PRINCIPLES OF MOBILE COMMUNICATION

PRINCIPLES OF MOBILE COMMUNICATION

by

Gordon L. Stüber
Georgia Institute of Technology

KLUWER ACADEMIC PUBLISHERS
Boston / Dordrecht / London

Distributors for North America:
Kluwer Academic Publishers
101 Philip Drive
Assinippi Park
Norwell, Massachusetts 02061 USA

Distributors for all other countries:
Kluwer Academic Publishers Group
Distribution Centre
Post Office Box 322
3300 AH Dordrecht, THE NETHERLANDS

Consulting Editor: Robert Gallager, Massachusetts Institute of Technology

Library of Congress Cataloging-in-Publication Data

A C.I.P. Catalogue record for this book is available
from the Library of Congress.

Printed on acid-free paper.

Printed in the United States of America

To my parents

CONTENTS

PREFACE

With the increasing importance of wireless systems and services in the telecommunications industry, courses in mobile communications are becoming very popular in many universities. When I first introduced a graduate level course in mobile communications at Georgia Tech in 1992, I had great difficulty in finding a textbook that i) provided an up-to-date treatment of the subject area, and ii) was suitable for instruction. In response to this need, I have compiled this book from a series of course notes that I developed at Georgia Tech from 1992 to 1995. My goal was to produce a textbook that would provide enough background material for a first-year graduate level course, and have enough advanced material to satisfy the more serious graduate students that would like to pursue research in the area. The book is intended to stress the *fundamentals* of mobile communications engineering that are important to *any* mobile system design, as opposed to providing a detailed description of existing and proposed wireless standards. This emphasis on fundamental issues should be of benefit not only to students taking formal instruction, but to practicing engineers who are likely to already have a detailed familiarity with the standards and are seeking to deepen their knowledge of the fundamentals of this important field.

Chapter 1 begins with an overview that is intended to introduce the broad array of issues relating to wireless communications. The remainder of the book is directed toward cellular and PCS issues, but many of the concepts apply to other types of wireless systems as well.

Chapter 2 treats propagation modeling and was inspired by the excellent reference by Jakes. It begins with a summary of propagation models for narrowband and wide-band channels, and provides a discussion of channel simulation techniques that are useful for radio link analysis. It concludes with a discussion of shadowing and path loss models. Chapter 3 is a related chapter that provides a detailed treatment of co-channel interference, the primary impairment in high capacity cellular systems.

Chapter 4 covers the various types of modulation schemes that are used in mobile communication systems along with their spectral characteristics. Chapter 5 discusses the performance of digital signal on narrow-band flat fading channels with a variety of receiver structures, and includes a discussion of diversity techniques.

Chapter 6 provides an extensive treatment of digital signaling on the fading ISI channels that are typical of mid-band land mobile radio systems. The chapter begins with the characterization of ISI channels and goes on to discuss techniques for combating ISI based on symbol-by-symbol equalization and sequence estimation.

Chapter 7 covers bandwidth efficient coding techniques. While convolutional and block codes find application in mobile communication systems, I have deliberately concentrated on the bandwidth efficient class of trellis codes. Included is a detailed discussion on the design and performance analysis of trellis codes for additive white Gaussian noise channels, interleaved flat fading channels, and equalized multipath fading ISI channels.

Chapter 8 is devoted to direct sequence code division multiple access (DS CDMA) techniques for cellular radio. The chapter includes a discussion of spreading sequences, RAKE receivers, and error probability approximations for DS CDMA systems. It concludes with a discussion of issues relevant to cellular CDMA, such as capacity estimation and power control.

Chapters 9 through 11 deal with resource management issues in FDMA and TDMA cellular systems. Chapter 9 considers frequency management schemes. Chapter 10 covers the important problem of link quality evaluation and handoff initiation in cellular systems, while Chapter 11 provides an overview of the various channel assignment techniques that have been proposed for FDMA and TDMA cellular systems.

The book contains far too much detail to be taught in a one-semester course. However, I believe that it can serve as a suitable text in most situations through the appropriate selection of material. My own preference for a one-semester course is to include the following in order: Chapter 1, Chapter 2, Sections 3.1 and 3.2, Chapter 4, Chapter 5, Sections 6.1 to 6.5, Chapter 9, and finally either Chapter 8 or Sections 10.1 to 10.4.

I would like to acknowledge all those who have contributed to the preparation of this book. The reviewers Vijay Bhargava at the University of British Columbia and Sanjiv Nanda at AT&T Bell Labs were very valuable in early stages of this project. The subsequent review by Upamanyu Madhow of the University ath Illinois and in particular the detailed review by Keith Chugg from the University of Arizona have been highly useful for improving this book. I am grateful to my doctoral students, past and present, who have contributed significantly to this book. The contributions of Wern-Ho Sheen, Khalid Hamied, Mark Austin, and Ming-Ju Ho are particularly noteworthy. Finally, I would like to thank BellSouth, GTE Labs, Motorola, Panasonic, Hitachi, Nortel, and NSF, for sustaining my research efforts in mobile communications over the past 6 or 7 years, from which much of the material in this book is drawn.

PRINCIPLES OF MOBILE COMMUNICATION

1

INTRODUCTION

The basic technological components of a cellular telephone system are radio transmission technology and computer technology. A cellular telephone system has two basic functions; it must locate and track the mobile stations (MSs), and it must always attempt to connect the MSs to the best available base station(s) (BS(s)). The latter task requires the continuous evaluation of the radio link quality with the serving BS(s), and the radio link quality with alternate BSs. This monitoring is performed by a computer system that uses knowledge of the link quality evaluations, in addition to the system topology and traffic flow, to decide upon the best BS(s) to serve a particular MS.

Unlike a traditional wire-line telephone system, a cellular telephone system allows for subscriber mobility, by using low power (less than 1 watt) radio communication between a MS and a grid of BSs [200]. MS movement, however, causes the radio link quality to be highly erratic and, therefore, careful monitoring and control are required to keep it acceptable. Radio link quality evaluation is based on a very large number of criteria, but at the core is a statistical measurement process based on *a priori* knowledge of the expected radio channel characteristics. The time required to measure the quality of a radio link, and the accuracy of the measurement, depends on the local propagation characteristics. Link quality measurements that require a long time will limit the ability of the cellular system to react to degradations in link quality, and compensate through changes in the allocation of power and bandwidth resources. If, on the other hand, the link quality measurements can be made very quickly, then the time required for the cellular system to process the link quality measurements, make decisions, and transmit desired changes to the network entities, including the MSs, will limit the adaptability of the cellular system. Limitations on the speed of monitoring and control essentially determine overall link quality and

1

the size and distribution of cells in modern cellular systems. The cell sizes, the ability radio links to withstand interference, and the ability of the cellular system to react to variations in traffic are the main factors in determining the spectral efficiency of the cellular system.

In cellular systems, the available spectrum is partitioned among the BSs, and a given frequency is reused at the closest possible distance that the radio link will allow. Smaller cells have a shorter distance between reused frequencies, and this results in an increased spectral efficiency and traffic carrying capacity. Dramatic improvement in spectral efficiency is the main reason for the intense research interest in very small cells, or microcells. Microcells, however, introduce their own set of challenges, because the microcellular propagation environment is highly erratic. Hence, microcellular resource allocation algorithms need to be fast to maintain high link quality, and distribution of control is needed to moderate the amount of network traffic.

The current trend in the cellular industry is toward spectrally efficient cellular systems with ubiquitous service coverage. These systems will require i) effective cellular architectures, ii) fast and accurate link quality measurements, iii) rapid control in all types of environments, iv) installation of BSs to provide radio coverage virtually everywhere, and v) power and bandwidth efficient radio links that can mitigate the harsh effects of the propagation environment and tolerate high levels of noise and interference.

Future personal communication services (PCSs) are envisioned that will allow low power (0.1-10 mW) global tetherless communication, where communication links can be established upon demand, at any time and any place [60, 61]. These systems will completely change our paradigm of the telephone system. The vision of PCS is to assign each subscriber a personal telephone number along with a small personal telephone. Currently, the success of using a telephone depends upon the knowledge of where a called party is located, with the result that 80% of today's calls never reach their intended party. With PCS, intelligent networks (INs) will be employed to assume the burden of locating a called party, leaving the subscribers free to roam anywhere in the world. The concept of just-in-time communications will emerge and enormously change the nature of business and personal interactions. Call management will become a necessity to ensure that PCS does not become a nuisance. That is, the subscribers must be able to control their availability for receiving calls.

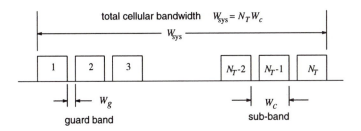

Figure 1.1 Typical FDMA bandwidth structure.

1.1 WIRELESS SYSTEMS AND STANDARDS

1.1.1 First Generation Cellular Systems

The early 1970s saw the emergence of the radio technology that was needed for the deployment of mobile radio systems in the 800/900 MHz band at a reasonable cost. In 1976, the World Allocation Radio Conference (WARC) approved frequency allocations for cellular telephones in the 800/900 MHz band, thus setting the stage for the commercial deployment of cellular systems. In the early 1980s, many countries deployed first generation cellular systems based on frequency division multiple access (FDMA) and analog FM technology. As shown in Fig. 1.1, FDMA divides the cellular frequency band into many sub-bands, each with its own distinct carrier frequency. The power spectral density of the modulated signals must be carefully controlled so that the radiated power into the adjacent band is 60 to 80 dB below that in the desired band; otherwise excessive adjacent channel interference is introduced, which will degrade performance. This spectral control creates **guard bands** between the adjacent channels. With FDMA there is a single channel per carrier. When a MS accesses the system it is actually assigned two carriers (channels), one for the forward (base-to-mobile) link and one for the reverse (mobile-to-base) link. Separation of the forward and reverse carrier frequencies is necessary to allow implementation of a **duplexer**, a complicated arrangement of filters that provides isolation of the forward and reverse frequencies.

In 1979, the first analog cellular system, the Nippon Telephone and Telegraph (NTT) system, became operational. In 1981, Ericsson Radio Systems AB fielded the Nordic Mobile Telephone (NMT) 900 system, and in 1983 AT&T fielded the Advanced Mobile Phone Service (AMPS) as a trial in Chicago. The NTT system operates in the 925-940/870-885 band with 25 kHz carrier spacings. The NMT 900 system operates in the 890-915/917-950 MHz band, and uses frequency interleaved carriers with a separation of 12.5 kHz such that overlapping carriers cannot be used with the same BS. Finally, the AMPS system operates in the 824-849/869-894 MHz band with 30 kHz carrier spacings. In all these systems, a separation of 45 MHz between the transmit and receive frequencies is used. Several other first generation analog systems were also deployed in the early 1980s including TACS, ETACS, NMT 450, C-450, RTMS, and Radiocom 2000 in Europe, and JTACS/NTACS in Japan.

Since these initial deployments, the cellular subscriber base has been growing 20% to 50% per year. In North America, there were only 25,000 subscribers in 1984 at the beginning of the Chicago AMPS trial. This grew to a phenomenal 20 million subscribers by the end of 1994, with an exponential-like growth rate as shown in Fig. 1.2. There are about an equal number of cellular subscribers outside North America, mostly in Europe and Japan. Current expectations are that 50% of the telephone traffic will use wireless links by the year 2000 [42]. With the prospect of obtaining no additional spectral allocations, cellular operators in largest cellular markets experienced capacity limitations in the late 1980s, and the need for spectrally efficient second generation cellular systems became intense.

1.1.2   Digital Cellular Techniques

Rapid developments in microelectronics have made second generation digital cellular systems viable. Digital cellular systems have many advantages over analog cellular systems including the provision of voice and data services, reduced RF transmit power, encryption for security, mobile assisted handoffs, and the ability to exploit incremental advances in low bit rate speech coding to further enhance spectral efficiency. Most, if not all, second generation digital cellular systems use either time division multiple-access (TDMA) or code-division multiple-access (CDMA). These access techniques allow the radio hardware at the BSs to be shared among multiple MSs. With TDMA and CDMA, digital modulation techniques are the only practical option.

As shown in Fig. 1.3, TDMA divides the usage of each carrier frequency into multiple time slots or channels. A pure TDMA system uses only one carrier frequency. The forward and reverse channels correspond to different time slots within each frame, an arrangement called time division duplexing (TDD). However, the use of a single carrier with TDMA is undesirable, because of the delay restrictions imposed by voice traffic and the tight timing tolerances

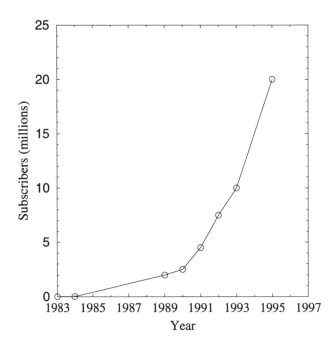

Figure 1.2 Number of cellular telephone subscribers in North America.

that would be required among the MSs. Therefore, cellular systems that employ TDMA actually use hybrid FDMA/TDMA, where there are multiple carriers with multiple channels per carrier. By staggering the transmit/receive time slots, TDMA systems have much less stringent duplexer requirements than FDMA systems.

As with FDMA, TDMA requires that suitable guard bands be established between adjacent carrier frequencies. As shown in Fig. 1.3, **guard times** must also be established between the time slots within a frame, because of variations in the propagation delay and variations in the delay spread (time dispersion) of the channel. Without guard times, the signals in adjacent time slots would overlap, causing severe distortion. The required guard time is approximately

$$\tau_g = \tau_p + \Delta_t \qquad (1.1)$$

where

$$\begin{aligned}
\tau_p &= \text{maximum differential propagation delay} \\
\Delta_t &= \text{delay spread of the channel.}
\end{aligned}$$

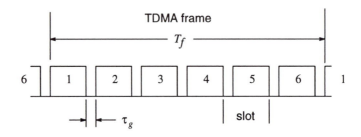

Figure 1.3   TDMA frame structure.

Radio waves in free space travel approximately 0.34 m/ns or 1 ft/ns, and the delay spread of a typical urban land mobile radio channel is about 5 μs. If the variation in radio path length due to MS movement is 10 km, then the required guard time is approximately $\tau_g = 29.4 + 5.0 = 34.4$ μs. In this case the effect of the delay spread is small compared to the effect of the propagation delay. The required guard time can be greatly reduced by adjusting the transmission times of the MSs, such that the more distance MSs begin their transmissions sooner. Also, as the cell sizes decrease, both the delay spread and the propagation delay will decrease and, therefore, smaller guard times can be used.

CDMA uses spread-spectrum signaling, where the bandwidth that is used to transmit a signal exceeds the minimum required. The band-spread is achieved by using pseudo-noise (PN) sequences or spreading codes, and users are distinguished by assigning them different spreading codes. A distinguishing feature of spread spectrum signals is that the spreading codes are independent of the information that is being transmitted. Therefore, modulation techniques such as wide-band FM are not spread spectrum techniques. A synchronized version of the code is used at the intended receiver to despread the received signal, allowing recovery of the transmitted information.

Two main types of CDMA systems have been studied extensively in the literature, based on direct-sequence (DS) and frequency-hopped (FH) spread-spectrum. With DS/CDMA the band-spread is achieved by using a high-rate PN sequence to introduce very rapid phase transitions in the carrier containing the data. With FH/CDMA the PN sequence is used to pseudo-randomly hop the carrier frequency throughout a large bandwidth. The earliest proposals for cellular CDMA systems favored FH/CDMA (e.g., [52]), however, most of the recent proposals favor DS/CDMA (e.g., [118]). Each type of CDMA system has its advantages and disadvantages.

Many comparisons between TDMA and CDMA systems have appeared in the literature. When compared to DS/CDMA, TDMA has the advantage of a much less stringent transmit power control requirement because the signals are truly orthogonal - something not achieved by CDMA with asynchronous users. Also, the time-slot structure allows a MS sufficient time to measure the signal quality associated with alternate time-slots, frequencies and BSs, thereby supporting mobile assisted handoffs. When compared to TDMA, DS/CDMA also has several advantages. First the need for guard bands/times can be greatly reduced, if not eliminated altogether, thereby improving spectral efficiency. Also, TDMA radios transmit using periodic pulses which presents a challenge to the design of portable units, while DS/CDMA radios do not (although some types of DS/CDMA systems do). DS/CDMA also has the potential of a greatly simplified frequency reuse plan, because all cells can use all carrier frequencies. Other benefits of DS/CDMA include; i) fine-time resolution, or the ability to resolve multipath which results in increased diversity, ii) the ease in which frequency-selective multipath is combined (e.g., a RAKE receiver vs. maximum likelihood sequence estimation (MLSE) for TDMA), iii) the ease of implementing soft-handoff, iv) low power flux density at the transmitter, and v) the ability to take advantage of reduced voice activity and the related soft-degradation properties.

The overall comparison of spectral efficiency between TDMA and DS/CDMA systems is quite difficult, because such comparisons are often made between systems that are at different evolutionary stages and with different constraints on deployment. However, a fair comparison between two suitably optimized TDMA and DS/CDMA systems without deployment constraints would probably show that their spectral efficiencies are roughly equal.

1.1.3 Second Generation Cellular Systems

European GSM and DCS1800

In 1982, the Conference of European Postal and Telecommunications Administrations (CEPT) established Groupe Special Mobile (GSM) with the mandate of defining standards for future Pan-European cellular radio systems. The GSM system (now "Global System for Mobile Communications") was developed to operate in a new frequency allocation, and made improved quality, Pan-European roaming, and the support of data services its primary objectives. GSM uses TDMA with 200 kHz carrier spacings, eight channels per carrier with a time slot (or burst) duration of 0.577 ms, and Gaussian mini-

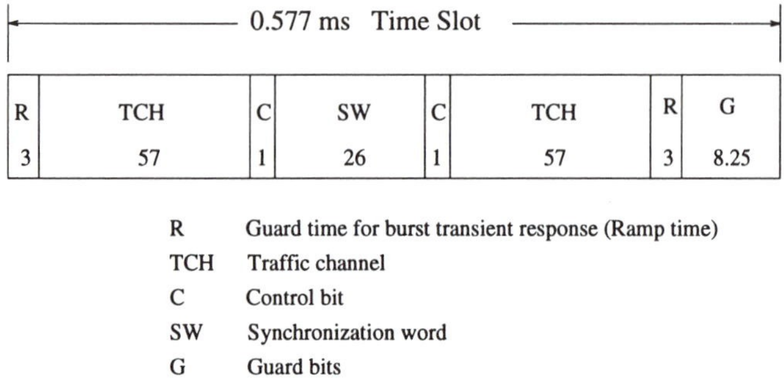

Figure 1.4   Time slot format for GSM. Units are in bits.

mum shift keying (GMSK) with a raw bit rate of 270.8 kb/s. The time slot format of the GSM traffic channels is shown in Fig. 1.4.

GSM was deployed in late 1992 as the world's first digital cellular standard. In its current version, GSM can support full-rate (8 slots/carrier) and half-rate (16 slots/carrier) operation, and provide various synchronous and asynchronous data services at 2.4, 4.8, and 9.6 kb/s that interface to voiceband modems (e.g., V.22bis or V.32) and ISDN. Additional details of the air interface specification for the GSM system are provided in [202, 263]. GSM has been a tremendous success. In late 1995, 156 regulators and network operators in 86 countries were firmly committed to GSM.

In Europe, the Digital Cellular System 1800 (DCS1800) has been developed by ETSI as a standard for personal communication networks (PCNs) [248]. DCS1800 is a derivative of the GSM system, but differs in a number of ways. First, DCS1800 operates in the 1710-1785 and 1805-1880 MHz bands, whereas GSM operates in the 900 MHz band. Second, DCS1800 is optimized for two classes of hand held portable units (rather than mobile units) with a peak power of 1 W and 250 mW, respectively. There are also some changes in the DCS1800 standard to support overlays of macrocells and microcells.

North American IS-54 and IS-95

In North America, second generation digital cellular systems were developed with the constraint of making a seamless transition from the AMPS system. That is, any second generation North American digital cellular standard must be reverse compatible with AMPS. While Europe has seen a convergence to a

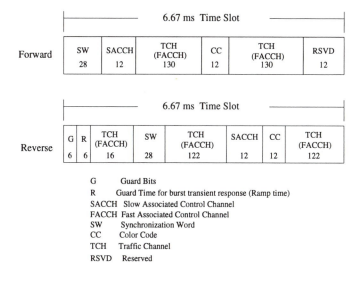

Forward	SW	SACCH	TCH (FACCH)	CC	TCH (FACCH)	RSVD
	28	12	130	12	130	12

Reverse	G	R	TCH (FACCH)	SW	TCH (FACCH)	SACCH	CC	TCH (FACCH)
	6	6	16	28	122	12	12	122

G   Guard Bits
R   Guard Time for burst transient response (Ramp time)
SACCH   Slow Associated Control Channel
FACCH   Fast Associated Control Channel
SW   Synchronization Word
CC   Color Code
TCH   Traffic Channel
RSVD   Reserved

Figure 1.5   Time slot format for North American IS-54. Units are in bits.

single cellular standard, North America has seen the emergence of two second generation standards IS-54 and IS-95. In 1990, a TDMA cellular system was adopted the IS-54 standard. IS-54 specifies dual-mode radio transceivers that are reverse compatible with AMPS, but are also capable of using a new digital signaling scheme based on TDMA with 30 kHz carrier spacings and $\pi/4$ phase-shifted quadrature differential phase shift keying ($\pi/4$-DQPSK) with a raw bit rate of 48.6 kb/s [78] . The time slot format for IS-54 is shown in Fig. 1.5. A key feature of IS-54 is the provision of an adaptive equalizer, although this requirement has been dropped in subsequent revisions of the standard. IS-54 offers at least three times the capacity of AMPS by using three slots (channels) per carrier. New advances in voice coding technology may permit six channels per carrier. IS-54 has now been deployed in the major cellular markets in the United States and the rate of customer adoption is growing.

In February 1990, just after the CTIA adopted IS-54, another second generation digital cellular system was proposed based on DS/CDMA [119]. This system, commonly called CDMA, is also reverse compatible with AMPS. In March 1992, CDMA was adopted as the IS-95 standard [79]. With IS-95, the basic user data rate is 9.6 kb/s, which is spread by using PN sequence with a chip (clock) rate of 1.2288 Mchips/s (a processing gain of 128). The forward channel supports coherent detection by using a pilot channel (code) for channel estimation. Information on the forward link is encoded by using a rate-1/2 convolutional code, interleaved, spread by using one of 64 Walsh codes, and transmitted in 20 ms bursts. Each MS in a cell is assigned a different code, thus providing complete orthogonality under conditions of low channel delay spread. Final spreading with a base-specific PN sequence of length 2^{15} is used to mitigate the multiple access interference to and from other cells. One of the major drawbacks of the IS-95 standard is that the downlink transmissions are not interleaved across bursts and, therefore, the signal is susceptible to fading.

The IS-95 reverse link uses noncoherent detection, and fast closed loop power control to combat the **near-far effect**. The information on the reverse link is encoded by using a rate-1/3 convolutional code, interleaved, and mapped onto one of 64 Walsh codes. Final spreading is achieved with a user-specific PN sequence of length $2^{42} - 1$. Both the BSs and the MSs use RAKE receivers to provide multipath diversity. An important feature of the IS-95 standard is the provision of **soft handoffs**, where the MS maintains a radio link with multiple BSs during the process of handing off between cells.

Initial capacity estimates of IS-95 suggested a 20 to 40 times capacity improvement over the AMPS system, a figure that raised high interest in the cellular industry. However, current estimates are more conservative and predict a 6 to 10 times capacity increase over AMPS. Ever since the introduction of IS-95, there has been a continued debate over the relative capacity of IS-54 and IS-95, an issue that will not be completely resolved until the both systems are widely deployed.

Japanese PDC

In Japan, development of a second generation digital cellular system began in 1989. In 1991, the Ministry of Posts and Telecommunications standardized the new system, named Personal Digital Cellular (PDC). Similar to IS-54, PDC uses TDMA with three channels per carrier, 25 kHz carrier spacings, and $\pi/4$-DQPSK modulation with a raw bit rate of 42 kb/s. The time slot format for the PDC traffic channels is shown in Fig. 1.6 and is considerably different from the IS-54 time slot format. Notice that the synchronization word appears near the center of the time slot for the PDC system in Fig. 1.6, and near the beginning of the time slot for the IS-54 system in Fig. 1.5. This feature better enables the PDC system to track channel variations over the time slot. Another key feature of PDC standard is the inclusion MS antenna diversity. Like IS-54, PDC suffers from degraded performance under conditions of low delay spread due to the loss of multipath diversity. However, antenna diversity in the PDC

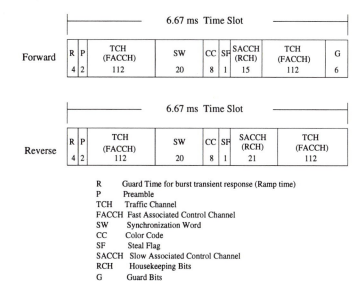

Figure 1.6 Time slot format for Japanese PDC. Units are in bits.

receivers maintains spatial diversity under these conditions. More details on the PDC system can be found in the complete standard [267].

1.1.4 Cordless Telephony and Mobile Satellite Systems

Cordless telephones find several applications including domestic telephones, telepoint (cordless phone booth), wireless PABX (private access business exchange), and wireless local loops or radio drops. Similar to cellular telephones, first generation cordless telephones were based on analog FM technology. Since their introduction, cordless telephones have demonstrated wide spread popularity. Currently, there are about 60 million cordless telephones in the United States. However, cordless telephones are becoming victims of their own success; the voice quality is unacceptable in high-density subscriber areas. The Europeans have remedied this problem by developing two digital cordless telephone systems, CT2 and DECT (Digital European Cordless Telephone) [307]. CT2 uses FDMA with 100 kHz carrier spacings and Gaussian minimum shift keying (GMSK) with a raw transmission rate of 72 kb/s. The forward and reverse links share the same carrier by using time division duplexing (TDD). Canada has developed a modification of CT2, called CT2Plus, that offers two-way call-ing, roaming, and enhanced data service capabilities. DECT employs TDMA and TDD, where each carrier supports 12 channels with GMSK having a raw transmission rate of 1.152 Mb/s. In Japan, the Personal Handyphone System (PHS) has been developed. PHS uses TDMA and TDD with 8 channels per carrier, along with $\pi/4$-DQPSK.

Several mobile satellite systems are also under development. These satellite systems fall under three categories. The first uses small low earth orbit (LEO) satellites that minimize path loss and propagation delays. The Motorola Irid-ium project stands out as the most comprehensive system of this type [132]. The Iridium system, named after an original plan to deploy 77 satellites, the same number of electrons in the Iridium atom, will consist of a deployment of 66 satellites equally divided into 6 polar orbital planes. The 24-satellite Loral-Qualcomm Globalstar system is another example of a LEO system. The second category of mobile satellite systems uses medium earth orbit (MEO) satellites to reduce the satellite count. An example of a MEO system is the TRW Odyssey system. The third category of mobile satellite systems uses large geostationary satellites with large spot beam antennas. Large antennas provide a high antenna gain, while spot beams provide good frequency reuse. The Personal Access Satellite System (PASS) proposed by the Jet Propulsion Laboratory is a good example of this type of satellite system. As compared to land mobile radio systems, global-mobile satellite systems generally suffer from severe Doppler or frequency offset as opposed to multipath fading, large round trip propagation delays, and high transmit powers (for geostationary systems).

1.1.5   Future Wireless Systems and Services

In March 1992, WARC approved a worldwide allocation in support of the Fu-ture Public Land Mobile Telephone System (FPLMTS) in the 1885-2200 MHz band. This new frequency allocation is leading to the development of a wide array of new wireless systems and services.

In Europe, the intermediate goal is the Universal Mobile Telecommunication System (UMTS), which is essentially a multi-operator system with mixed cell architectures and multimedia capabilities. The basic idea of UMTS is to provide the same service everywhere, but with data rates that depend upon the local network loading and propagation conditions. Some effort is being made to extend current standards such as GSM, DCS1800, and DECT for this purpose. Other projects such as Advanced TDMA (ATDMA) and Code

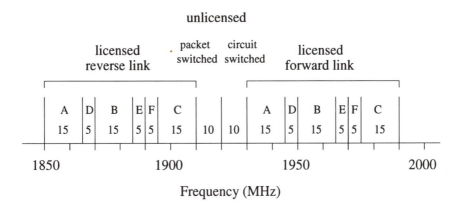

Figure 1.7 North American PCS frequency allocations.

Division Testbed (CODIT) are taking more radical approaches that are based on TDMA and CDMA technology.

In North America, the 1885-2200 MHz band allocated for FPLMTS is currently occupied by private microwave radio systems. The FCC has recently reallocated 140 MHz of spectrum in the 1850-1990 MHz band to support PCS. according to the plan shown in Fig. 1.7. Blocks A and B correspond to major trading areas (MTAs) while blocks C through F correspond to basic trading areas (BTAs). There are 51 MTAs and 492 BTAs in the United States. In addition to these allocations, 20 MHz was reserved for unlicensed use according to FCC Part 15 rules. Of this 20 MHz, 10 MHz is for packet switched applications and 10 MHz is for circuit switched applications. The FCC has awarded licenses in the PCS band through "auctioning." Once a license is purchased, an operator is free to deploy any system they wish, provided that they comply with FCC rules. Potential PCS standards fall into two categories: "high tier" (cellular) and "low tier" (cordless telephony). At this point there are seven proposed standards: Personal Access Communication Services (PACS) and DECT (low tier), DCS1800 and proposals based on IS-54 and IS-95 (high tier), and hybrid TDMA/CDMA and wide-band CDMA (WCDMA). The private microwave users that currently occupy the PCS band will be relocated to other frequency bands. The cost of these relocations must be assumed by the PCS licensees.

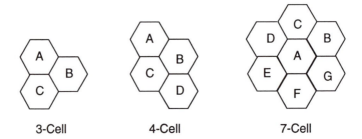

3-Cell   **4-Cell**   **7-Cell**

Figure 1.8   Commonly used cellular reuse clusters.

1.2 FREQUENCY REUSE AND THE CELLULAR CONCEPT

Cellular mobile radio systems that use TDMA and FDMA rely upon **frequency reuse**, where users in geographically separated cells simultaneously use the same carrier frequency. The cellular layout of a conventional macrocellular system is quite often described by a uniform grid of hexagonal cells or radio coverage zones. In practice the cells are not regular hexagons, but instead are distorted and overlapping areas. The hexagon is an ideal choice for representing macrocellular coverage areas, because it closely approximates a circle and offers a wide range of tessellating reuse cluster sizes. A tessellating reuse cluster of size N can be constructed if [244]

$$N = i^2 + ij + j^2 \qquad (1.2)$$

where i and j are non-negative integers, and $i \geq j$. It follows that the allowable cluster sizes are $N = 1, 3, 4, 7, 9, 12, \ldots$ Examples of 3-, 4-, and 7-cell reuse clusters are shown in Fig. 1.8. The reuse clusters are tessellated to form a frequency plan. A simplified 7-cell frequency reuse plan is shown in Fig. 1.9, where similarly marked cells use identical sets of carrier frequencies.

For microcellular systems with lower BS antenna heights, regular hexagons are no longer appropriate for approximating the radio coverage zones. Typical microcell BSs use an antenna height of about 15 m, well below the skyline of any buildings that might be present, and acceptable link quality can be obtained anywhere within 200-500 m of the BS. For microcells, the choice of cell shape depends greatly upon the particular deployment. For example, the linear cells shown in Fig. 1.10 may provide a more accurate model of **highway microcells** that are deployed along a highway with directional antennas. In an area with urban canyons, the buildings act as waveguides to channel the signal energy

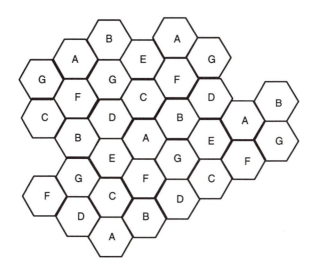

Figure 1.9 Macrocellular deployment using 7-cell reuse pattern.

A	B	C	A	B	C	A	B	C	A

Figure 1.10 Microcellular deployment along a highway with a 3-cell reuse pattern.

along the street corridors. Fig. 1.11 shows a typical **Manhattan microcell** deployment that is often used to model microcells that are deployed in city centers.

1.3 MOBILE RADIO PROPAGATION ENVIRONMENT

Radio signals generally propagate according to three mechanisms ; **reflection**, **diffraction**, and **scattering**. Reflections arise when the plane waves are incident upon a surface with dimensions that are very large compared to the wavelength. Diffraction occurs according to Huygen's principle when there is an obstruction between the transmitter and receiver antennas, and secondary waves are generated behind the obstructing body. Scattering occurs when the plane waves are incident upon an object whose dimensions are on the order of

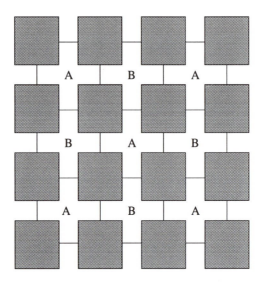

Figure 1.11   Microcellular deployment in an urban canyon. Base stations are deployed at every intersection in a dense urban area with a 2-cell reuse pattern.

a wavelength or less, and causes the energy to be redirected in many directions. The relative importance of these three propagation mechanisms depends on the particular propagation scenario.

As a result of the above three mechanisms, macrocellular radio propagation can be roughly characterized by three nearly independent phenomenon; **path loss** variation with distance, slow log-normal **shadowing**, and fast **multipath-fading**. Each of these phenomenon is caused by a different underlying physical principle and each must be accounted for when designing and evaluating the performance of a cellular system. Multipath-fading results in rapid variations in the envelope of the received signal and is caused when plane waves arrive from many different directions with random phases and combine vectorially at the receiver antenna. Typically, the received envelope can vary by as much as 30 to 40 dB over a fraction of a wavelength due to constructive and destructive addition. Multipath also causes time dispersion, because the multiple replicas of the transmitted signal propagate over different transmission paths and reach the receiver antenna with different time delays. Time dispersion may require equalization in TDMA systems and RAKE reception in CDMA systems.

It is well known that the intensity of an electromagnetic wave in free space decays with the square of the radio path length, d, such that the received power is

$$\Omega(d) = P \left(\frac{\lambda_c}{4\pi d} \right)^2 \tag{1.3}$$

where P is the transmitted power and λ_c is the wavelength. Although it may seem counter-intuitive, path loss is essential in high capacity cellular systems, the reason being that a rapid attenuation of signal strength with distance permits a small co-channel reuse distance and, therefore, a high spectral efficiency. The 800-900 MHz UHF band was chosen for first generation cellular systems, partly because of its relatively short range radio propagation characteristics. Of course if a large radio coverage area is desired, as is the case with low capacity emergency communication systems (police, fire, etc..), then a small path loss is preferred. For this reason the VHF band is preferred for these applications which results in a smaller attenuation with distance.

Free space propagation does not apply in a mobile radio environment and the propagation path loss depends not only on the distance and wavelength, but also on the antenna heights of the MSs and the BSs, and the local terrain characteristics such as buildings and hills (in macrocells). The site specific nature of radio propagation makes the theoretical prediction of path loss difficult and there are no easy solutions. The simplest path loss model assumes that the received power is

$$\Omega_{(\text{dB})}(d) = \Omega_{(\text{dB})}(d_o) - 10\beta \log_{10}(d/d_o) + \epsilon_{(\text{dB})} \quad \text{dBm} \tag{1.4}$$

where the term $\Omega_{(\text{dB})}(d_o)$ gives the received signal power (in dBm) at a known reference distance that is in the far field of the transmitting antenna. Typically, d_o is 1 km for macrocells, 100 m for outdoor microcells, and 1 m for indoor picocells. The value of $\Omega_{(\text{dB})}(d_o)$ depends on the frequency, antenna heights and gains, and other factors. The parameter β is called the **path loss exponent** and is a key parameter that affects the spectral efficiency of a cellular system. This parameter is strongly dependent on the cell size and local terrain characteristics. The path loss exponent ranges from 3 to 4 for a typical urban macrocellular environment, and from 2 to 8 for a microcellular environment. Usually, the path loss exponents are determined by empirical measurements.

The parameter $\epsilon_{(\text{dB})}$ in (1.4) is a zero-mean Gaussian random variable (in dB) that represents the error between the actual and estimated path loss. This statistical variation in $\Omega_{(\text{dB})}$ is caused by shadowing. Shadows are generally modeled as being log-normally distributed, meaning that the probability density function of $\Omega_{(\text{dB})}$ is

$$p_{\Omega_{(\text{dB})}}(x) = \frac{1}{\sqrt{2\pi}\sigma_\Omega} \exp \left\{ -\frac{(x - \mu_\Omega)^2}{2\sigma_\Omega^2} \right\} \tag{1.5}$$

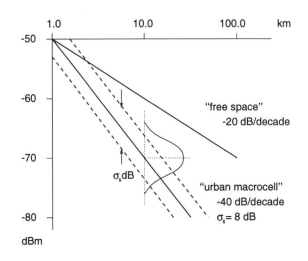

Figure 1.12 Path loss in free space and typical urban macrocellular environments; $\beta = 4$, $\sigma_\Omega = 8$ dB.

where

$$\mu_{\Omega(d)} = \mu_{\Omega(d_o)} - 10\beta \log_{10}(d/d_o) \ \ \text{dBm} \qquad (1.6)$$

and $\mu_\Omega = \mathrm{E}[\Omega_{(\mathrm{dB})}]^1$.

The parameter σ_Ω is the **shadow standard deviation**. A more accurate path loss model results in a smaller σ_Ω. For macrocells, σ_Ω typically ranges from 5 to 12 dB, with $\sigma_\Omega = 8$ dB being a typical value. Furthermore, σ_Ω has been observed to be nearly independent of the radio path length d. Fig. 1.12 illustrates the above concepts by plotting the received signal strength as a function of the radio path length for both free space and a typical urban macrocellular environment.

One of the major difficulties in the accurate prediction of path loss arises from variations in the statistical characteristics of the radio propagation environment with location. This has lead to considerable research into site specific radio propagation methods that can accurately predict path loss by using ray tracing techniques. Ray tracing techniques, however, require detailed knowledge of the location, texture, and composition of the reflecting, scattering and diffracting surfaces in the propagation environment with respect to the physical locations of the BS and MS antennas.

---

[1]Note that here and throughout the text, the units of μ_Ω are dBm and the units of σ_Ω are in dB.

1.4 CO-CHANNEL INTERFERENCE AND NOISE

Frequency reuse in FDMA/TDMA cellular systems introduces **co-channel interference**, one of the major factors that limits the capacity of cellular systems. Co-channel interference arises when the same carrier frequency is used in different cells. In this case, the power density spectra of the desired and interfering signals completely overlap. Frequency reuse also introduces **adjacent channel interference**. This type of interference arises when neighboring cells use carrier frequencies that are spectrally adjacent to each other. In this case the power density spectrum of the desired and interfering signals partially overlap.

Cellular radio links quite often exhibit a **threshold effect**, such that the link quality will be acceptable provided that both the average received carrier-to-noise ratio Γ and the average carrier-to-interference ratio Λ exceed certain thresholds, denoted by Γ_{th} and Λ_{th}, respectively [96].[2] These thresholds depend on many parameters of the radio link, including the particular modulation and coding scheme that is employed, the receiver structure, the measure of transmission quality, the propagation environment, and the MS velocity. For fast moving MSs, path loss and shadowing determine the link quality once Γ_{th} and Λ_{th} have been specified. Conversely, for slow moving MSs, the link quality may also become unacceptable when the channel exhibits a deep fade. However, for this introductory section we will restrict our attention to fast moving MSs, where the receiver can "average" over the effects of envelope fading.

Here we introduce two quantities that measure the performance of a cellular system. The first is the **probability of thermal noise** (TN) defined as

$$O = \Pr(\Gamma < \Gamma_{th}) \tag{1.7}$$

and the second is the probability of co-channel interference (CCI). , defined as

$$O = \Pr(\Lambda < \Lambda_{th}) . \tag{1.8}$$

In the basic design of a macrocellular system, two parameters must be specified; the required transmitter power and the **co-channel reuse factor**, D/R, defined as the ratio of the co-channel reuse distance D between cells using the same set of carrier frequencies and the radius of the cells R^3. The reuse cluster

---

[2] For the time being, the effect of adjacent channel interference will be neglected.

[3] For hexagonal cells, R is the distance from the center to the corner of a cell.

size N and the co-channel reuse factor D/R are related by (see Problem 1.2)

$$D/R = \sqrt{3N} \ . \tag{1.9}$$

The required transmitter power can be determined once the acceptable probability of TN is known. Likewise, the required co-channel reuse factor can be determined once the acceptable probability of CCI is known. Note that co-channel reuse distance actually depends on both the level of thermal noise and the level of co-channel interference. However, in high capacity cellular systems the thermal noise can usually be neglected in difference to the typically dominant effect of the co-channel interference.

1.4.1   Determining the Minimum Transmitter Power

Consider a cellular system described by a uniform grid of hexagonal cell of radius R, and suppose that Γ_{th} is known. We first note that the received carrier-to-noise ratio Γ and the received field strength are related by $\Gamma = \Omega/N_o$, where N_o is the one-sided power spectral density of the thermal noise. Let $\mu_{\Gamma(R)}$ be the value of $\mu_\Gamma = \mu_\Omega/N_o$ (in dB) on the cell fringe. The probability of TN that a MS will experience on a cell fringe is

$$
\begin{aligned}
O(R) &= \Pr(\Gamma(R) < \Gamma_{th}) \\
&= \int_{-\infty}^{\Gamma_{th(dB)}} \frac{1}{\sqrt{2\pi}\sigma_\Omega} \exp\left\{ -\frac{(x - \mu_{\Gamma(R)})^2}{2\sigma_\Omega^2} \right\} dx \\
&= Q\left(\frac{\mu_{\Gamma(R)} - \Gamma_{th(dB)}}{\sigma_\Omega} \right) \ .
\end{aligned} \tag{1.10}
$$

In equation (1.10), the quantity $M_\Gamma = \mu_{\Gamma(R)} - \Gamma_{th(dB)}$ is the minimum carrier-to-noise ratio margin required to sustain an probability of TN equal to $O(R)$ on the cell fringe. The probability of TN $O(R)$ is plotted against M_Γ in Fig. 1.13 for various shadow standard deviations.

To obtain a relationship between the probability of TN on a cell fringe and the area averaged probability of TN, requires models for the propagation path loss and spatial density of the MSs. For macrocells it is reasonable to assume that the MSs are uniformly distributed throughout the cell area. Assuming a uniform spatial density along with the path loss model in (1.6) yields the area averaged proability of TN [96]

$$O = \frac{1}{\pi R^2} \int_0^R O(r) 2\pi r \ dr$$

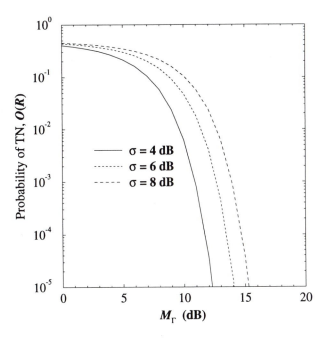

Figure 1.13 Probability of TN and required carrier-to-noise ratio margin.

$$= Q(X) - \exp\left\{XY + Y^2/2\right\} Q(X + Y) \tag{1.11}$$

where

$$X = \frac{\mu_{\Gamma(R)} - \Gamma_{\text{th(dB)}}}{\sigma_\Omega}, \qquad Y = \frac{2\sigma_\Omega}{\beta\xi} \tag{1.12}$$

where $\xi = 10/\ln 10$. The first term of this expression is equal to the probability of TN on a cell fringe, $O(R)$, while the second term is a correction factor.

Once O or $O(R)$ is specified, (1.10) or (1.11) can be easily solved for $\mu_{\Gamma(R)} - \Gamma_{\text{th(dB)}}$. If Γ_{th} is known from theory, experiments, or simulation, then $\mu_{\Gamma(R)}$ can be obtained. Once $\mu_{\Gamma(R)}$ is known, the minimum required transmitter power can be determined by using knowledge of the path loss, antenna gains, and received noise power. Note that it is desirable to design radio links having Γ_{th} as small as possible.

1.4.2   Determining the Co-channel Reuse Distance

Consider the situation shown in Fig. 1.14, depicting the co-channel interference on the forward channel at a MS. The MS is at distance d_0 from the serving BS and at distances $d_k, k = 1, 2, \cdots, N_I$ from the first tier of N_I interfering co-channel BSs. Suppose that the desired signal and co-channel signals are characterized by independent log-normal shadowing with the same shadow standard deviation. The received signal power, Ω, for each of the received signals has the log-normal distribution in (1.5). Let $\mathbf{d} = (d_0, d_1, \cdots, d_{N_I})$ denote the vector of distances at a particular MS location. Then the average downlink carrier-to-interference ratio as a function of $\mathbf{d}$ is

$$\Lambda_{(\mathrm{dB})}(\mathbf{d}) = \Omega_{(\mathrm{dB})}(d_0) - 10\log_{10}\left\{\sum_{k=1}^{N_I} 10^{\Omega_{(\mathrm{dB})}(d_k)/10}\right\} . \qquad (1.13)$$

The probability of CCI is then given by

$$O(\mathbf{d}) = \mathrm{P_r}\left(\Lambda_{(\mathrm{dB})}(\mathbf{d}) < \Lambda_{\mathrm{th}(\mathrm{dB})}\right) . \qquad (1.14)$$

A major difficulty arises from the fact that although each of the $\Omega_{(\mathrm{dB})}(d_i)$ are normally distributed, $\Lambda_{(\mathrm{dB})}(\mathbf{d})$ is not normally distributed except for $N_I = 1$. Unfortunately, there is no simple analytical formula for the density of $\Lambda(\mathbf{d})$ (and $\Lambda_{(\mathrm{dB})}(\mathbf{d})$) for $N_I > 1$, but several very useful approximations have been derived. These approximations are explored in Chapter 3, where the probability of CCI is covered in detail.

Finally, Fig. 1.15 depicts the co-channel interference on the reverse channel at the serving BS. Note that the co-channel interference may not be exactly the same on the forward and reverse channels, because the vector $\mathbf{d}$ is different in each direction. This phenomenon is known as **link imbalance**.

1.5   MODULATION TECHNIQUES

Modulation is the process where the message information is added to the radio carrier. Most first generation cellular systems such as AMPS use analog FM, because analog technology was very mature when these systems were developed. However, digital modulation schemes are the obvious choice for future wireless systems, especially if data services such as wireless multimedia are to be supported. Digital modulation can also improve spectral efficiency, because

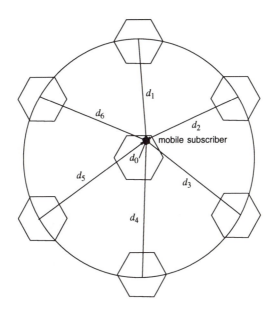

Figure 1.14 Co-channel interference on the forward channel at a desired MS. There are six interfering BSs.

the digital signals are more robust against channel impairments. Spectral efficiency is a key attribute of wireless systems that must operate in a crowded radio frequency spectrum.

To achieve high spectral efficiency, modulation schemes for FDMA and TDMA systems must be selected that have a high **bandwidth efficiency**, measured in units of bits per second per Hertz of bandwidth (bits/s/Hz). As mentioned earlier, the link quality in many wireless systems is limited by co-channel interference. Hence, modulation schemes must be identified that are both bandwidth efficient and capable of tolerating high levels of co-channel interference. More specifically, digital modulation techniques are chosen for FDMA and TDMA wireless systems that satisfy the following properties:

- *Compact Power Density Spectrum:* To minimize the effect of adjacent channel interference, it is desirable that the power radiated into the adjacent band be 60 to 80 dB below that in the desired band. Hence, modulation techniques with a narrow main lobe and fast roll-off of side-lobes are needed.

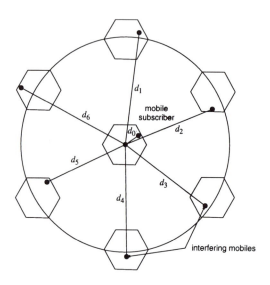

Figure 1.15 Co-channel interference on the reverse channel at a desired BS. There are six interfering MSs.

■ *Good Bit Error Rate Performance:* A low bit error probability should be achieved in the presence of co-channel interference, adjacent channel interference, thermal noise, and other channel impairments such as fading and intersymbol interference.

■ *Envelope Properties:* Portable and mobile applications typically employ non-linear (Class-C) power amplifiers to minimize battery drain. Nonlinear amplification may degrade the bit error rate performance of modulation schemes that transmit information in the amplitude of the carrier. Also, spectral shaping is usually performed prior to up-conversion and non-linear amplification. To prevent the regrowth of spectral side-lobes during non-linear amplification, relatively constant envelope modulation schemes are preferred.

A variety of digital modulation techniques are currently being used in wireless communication systems. Two of the more widely used digital modulation techniques for cellular mobile radio are $\pi/4$-DQPSK and GMSK. The former is used in the North American IS-54 and Japanese PDC and PHS systems, while the latter is used in the European GSM, DCS1800, DECT, and CT2 systems.

1.6   HANDOFFS AND CHANNEL ASSIGNMENT

When a new call arrives, the MS must be connected to a suitable BS. Also, when a MS traverses a cell boundary an **intercell handoff** is required so that an acceptable link quality can be maintained without causing unnecessary co-channel and adjacent channel interference. Failure to handoff a MS at an established cell boundary also tends to increase blocking, because some cells will carry more traffic than planned. Sometimes an **intracell handoff** is desirable when the link with the serving BSis affected by excessive interference, while another link with the same BS can provide better quality. The handoff process consists of two stages: i) link quality evaluation and handoff initiation, ii) allocation of radio and network resources.

In general, cellular systems with smaller cell sizes require faster and more reliable link quality evaluation and handoff algorithms. Labedz [169] has shown that the number of cell boundary crossings is inversely proportional to the cell size. Furthermore, Nanda [231] has shown that the handoff rate increases with only the square-root of the call density in macrocells, but linearly with the call density in microcells. Since the MS has a certain probability of handoff failure each time a handoff is attempted, it is clear that handoff algorithms must become more robust and reliable as the cell sizes decrease.

1.6.1   Link Quality Evaluation and Control

One of the major tasks in a cellular system is to monitor the link quality and determine when handoff is required. If a handoff algorithm does not detect poor signal quality fast enough, or makes too many handoffs, then capacity is diminished due to excessive co-channel interference and/or an undue switching load. Based on the roles that the BSs and MSs perform in the process of link quality evaluation and handoff initiation, there are three categories of handoff algorithms . The first is a **network-controlled handoff** (NCHO) algorithm which has been widely used in first generation analog cellular systems, such as AMPS. With a NCHO algorithm, the link quality is only monitored by the serving BS and the surrounding BSs. The handoff decision is made under the centralized control of a mobile telephone switching office (MTSO). Typically, NCHO algorithms only support only intercell handoffs, have handoff network delays of several seconds, and have relatively infrequent updates of the link quality estimates from the alternate BSs.

The second type of handoff algorithm is the **mobile-assisted handoff** (MAHO) algorithm which is widely used in many second generation digital cellular systems, such as IS-54 and GSM. MAHO algorithms use both the serving BS and the MS to measure link quality of the serving BS; however, link quality measurements of the alternate BSs are only obtained by the MS. The MS periodically relays the link quality measurements back to the serving BS, and the handoff decision is still made the serving BS along with the MTSO. MAHO algorithms typically support both intracell and intercell handoffs, have network delays on the order of one to two seconds, an use relatively frequent updates of the link quality measurements.

The third type of handoff algorithm is a **mobile-controlled handoff** (MCHO) algorithm, a decentralized strategy that is used in some of the more recent digital cordless telephone systems, such as DECT. With MCHO algorithms the link quality with the serving BS is measured by both the serving BS and the MS. Like a MAHO algorithm, the measurements of link quality for alternate BSs are done at the MS, and both intracell and intercell handoffs are supported. However, unlike the MAHO algorithms, the link measurements at the serving BS are relayed to the MS, and the handoff decision is made by the MS. MCHO algorithms typically have the lowest handoff network delays (usually about 100 ms) and are the most reliable.

Once the handoff process is initiated, handoff algorithms can also differ in the way that a call transferred to a new link. Handoff algorithms can be categorized into forward and backward types. **Backward handoff** algorithms initiate the handoff process through the serving BS, and no access to the "new" channel is made until the control entity of the new channel has confirmed the allocation of resources. The advantage of backward algorithms is that the signaling information is transmitted through an existing radio link and, therefore, the establishment of a new signaling channel is not required during the initial stages of the handoff process. The disadvantage is that the algorithm may fail in conditions where the link quality with the serving BS is rapidly deteriorating. This type of handoff is used in most of the TDMA cellular systems such as GSM. **Forward handoff** algorithms initiate the handoff process via a channel the the target BS without relying on the "old" channel during the initial phase of the handoff process. The advantage is a faster handoff process, but the disadvantage is a reduction in handoff reliability. This type of handoff is used in digital cordless telephone systems such as DECT.

Handoff can also be distinguished according to **hard handoffs** and **soft handoffs** . Hard handoffs release the radio link with the old BS at the same time that the radio link with the new BS is established. This type of handoff is used in most TDMA cellular systems such as IS-54, PDC, and GSM. Soft handoffs maintain a radio link with at least two BSs in a handoff region, and a link is dropped only when the signal level drops below a certain threshold. This type of handoff is used in CDMA cellular systems such as IS-95.

Classical handoff algorithms use received signal strength power as the link quality criterion. These algorithms execute a handoff if the received signal strength with an alternate BS, measured over a T second interval, exceeds that of the serving BS by H dB, where H is a handoff **hysteresis** that prevents excessive handoffs due to "ping-ponging" between BSs. The best choice of T and H depends on the propagation environment. Usually, the averaging interval T is chosen to correspond to 20 to 40 wavelengths, and the hysteresis H is chosen on the order of the shadow standard deviation. One of the main drawbacks of signal strength based handoff algorithms is their inability to distinguish between the received carrier power, C, from the serving BS and the total interfering power, I, from the co-channel BSs. Hence, a link having a large received signal power $C + I$ may have a small C/I. This can be remedied by using other link quality measures such as the received C/I, the bit error rate, or the distance from the serving BS.

Although the best handoff algorithm is the one that maximizes the capacity of the network, there are many criterion to judge the performance of a handoff algorithm. These include the probability of handoff initiation, probability of dropped call, the mean number of handoff requests as a MS traverses over a handoff route, and the delay before a handoff is initiated after a MS crosses an established cell boundary. These quantities depend on the measure of link quality and the propagation environment. Finally, network parameters such as the probabilities of new call blocking, the probability of forced termination, and handoff queuing time are important. Note that we may also wish to distinguish between dropped calls that are due to a failed handoff mechanism, and forced terminations that are due to the lack of an unavailable channel in the target cell after successful initiation of the handoff process.

1.6.2 Channel Assignment Techniques

There are many methods of allocating a channel upon a new call arrival or handoff attempt. A good channel allocation algorithm is the one that yields high spectral efficiency for a specified **grade of service** (including link quality, probability of new call blocking, and the probability of forced termination) and given degree of computational complexity. It also keeps the planned cell

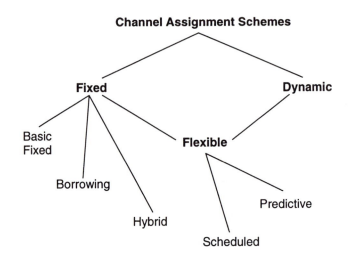

Figure 1.16   Basic classifications of channel assignment schemes, from [299].

boundaries intact, allocates a channel to a MS quickly, maintains the best speech quality for the MS at any instant, and relieves undesired network congestion. As shown in Fig. 1.16, there are three basic types of channel assignment algorithms, **fixed**, **flexible**, and **dynamic** [299].

Fixed Channel Assignment (FCA)

With fixed channel assignment (FCA), each cell is permanently assigned a fixed set of channels. A new call arrival or handoff attempt is blocked if all channels are busy. These blocking probabilities can be reduced by using various schemes that borrow channels from neighboring cells. The most basic scheme is simple borrowing, where a MS can be allocated a channel from a neighboring cell, provided that it does not degrade the link quality of other calls by introducing excessive co-channel interference. Once a channel is borrowed, all other cells that are within the co-channel reuse distance are prohibited from using the channel. The efficiency of this borrowing strategy tends to degrade in heavy traffic and the channel utilization is worse than FCA. This problem can be partially solved by using a hybrid channel assignment scheme, where the channels assigned to a cell are divided into two groups; the channels in one group are owned by the cell, while the channels in the other group may be borrowed. There are several variations of this theme. The ratio of the number of owned-to-borrowable channels can be dynamically varied to compensate for traffic changes.

Dynamic Channel Assignment (DCA)

With dynamic channel assignment (DCA) channels (or carriers) are not permanently assigned to cells. Any cell can use any channel (or carrier) that does not violate the co-channel (of frequency) reuse constraint. We distinguish between channels and carriers, because in TDMA systems there are multiple channels per carrier. Hence, the channels can be assigned individually or in carrier groups. DCA schemes tend to be more efficient than FCA schemes in conditions of light, non-homogeneous, and time-varying traffic.

DCA schemes differ in the distribution of control and the communication required among the BSs. **Centralized DCA** schemes such as maximum packing (MP) require system-wide information [88]. With MP a call is blocked only if there is no global rearrangement of calls to channels that will accommodate the call. **Fully decentralized DCA** schemes are the other extreme, where no communication is required among BSs [92], [241], [5]. These schemes are ideal for cordless telephone systems that use MCHO, such as DECT, and rely only upon local interference monitoring to make channel assignments.

Decentralized DCA schemes require limited communication among local clusters of BSs. One DCA scheme is dynamic resource acquisition (DRA) [230]. With DRA, the channel (or carrier) that is acquired due to a new call arrival or handoff is chosen to minimize a cost function, and the channel (or carrier) that is released due a call completion or handoff is chosen to maximize a reward function. The cost and reward functions can be selected to maximize the spectral efficiency of the cellular network for a specified grade of service. The computation of the cost and reward functions for a given cell depends on the usages of the channels (or carriers) in the set of surrounding cells called the DRA neighborhood [230]. Another distributed DCA scheme is simple dynamic channel assignment (SDCA) [334]. SDCA performs slightly worse than DRA, but requires communication among a smaller set of cells called the interference neighborhood [334].

Flexible Channel Assignment

Flexible channel assignment algorithms combine aspects of fixed and dynamic channel assignment schemes. Each cell is assigned a fixed set of channels, but a pool of channels is reserved for flexible assignment. The assignment of flexible channels can be either scheduled or predictive [297]. Scheduled assignment schemes rely on known changes in traffic patterns. The flexible channels are assigned to the cells on a scheduled basis to account for these foreseeable changes in traffic patterns. With predictive assignment, the traffic load is continuously or periodically measured at every BS, and the flexible channels are assigned to the cells according to these measurements.

Hand-off Priority

Forced terminations are generally perceived to severely degrade the quality of service. For this reason, **handoff priority** schemes are usually employed to allocate channels to handoff requests more readily than to new call arrivals. Handoff priority reduces the probability of forced termination at the expense of a (slight) increase in the probability of new call blocking. Practical cellular systems are designed to have a probability of new call blocking less that 5%, with a probability of forced termination perhaps an order of magnitude smaller.

The use of **guard channels** is one method of achieving handoff priority, where the channels are divided into two groups; one group is for new calls and handoff requests, and the other group is reserved for handoff requests only [149]. Another method is to queue the handoff requests (but not the new call arrivals) [149], [110]. This method can be combined with guard channels.

1.7 SPECTRAL EFFICIENCY AND GRADE OF SERVICE

Spectral efficiency is of primary concern to cellular system planners. There are a variety of definitions for spectral efficiency, but an appropriate definition measures spectral efficiency in terms of the spatial traffic density per unit bandwidth. For a cellular system that consists of a deployment of uniform cells, the spectral efficiency can be expressed in terms of the following parameters:

$$
\begin{aligned}
G_c &= \text{offered traffic per channel (Erlangs/channel)} \\
N_c &= \text{number of channels per cell} \\
W_{\text{sys}} &= \text{total system bandwidth (Hz)} \\
A &= \text{area per cell (m}^2) \quad .
\end{aligned}
$$

One Erlang is the traffic intensity in a channel that is continuously occupied, so that a channel occupied for $x\%$ of the time carriers $x/100$ Erlangs. The spectral efficiency is defined as

$$
\eta_{\text{S}} = \frac{N_c \cdot G_c}{W_{\text{sys}} \cdot A} \quad \text{Erlangs/m}^2/\text{Hz} \quad . \tag{1.15}
$$

Suppose that the cellular deployment consists of N-cell reuse clusters. Then the number of channels per cell with FDMA is

$$N_c = \frac{W_{\text{sys}}}{W_c \cdot N} \qquad (1.16)$$

where W_c is the bandwidth per channel. If TDMA is used, then W_c is the bandwidth per carrier divided by the number of channels per carrier. The spectral efficiency can be written as the product of three efficiencies, viz.,

$$\begin{aligned} \eta_{\text{S}} &= \frac{1}{W_c} \cdot \frac{1}{NA} \cdot G_c \\ &= \eta_{\text{B}} \cdot \eta_{\text{C}} \cdot \eta_{\text{T}} \, . \end{aligned} \qquad (1.17)$$

The first factor η_B is the bandwidth efficiency. High bandwidth efficiency can be achieved by using low bit rate voice coding and bandwidth efficient modulation.

The second factor η_C is the spatial efficiency. High spatial efficiency can be achieved by i) minimizing the area per cell, and ii) minimizing the co-channel reuse distance. The first of these explains the intense interest in microcellular systems, where cell radii on the order of 200-500 m are used. The co-channel reuse distance D/R is minimized by designing the radio links to minimize Λ_{th}. In practice, this is achieved by using error control coding, antenna diversity, adaptive equalization, and other such receiver techniques. It is also possible to minimize the co-channel reuse distance by using techniques such as cell sectoring, adaptive transmitter power control, discontinuous transmission, effective hand-off algorithms, macroscopic BS diversity, and others, to control co-channel interference.

Finally, $\eta_T = G_c$ is the trunking efficiency. High trunking efficiency can be achieved by using channel assignment schemes that maximize channel utilization. There is usually a trade-off between trunking efficiency (or offered traffic per channel) and grade of service in terms of new call and handoff blocking probabilities. Various fundamental formula were developed by Erlang, who laid the foundations of modern teletraffic theory. One of his most famous results is the **Erlang-B formula**, first derived in 1917, that gives the probability that a new call attempt will not find an available channel in a trunk of channels and is lost. Sometimes this policy is called the **blocked calls cleared** queueing discipline and it is widely used to model wireline telephone traffic. The Erlang-B formula is not really applicable to cellular systems, because it does not account for handoff traffic. Furthermore, the total offered traffic per cell is time-varying due to the spatial movement of the subscribers, whereas the offered traffic in the Erlang-B formula is assumed to be constant. Nevertheless,

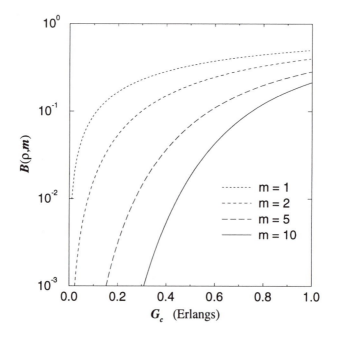

Figure 1.17 erlang-B blocking probability $B(\rho, m)$ vs. offered traffic per channel $G_c = \rho/m$. Trunking is shown to improve the spectral efficiency.

it provides useful insight. The Erlang-B formula is

$$B(\rho, m) = \frac{\rho^m}{m! \sum_{k=0}^{m} \frac{\rho^k}{k!}} \qquad (1.18)$$

where m is the total number of channels in the trunk and $\rho = \lambda\mu$ is the total offered traffic (λ is the call arrival rate and μ is the mean call duration). The Erlang-B formula is derived under the assumption of an infinite subscriber population, Poisson call arrivals with rate λ calls/s, and exponentially distributed call durations with a mean call duration μ s/call. Fig. 1.17 plots the blocking probability $B(\rho, m)$ as a function of the offered traffic per channel $G_c = \rho/m$. The benefits from **trunking** is obvious, since the offered traffic per channel, G_c, increases as the number of trunked channels increases, for any blocking probability. However, diminishing returns are obtained as the number of trunked channels becomes larger. In general, the design of high capacity spectrally efficient cellular systems is a very difficult task because there are many complex inter-relationships and trade-offs between the factors that determine the spectral efficiency and grade of service.

Problems

1.1. Show that the area averaged probability of TN is given by (1.11).

1.2. By using geometric arguments, show that the co-channel reuse factor for cellular deployments based on hexagonal cells is given by (1.9).

1.3. Consider a regular hexagonal cell deployment, where the MSs and BSs use omnidirectional antennas. Suppose that we are interested in the forward channel performance and consider only the first tier of co-channel interferers as shown in Fig. 1.14. Ignore the effects of shadowing and multipath fading, and assume that the propagation path loss is described by the inverse β law in (1.6).

a) Determine the worst case carrier-to-interference ratio, Λ, as a function of the reuse cluster size N, for $\beta = 3, 3.5$, and 4.

b) What is the minimum cluster size that is needed if the radio receivers have $\Lambda_{th} = 18$ dB?

c) Referring to Fig. 1.15, repeat a) and b) for the reverse channel.

1.4. Suppose that a single isolated cell is used to provide cellular telephone service to a small town. We make the following assumptions; the number of customers is effectively infinite, the call interarrival times and durations are exponentially distributed, and the call set-up times are negligible. An arriving call that does not find an idle channel is allowed to remain in a queue for t_q seconds and is dropped from the queue (blocked) if no channel becomes available in that time. The queue is serviced using a "first come first served" discipline. If m is the total number of channels in the trunk and ρ is the total offered traffic, then the probability of queueing is given by the famous **Erlang-C formula**

$$C(\rho, m) = \frac{\rho^m}{\rho^m + m! \left(1 - \frac{\rho}{m}\right) \sum_{k=0}^{m-1} \frac{\rho^k}{k!}}$$

The probability that a queued call will have to wait more than t_q seconds in the queue is

$$P(W > t_q) = \exp\left\{-\frac{(m - \rho)t_q}{\mu}\right\}$$

where μ is the mean call duration. Assuming that $\mu = 120$ s and $t_q = 5$ s, plot the blocking probability against the normalized offered traffic per channel $G_c = \rho/m$, for various values of m.

2

PROPAGATION MODELING

A typical cellular radio system consists of a collection of base stations (BSs) that are relatively free from local scatterers. The BS antenna height and placement affects the proximity of local scatterers. In a macrocellular environment, the BS antennas are usually well elevated above the local terrain. No direct line-of-sight (LOS) path exists between the BS and mobile station (MS) antennas, because of the natural and man-made objects that are in the immediate vicinity of the MS. As a consequence of reflections, scattering and diffraction, multiple plane waves arrive at a MS from many different directions and with different delays, as shown in Fig. 2.1. This property is called multipath propagation. The multiple plane waves combine vectorially at the receiver antenna to produce a composite received signal.

The carrier wavelength used in UHF mobile radio applications ranges from 15 to 60 cm. Therefore, small changes in the differential delays due to MS mobility will cause large changes in the phases of the arriving plane waves. These phase differences cause constructive and destructive addition of the arriving plane waves which manifests itself as large variations in the amplitude and phase of the composite received signal. Since the MS is moving through space, the spatial variations in the envelope and phase of the composite received signal manifest themselves as time variations, a phenomenon called envelope fading. As we will see later, the fading rate depends on the velocity of the MS.

Radio channels are reciprocal in the sense that if a propagation path exists, it carries energy equally well in both directions. However, the direction of arriving plane waves may be significantly different in each direction. A MS in a typical macrocellular environment is usually surrounded by local scatterers so that the plane waves will arrive from many directions without a direct LOS

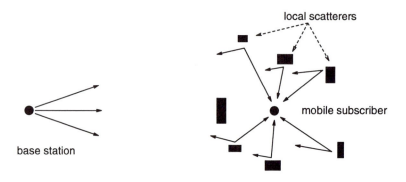

Figure 2.1 Typical macrocellular radio propagation.

component. Isotropic scattering where the arriving plane waves arrive from all directions with equal probability is a reasonable modeling assumptions at a MS in a macrocellular system. However, the BSs are relatively free from local scatterers so that the plane waves tend to arrive from one general direction as shown in Fig. 2.1. These differences in the scattering environment for the forward and reverse channels causes differences in the spatial correlation of their respective faded envelopes.

In a microcellular environment, the BS antennas are only moderately elevated above the local scatterers. Sometimes a LOS path exists between the MS and BS, while at others times there is no LOS component. Therefore, the received signal will still experience fading. However, the scattering is usually non-isotropic [12] and Ricean distributed envelope fading is often assumed [130], [346], [337].

Multipath-fading is sometimes called fast fading to distinguish it from shadowing, which manifests itself as a slow variation in the mean envelope (or mean square-envelope) over a distance corresponding to several tens of wavelengths. Experimental observations have confirmed that shadows are log-normally distributed with the pdf in (1.5) for both macrocellular [174], [326] and microcellular systems [211], [213], [131].

Path loss predicts how the mean signal strength power decays with distance from a BS. Early studies by Okumura [240] and Hata [141] resulted in path loss models for urban, suburban, and rural areas that are accurate to within 1 dB for distances ranging from 1 to 20 km. Since these studies concentrated on macrocellular systems, recent work has been directed to path loss prediction in microcells. The COST231 study [56] has resulted in the COST231-Hata and COST231-Walfish-Ikegami models for urban microcellular path loss prediction.

The remainder of this chapter presents the fundamentals of radio propagation. Section 2.1 introduces the mechanism of multipath-fading. Various properties of the faded envelope are then derived in Sections 2.1.1 through 2.1.5. Section 2.2 treats the characterization of wide-band multipath-fading channels. Laboratory simulation of fading channels is covered in Section 2.3. Shadowing models and simulation techniques are discussed in Section 2.4. Finally, Section 2.5 treats theoretical and empirical path loss models for macrocellular and microcellular systems.

2.1 FREQUENCY-NON-SELECTIVE (FLAT) MULTIPATH-FADING

Fig. 2.2 depicts a horizontal $x-y$ plane with a MS moving along the x-axis with velocity v. In portable and mobile radio applications the transmitted signals are usually vertically polarized and, therefore, the electric field vector is aligned with the z-axis. The nth plane wave arrives at the MS antenna with an angle of incidence $\theta_n(t)$. The MS movement introduces a **Doppler**, or frequency, shift into the incident plane wave, given by

$$f_{D,n}(t) = f_m \cos \theta_n(t) \quad \text{Hz} \tag{2.1}$$

where $f_m = v/\lambda_c$ and λ_c is the wavelength of the arriving plane wave. Plane waves arriving from the direction of motion will experience a positive Doppler shift, while those arriving opposite from the direction of motion will experience a negative Doppler shift.

Consider the transmission of the band-pass signal

$$s(t) = \text{Re}\left\{u(t)e^{j2\pi f_c t}\right\} \tag{2.2}$$

where $u(t)$ is the complex low-pass signal, f_c is the carrier frequency, and $\text{Re}\{z\}$ denotes the real part of z. If the channel is comprised of N paths, then the received band-pass waveform is

$$x(t) = \text{Re}\left\{r(t)e^{j2\pi f_c t}\right\} \tag{2.3}$$

where the received complex low-pass signal $r(t)$ is given by

$$r(t) = \sum_{n=1}^{N} \alpha_n(t)e^{-j2\pi[(f_c+f_{D,n}(t))\tau_n(t)-f_{D,n}(t)t]}u(t-\tau_n(t)) \tag{2.4}$$

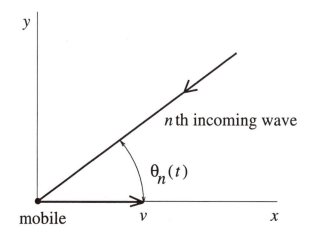

Figure 2.2   A typical plane wave component incident on a MS receiver.

and $\alpha_n(t)$ and $\tau_n(t)$ are the amplitude and time delay, respectively, associated with the nth path. The received complex low-pass signal can be rewritten as

$$r(t) = \sum_{n=1}^{N} \alpha_n(t) e^{-j\phi_n(t)} u(t - \tau_n(t)) \qquad (2.5)$$

where

$$\phi_n(t) = 2\pi \left\{ (f_c + f_{D,n}(t)) \tau_n(t) - f_{D,n}(t)t \right\} \qquad (2.6)$$

is the phase associated with the nth path. From (2.5), the channel can be modeled by a time-variant linear filter having the complex low-pass impulse response

$$c(\tau, t) = \sum_{n=1}^{N} \alpha_n(t) e^{-j\phi_n(t)} \delta(\tau - \tau_n(t)) \qquad (2.7)$$

where $c(\tau, t)$ response channel response at time t to an impulse applied at time $t - \tau$, and $\delta(\cdot)$ is the dirac delta function.

From (2.5) and (2.6), several interesting observations can be made. Since $f_c + f_{D,n}(t)$ is very large, a small change in the path delay $\tau_n(t)$ causes a large change in the phase $\phi_n(t)$. At any time t these random phases may result in the constructive or destructive addition of the components. The amplitudes $\alpha_n(t)$ depend on the cross sectional area of the nth scatterer or the length of the nth diffracting surface. However, these quantities do not change significantly over small spatial distances. Therefore, fading is primarily due to time variations in the random phases $\phi_n(t)$ that are caused by the Doppler shifts $f_{D,n}(t)$.

Sometimes the channel is characterized by either a LOS path or a specular component from a strong (fixed) local scatterer. In this case, the amplitude $\alpha_0(t)$ is significantly larger than the other $\alpha_n(t)$.

2.1.1 Received Signal Correlation and Spectrum

It is apparent that the different frequency components in a signal will be affected differently by the multipath-fading channel. However, for narrow-band signals where the signal bandwidth is very small compared to the carrier frequency, it suffices to derive the characteristics of the received complex low-pass signal by considering the transmission of an unmodulated carrier. For an unmodulated carrier, the received complex low-pass signal is

$$r(t) = \sum_{n=1}^{N} \alpha_n(t) e^{-j\phi_n(t)} \ . \tag{2.8}$$

Using (2.3), the received band-pass signal can be expressed in the quadrature form

$$x(t) = r_I(t) \cos 2\pi f_c t \ - \ r_Q(t) \sin 2\pi f_c t \tag{2.9}$$

where

$$r_I(t) \ = \ \sum_{n=1}^{N} \alpha_n(t) \cos \phi_n(t) \tag{2.10}$$

$$r_Q(t) \ = \ \sum_{n=1}^{N} \alpha_n(t) \sin \phi_n(t) \tag{2.11}$$

and $r(t) = r_I(t) + jr_Q(t)$. For large N, the central limit theorem can be invoked so that the quadrature components $r_I(t)$ and $r_Q(t)$ can be treated as independent Gaussian random processes. Assuming that these random processes are wide sense stationary (i.e., $f_{D,n}(t) = f_{D,n}$, $\alpha_n(t) = \alpha_n$, and $\tau_n(t) = \tau_n$), and assuming that $x(t)$ is wide sense stationary, the autocorrelation of $x(t)$ is

$$
\begin{aligned}
\phi_{xx}(\tau) \ &= \ \mathrm{E}[x(t)x(t+\tau)] \\
&= \ \mathrm{E}[r_I(t)r_I(t+\tau)] \cos 2\pi f_c \tau - \mathrm{E}[r_Q(t)r_I(t+\tau)] \sin 2\pi f_c \tau \\
&= \ \phi_{r_I r_I}(\tau) \cos 2\pi f_c \tau - \phi_{r_Q r_I}(\tau) \sin 2\pi f_c \tau \ . \tag{2.12}
\end{aligned}
$$

Note that

$$\phi_{r_I r_I}(\tau) = \phi_{r_Q r_Q}(\tau) \tag{2.13}$$

$$\phi_{r_I r_Q}(\tau) = -\phi_{r_Q r_I}(\tau) \ . \tag{2.14}$$

It is reasonable to assume that the phases $\phi_n(t)$ and $\phi_m(t)$ are independent for $n \neq m$ since their associated delays and Doppler shifts are independent. Furthermore, the phases $\phi_n(t)$ can be assumed to be uniformly distributed over $[-\pi, \pi]$, since $f_c \tau_n \gg 1$. By using these properties, it is straightforward to obtain the autocorrelation $\phi_{r_I r_I}(\tau)$ from (2.10) and (2.1) as follows:

$$\begin{aligned}
\phi_{r_I r_I}(\tau) &= \mathrm{E}[r_I(t) r_I(t + \tau)] \\
&= \frac{\Omega_p}{2} \mathrm{E}[\cos 2\pi f_{D,n} \tau] \\
&= \frac{\Omega_p}{2} \mathrm{E}_\theta[\cos(2\pi f_m \tau \cos\theta)]
\end{aligned} \tag{2.15}$$

where

$$\frac{\Omega_p}{2} = \mathrm{E}[x^2(t)] = \mathrm{E}[r_I^2(t)] = \mathrm{E}[r_Q^2(t)] = \frac{1}{2} \sum_{n=1}^{N} \mathrm{E}[\alpha_n^2] \tag{2.16}$$

is the total average received power from all multipath components.

Likewise, the crosscorrelation $\phi_{r_I r_Q}(\tau)$ is

$$\begin{aligned}
\phi_{r_I r_Q}(\tau) &\overset{\Delta}{=} \mathrm{E}[r_I(t) r_Q(t + \tau)] \\
&= \frac{\Omega_p}{2} \mathrm{E}_\theta[\sin(2\pi f_m \tau \cos\theta)] \ .
\end{aligned} \tag{2.17}$$

Evaluation of the expectations in (2.15) and (2.17) requires that we specify the probability density function for the angle of incidence of the arriving plane waves, $p(\theta)$. For macrocellular applications, it is reasonable to assume that the plane waves arrive at the MS antenna arrive from all directions in the (x, y) plane with equal probability, i.e., θ is uniformly distributed over $[-\pi, \pi]$. This model was first suggested by Clarke [51], and is commonly referred to as Clarke's two-dimensional **isotropic scattering** model. With isotropic scattering, the expectation in (2.15) becomes

$$\begin{aligned}
\phi_{r_I r_I}(\tau) &= \frac{\Omega_p}{2} \frac{1}{2\pi} \int_{-\pi}^{\pi} \cos\left(2\pi f_m \tau \cos\theta\right) d\theta \tag{2.18} \\
&= \frac{\Omega_p}{2} \frac{1}{\pi} \int_{0}^{\pi} \cos\left(2\pi f_m \tau \sin\theta\right) d\theta \tag{2.19} \\
&= \frac{\Omega_p}{2} J_0(2\pi f_m \tau) \tag{2.20}
\end{aligned}$$

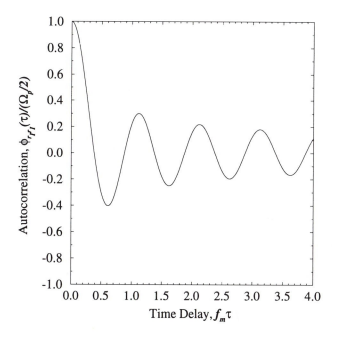

Figure 2.3 Autocorrelation of the inphase and quadrature components of the received complex low-pass signal for isotropic scattering channel.

where $J_0(x)$ is the zero-order Bessel function of the first kind. Likewise, (2.17) becomes

$$
\begin{aligned}
\phi_{r_I r_Q}(\tau) &= \frac{\Omega_p}{2} \frac{1}{2\pi} \int_{-\pi}^{\pi} \sin\left(2\pi f_m \tau \cos\theta\right) d\theta \\
&= 0 .
\end{aligned}
\tag{2.21}
$$

The normalized autocorrelation $\phi_{r_I r_I}(\tau)/(\Omega_p/2)$ is plotted against the normalized time delay $f_m \tau$ in Fig. 2.3.

The power density spectrum (psd) of $r_I(t)$ and $r_Q(t)$ is the Fourier transform of $\phi_{r_I r_I}(\tau)$ or $\phi_{r_Q r_Q}(\tau)$, i.e.,

$$
\begin{aligned}
S_{r_I r_I}(f) &= \mathcal{F}[\phi_{r_I r_I}(\tau)] \\
&= \begin{cases} \dfrac{\Omega_p}{4\pi f_m} \dfrac{1}{\sqrt{1-(f/f_m)^2}} & |f| \leq f_m \\ 0 & \text{otherwise} \end{cases} .
\end{aligned}
\tag{2.22}
$$

The autocorrelation of the received complex low-pass signal $r(t) = r_I(t) + jr_Q(t)$ is

$$
\begin{aligned}
\phi_{rr}(\tau) &= \frac{1}{2}\mathrm{E}[r^*(t)r(t+\tau)] \\
&= \phi_{r_I r_I}(\tau) + j\phi_{r_I r_Q}(\tau) \ .
\end{aligned}
\tag{2.23}
$$

From (2.12) we have

$$
\phi_{xx}(\tau) = \mathrm{Re}\left[\phi_{rr}(\tau)e^{j2\pi f\tau}\right] \ .
\tag{2.24}
$$

Since $\phi_{rr}(\tau) = \phi_{rr}^*(-\tau)$, it follows that the psd of the band-pass waveform $x(t)$ is

$$
S_{xx}(f) = \frac{1}{2}[S_{rr}(f - f_c) + S_{rr}(-f - f_c)] \ .
\tag{2.25}
$$

With isotropic scattering $\phi_{r_I r_Q}(\tau) = 0$ so that

$$
S_{xx}(f) = \frac{1}{2}[S_{r_I r_I}(f - f_c) + S_{r_I r_I}(-f - f_c)]
\tag{2.26}
$$

where $S_{r_I r_I}(f)$ is given by (2.22).

The psd in (2.26) can be derived by using a different approach that is perhaps more insightful and sometimes more useful. As $N \to \infty$, the incident power on the receiver antenna as a function of the angle of incidence θ approaches a continuous distribution, denoted by $p(\theta)$. The fraction of the total incoming power that arrives between θ and $\theta + d\theta$ is $p(\theta)d\theta$. If the antenna has a gain of $G(\theta)$ at angle θ, then the corresponding received power is $G(\theta)p(\theta)d\theta$. Therefore, the psd of the received signal can be expressed as

$$
S_{xx}(f)|df| = \frac{\Omega_p}{2}\{G(\theta)p(\theta) + G(-\theta)p(-\theta)\}|d\theta| \ .
\tag{2.27}
$$

From Fig. 2.2, the frequency of the incident plane wave arriving at angle θ is

$$
f = f_m \cos\theta + f_c \ ,
\tag{2.28}
$$

where $f_m = v/\lambda_c$ is the maximum Doppler shift and, hence,

$$
|df| = f_m| - \sin\theta d\theta| = \sqrt{f_m^2 - (f - f_c)^2}|d\theta| \ .
\tag{2.29}
$$

Therefore,

$$
S_{xx}(f) = \frac{\Omega_p/2}{\sqrt{f_m^2 - (f - f_c)^2}}\{G(\theta)p(\theta) + G(-\theta)p(-\theta)\}
\tag{2.30}
$$

where

$$\theta = \cos^{-1}\left(\frac{f - f_c}{f_m}\right) . \tag{2.31}$$

To proceed further requires assumptions about the density $p(\theta)$ and the antenna gain pattern $G(\theta)$. If a vertical monopole antenna is used, then $G(\theta) = 3/2$. Furthermore, with isotropic scattering $p(\theta0 = 1/2\pi, -\pi \leq \theta \leq \pi$, so that

$$S_{xx}(f) = \begin{cases} \frac{3\Omega_p}{4\pi f_m} \dfrac{1}{\sqrt{1 - \left(\frac{|f - f_c|}{f_m}\right)^2}} & |f - f_c| \leq f_m \\ 0 & \text{otherwise} \end{cases} \tag{2.32}$$

Except for the factor of $3/2$ that results from the antenna gain, (2.32) is same result derived in (2.22) and (2.25). If a more complicated antenna gain pattern is used, or if the scattering is not isotropic, then (2.30) can be easily used to compute the psd and autocorrelation of the band-pass received signal.

The normalized psd $S_{xx}(f)/(3\Omega_p/4\pi f_m)$ is plotted against the normalized frequency difference $(f - f_c)/f_m$ in Fig. 2.4. Notice that $S_{xx}(f)$ is limited to the range of frequencies $|f - f_c| \leq f_m$ or twice the maximum Doppler frequency, and $S_{xx}(f) = \infty$ at $f = f_c \pm f_m$. In reality the psd will not goto infinity, and the reason for this behavior is that the plane waves were assumed to propagate in a 2-D plane, whereas in reality the propagation is actually in 3-D space. Aulin [12] modified Clarke's model to include 3-D propagation. The psd that Aulin obtained was very similar to Fig. 2.4, except that it remained finite at $f = f_c \pm f_m$.

In many cases, a LOS or specular component will often be present in the received signal. For example, consider the scattering environment shown in Fig. 2.5, where a LOS or specular component arrives at angle θ_0, but the scattering is otherwise isotropic. By using (2.30) the psd has the form

$$S_{xx}(f) = \begin{cases} \dfrac{1}{K+1} \cdot \dfrac{3\Omega_p}{4\pi f_m} \dfrac{1}{\sqrt{1 - \left(\frac{|f - f_c|}{f_m}\right)^2}} \\ \quad + \dfrac{K}{K+1} \dfrac{3\Omega_p}{4} \delta(f - f_c - f_m \cos\theta_0) & |f - f_c| \leq f_m \\ 0 & \text{otherwise} \end{cases} \tag{2.33}$$

The parameter K is called the **Rice factor** and is defined as the ratio of the power in the specular and scatter components of the received signal. The psd in (2.33) is the same as Fig. 2.4, except for a discrete tone at $f_c + f_m \cos\theta_0$.

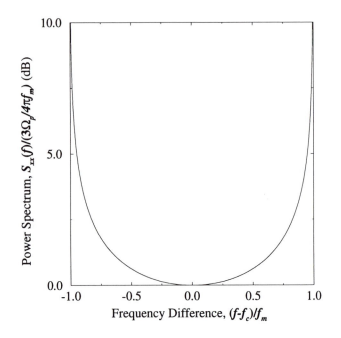

Figure 2.4 Psd of the received band-pass signal for an isotropic scattering channel.

The autocorrelation functions $\phi_{r_I r_I}(\tau)$ and $r_{r_I r_Q}(\tau)$ corresponding to (2.33) can be readily obtained from (2.18) and (2.21) as

$$\phi_{r_I r_I}(\tau) = \frac{1}{K+1}\frac{\Omega_p}{2}J_0(2\pi f_m \tau) + \frac{K}{K+1}\frac{\Omega_p}{2}\cos(2\pi f_m \tau \cos\theta_0) \quad (2.34)$$

$$\phi_{r_I r_Q}(\tau) = \frac{K}{K+1}\frac{\Omega_p}{2}\sin(2\pi f_m \tau \cos\theta_0) \ . \quad (2.35)$$

For microcells that are deployed in dense urban areas, the plane waves may be channeled by the buildings along the streets and arrive at the receiver antenna from just one direction, as shown in Fig. 2.6. Clearly, the scattering is non-isotropic. In this case, a variety of models may be used for distribution of arriving plane waves. One plausible distribution is

$$p(\theta) = \begin{cases} \frac{\pi}{4|\theta_m|}\cos\left(\frac{\pi}{2}\cdot\frac{\theta}{\theta_m}\right) & , \ |\theta| \leq |\theta_m| \leq \frac{\pi}{2} \\ 0 & , \ \text{elsewhere} \end{cases} \quad (2.36)$$

The parameter θ_m determines the directivity of the incoming waves. Fig. 2.7 shows a polar plot of $p(\theta)$ against the angle of arrival of the plane waves for

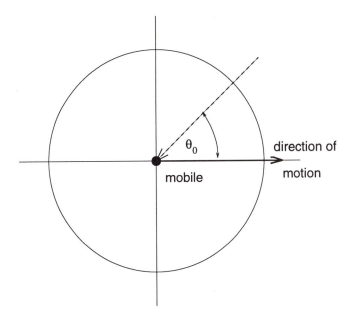

Figure 2.5 Arriving plane waves with isotropic scattering plus a specular component arriving at angle θ_0.

$\theta_m = 30°$, $60°$, and $90°$. If we assume an omnidirectional antenna, then the psd of the received band-pass signal $S_{xx}(f)$ can be readily obtained by substituting (2.36) into (2.30). Likewise, the autocorrelation $\phi_{r_I r_I}(\tau)$ and crosscorrelation $\phi_{r_I r_Q}(tau)$ can be obtained by evaluating the expectations in (2.15) and (2.17), respectively, with the density in (2.36).

2.1.2 Received Envelope and Phase Distribution

Rayleigh Fading

When the composite received signal consists of a large number of plane waves, the received complex low-pass signal $r(t) = r_I(t) + jr_Q(t)$ can be modeled as a

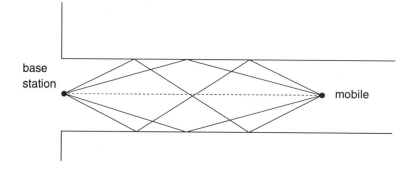

Figure 2.6 An urban microcellular propagation environment characterized by unidirectional scattering.

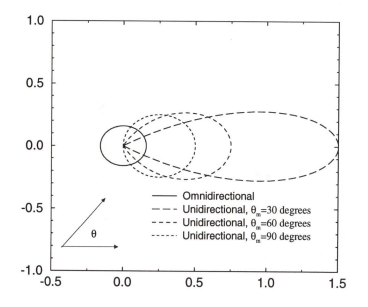

Figure 2.7 Arriving planes waves with unidirectional scattering.

complex Gaussian random process. In the absence of a LOS or specular component, $r_I(t)$ and $r_Q(t)$ have zero-mean. By using a bivariate transformation (see Appendix 2A), the received complex envelope $z(t) = |r(t)|$ has a Rayleigh distribution at any time t, i.e.,

$$p_z(x) = \frac{x}{\sigma^2} \exp\left\{-\frac{x^2}{2\sigma^2}\right\} \ . \tag{2.37}$$

For a Rayleigh distributed envelope, the average power is $E[z^2] = \Omega_p = 2\sigma^2$ so that

$$p_z(x) = \frac{2x}{\Omega_p} \exp\left\{-\frac{x^2}{\Omega_p}\right\} \qquad x \geq 0 . \tag{2.38}$$

This type of fading is called **Rayleigh fading** and agrees very well with empirical observations for macrocellular applications. Rayleigh fading usually applies to any scenario where there is no LOS path between the transmitter and receiver antennas. By using a transformation of random variables, the squared-envelope $z^2(t) = |r(t)|^2$ is exponentially distributed with density

$$p_{z^2}(x) = \frac{1}{\Omega_p} \exp\left\{-\frac{x}{\Omega_p}\right\} . \tag{2.39}$$

Ricean Fading

For a multipath-fading channel containing a specular or LOS component, $r_I(t)$ and $r_Q(t)$ have non-zero mean and the complex envelope has a Ricean distribution at any time t, i.e.,

$$p_z(x) = \frac{x}{\sigma^2} \exp\left\{-\frac{x^2 + s^2}{2\sigma^2}\right\} I_o\left(\frac{xs}{\sigma^2}\right) \qquad x \geq 0 , \tag{2.40}$$

where

$$s^2 = \alpha_0^2 \cos^2\theta_0 + \alpha_0^2 \sin^2\theta_0 = \alpha_0^2 \tag{2.41}$$

is the non-centrality parameter. This type of fading is called **Ricean fading** and is very often observed in microcellular applications. The Rice factor or K factor is the ratio of the power in the specular and scattered components, i.e., $K = s^2/2\sigma^2$. When $K = 0$ the channel exhibits Rayleigh fading, and when $K = \infty$ the channel does not exhibit fading. The envelope distribution can be written in terms of the Rice factor $K = s^2/(2\sigma^2)$. For a Ricean distributed envelope, the average power is $E[z^2] = \Omega_p = s^2 + 2\sigma^2$ and

$$s^2 = \frac{K\Omega_p}{K+1}, \qquad 2\sigma^2 = \frac{\Omega_p}{K+1} . \tag{2.42}$$

Hence,

$$p_z(x) = \frac{2x(K+1)}{\Omega_p} \exp\left\{-K - \frac{(K+1)x^2}{\Omega_p}\right\} I_o\left(2x\sqrt{\frac{K(K+1)}{\Omega_p}}\right), \qquad x \geq 0 . \tag{2.43}$$

Once again, by using a transformation of random variables, the squared-envelope has the following non-central chi-square distribution with two degrees of freedom

$$p_{z^2}(x) = \frac{(K+1)}{\Omega_p} \exp\left\{-K - \frac{(K+1)x}{\Omega_p}\right\} I_o\left(2\sqrt{\frac{K(K+1)x}{\Omega_p}}\right), \quad x \geq 0 \; .$$

$$(2.44)$$

At time t, the phase $\phi(t)$ of the received complex low-pass signal is

$$\phi(t) = \mathrm{Tan}^{-1}\left(\frac{x_Q(t)}{x_I(t)}\right) \; . \tag{2.45}$$

If $x_I(t)$ and $x_Q(t)$ are uncorrelated Gaussian random processes, then it follows from Appendix 2A that the phase is uniformly distributed over the interval $[-\pi, \pi]$, i.e.,

$$p_\phi(x) = \frac{1}{2\pi} \qquad -\pi \leq x \leq \pi \; . \tag{2.46}$$

This result applies to both Ricean and Rayleigh fading channels.

Nakagami Fading

The Nakagami distribution was introduced by Nakagami in the early 1940's to characterize rapid fading in long distance HF channels [229]. The Nakagami distribution was selected to fit empirical data, and is known to provide a closer match to some experimental data than either the Rayleigh, Ricean, or log-normal distributions [36].

In essence, the Nakagami distribution describes the received envelope $z(t) = |r(t)|$ by a central chi-square distribution with m degrees of freedom, i.e.,

$$p_z(x) = \frac{2m^m x^{2m-1}}{\Gamma(m)\Omega_p^m} \exp\left\{-\frac{mx^2}{\Omega_p}\right\} \qquad m \geq \frac{1}{2} \tag{2.47}$$

where $\Omega_p = \mathrm{E}[z^2]$. Fig. 2.8 shows the Nakagami distribution for several values of m. The Nakagami distribution is often used to model multipath-fading for the following reasons. First, the Nakagami distribution can model fading conditions that are either more or less severe than Rayleigh fading. When $m = 1$, the Nakagami distribution becomes the Rayleigh distribution, when $m = 1/2$ it becomes a one-sided Gaussian distribution, and when $m \to \infty$ the distribution becomes an impulse (no fading). Second, the Rice distribution (which does have physical significance) can be closely approximated by using

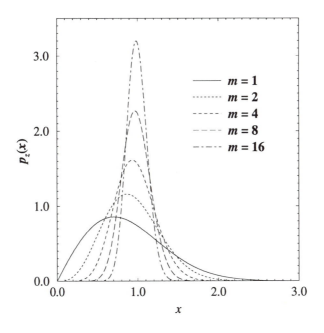

Figure 2.8   The Nakagami pdf for several values of m with $\Omega_p = 1$.

the following relation between the Rice factor K and the Nakagami parameter m [229];

$$K \;=\; \frac{\sqrt{m^2 - m}}{m - \sqrt{m^2 - m}} \qquad m > 1 \qquad\qquad (2.48)$$

$$m \;=\; \frac{(K+1)^2}{(2K+1)} \;. \qquad\qquad (2.49)$$

Since the Rice distribution contains a Bessel function while the Nakagami distribution does not, the Nakagami distribution often leads to closed form analytical expressions and insights that are otherwise unattainable.

By using a transformation of random variables, the squared-envelope has the Gamma density

$$p_{z^2}(x) = \left(\frac{m}{\Omega_p}\right)^m \frac{x^{m-1}}{\Gamma(m)} \exp\left\{-\frac{mx}{\Omega_p}\right\} \;. \qquad\qquad (2.50)$$

Finally, by using the relationship between the K factor and the parameter m in (2.48), the cumulative distribution function (cdf), $F_{z^2}(x) = \mathrm{P_r}[z^2 \leq x]$ of

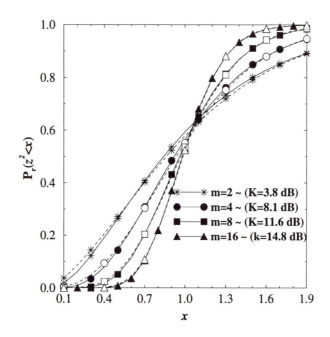

Figure 2.9 Comparison of the cdfs for the Ricean and Nakagami distributions.

the squared-envelope with Nakagami and Ricean fading is plotted in Fig. 2.9. It is apparent from Fig. 2.9 that for larger values of m, a Gamma distribution can well approximate a non-central chi-square distribution.

2.1.3 Envelope Correlation and Spectra

The autocorrelation of the envelope $z(t) = |r(t)|$ of a complex Gaussian random process can be expressed in terms of the hypergeometric function $F[\cdot, \cdot; \cdot, \cdot]$ as [65, p. 170]

$$
\begin{aligned}
\phi_{zz}(\tau) &= \mathrm{E}[z(t)\, z(t+\tau)] \\
&= \frac{\pi}{2}|\phi_{rr}(0)|F\left[-\frac{1}{2}, -\frac{1}{2}; 1, \frac{|\phi_{rr}(\tau)|^2}{|\phi_{rr}(0)|^2}\right]
\end{aligned}
\tag{2.51}
$$

where

$$
|\phi_{rr}(\tau)|^2 = \phi_{r_I r_I}^2(\tau) + \phi_{r_I r_Q}^2(\tau) \ .
\tag{2.52}
$$

The above expression is analytically cumbersome, but fortunately a useful approximation can be obtained by expanding the hypergeometric function into the following infinite series:

$$F\left[-\frac{1}{2}, -\frac{1}{2}; 1, x\right] = 1 + \frac{1}{4}x + \frac{1}{64}x^2 + \cdots \tag{2.53}$$

Neglecting the terms beyond second order, the approximation becomes

$$\phi_{zz}(\tau) \doteq \frac{\pi}{2}|\phi_{rr}(0)|\left[1 + \frac{1}{4}\frac{|\phi_{rr}(\tau)|^2}{|\phi_{rr}(0)|^2}\right] . \tag{2.54}$$

At $\tau = 0$, the approximation gives $\phi_{zz}(0) = 5\pi\Omega_p/32$, whereas the true value is $\phi_{zz}(0) = \Omega_p/2$. Hence, the relative error in the signal power is only 1.86%, leading us to believe that the approximation is probably very good.

The psd of the received envelope can be obtained by taking the Fourier transform of $\phi_{zz}(\tau)$. In general, the psd will include both discrete and continuous components. The dc component of the received envelope gives rise to the discrete spectral component. However, we are primarily interested in the continuous portion of the psd. Therefore, the autocovariance function $\phi_{zz}(\tau)$ is used, where

$$
\begin{aligned}
\mu_{zz}(\tau) &= E[z(t)z(t+\tau)] - E[z(t)]E[z(t+\tau)] \\
&= \frac{\pi}{2}|\phi_{rr}(0)|\left[1 + \frac{1}{4}\frac{|\phi_{rr}(\tau)|^2}{|\phi_{rr}(0)|^2}\right] - \frac{\pi}{2}|\phi_{rr}(0)| \\
&= \frac{\pi}{8|\phi_{rr}(0)|}|\phi_{rr}(\tau)|^2 .
\end{aligned}
\tag{2.55}
$$

With isotropic scattering $|\phi_{rr}(\tau)|^2 = \phi_{r_I r_I}^2(\tau)$ and, therefore,

$$\mu_{zz}(\tau) = \frac{\pi\Omega_p}{16}J_0^2(2\pi f_m\tau) . \tag{2.56}$$

Fig. 2.10 plots the normalized envelope auto-covariance $\mu_{zz}(\tau)/(\pi\Omega_p/16)$ against the normalized time delay $f_m\tau$ for the case of isotropic scattering. The Fourier transform of $\mu_{zz}(\tau)$ can be calculated by using the fact that $|\phi_{rr}(\tau)|^2 = \phi_{rr}(\tau)\phi_{rr}^*(\tau)$ and $\phi_{rr}(\tau) = \phi_{rr}^*(-\tau)$ to write

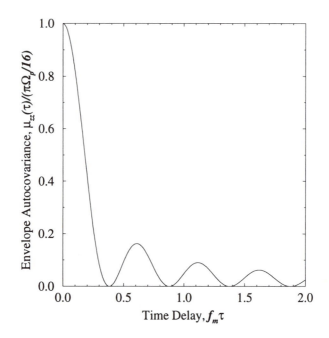

Figure 2.10 Envelope autocovariance against the time delay $f_m\tau$ for an isotropic scattering channel.

$$
\begin{aligned}
S_{zz}(f) &= \frac{\pi}{8|\phi_{rr}(0)|} S_{rr}(f) * S_{rr}(f) \\
&= \frac{\pi}{8|\phi_{rr}(0)|} \int_{-\infty}^{\infty} S_{rr}(x)S_{rr}(x-f)\, dx \\
&= \frac{\pi}{8|\phi_{rr}(0)|} \int_{-f_m}^{f_m-|f|} S_{rr}(x)S_{rr}(x+|f|)\, dx \qquad 0 \le |f| \le 2f_m \ .
\end{aligned}
$$

$$(2.57)$$

Note that $S_{zz}(f)$ is always real, positive, and even. It is centered about $f = 0$ with a spectral width of $4f_m$, where f_m is the maximum Doppler frequency. To proceed further, we need to specify $S_{rr}(f)$. With isotropic scattering $\phi_{rr}(\tau) = \phi_{r_I r_I}(\tau)$ so that $S_{rr}(f) = S_{r_I r_I}(f)$, where $S_{r_I r_I}(f)$ is given by (2.22). The result from evaluating (2.57) is

$$
S_{zz}(f) = \frac{\Omega_p}{64\pi} \frac{1}{f_m} K\left(\sqrt{1 - \left(\frac{f}{2f_m}\right)^2} \right) \qquad 0 \le |f| \le 2f_m \qquad (2.58)
$$

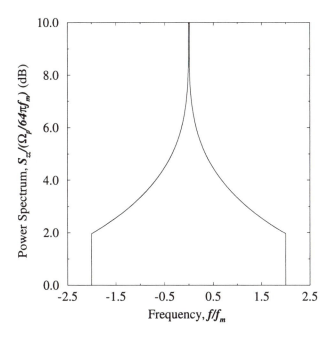

Figure 2.11 Envelope psd against the frequency f/f_m for an isotropic scattering channel.

where $K(\,\cdot\,)$ is the complete elliptic integral of the first kind, defined by any of the following equivalent forms

$$K(\gamma) \;=\; \int_0^\infty \frac{dx}{\sqrt{(1+x^2)(1+\gamma^2 x^2)}} \tag{2.59}$$

$$=\; \int_0^1 \frac{dx}{\sqrt{(1-x^2)(1-(1-\gamma^2)x^2)}} \tag{2.60}$$

$$=\; \int_0^{\pi/2} \frac{d\phi}{\sqrt{\cos^2\phi + (1-\gamma^2)\sin^2\phi}} \;. \tag{2.61}$$

The normalized psd $S_{zz}(f)/(\Omega_p/64\pi f_m)$ is plotted against the normalized frequency f/f_m in Fig. 2.11. The psd of the complex envelope for a non-isotropic scattering channel can be obtained with some minor modifications to the above development. For example, consider the particular scattering environment shown in Fig. 2.5. In this case, the psd of $r(t)$ can be obtained from (2.23), (2.34), and (2.35) as

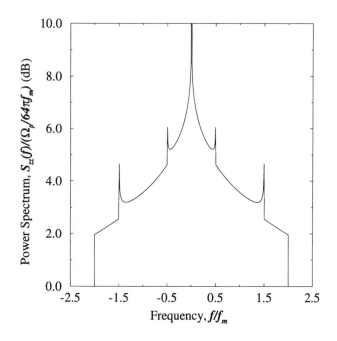

Figure 2.12 Envelope psd against the frequency f/f_m for the scattering environment shown in Fig. 2.5; $2\pi f_m K = 0.1$ and $\theta_0 = \pi/3$.

$$S_{rr}(f) = \frac{1}{K+1} \frac{\Omega_p}{4\pi f_m} \frac{1}{\sqrt{1-(f/f_m)^2}} + \frac{K}{K+1} \frac{\Omega_p}{2} \delta(f - f_m \cos\theta_0) \quad |f| \leq f_m$$

$$(2.62)$$

where K is the Rice factor. Note that the psd of the complex low-pass waveform $r(t)$ is asymmetrical. To obtain the psd of the complex envelope $z(t)$, we substitute (2.62) into (2.57) to obtain

$$
\begin{aligned}
S_{zz}(f) &= \left(\frac{1}{K+1}\right)^2 \frac{\Omega_p}{64\pi f_m} K\left(\sqrt{1-\left(\frac{f}{2f_m}\right)^2}\right) \\
&\quad + \frac{K}{(K+1)^2} \frac{\pi}{8} \left(S_{r_I r_I}(f + f_m \cos\theta_0) \right. \\
&\quad \left. + S_{r_I r_I}(f - f_m \cos\theta_0) + \frac{K\Omega_p}{2} \delta(f)\right)
\end{aligned}
$$

$$(2.63)$$

where $S_{r_I r_I}(f)$ is given by (2.22). Fig. 2.12 plots the normalized envelope psd $S_{zz}(f)/(\Omega_p/64\pi f_m)$ against the normalized frequency f/f_m.

Squared-Envelope Correlation and Spectra

Sometimes the squared-envelope $z^2(t) = |r(t)|^2$ is of interest; for example, the instantaneous received signal power is proportional to $z^2(t)$. The autocorrelation of the squared-envelope is

$$\phi_{z^2z^2}(\tau) = E[z(t)^2 z(t+\tau)^2] . \qquad (2.64)$$

Since $z(t)^2 = r_I^2(t) + r_Q^2(t)$, it follows that

$$\begin{aligned}\phi_{z^2z^2}(\tau) &= 2E[r_I^2(t)r_I^2(t+\tau)] + E[r_Q^2(t)r_Q^2(t+\tau)] \\ &\quad + 2E[r_I^2(t)r_Q^2(t+\tau)] + E[r_Q^2(t)r_I^2(t+\tau)] .\end{aligned} \qquad (2.65)$$

First consider the case where the propagation environment is characterized by diffuse scattering, so that $r_I(t)$ and $r_Q(t)$ have zero mean. Then, from Problem 2.7, we have

$$\phi_{z^2z^2}(\tau) = 4\phi_{r_Ir_I}^2(0) + 4\phi_{r_Ir_I}^2(\tau) + 4\phi_{r_Ir_Q}^2(\tau) . \qquad (2.66)$$

Finally, the squared-envelope autocovariance is

$$\begin{aligned}\mu_{z^2z^2}(\tau) &= \phi_{z^2z^2}(\tau) - E^2[z(t)^2] \\ &= 4\phi_{r_Ir_I}^2(\tau) + 4\phi_{r_Ir_Q}^2(\tau) \\ &= 4|\phi_{rr}(\tau)|^2 .\end{aligned} \qquad (2.67)$$

With isotropic scattering the above expression reduces to

$$\mu_{z^2z^2}(\tau) = \Omega_p^2 J_0^2(2\pi f_m \tau) . \qquad (2.68)$$

By comparing (2.55) and (2.67), we observe that the *approximate* autocorrelation and psd of the envelope and the exact autocorrelation and psd of the squared-envelope are identical, except for a multiplicative constant. If the propagation environment is characterized by a specular or LOS component (Ricean fading), then $r_I(t)$ and $r_Q(t)$ have non-zero mean and the autocovariance of the squared-envelope assumes a more complicated form. For Ricean fading, Aulin [12] has shown that

$$\mu_{z^2z^2}(\tau) = r\phi_{r_Ir_I}(\tau)[\phi_{r_Ir_I}(\tau) + s^2 \cos(2\pi f_m \cos\theta_0)] \qquad (2.69)$$

where s^2 is the power in the specular component and θ_0 is the angle that the specular component makes with the MS direction of motion. For the scattering environment shown in Fig. 2.5, $\phi_{r_Ir_I}(\tau)$ in (2.69) is given by (2.20) so that

$$\mu_{z^2z^2}(\tau) = \left(\frac{\Omega_p}{K+1}\right)^2 \left(J_0^2(2\pi f_m \tau) + 2K J_0(2\pi f_m \tau)\cos(2\pi f_m \tau \cos\theta_0)\right) \qquad (2.70)$$

where K is the Rice factor. The corresponding normalized squared-envelope

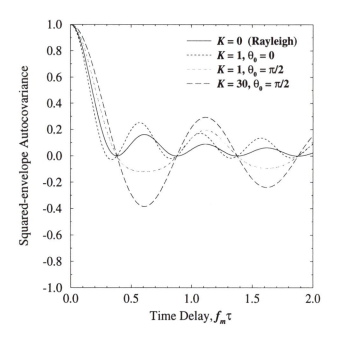

Figure 2.13   Squared-envelope auto-covariance against the time delay $f_m\tau$ for the scattering environment shown in Fig. 2.5.

autocovariance

$$\left(\frac{K+1}{\Omega_p}\right)^2 \frac{1}{1+2K} \cdot \mu_{z^2z^2}(\tau) \tag{2.71}$$

is plotted in Fig. 2.13 as a function of the normalized time delay $f_m\tau$, for various values of K and θ_0.

2.1.4   Level Crossing Rates and Fade Durations

Two important second order statistics associated with envelope fading are the **level crossing rate** (how often the envelope crosses a specified level) and the **average fade duration** (how long the envelope remains below a specified level). These quantities are second order statistics, because they are not only affected by the scattering environment but also by the velocity of the MS. For the case of Ricean (and Rayleigh) fading, closed form expressions can be derived for these parameters.

Envelope Level Crossing Rate

The envelope level crossing rate at a specified level R, L_R, is defined as the rate at which the envelope crosses level R in the positive (or negative) going direction. Obtaining the level crossing rate requires the joint pdf, $p(z, \dot{z})$, of the envelope level $z = |r|$ and the envelope slope $\dot{z} = |\dot{r}|$. In terms of the joint pdf $p(z, \dot{z})$, the expected amount of time spent in the interval $(R, R + dz)$ for a given envelope slope $\dot{z}$ and time duration dt is

$$p(R, \dot{z})dzd\dot{z}dt \ . \tag{2.72}$$

The time required to cross the level z once for a given envelope slope $\dot{z}$, in the interval $(R, R + dz)$ is

$$dz/\dot{z} \ . \tag{2.73}$$

The ratio of these two quantities is the expected number of crossings of the envelope z within the interval $(R, R + dz)$ for a given envelope slope $\dot{z}$ and time duration dt, i.e.,

$$\dot{z}p(R, \dot{z})d\dot{z}dt \ . \tag{2.74}$$

The expected number of crossings of the envelope level R for a given envelope slope $\dot{z}$ in a time interval of duration T is

$$\int_0^T \dot{z}p(R, \dot{z})d\dot{z}dt = \dot{z}p(R, \dot{z})d\dot{z}T \ . \tag{2.75}$$

The expected number of crossings of the envelope level R with a positive slope is

$$N_R = T \int_0^\infty \dot{z}p(R, \dot{z})d\dot{z} \ . \tag{2.76}$$

Finally, the expected number of crossings of the envelope level R per second, or the level crossing rate, is

$$L_R = \int_0^\infty \dot{z}p(R, \dot{z})d\dot{z} \ . \tag{2.77}$$

This is a general result that applies to any random process. For the case of a non-zero mean complex Gaussian random process (Ricean fading), Rice [270] has derived the joint pdf of $p(z, \dot{z})$ as

$$p(z, \dot{z}) = \frac{z(2\pi)^{-3/2}}{\sqrt{Bb_0}} \int_{-\pi}^\pi d\theta \tag{2.78}$$

$$\times \exp\left\{ -\frac{1}{2Bb_0} \left[B\left(z^2 - 2zs\cos\theta + s^2\right) + (b_0\dot{z} + b_1 s \sin\theta)^2 \right] \right\}$$

where s is the non-centrality parameter in the Rice distribution, and $B = b_0 b_2 - b_1^2$, where b_0, b_1, and b_2 are constants that are derived from the psd of the *diffuse* component of the faded envelope. In particular [326],

$$b_n = (2\pi)^n \int_{f_c-f_m}^{f_c+f_m} \hat{S}_{xx}(f)(f-f_c)^n df \qquad (2.79)$$

$$= (2\pi f_m)^n b_0 \int_0^{2\pi} \hat{p}(\theta) \cos^n \theta d\theta \qquad (2.80)$$

where $\hat{S}_{xx}(f)$ is the psd of the *diffuse* component of the received band-pass signal and $\hat{p}(\theta)$ is its density. The above expression is general enough to apply to the case where the diffuse component experiences non-isotropic scattering. However, if $\hat{S}_{xx}(f)$ is symmetrical about the carrier frequency f_c, then $b_1 = 0$ and the integration in (2.78) leads to the product form

$$p(z,\dot{z}) = \sqrt{\frac{1}{2\pi b_2}} \exp\left\{-\frac{\dot{z}^2}{2b_2}\right\} \cdot \frac{z}{b_0} \exp\left\{-\frac{(z^2+s^2)}{2b_0}\right\} I_0\left(\frac{zs}{b_0}\right) \qquad (2.81)$$

$$= p(\dot{z}) \cdot p(z) .$$

An equivalent condition is that the density $\hat{p}(\theta)$ be symmetrical about a line that is perpendicular to the direction of motion of the MS. Since $p(z,\dot{z}) = p(\dot{z}) \cdot p(z)$ in (2.81), it follows that z and $\dot{z}$ are independent. When the diffuse component experiences isotropic scattering as in Fig. 2.5, a closed form expression for the envelope level crossing rate can be obtained. Substituting (2.32) into (2.79) gives

$$b_n = \begin{cases} b_0(2\pi f_m)^n \frac{1 \cdot 3 \cdot 5 \cdots (n-1)}{2 \cdot 4 \cdot 6 \cdots n} & n \text{ even} \\ 0 & n \text{ odd} \end{cases} . \qquad (2.82)$$

Therefore, $b_1 = 0$ and $b_2 = b_0(2\pi f_m)^2/2$, where b_0 is the power in the diffuse component of the received signal, given by

$$b_0 = \frac{\Omega_p}{2(K+1)} . \qquad (2.83)$$

Substituting the joint density in (2.81) into (2.77) gives the envelope level crossing rate

$$L_R = \sqrt{2\pi(K+1)} f_m \rho e^{-K-(K+1)\rho^2} I_0\left(2\rho\sqrt{K(K+1)}\right) \qquad (2.84)$$

where

$$\rho = \frac{R}{\sqrt{\Omega_p}} = \frac{R}{R_{\text{rms}}} \qquad (2.85)$$

and $\sqrt{\Omega_p} \triangleq R_{\text{rms}}$ can be interpreted as the *rms* envelope level. For Rayleigh fading ($K = 0$), the above expression simplifies to

$$L_R = \sqrt{2\pi} f_m \rho e^{-\rho^2} . \qquad (2.86)$$

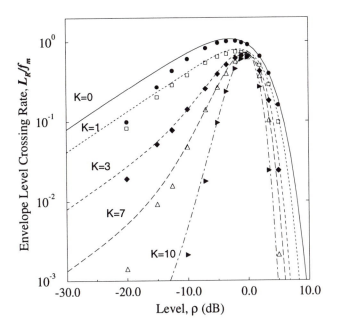

Figure 2.14 Envelope level crossing rate for the scattering environment shown in Fig. 2.5. Lines denote theoretical results, while points denote simulation results.

The normalized envelope level crossing rate L_R/f_m is plotted in Fig. 2.14 as a function of ρ and K. The maximum LCR can be found by taking the derivative of (2.84) with respect to ρ and solving

$$I_0\left(2\rho\sqrt{K(K+1)}\right)\left(1-2(K+1)\rho^2\right)+2\rho\sqrt{K(K+1)}I_1\left(2\rho\sqrt{K(K+1)}\right)=0 \tag{2.87}$$

for ρ as a function of K. Fig. 2.15 plots the maximum envelope level crossing rate as a function of K. Finally, we note that the envelope level crossing rate around $\rho = 0$ dB is nearly independent of K. This attractive property will be exploited in Chapter 10 we use the envelope level crossing rate to estimate the MS velocity. The simulation results in Fig. 2.14 were obtained with a fading simulator that will be described in Section 2.3.

Zero Crossing Rate

Recall that received low-pass signal $r(t) = r_I(t) + r_Q(t)$ is a complex Gaussian random process. If the channel is characterized by a specular component with

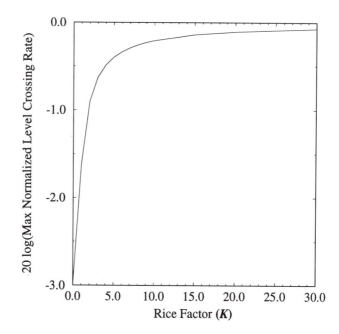

Figure 2.15 Maximum normalized envelope level crossing rate L_R/f_m for the scattering environment shown in Fig. 2.5.

amplitude α_0 and angle of arrival θ_0, then $r_I(t)$ and $r_Q(t)$ have mean values $\alpha_0 \cos \theta_0$ and $\alpha_0 \sin \theta_0$, respectively. Here we are interested in the **zero crossing rate** of the zero-mean Gaussian random processes $r_I(t) - \alpha_0 \cos \theta_0$ and $r_Q(t) - \alpha_0 \sin \theta_0$. Rice [270] has derived this zero crossing rate as

$$L_Z = \frac{1}{\pi}\sqrt{\frac{b_2}{b_0}} = 2f_m\sqrt{\int_0^{2\pi} \hat{p}(\theta)\cos^2\theta d\theta} \ . \tag{2.88}$$

When the diffuse component experience isotropic scattering, as in Fig. 2.5, the above expression reduces to

$$L_Z = \sqrt{2}f_m \ . \tag{2.89}$$

Average Envelope Fade Duration

Another quantity of interest is the average duration that the envelope level remains below a specified level R. Although the pdf of the envelope fade duration is unknown, the average fade duration can be calculated.

Consider a very long time interval of length T and let t_i be the duration of the ith fade below the level R. The probability of the received envelope level being less than R is

$$\Pr[z \leq R] = \frac{1}{T} \sum_i t_i \ . \tag{2.90}$$

The average envelope fade duration is equal to

$$\bar{t} = \frac{1}{T L_R} \sum_i t_i = \frac{\Pr[z \leq R]}{L_R} \ . \tag{2.91}$$

If the envelope has the Rice distribution in (2.40), then

$$\Pr[z \leq R] = \int_0^R p(z)dz = 1 - Q\left(\sqrt{2K}, \sqrt{2(K+1)\rho^2}\right) \tag{2.92}$$

where $Q(a, b)$ is the Marcum Q function. Therefore,

$$\bar{t} = \frac{1 - Q\left(\sqrt{2K}, \sqrt{2(K+1)\rho^2}\right)}{\sqrt{2\pi(K+1)}f_m \rho e^{-K-(K+1)\rho^2} I_0\left(2\rho\sqrt{K(K+1)}\right)} \ . \tag{2.93}$$

If the envelope is Rayleigh distributed, then

$$\Pr[z \leq R] = \int_0^R p(z)dz = 1 - e^{-\rho^2} \tag{2.94}$$

and, therefore,

$$\bar{t} = \frac{e^{\rho^2} - 1}{\rho f_m \sqrt{2\pi}} \ . \tag{2.95}$$

The normalized average envelope fade duration $\bar{t} f_m$ is plotted in Fig. 2.16 as a function of ρ.

Note that the level crossing rate, zero crossing rate, and the average fade duration all depend on the velocity of the MS ($f_m = v/\lambda_c$). Very deep fades tend to occur infrequently and do not last very long. For example, at 60 mi/hr and 900 MHz, the maximum Doppler frequency is $f_m = 88$ Hz. Therefore, with isotropic scattering and Rayleigh fading ($K = 0$) there are $L_R = 81$ fades/s at $\rho = 0$ dB with an average fade duration of 7.8 ms. However, at $\rho = -20$ dB there are only 2.2 fades/s with an average fade duration of 45 μs.

Observe from Fig. 2.14 that the fades are shallower when the Rice factor, K,

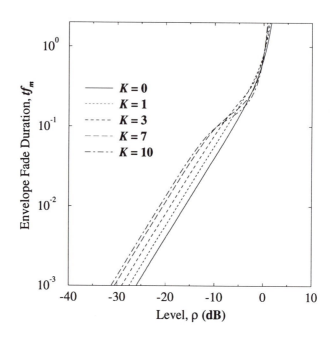

Figure 2.16   Average envelope fade duration for the scattering environment shown in Fig. 2.5.

is larger. Furthermore, we see from Fig. 2.16 that the average fade duration tends to be smaller with larger Rice factors.

2.1.5   Spatial Correlation

Many mobile radio systems employ antenna diversity, where spatially separated antennas are used to provide multiple faded replicas of the same information bearing signal. A fundamental question that arises is the antenna separation needed to provide uncorrelated antenna diversity branches. This question can be answered by using our previously derived results along with the distance-time transformation $\ell = v\tau$, where v is the MS velocity. This transformation results in $f_m\tau = \ell/\lambda_c$. For the case of isotropic scattering (2.20) and (2.56) become, respectively

$$\phi_{r_I r_I}(\ell) = \frac{\Omega_p}{2} J_o(2\pi\ell/\lambda_c) \qquad (2.96)$$

$$\mu_{zz}(\ell) = \frac{\pi\Omega_p}{16} J_o^2(2\pi\ell/\lambda_c) . \qquad (2.97)$$

It follows that Fig. 2.10 also shows the normalized envelope autocovariance $\mu_{zz}(\tau)/(\pi\Omega_p/16)$ against the distance ℓ/λ_c. The autocovariance is zero at $\ell = 0.38\lambda_c$ and is less than 0.3 for $\ell > 0.38\lambda_c$. Therefore, in practice, uncorrelated diversity branches can be obtained at the MS by spacing the antenna elements about a half-wavelength apart.

Received Signal at the Base Station

Multipath-fading channels are reciprocal in the sense that if a propagation path exists, it will carry energy equally well in either direction. That is the energy in either direction propagates by exactly the same set of scatterers. Since the MS moves and the BS is stationary, the temporal autocorrelations and signal spectrum will be identical at both the MS and the BS. However, the BS antennas are usually elevated and free of local obstructions so as to provide the best possible radio coverage. Therefore, most of the scatters in the medium will be in the vicinity of the MS rather than the BS. Consequently, the probability density function of the arriving plane waves at the BS tends to be concentrated in a narrow angle of arrival. Two antennas located at a BS will view the scattered volume around a MS from only a slightly different angle. It is therefore expected that a much larger distance is required for spatial decorrelation at the BS as compared to the MS. The analysis is much more complicated, but it suffices to state here that horizontally separated antennas generally provide more spatial decorrelation than vertically separated antennas. Usually, an antenna separation of about $20\lambda_c$ is required to obtain a correlation of about 0.7. Even with this high value of correlation a significant diversity improvement can still be realized.

2.2 FREQUENCY-SELECTIVE MULTIPATH-FADING

To this point we have considered channel models that are appropriate for narrow-band transmission, where the inverse signal bandwidth is much greater than the time spread of the propagation path delays. For digital communication systems this means that the duration of a modulated symbol is much greater than the time spread of the propagation path delays. Under this condition all frequencies in the transmitted signal will experience the same random attenuation and phase shift due to multipath-fading. Such a channel introduces very little or no distortion into the received signal and is said to exhibit **flat fading** .

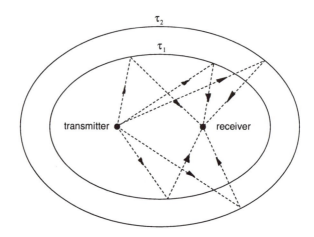

Figure 2.17   Path geometry for multipath-fading channels. Signals will arrive at the receiver antenna at the same time if they reflect off scatterers that are located on the same ellipse.

If the range in the propagation path delays is large compared to the inverse signal bandwidth, then the frequency components in the transmitted signal will experience different phase shifts along the different paths. As the differential path delays become large, even closely separated frequencies in the transmitted signal can experience significantly different phase shifts. Under this condition the channel introduces amplitude and phase distortion into the message waveform. Such a channel is said to exhibit **frequency-selective fading**. The path geometry for a multipath-fading channel is shown in Fig. 2.17. Considering only single reflections, all scatterers that are associated with a particular path length are located on an ellipse with the transmitter and receiver located at the foci. Different delays correspond to different confocal ellipses. Flat fading channels have their scatterers located on ellipses corresponding to differential delays that are small compared to the duration of a modulated symbol. Frequency selective channels have strong scatterers that are located on several ellipses that correspond to differential delays that are significant compared to a symbol duration. In urban and suburban macrocellular systems, these strong scatterers usually correspond to high-rise buildings or perhaps large distant terrain features such as mountains.

Multipath-fading channels can be modeled as time-variant linear filters, whose inputs and outputs can be described in both the time and frequency domains. This leads to four possible **transmission functions** [24]; the input delay-spread function $c(\tau,t)$, the output Doppler-spread function $H(f,\nu)$, the time-variant transfer function $T(f,t)$, and the delay Doppler-spread function $S(\tau,\nu)$.

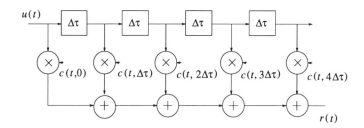

Figure 2.18 Discrete-time tapped delay line model for a multipath-fading channel, from [243].

The complex low-pass impulse response relates the complex low-pass input and output time waveforms, $u(t)$ and $r(t)$, respectively, through the convolution

$$r(t) = \int_{-\infty}^{\infty} u(t - \tau)c(t, \tau)d\tau \ . \tag{2.98}$$

Bello called the low-pass impulse response $c(t, \tau)$ the **input delay-spread function** [24]. In physical terms, $c(t, \tau)$ can be interpreted as the channel response at time t due to an impulse applied at time $t - \tau$. Since a physical channel cannot have an output before an input is applied $c(t, \tau) = 0$ for $\tau < 0$ and the lower limit of integration in (2.98) becomes 0. If the convolution in (2.98) is written as a discrete sum, then

$$r(t) = \sum_{m=0}^{n} u(t - m\Delta\tau)c(t, m\Delta\tau)\Delta\tau \ . \tag{2.99}$$

This representation allows us to visualize the channel as a transversal filter with tap spacing $\Delta\tau$ and time-varying tap gains $c(t, m\Delta\tau)$ as shown in Fig. 2.18.

The second transmission function relates the input and output spectra, $U(f)$ and $R(f)$, respectively, through the integral equation

$$R(f) = \int_{-\infty}^{\infty} U(f - \nu)H(f - \nu, \nu)d\nu \ . \tag{2.100}$$

Bello called the function $H(f - \nu, \nu)$ the **output Doppler-spread function** [24]. This function explicitly shows the effect of Doppler shift or spectral broadening on the output spectrum. In physical terms, the frequency-shift variable ν can be interpreted as the Doppler shift that is introduced by the channel. Once again, the integral in (2.100) can be approximated by the discrete sum

$$R(f) = \sum_{m=0}^{n} U(f - m\nu)H(f - m\nu, m\nu)\Delta\nu \ . \tag{2.101}$$

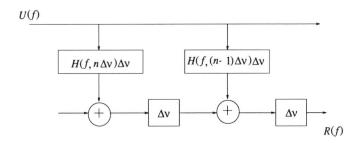

Figure 2.19 Frequency conversion model for a multipath-fading channel, from [243].

This allows the channel to be represented by a bank of filters with transfer functions $H(f-m\nu, m\nu)$ followed by a frequency conversion chain that produces the Doppler shifts.

The third transmission function relates the output time waveform to the input spectrum through the integral equation

$$r(t) = \int_{-\infty}^{\infty} U(f)T(f,t)e^{j2\pi ft}df \ . \tag{2.102}$$

Zadeh called the function $T(f,t)$ the **time-variant transfer function** [351]. The time-variant transfer function is the Fourier transform of the Doppler-spread function with respect to the delay variable τ, i.e.,

$$c(t,\tau) \Longleftrightarrow T(f,t) \ . \tag{2.103}$$

Also the time-variant transfer function is the inverse Fourier transform of the output Doppler-spread function with respect to the Doppler shift variable ν, i.e.,

$$H(f,\nu) \Longleftrightarrow T(f,t) \ . \tag{2.104}$$

The final description relates the input and output time waveforms through the double integral

$$r(t) = \int_{-\infty}^{\infty} \int_{-\infty}^{\infty} u(t-\tau)S(\tau,\nu)e^{-j2\pi f\tau} \ d\nu \ d\tau \ . \tag{2.105}$$

The function $S(\tau,\nu)$ is called the **delay Doppler-spread function** [24]. The time-variant transfer function is the inverse Fourier transform of the delay Doppler-spread function with respect to the Doppler shift variable ν, i.e.,

$$c(t,\tau) \Longleftrightarrow S(\tau,\nu) \ . \tag{2.106}$$

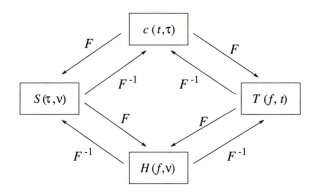

Figure 2.20 Fourier transform relations between the transmission functions, from [243].

The delay Doppler-spread function provides a measure of the scattering amplitude of the channel in terms of the time delay τ and Doppler frequency ν.

The various Fourier transform relations between the above transmission functions are shown in Fig. 2.20. In each transform pair there is always a fixed variable, so that the transform involves the other two variables.

2.2.1 Statistical Channel Correlation Functions

Recall the channel impulse response $c(\tau, t) = c_I(\tau, t) + jc_Q(\tau, t)$ can be modeled as a complex Gaussian random process, where the quadrature components $c_I(\tau, t)$ and $c_Q(\tau, t)$ are uncorrelated Gaussian random processes. In general, all of the transmission functions defined in the last section are random processes. A thorough characterization of a channel requires knowledge of the joint pdf of all the transmission functions. Since this is formidable, a more reasonable approach is to obtain statistical correlation functions for the individual transmission functions. If the underlying process is Gaussian, then a complete statistical description is provided by the means and autocorrelation functions. In the following discussion, we assume zero-mean Gaussian random processes so that only the autocorrelation functions are of interest. Since there are four transmission functions, four autocorrelation functions can be defined as follows [243], [257]:

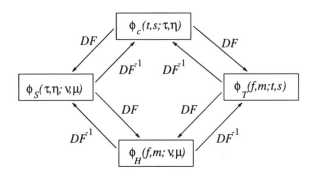

Figure 2.21 Double Fourier transform relations between the channel auto-correlation functions, from [243].

$$\phi_c(t, s; \tau, \eta) = \frac{1}{2}\mathrm{E}[c(t, \tau)c^*(s, \eta)] \qquad (2.107)$$

$$\phi_T(f, m; t, s) = \frac{1}{2}\mathrm{E}[T(f, t)T^*(m, s)] \qquad (2.108)$$

$$\phi_H(f, m; \nu, \mu) = \frac{1}{2}\mathrm{E}[H(f, \nu)H^*(m, \mu)] \qquad (2.109)$$

$$\phi_S(\tau, \eta; \nu, \mu) = \frac{1}{2}\mathrm{E}[S(\tau, \nu)S^*(\eta, \mu)] . \qquad (2.110)$$

These autocorrelation functions are related to each other through double Fourier transform pairs. For example,

$$\phi_S(\tau, \eta; \nu, \mu) = \int_{-\infty}^{\infty}\int_{-\infty}^{\infty} \phi_c(t, s; \tau, \eta)e^{j2\pi(\nu t - \mu s)}\, dt\, ds \qquad (2.111)$$

$$\phi_c(t, s; \tau, \eta) = \int_{-\infty}^{\infty}\int_{-\infty}^{\infty} \phi_S(\tau, \eta; \nu, \mu)e^{-j2\pi(\nu t - \mu s)}\, d\nu\, d\mu . \qquad (2.112)$$

The complete set of such relationships is summarized in Fig. 2.21.

2.2.2 Classification of Channels

Wide sense stationary (WSS) channels have fading statistics that remain constant over short periods of time or small spatial distances. This implies that the channel correlation functions depend on the time variables t and s only through the time difference $\Delta t = t - s$. It can be demonstrated (see Problem 2.3) that WSS channels give rise to scattering with uncorrelated Doppler shifts.

This behavior suggests that the attenuations and phase shifts associated with signal components having different Doppler shifts are uncorrelated. Hence for WSS channels, the correlation functions become

$$\phi_c(t, t + \Delta t; \tau, \eta) = \phi_c(\Delta t; \tau, \eta) \tag{2.113}$$

$$\phi_T(f, m; t, t + \Delta t) = \phi_T(f, m; \Delta t) \tag{2.114}$$

$$\phi_H(f, m; \nu, \mu) = \psi_H(f, m; \nu)\delta(\nu - \mu) \tag{2.115}$$

$$\phi_S(\tau, \eta; \nu, \mu) = \psi_S(\tau, \eta; \nu)\delta(\nu - \mu) \tag{2.116}$$

where

$$\psi_H(f, m; \nu) = \int_{-\infty}^{\infty} \phi_T(f, m; \Delta t)e^{-j2\pi\nu\Delta t}d\Delta t \tag{2.117}$$

$$\psi_S(\tau, \eta; \nu) = \int_{-\infty}^{\infty} \phi_c(\Delta t; \tau, \eta)e^{-j2\pi\nu\Delta t}d\Delta t \tag{2.118}$$

are Fourier transform pairs. **Uncorrelated scattering** (US) channels are characterized by an uncorrelated attenuation and phase shift with paths of different delays. Bello showed that US channels are wide sense stationary in the frequency variable so that the correlation functions depend on the frequency variables f and m only through the frequency difference $\Delta f = f - m$ [24]. Analogous to (2.115) and (2.116), the channel correlation functions can be shown (see Problem 2.4) to be singular in the time-delay variable. For US channels, the channel correlation functions become

$$\phi_c(t, s; \tau, \eta) = \psi_c(t, s; \tau)\delta(\eta - \tau) \tag{2.119}$$

$$\phi_T(f, f + \Delta f; t, s) = \phi_T(\Delta f; t, s) \tag{2.120}$$

$$\phi_H(f, f + \Delta f; \nu, \mu) = \phi_H(\Delta f; \nu, \mu) \tag{2.121}$$

$$\phi_S(\tau, \eta; \nu, \mu) = \psi_S(\tau; \nu, \mu)\delta(\eta - \tau) \tag{2.122}$$

where

$$\psi_c(t, s; \tau) = \int_{-\infty}^{\infty} \phi_T(\Delta f; t, s)e^{j2\pi\Delta f\tau}d\Delta f \tag{2.123}$$

$$\psi_S(\tau; \nu, \mu) = \int_{-\infty}^{\infty} \phi_H(\Delta f; \nu, \mu)e^{j2\pi\Delta f\tau}d\Delta f \ . \tag{2.124}$$

Wide sense stationary uncorrelated scattering (WSSUS) channels are a very special type of multipath-fading channel. These channel display uncorrelated scattering in both the time-delay and Doppler shift. Fortunately, many radio channels can be accurately modeled as WSSUS channels. For WSSUS channels, the correlation functions have singular behavior in both the time delay and Doppler shift variables, and reduce to the following simple forms:

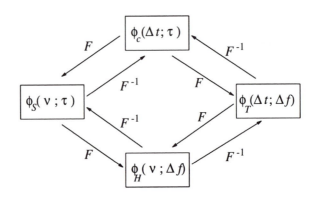

Figure 2.22 Fourier transform relations between the channel correlation functions for WSSUS channels, from [243].

$$\phi_c(t, t + \Delta t; \tau, \eta) = \psi_c(\Delta t; \tau)\delta(\eta - \tau) \tag{2.125}$$
$$\phi_T(f, f + \Delta f; t, t + \Delta t) = \phi_T(\Delta f; \Delta t) \tag{2.126}$$
$$\phi_H(f, f + \Delta f; \nu, \mu) = \psi_H(\Delta f; \nu)\delta(\nu - \mu) \tag{2.127}$$
$$\phi_S(\tau, \eta; \nu, \mu) = \phi_S(\tau, \nu)\delta(\eta - \tau)\delta(\nu - \mu) . \tag{2.128}$$

These correlation functions are related through the Fourier transform pairs shown in Fig. 2.22.

The function $\phi_c(0; \tau) \equiv \phi_c(\tau)$ is called the **multipath intensity profile** or **power delay profile** and gives the average power at the channel output as a function of the time delay τ. It can be viewed as the scattering function averaged over all Doppler shifts. A typical power delay profile is shown in Fig. 2.23. One quantity of interest is the **average delay** , defined as

$$\mu_\tau = \frac{\int_0^\infty \tau \phi_c(\tau) d\tau}{\int_0^\infty \phi_c(\tau) d\tau} . \tag{2.129}$$

Note that the normalization $\int_0^\infty \phi_c(\tau) d\tau$ is applied because $\phi_c(\tau)$ is not a pdf. Another quantity of interest is the *rms* **delay spread** , defined as

$$\sigma_\tau = \sqrt{\frac{\int_0^\infty (\tau - \mu_\tau)^2 \phi_c(\tau) d\tau}{\int_0^\infty \phi_c(\tau) d\tau}} . \tag{2.130}$$

There are other quantities that can also be used to describe the power delay profile. One is the width, W_x, of the middle portion of the power delay profile that contains $x\%$ of the total power in the profile. Referring to Fig. 2.23

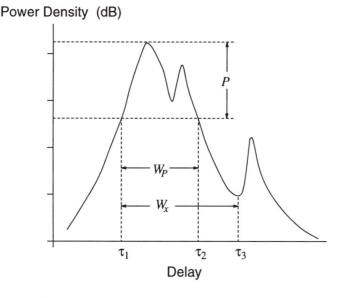

Figure 2.23 A typical power delay profile.

$$W_x = \tau_3 - \tau_1 \qquad (2.131)$$

where τ_1 and τ_3 are chosen so that

$$\int_0^{\tau_1} \phi_c(\tau)d\tau = \int_{\tau_3}^{\infty} \phi_c(\tau)d\tau \qquad (2.132)$$

and

$$\int_{\tau_1}^{\tau_3} \phi_c(\tau)d\tau = x \int_0^{\infty} \phi_c(\tau)d\tau . \qquad (2.133)$$

Another quantity is the difference in delays where the delay profile rises to a value P dB below its maximum value and where the delay profile drops to a value P dB below its maximum value for the last time. This quantity is denoted by W_P and is also illustrated in Fig. 2.23, where $W_P = \tau_2 - \tau_1$. Power delay profiles play a key role in determining whether or not an adaptive equalizer is required at the receiver. If the excess delay spread exceeds 10% to 20% of the symbol duration, then an adaptive equalizer may be required. In general, the average delay and delay spread of the channel will diminish with decreasing cell size, the reason being that the radio path lengths are shorter. While the delay spread in a typical macrocellular application may be on the order of 1 to 10 μs, the delay spreads in a typical microcellular applications are much less. Delay spreads within buildings range can anywhere from 30 to 60 ns in buildings with interior walls and little metal, to 300 ns in buildings with open plans and a significant amount of metal.

The function $\phi_T(\Delta t; \Delta f)$ is called the spaced-frequency spaced-time correlation function. The function $\phi_T(0; \Delta f) \equiv \phi_T(\Delta f)$ measures the frequency correlation of the channel. The **coherence bandwidth**, B_c, of the channel is defined as the smallest value of Δf for which $\phi_T(\Delta f)$ equals some suitable correlation coefficient such as 0.5. As a result of the Fourier transform relation between $\phi_c(\tau)$ and $\phi_T(\Delta f)$, the reciprocal of either the average delay or the delay spread is a measure of the coherence bandwidth of the channel. i.e.,

$$B_c \propto \frac{1}{\mu_\tau} \quad \text{or} \quad B_c \propto \frac{1}{\sigma_\tau} . \tag{2.134}$$

The function $\phi_H(\nu; 0) \equiv \phi_H(\nu)$ is called the Doppler psd and gives the average power at the channel output as a function of the Doppler frequency ν. The range of values over which $\phi_H(\nu)$ is significant is called the Doppler spread and is denoted by B_d. Since $\phi_H(\nu)$ and $\phi_T(\Delta t)$ are a Fourier transform pair, it follows that the inverse of the Doppler spread gives a measure of the **coherence time**, T_c, of the channel, i.e.,

$$T_c \approx \frac{1}{B_d} . \tag{2.135}$$

The coherence time of the channel is important for evaluating the performance of coding and interleaving techniques that try to exploit the inherent time diversity of the channel. Note that the Doppler spread and, hence, the coherence time depend directly on the velocity of a moving MS. Therefore, any scheme that exploits the time diversity of the channel must be evaluated over the complete range of expected MS velocities.

The function $\phi_S(\nu, \tau)$ is called the **scattering function** and gives the average power output of the channel as a function of the time delay τ and the Doppler shift ν. The scattering function is widely used in the characterization of multipath-fading channels.

2.2.3 Channel Output Autocorrelation

The autocorrelation of the channel output can be expressed in terms of the transmission functions. For example, from (2.98) we have

$$\phi_{rr}(t,s) = \int_{-\infty}^{\infty}\int_{-\infty}^{\infty} u(t-\tau)u^*(s-\eta)\frac{1}{2}\mathrm{E}[c(t,\tau)c^*(s,\eta)]d\tau d\eta$$

$$= \int_{-\infty}^{\infty}\int_{-\infty}^{\infty} u(t-\tau)u^*(s-\eta)\phi_c(t,s;\tau,\eta)d\tau d\eta \ .$$

$$(2.136)$$

For WSSUS channels, the above expression reduces to

$$\phi_{rr}(t,t+\Delta t) = \int_{-\infty}^{\infty}\int_{-\infty}^{\infty} u(t-\tau)u^*(t+\Delta t-\eta)\phi_c(\Delta t;\tau)\delta(\eta-\tau)d\tau d\eta$$

$$= \int_{-\infty}^{\infty} u(t-\tau)u^*(t+\Delta t-\tau)\phi_c(\Delta t;\tau)d\tau \ . \quad (2.137)$$

The channel output autocorrelation can also be expressed in terms of the scattering function by substituting the double inverse Fourier transform in (2.112) into (2.135). For WSSUS channels, we can use (2.128) to write

$$\phi_{rr}(t,t+\Delta t) = \int_{-\infty}^{\infty}\int_{-\infty}^{\infty} u(t-\tau)u^*(t+\Delta t-\tau)\phi_S(\tau;\nu)e^{j2\pi\nu\Delta t}d\tau d\nu \ . \quad (2.138)$$

2.3 LABORATORY SIMULATION OF MULTIPATH-FADING CHANNELS

The most accurate method of simulating multipath-fading channels is to use recorded strips of actual wide band channel measurements. Unfortunately, it is impossible to control the channel characteristics with this method, except by using different recorded strips of measurements. Since it is desirable to assess the system performance over all ranges of expected channel conditions, control of the channel characteristics is essential. For this reason, channel simulators are of interest that are derived from theoretical principles.

2.3.1 Filtered Gaussian Noise

A straightforward method of constructing a fading simulator is to amplitude modulate the inphase and quadrature components of a carrier with a low-pass filtered Gaussian noise source as shown in Fig. 2.24. If the Gaussian noise sources have zero-mean, then this method produces a Rayleigh faded envelope;

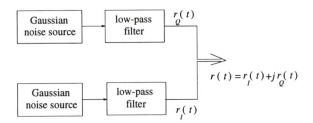

Figure 2.24 Fading simulator that uses low-pass filtered white Gaussian noise.

otherwise a Ricean faded envelope is produced. Here, we consider only the Rayleigh fading case. The two different noise sources must have the same psd to produce a fading process that is stationary, and the psd of the inphase and quadrature components of the faded carrier will be that of the filtered white Gaussian noise. The main limitation with this approach is that only rational forms of the fading spectra can be produced. In reality, the fading spectrum is typically non-rational as shown for example in Fig. 2.4. In order to approximate the spectrum in Fig. 2.4, a high order pole-zero filter is required. Unfortunately, a high order filter has a long impulse response, and this will significantly increase the run times if this simulator is implemented in software.

The simplest simulator of this type uses a first-order low-pass filter, implying that the fading process is a Markov process. To describe this simulator further, let $r_{I,k} \equiv r_I(kT)$ and $r_{Q,k} \equiv r_Q(kT)$ represent the amplitudes of the in-phase and quadrature components of the fading signal at epoch k, where T is the simulation step size. Then $r_{I,k}$ and $r_{Q,k}$ are Gaussian random variables with the state equation

$$(r_{I,k+1}, r_{Q,k+1}) = \zeta(r_{I,k}, r_{Q,k}) + (1 - \zeta)(w_{1,k}, w_{2,k}) \qquad (2.139)$$

where $w_{1,k}$ and $w_{2,k}$ are independent Gaussian random variables with zero mean and correlation $\mathrm{E}[w_{i,k}w_{i,\ell}] = \sigma^2\delta_{k\ell}$, $i = 1, 2$. The envelope $\sqrt{r_{I,k}^2 + r_{Q,k}^2}$ is Rayleigh distributed. It can be shown that the discrete autocorrelation of $r_{I,k}$ and $r_{Q,k}$ is

$$
\begin{aligned}
\phi_{r_Q r_Q}(n) = \phi_{r_I r_I}(n) &= \mathrm{E}[r_{I,k} r_{I,k+n}] \\
&= \frac{1 - \zeta}{1 + \zeta}\sigma^2 \zeta^{|n|} \qquad (2.140)
\end{aligned}
$$

With isotropic scattering, for example, the desired autocorrelation is, from (2.20),

$$\phi_{r_I r_I}(n) = \frac{\Omega_p}{2} J_o(2\pi f_m nT) \tag{2.141}$$

where, again, T is the simulation step size. By using the first-order low-pass filter we are essentially approximating the Bessel function with a negative exponential function. The parameters σ^2 and ζ must be specified to complete the model. To determine appropriate values for σ^2 and ζ, we first note that the noise spectral density at the output of the filter is

$$S_{nn}(f) = \frac{\sigma^2}{1 + \zeta^2 - 2\zeta \cos 2\pi fT} \cdot \tag{2.142}$$

This is to be compared with the psd of a faded carrier shown in Fig. 2.4. Observe from Fig. 2.4 that the spectrum is confined to the interval $f_c - f_m \leq f \leq f_c + f_m$. Therefore, one possibility is to arbitrarily set the corner frequency of the first-order low-pass filter to $f_m/4$, where $f_m = v/\lambda_c$. This gives

$$\zeta = 2 - \cos(2\pi f_m T) - \sqrt{(2 - \cos 2\pi f_m T)^2 - 1} \ . \tag{2.143}$$

To normalized the mean square envelope to 0 dB, the value of σ^2 is chosen as

$$\sigma^2 = \frac{1 + \zeta}{2(1 - \zeta)} \ . \tag{2.144}$$

Fig. 2.25 plots the received envelope level in decibels against normalized time, t/T, that is obtained by using this simulator. Note that the slow roll-off of the first order low-pass filter causes some high frequency components in the faded envelope. Some improvement can be obtained by using a higher order filter, but as explained earlier, this will increase the complexity of the simulator. One big advantage of using low-pass filtered white Gaussian noise is the ease by which multiple uncorrelated fading waveforms can be generated. We just need to use uncorrelated noise sources.

2.3.2 Jakes' Method

Another very effective channel simulator has been suggested by Jakes. The description of this method begins with (2.8) and (2.6), and assumes stationarity ($f_{D,n}(t) = f_{D,n}$) and equal strength multipath components ($\alpha_n = 1$). In this case the received complex low-pass envelope is

$$r(t) = \sum_{n=1}^{N} e^{-j(\hat{\phi}_n + 2\pi f_m t \cos \theta_n)} \ , \tag{2.145}$$

where

$$\hat{\phi}_n = 2\pi(f_c + f_m)\tau_n \ . \tag{2.146}$$

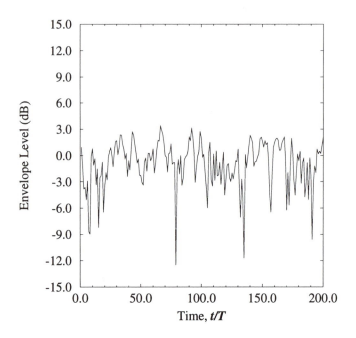

Figure 2.25   Faded envelope generated by filtering white Gaussian noise with a first-order low-pass filter; $f_m T = 0.1$.

Suppose that we wish to model an isotropic scattering channel. Then the N components are uniformly distributed in angle, i.e.,

$$\theta_n = \frac{2\pi n}{N} \ , \quad n = 1,\, 2, \ldots,\, N \ . \tag{2.147}$$

If $N/2$ is an odd integer, then (2.145) can be rearranged into the form

$$
\begin{aligned}
r(t) \ = \ & \sum_{n=1}^{N/2-1} \left[e^{-j(\hat{\phi}_{-n}+2\pi f_m t \cos\theta_n)} + e^{j(\hat{\phi}_n+2\pi f_m t \cos\theta_n)} \right] \\
& + e^{-j(\hat{\phi}_{-N}+2\pi f_m t)} + e^{j(\hat{\phi}_N+2\pi f_m t)} \\
= \ & \sqrt{2} \sum_{n=1}^{M} \left[e^{-j(\hat{\phi}_{-n}+2\pi f_m t \cos\theta_n)} + e^{j(\hat{\phi}_n+2\pi f_m t \cos\theta_n)} \right] \\
& e^{-j(\hat{\phi}_{-N}+2\pi f_m t)} + e^{j(\hat{\phi}_N+2\pi f_m t)}
\end{aligned}
\tag{2.148}
$$

where

$$M = \frac{1}{2}\left(\frac{N}{2}-1\right) \ . \tag{2.149}$$

The last equality in (2.148) follows because the Doppler shifts progress from $+2\pi f_m \cos(2\pi/N)$ to $-2\pi f_m \cos(2\pi/N)$ as n progresses from 1 to $N/2 - 1$ in the first sum, while the Doppler shifts progress from $-2\pi f_m \cos(2\pi/N)$ to $+2\pi f_m \cos(2\pi/N)$ as n progresses from 1 to $N/2 - 1$ in the second sum. Therefore, the frequencies overlap and only half of them are required. The factor $\sqrt{2}$ is included so that the total power will remain unchanged. By using some trigonometric identities and further algebra, (2.148) can be expressed in the form

$$r(t) = \sqrt{2} \left\{ \left[2\sum_{n=1}^{M} \cos \beta_n \cos 2\pi f_n t + \sqrt{2} \cos \alpha \cos 2\pi f_m t \right] \right. \tag{2.150}$$
$$\left. + j \left[\sum_{n=1}^{M} \sin \beta_n \cos 2\pi f_n t + \sqrt{2} \sin \alpha \cos 2\pi f_m t \right] \right\}$$

where

$$\alpha = \frac{\hat{\phi}_N - \hat{\phi}_{-N}}{2} \tag{2.151}$$

is an arbitrary phase. The $\{\beta_n\}$ are gains that are appropriately chosen as discussed below. From the above development, the Rayleigh fading simulator shown in Fig. 2.26 can be constructed. With this simulator, M low-frequency oscillators with frequencies $f_n = f_m \cos(2\pi n/N)$, $n = 1, 2, \ldots, M$, where $M = \frac{1}{2} \left(\frac{N}{2} - 1 \right)$, and with one oscillator at frequency f_m are used to generate waveforms that are added together to produce $r_I(t)$ and $r_Q(t)$. The amplitudes of the oscillators are all unity except for the oscillator at frequency f_m which has amplitude $1/\sqrt{2}$. It follows from Fig. 2.26 that

$$r_I(t) = 2\sum_{n=1}^{M} \cos \beta_n \cos 2\pi f_n t + \sqrt{2} \cos \alpha \cos 2\pi f_m t \tag{2.152}$$

$$r_Q(t) = 2\sum_{n=1}^{M} \sin \beta_n \cos 2\pi f_n t + \sqrt{2} \sin \alpha \cos 2\pi f_m t . \tag{2.153}$$

Note that (2.152) and (2.153) are equal the quadrature components of (2.150), except for a factor of $\sqrt{2}$.

It is desirable that the phase of $r(t) = r_I(t) + j r_Q(t)$ be uniformly distributed. This can be accomplished by choosing the parameters α and β_n so that $< r_I^2(t) >=< r_Q^2(t) >$ and $< r_I(t)r_Q(t) >= 0$, where $< \cdot >$ is a time average operator. From Fig. 2.26

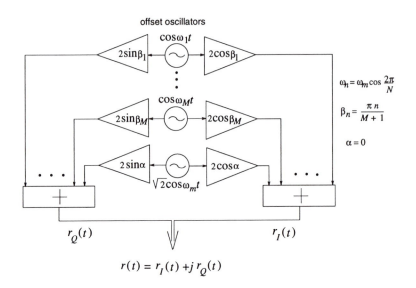

Figure 2.26 Jakes' Rayleigh fading simulator that generates a faded envelope
by using a number of low frequency oscillators. The choices $\alpha = 0$ and $\beta_n =$
$\pi n/M$ will yield $< r_Q^2(t) >= M$, $< r_I^2(t) >= M + 1$, and $< r_I(t)r_Q(t) >= 0$.

$$
\begin{aligned}
< r_I^2(t) > &= 2 \sum_{n=1}^{M} \cos^2 \beta_n + \cos^2 \alpha \\
&= M + \cos^2 \alpha + \sum_{n=1}^{M} \cos 2\beta_n \qquad (2.154)
\end{aligned}
$$

$$
\begin{aligned}
< r_Q^2(t) > &= 2 \sum_{n=1}^{M} \sin^2 \beta_n + \sin^2 \alpha \\
&= M + \sin^2 \alpha - \sum_{n=1}^{M} \cos 2\beta_n \qquad (2.155)
\end{aligned}
$$

$$
< r_I(t)r_Q(t) >= 2 \sum_{n=1}^{M} \sin \beta_n \cos \beta_n + \sin \alpha \cos \alpha \ . \qquad (2.156)
$$

By choosing $\alpha = 0$ and $\beta_n = \pi n/M$, the above equations reduce to $< r_Q^2(t) >=$
M, $< r_I^2(t) >= M + 1$, and $< r_I(t)r_Q(t) >= 0$. The mean square values
$< r_I^2(t) >$ and $< r_Q^2(t) >$ can be scaled to any desired value. A typical Rayleigh

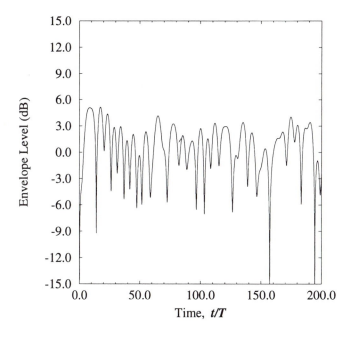

Figure 2.27 Faded envelope generated by using Jakes' fading simulator with 8 oscillators; $f_m T = 0.1$.

faded envelope, obtained by using $N = 34$ or $(M = 8)$ is shown in Fig. 2.27. The normalized autocorrelation function

$$\phi_{rr}(\tau) = \frac{E[r^*(t)r(t+\tau)]}{E[z(t)^2]} \qquad (2.157)$$

is plotted against the normalized time delay $f_m \tau$ in Fig. 2.28. Observe that the autocorrelation tends to deviate from the desired values at large lags. This can be improved upon by increasing the number of oscillators that are used in the simulator. For example, Fig. 2.29 shows the normalized autocorrelation function when the number of oscillators is doubled from 8 to 16. One of the advantages of using Jakes' method is that the autocorrelation and, hence, the psd of the inphase and quadrature components of the received signal can be generated to reflect an isotropic scattering environment, with a simulator of reasonable complexity.

In many cases it is desirable to generate multiple uncorrelated faded carriers. Jakes' method may be extended to provide up to M fading signals by using the same low frequency oscillators. This is accomplished by giving the nth oscillator the additional phase shift $\gamma_{nj} + \beta_n$, $1 \leq j \leq M$. The appropriate

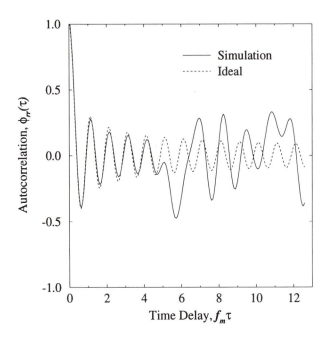

Figure 2.28 Autocorrelation of inphase and quadrature components obtained by using Jakes' fading simulator with 8 oscillators.

values of γ_{nj} and β_n are determined by imposing the additional constraint that the complex fading gains be uncorrelated (or as nearly uncorrelated as possible). By using two quadrature low-frequency oscillators per offset, rather than a single oscillator, the use of phase shifters to perform the phase shift $\gamma_{nj} + \beta_n$ can be eliminated. This leads to the fading generator shown in Fig. 2.30.

Consider the following choice for β_n and γ_{nj} with the objective yielding uncorrelated waveforms

$$\beta_n = \frac{\pi n}{M} \quad n = 1, 2, \ldots, M \tag{2.158}$$

$$\gamma_{nj} = \frac{2\pi(j-1)n}{M} \quad n = 1, 2, \ldots, M . \tag{2.159}$$

By using these values, the crosscorrelations between the different faded envelopes can be computed. Fig. 2.31 plots the typical normalized crosscorrelation

$$\phi_{r_i r_j}(\tau) = \frac{E[r_i^*(t)r_j(t+\tau)]}{\sqrt{E[|r_i(t)|^2]E[|r_j(t)|^2]}} \tag{2.160}$$

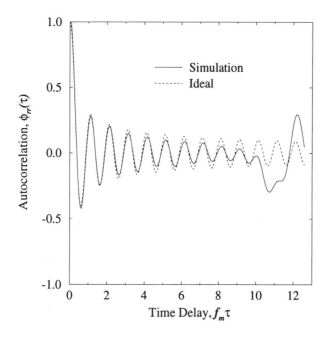

Figure 2.29 Autocorrelation of inphase and quadrature components obtained by using Jakes' fading simulator with 16 oscillators.

against the normalized time delay $f_m \tau$. Observe that the crosscorrelations are not always close to zero. Hence, one problem with Jakes' approach is that there is significant crosscorrelation between the different faded envelopes that are generated. One method to solve this problem has been suggested by Dent *et. al.* [68], see Problem 2.16.

2.3.3 Simulation of Wide-band Multipath-Fading Channels

τ-spaced model: The τ-spaced channel model assumes that the channel consists of a number of discrete paths at different delays. The channel is modeled by ℓ discrete multipath components having random complex gains and different delays, so that the received signal is

$$r(t) = \sum_{i=0}^{\ell} r_i(t) u(t - \tau_i(t)) \tag{2.161}$$

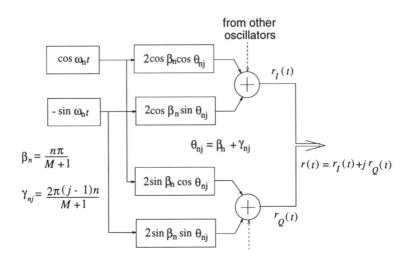

Figure 2.30   Jakes' method of obtaining multiple Rayleigh fading envelopes.

where the $r_i(t)$ and $\tau_i(t)$ are the gains and delays associated with the discrete multipath components. Although the $\tau_i(t)$ are time-varying, they are often assumed to be randomly static. The τ-spaced channel is described by the tap gain vector

$$\mathbf{r}(t) = (r_0(t),\ r_1(t),\ \ldots,\ r_\ell(t)) \tag{2.162}$$

and the delay vector

$$\tau(t) = (\tau_0(t), \tau_1(t),\ \ldots, \tau_\ell(t))\ . \tag{2.163}$$

Sometimes it is convenient if the differential path delays are multiples of some small number τ, leading to the τ-spaced tapped delay line channel model shown in Fig. 2.32. Note that many of the tap coefficients in the tapped delay line are zero, reflecting the fact that no energy is received at these delays.

The time varying channel tap coefficients $\{r_k(t)\}$ can be generated by using either of the two approaches described earlier, i.e., filtered white noise or Jakes' method. The filtered white noise approach has the advantage that the taps are uncorrelated, and the disadvantage that a very high filter order is needed to obtain the desired tap autocorrelations. Jakes' method provides the correct tap autocorrelations, but the tap coefficients are somewhat correlated.

If we wish to model an isotropic scattering channel, then an attempt is made to generate a tap gain vector $\mathbf{r}$ having the covariance matrix [326], [145]

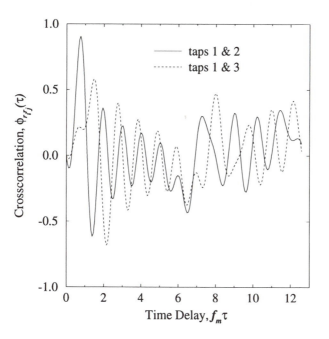

Figure 2.31 Crosscorrelation between the faded envelopes that are obtained by using Jakes' fading simulator with 8 oscillators.

$$\Phi_{\mathbf{r}}(\tau) = \frac{1}{2}\mathrm{E}[\mathbf{r}^H(t+\tau)\mathbf{r}(t)] = J_0(2\pi f_m\tau)\mathbf{\Sigma}^2 \qquad (2.164)$$

where H denotes Hermitian transposition, $J_0(\,\cdot\,)$ is the zero-order Bessel function of the first kind, $\mathbf{\Sigma} \triangleq \mathrm{diag}[\sigma_{r_0}, \sigma_{r_1}, \ldots, \sigma_{r_\ell}]$, and f_m is the maximum Doppler frequency. The set of tap variances $\{\sigma_{r_i}^2\}$ determines the fractional power at each lag. Typical 6-ray urban and bad urban multipath delay profiles have been defined in the COST207 study [54] and are shown in Table 2.1 and Fig. 2.33. Sometimes it is desirable to improve the accuracy by including more rays, because in reality the propagation medium is characterized by continuum of multipaths. Table 2.2 and Fig. 2.34 show 12-ray models for the typical urban and bad urban channel, as defined by COST207 [54]. The COST207 channel models were developed for macrocellular applications. The reader is cautioned that the delay spread experienced in microcells will be much smaller due to the shorter radio path lengths.

T-**spaced model:** The *T*-spaced channel model is similar to the *τ*-spaced model, except that the taps are symbol-spaced. The taps in the *T*-spaced model

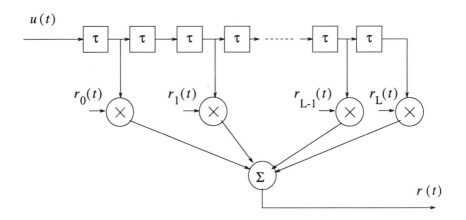

Figure 2.32 Wide-band multipath-fading channel model with discrete multipath components.

Typical Urban		Bad Urban	
delay, μs	Fractional Power	delay, μs	Fractional Power
0.0	0.189	0.0	0.164
0.2	0.379	0.3	0.293
0.5	0.239	1.0	0.147
1.6	0.095	1.6	0.094
2.3	0.061	5.0	0.185
5.0	0.037	6.6	0.117

Table 2.1 Typical macrocellular urban and bad urban 6-ray power delay profiles, from [54].

are usually all non-zero and correlated. The tap correlations often lead to analytical difficulties. To overcome these difficulties, a widely used approximation is to treat the taps as being uncorrelated [66], [82], [120], [191], [301], [173]. However, this approximation does not produce accurate results when the symbol duration is much greater than the delay spread. Crosscorrelation between the taps in the T-spaced model arises when the actual channel rays are not T-spaced, as in the τ-spaced model. The crosscorrelation is introduced by the receiver filter. Fig 2.35 shows a method for generating the T-spaced tap coefficients with the proper crosscorrelations when a linear modulation scheme is used and the underlying channel model is τ-spaced. A pulse generator produces

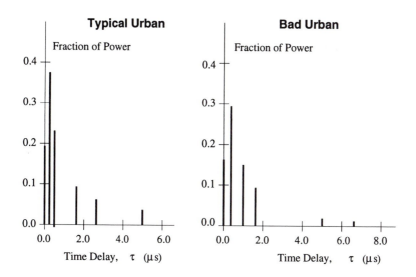

Figure 2.33 Typical macrocellular urban and bad urban 6-ray power delay profiles, from [54].

Typical Urban		Bad Urban	
delay, μs	Fractional Power	delay, μs	Fractional Power
0.0	0.092	0.0	0.033
0.1	0.115	0.1	0.089
0.3	0.231	0.3	0.141
0.5	0.127	0.7	0.194
0.8	0.115	1.6	0.114
1.1	0.074	2.2	0.052
1.3	0.046	3.1	0.035
1.7	0.074	5.0	0.140
2.3	0.051	6.0	0.136
3.1	0.032	7.2	0.041
3.2	0.018	8.1	0.019
5.0	0.025	10.0	0.006

Table 2.2 Typical macrocellular urban and bad urban 12-ray power delay profiles, from [54].

pulses having a shape that is determined by the combination of the transmitter and receiver filter, e.g., a raised cosine pulse. After passing the pulse through the τ-spaced channel, T-spaced samples are extracted. The T-spaced samples

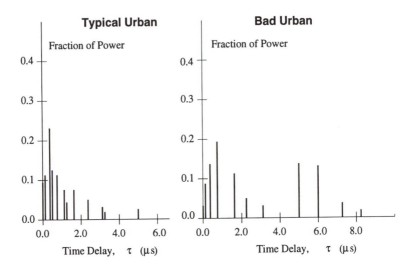

Figure 2.34 Typical macrocellular urban and bad urban 12-ray power delay profiles, from [54].

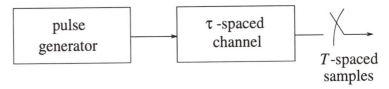

Figure 2.35 Method for generating correlated tap coefficients in a T-spaced channel model.

are just a linear combination of the tap gains in the τ-spaced model. That is, if $r_i^T(t)$ denotes the ith tap coefficient in the T-spaced model, then

$$r_i^T(t) = a_1 r_1(t) + a_2 r_2(t) + \cdots + a_L r_L(t) \qquad (2.165)$$

where the coefficients $\{a_i\}$ are determined by the overall pulse response of the transmitter and receiver filters along with the relative delays of the rays in the τ-spaced model. Note that the $\{a_i\}$ only have to be generated once, provided that the relative delays of the rays in the τ-spaced channel do not change.

For example, suppose that the overall pulse is a raised cosine pulse[1] having a roll-off factor of 0.35, and the channel is characterized by two equal strength rays, i.e., $\mathbf{r}(t) = (r_0(t), r_1(t))$, where $\sigma_0^2 = \sigma_1^2$. Then, Figs. 2.36 and 2.37 plot the normalized autocorrelation function

---
[1] See Chapter 4 for a discussion of raised cosine pulse shaping.

$$\phi_{r_i^T r_i^T}(\nu) = \frac{E[r_i^T(t) r_i^{T^*}(t+\nu)]}{E[|r_i^T(t)|^2]} \quad i = 1, 2 \tag{2.166}$$

and the normalized crosscorrelation function

$$\phi_{r_0^T r_1^T}(\nu) = \frac{E[r_0^T(t) r_1^{T^*}(t+\nu)]}{\sqrt{E[|r_0^T(t)|^2] E[|r_1^T(t)|^2]}} \tag{2.167}$$

of the two main taps in the T-spaced model when the differential delay $\tau = |\tau_1(t) - \tau_0(t)|$ is equal to $T/4$, $T/2$, $3T/4$, and T. Observe, from Fig. 2.37 that $\phi_{r_0^T r_1^T}(0) \neq 0$ when $\tau \neq T$. The large non-zero values of $\phi_{r_0^T r_1^T}(\nu)$ at larger values of ν is due to the aforementioned limitations of Jakes's fading simulator. Fortunately, for most practical applications $f_m T \ll 1$ so that these large non-zero crosscorrelations have little effect on the performance of the various systems that we will be examining. For example, using the typical IS-54 parameters of $T = 41.667$ μs and a carrier frequency of 850 MHz yields $f_m T = 0.0013$ for a MS velocity of 40 km/h.

2.4 SHADOWING

Let $\Omega_v = E[z(t)]$ denote the mean envelope level, where the expectation is taken over the pdf of the received envelope, e.g., a Rayleigh or Rice distribution. Sometimes Ω_v is called the **local mean** because it represents the envelope level averaged over a distance of a few wavelengths. The local mean Ω_v itself is a random variable due to shadow variations that are caused by large terrain features between the BS and MS, such as buildings and hills in macrocells and smaller objects such as vehicles in microcells. Empirical studies have shown that Ω_v follows a log-normal distribution, such that the pdf of Ω_v is

$$p(\Omega_v) = \frac{2\xi}{\Omega_v \sigma_\Omega \sqrt{2\pi}} \exp\left\{ -\frac{\left(10\log_{10}\Omega_v^2 - \mu_{\Omega_v}\right)^2}{2\sigma_\Omega^2} \right\} \tag{2.168}$$

where $\mu_{\Omega_v} = E[\Omega_v \text{ (dB)}]$ denotes the mean value of Ω_v (dB) and $\xi = 10/\ln 10$. The value of μ_{Ω_v} is determined by the propagation path loss between the BS and MS. A transformation of random variables gives the following Gaussian pdf for Ω_v (dB) $= 10\log_{10}\Omega_v$;

$$p(\Omega_v \text{ (dB)}) = \frac{1}{\sqrt{2\pi}\sigma_\Omega} \exp\left\{ -\frac{(\Omega_v \text{ (dB)} - \mu_{\Omega_v})^2}{2\sigma_\Omega^2} \right\} . \tag{2.169}$$

Some confusion may arise in the description of log-normal shadow distribution, because some authors [222], [102], [103] treat the mean envelope $\Omega_v = E[z(t)]$

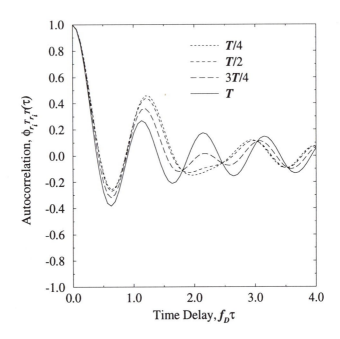

Figure 2.36   Autocorrelation function for one of the main taps in the overall
T-spaced discrete-time channel impulse response when the channel consists of
two equal strength Rayleigh faded rays at various differential delays.

as being log-normally distributed with standard deviation σ_Ω, while other authors [192], [212], [254], [109] treat the mean square-envelope $\Omega_p = \mathrm{E}[z(t)^2]$ as
being log-normally distributed with the same value of σ_Ω. Clearly, these two
quantities are not the same. It is shown in Appendix 2A that the standard
deviation σ_Ω is the same in each case. However, with Ricean fading the means
differ by

$$\mu_{\Omega_p} = \mu_{\Omega_v} + 10\log_{10}C(K) \tag{2.170}$$

where

$$C(K) = \frac{4e^{2K}(K+1)}{\pi_1 F_1^2(3/2, 1; K)} \tag{2.171}$$

and $_1F_1(\cdot, \cdot; \cdot)$ denotes the confluent hypergeometric function. The standard
deviation of log-normal shadowing ranges from 5 to 12 dB with $\sigma_\Omega = 8$ dB
being a typical value for macrocellular applications. The standard deviation
increases slightly with frequency (0.8 dB higher at 1800 MHz than at 900 MHz),
but has been observed to be nearly independent radio path length, even for
distances that are very close to the transmitter [212]. The standard deviation
that is observed in microcells varies between 4 and 13 dB [265], [25], [203],
[124], [126], [213]. Mogensen [213] has reported $\sigma_\Omega = 6.5$ to 8.2 dB at 900 MHz

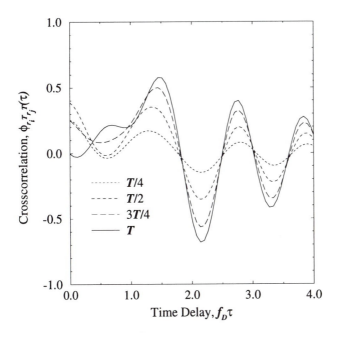

Figure 2.37 Crosscorrelation function for the two main taps in the overall *T*-spaced discrete-time channel impulse response when the channel consists of two equal strength Rayleigh faded rays at various differential delays.

in urban areas, while Mockford *et. al.* [212] report a value of 4.5 dB for urban areas. Berg [25] and Goldsmith and Greenstein [126] report that σ_Ω is around 4 dB for a spatial averaging window of 20 wavelengths and BS antenna heights of about 10 (m). Several studies suggest that σ_Ω decreases with an increase in the degree of urbanization or density of scatters. For example, the results presented by Mockford *et. al.* [212] suggest that σ_Ω is 1.3 to 1.8 dB higher in a suburban environment than in an urban environment.

2.4.1 Laboratory Simulation of Shadowing

One of the challenges when constructing a shadow simulator is to account for the spatial correlation of the shadows. Several studies have investigated the spatial correlation of shadows [141], [134], [203], [153], [135]. One simple, convenient and realistic model has been suggested by Gudmundson [135], where log-normal shadowing is modeled as a Gaussian white noise process that is filtered with a first-order low-pass filter. With this model

$$\Omega_{k+1 \text{ (dB)}} = \xi \Omega_{k \text{ (dB)}} + (1 - \xi) v_k \tag{2.172}$$

where $\Omega_{k \text{ (dB)}}$ is the mean envelope level or mean square-envelope level (in dB) that is experienced at location k, ξ is a parameter that controls the spatial decorrelation of the shadowing, and v_k is a zero-mean Gaussian random variable with $\phi_{vv}(n) = \tilde{\sigma}^2 \delta(n)$. From equation (2.140), it immediately follows that the spatial autocorrelation function of $\Omega_{k \text{ (dB)}}$ is

$$\phi_{\Omega_{(dB)}\Omega_{(dB)}}(n) = \frac{1 - \xi}{1 + \xi} \tilde{\sigma}^2 \xi^{|n|} . \tag{2.173}$$

Since the variance of log-normal shadowing is

$$\sigma_\Omega^2 = \phi_{\Omega_{(dB)}\Omega_{(dB)}}(0) = \frac{1 - \xi}{1 + \xi} \tilde{\sigma}^2 \tag{2.174}$$

we can express the autocorrelation of Ω_k as

$$\phi_{\Omega_{(dB)}\Omega_{(dB)}}(n) = \sigma_\Omega^2 \xi^{|n|} . \tag{2.175}$$

To express the above result in another form, suppose that the signal strength is sampled every T seconds. Then the time autocorrelation of the shadowing is

$$\phi_{\Omega_{(dB)}\Omega_{(dB)}}(k) \equiv \phi_{\Omega_{(dB)}\Omega_{(dB)}}(kT) = \sigma_\Omega^2 \xi_D^{(vT/D)|k|} \tag{2.176}$$

where ξ_D is the correlation between two points separated by a spatial distance of D (m) and v (m/s) is the velocity of the MS. The parameter ξ_D can be chosen to adjust the correlation of the shadows. For typical suburban propagation at 900 MHz, it has been experimentally verified by Gudmundson [133] that $\sigma_\Omega \approx 7.5$ dB with a correlation of approximately 0.82 at a distance of 100 m. For typical microcellular propagation at 1700 MHz, Gudmundson has also reported $\sigma_\Omega = 4.3$ dB with a correlation of 0.3 at a distance of 10 m.

2.4.2 Composite Shadowing-Fading Distributions

Sometimes it is desirable to know the composite distribution due to shadowing and multipath fading. This may be particularly true for the case of slow moving or stationary MSs, where the receiver is unable to average over the effects of fading and a composite distribution is necessary for evaluating link performance and other quantities.

Two different approaches have been suggested in the literature for obtaining the composite distribution. The first approach is to express the envelope (or squared-envelope) as a conditional density on Ω_v (or Ω_p), and then integrate over the density of Ω_v (or Ω_p) to obtain the composite distribution

$$p_{z_c}(x) = \int_0^\infty p_{z|\,\Omega_v}(x|w)p_{\Omega_v}(w)dw \ . \tag{2.177}$$

For the case of Rayleigh fading

$$\Omega_v = E[z(t)] = \sqrt{\frac{\pi}{2}}\sigma \tag{2.178}$$

and, hence,

$$p_{z|\,\Omega_v}(x|w) = \frac{\pi x}{2w^2}\exp\left\{-\frac{\pi x^2}{4w^2}\right\} \ . \tag{2.179}$$

The composite envelope distribution with Rayleigh fading and log-normal shadowing is

$$\begin{aligned} p_{z_c}(x) &= \int_0^\infty \frac{\pi x}{2w^2}\exp\left\{-\frac{\pi x^2}{4w^2}\right\} \\ &\times \frac{2\xi}{w\sigma_\Omega\sqrt{2\pi}}\exp\left\{-\frac{\left(10\log_{10}w^2 - \mu_{\Omega_v}\right)^2}{2\sigma_\Omega^2}\right\}dw \ . \end{aligned} \tag{2.180}$$

Sometimes this distribution is called a Susuki distribution, after the original work by Susuki [292].

The second approach, originally suggested by Lee and Yeh [182], is to express the composite received signal as the product of the short term multipath fading and the long term shadow fading. Hence, at any time t, the envelope of the composite signal has the form

$$\hat{z}_c(t) = z(t) \cdot \Omega_v(t) \tag{2.181}$$

and the squared-envelope of the composite signal has the form

$$\hat{z}_c(t)^2 = z(t)^2 \cdot \Omega_p(t) \ . \tag{2.182}$$

Under the assumption that the fading and shadowing are independent random processes, we now demonstrate that both approaches lead to identical results.

The density function of envelope in (2.181) can be obtained by using a bivariate transformation and then integrating to obtain the marginal density. This leads to the density

$$p_{\hat{z}_c}(x) = \int_0^\infty \frac{1}{w} p_z \left(\frac{x}{w}\right) p_{\Omega_v}(w)dw \ .$$

(2.183)

Again, consider the case of log-normal shadowing and Rayleigh fading. Using (2.38) and (2.168) gives

$$
\begin{aligned}
p_{\hat{z}_c}(x) &= \int_0^\infty \frac{x}{(w\sigma)^2} \exp\left\{-\frac{x^2}{2(w\sigma)^2}\right\} \\
&\quad \times \frac{2\xi}{w\sigma_\Omega\sqrt{2\pi}} \exp\left\{-\frac{\left(10\log_{10}w^2 - \mu_{\Omega_v}\right)^2}{2\sigma_\Omega^2}\right\} dw \ .
\end{aligned}
$$

(2.184)

Observe that (2.180) and (2.184) are related by

$$p_{z_c}(x) = \sqrt{\frac{\pi}{2}}\sigma p_{\hat{z}_c} \left(\sqrt{\frac{\pi}{2}}\sigma x\right)$$

(2.185)

It follows that the random variables $\hat{z}_c$ and z_c are simply related through the linear transformation

$$z_c = \sqrt{\frac{2}{\pi}}\frac{1}{\sigma}\hat{z}_c \ .$$

(2.186)

Note, however, that $\sqrt{\pi/2}\sigma$ is just the mean of the Rayleigh distribution. Therefore, if we normalize $z(t)$ to have unit mean, then z_c and $\hat{z}_c$ have the exact same distribution!

Composite Gamma-log-normal Distribution

It is sometimes very useful to model the radio propagation environment as a shadowed Nakagami fading channel, because the Nakagami distribution is mathematically convenient and can closely approximate a Ricean distribution which is often used to model a LOS propagation environment. The composite distribution of the squared-envelope due to Nakagami fading and log-normal shadowing has the *Gamma*-log-normal density function

$$
\begin{aligned}
p_{z_c^2}(x) &= \int_0^\infty \left(\frac{m}{w}\right)^m \frac{x^{m-1}}{\Gamma(m)} \exp\left\{-\frac{mx}{w}\right\} \\
&\quad \times \frac{\xi}{\sqrt{2\pi}\sigma_\Omega w} \exp\left\{-\frac{(10\log_{10}w - \mu_{\Omega_p})^2}{2\sigma_\Omega^2}\right\} dw \ .
\end{aligned}
$$

(2.187)

As shown in Appendix 2B, the composite Gamma-log-normal distribution in (2.187) can be accurately approximated by a simple log-normal distribution with mean and standard deviation

$$\mu_{(\text{new})} = \xi[\psi(m) - \ln(m)] + \mu_{\Omega_p}$$
$$\sigma^2_{(\text{new})} = \xi^2 \zeta(2, m) + \sigma^2_\Omega \qquad (2.188)$$

where $\psi(\cdot)$ is the Euler psi function and $\zeta(\cdot)$ is Riemann's zeta function as defined in Appendix 2B. When $m = 1$ the approximation is valid for $\sigma_\Omega > 6$ dB, and for $m > 2$ the approximation is valid for all ranges of σ_Ω of interest [143]. The effect of Nakagami fading in (2.187) is to decrease the mean and increase the variance. However, this affect decreases as the shape factor m increases (corresponding to less severe fading). For example, with $m = 1$ (Ralyeigh fading) we have $\mu_{(\text{new})} = \mu - 2.50675$ and $\sigma^2_{(\text{new})} = \sigma^2 + 31.0215$, while with $m = 8$ we have $\mu_{(\text{new})} = \mu - 0.277$ and $\sigma^2_{(\text{new})} = \sigma^2 + 2.50972$. We conclude that it is important to include the effects of Nakagami fading when the shape factor m and the shadow standard deviation σ are small.

2.5 PATH LOSS MODELS

It is well known that the received signal power decays with the square of the path length in free space. That is, the received signal power is [243]

$$\mu_\Omega = 10\log_{10}\left(\Omega_t G_T G_R \left(\frac{\lambda_c}{4\pi d}\right)^2\right) \quad \text{dBm} \qquad (2.189)$$

where Ω_t is the transmitted power, G_T and G_R are the transmitter and receiver antenna gains, and d is the radio path length. The signals in land mobile radio applications, however, do not experience free space propagation. A more appropriate theoretical model assumes propagation over a flat reflecting surface (the earth) as shown in Fig. 2.38. In this case, the received signal power is [243]

$$\mu_\Omega = 10\log_{10}\left(\Omega_t 4\left(\frac{\lambda_c}{4\pi d}\right)^2 G_T G_R \sin^2\left(\frac{2\pi h_b h_m}{\lambda_c d}\right)\right) \quad \text{dBm} \qquad (2.190)$$

where h_b and h_m are the heights of the BS and MS antennas, respectively. Under the condition that $d \gg h_b h_m$, (2.190) reduces to

$$\mu_\Omega = 10\log_{10}\left(\Omega_t G_T G_R \left(\frac{h_b h_m}{d^2}\right)^2\right) \quad \text{dBm} \qquad (2.191)$$

where we have invoked the approximation $\sin x \approx x$ for small x. Observe that the propagation over a flat reflecting surface differs from free space propagation in two ways when $d \gg h_b h_m$. First, the path loss is not frequency dependent and, second, the signal power decays with the 4th power of the distance.

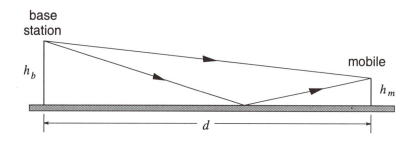

Figure 2.38   Radio propagation over a flat reflecting surface.

Fig. 2.39 plots the path loss

$$
\begin{aligned}
L_p &= 10\log_{10}\left\{\frac{\Omega_t G_T G_R}{\mu_\Omega}\right\} \\
&= -10\log_{10}\left\{4\left(\frac{\lambda_c}{4\pi d}\right)^2 \sin^2\left(\frac{2\pi h_b h_m}{\lambda_c d}\right)\right\} \quad \text{dB} \qquad (2.192)
\end{aligned}
$$

against the distance d. Notice that the received signal power has alternate minima and maxima when the MS is close to the BS. This property has been noted before by Milstein *et. al.* [210]. The last local minima in the path loss occurs when

$$
\frac{2\pi h_b h_m}{\lambda_c d} = \frac{\pi}{2} \ .
$$

2.5.1   Path Loss in Macrocells

Several highly useful empirical models for macrocellular systems have been obtained by curve fitting experimental data. Two of the more useful models for 900 MHz cellular systems are Hata's model [240] based on Okumura's prediction method [141], and Lee's model [176].

Okumura and Hata's Model

Hata's empirical model [141] is probably the simplest to use, and can distinguish between man-made structures. The empirical data for this model was collected by Okumura [240] in the city of Tokyo. Be cautioned, however, that the path losses for Japanese suburban areas do not match North American suburban areas very well. The latter are more like the quasi-open areas in Japan.

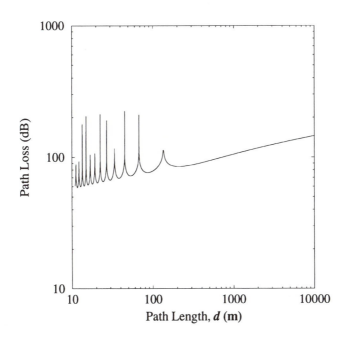

Figure 2.39 Propagation path loss with distance over a flat reflecting surface; $h_b = 7.5$ m, $h_m = 1.5$ m, $f_c = 1800$ MHz.

Okumura and Hata's model is expressed in terms of the carrier frequency $450 \leq f_c \leq 1000$ (MHz), BS antenna height $30 \leq h_b \leq 200$ (m), the MS antenna height $1 \leq h_m \leq 10$ (m), and the distance $1 \leq d \leq 20$ (km) between the BS and MS. The model is known to be accurate to within 1 dB for distances ranging from 1 to 20 km. With Okumura and Hata's model, the path loss (in dB) is

$$L_p = \begin{cases} A + B \log_{10}(d) & \text{for urban area} \\ A + B \log_{10}(d) - C & \text{for suburban area} \\ A + B \log_{10}(d) - D & \text{for open area} \end{cases} \qquad (2.193)$$

where

$$
\begin{aligned}
A &= 69.55 + 26.16 \log_{10}(f_c) - 13.82 \log_{10}(h_b) - a(h_m) \\
B &= 44.9 - 6.55 \log_{10}(h_b) \\
C &= 5.4 + 2 \left[\log_{10}(f_c/28) \right]^2 \\
D &= 40.94 + 4.78 \left[\log_{10}(f_c) \right]^2 - 19.33 \log_{10}(f_c)
\end{aligned}
$$

and

$$
a(h_m) = \begin{cases} (1.1\log_{10}(f_c) - 0.7)\,h_m - (1.56\log_{10}(f_c) - 0.8) \\[2mm] \qquad\qquad\qquad\qquad\qquad\quad \text{for medium or small city} \\[3mm] \begin{cases} 8.28\,(\log_{10}(1.54 h_m))^2 - 1.1 & \text{for } f_c \geq 400 \text{ MHz} \\ 3.2\,(\log_{10}(11.75 h_m))^2 - 4.97 & \text{for } f_c < 400 \text{ MHz} \end{cases} \\[2mm] \qquad\qquad\qquad\qquad\qquad\quad \text{for large city} \end{cases}
$$

Lee's Model

Lee's empirical path loss prediction model [176] is also accurate and easy to use. The method is generally used to predict a path loss over flat terrain. If the actual terrain is not flat, e.g., hilly, there will be large prediction errors. Two parameters are required for Lee's path loss prediction model; the power at a 1 mile (1.6 km) point of interception, $\mu_{\Omega_{d_o}}$, and the path-loss exponent, β. The received signal power can be expressed as

$$
\mu_\Omega = 10\log_{10}\left(\mu_{\Omega_{d_o}} \left(\frac{d_o}{d}\right)^\beta \left(\frac{f_c}{f}\right)^n \alpha_o \right) \quad \text{dBm} \qquad (2.194)
$$

where d is in kilometers and $d_o = 1.6$ km. The parameter α_o is a correction factor used to account for different BS and MS antenna heights, transmit powers, and antenna gains. The following set of *nominal* conditions are assumed in Lee's path loss model:

- frequency $f_c = 900$ MHz

- BS antenna height $= 30.48$ m

- BS transmit power $= 10$ watts

- BS antenna gain $= 6$ dB above dipole gain

- MS antenna height $= 3$ m

- MS antenna gain $= 0$ dB above dipole gain

If the actual conditions are different from those listed above, then we compute the following parameters:

Terrain	$\mu_{\Omega_{r_o}}$ (dBm)	β
Free Space	-45	2
Open Area	-49	4.35
North American Suburban	-61.7	3.84
North American Urban (Philadelphia)	-70	3.68
North American Urban (Newark)	-64	4.31
Japanese Urban (Tokyo)	-84	3.05

Table 2.3 Parameters for the Lee's path loss prediction model in various propagation environments, from [176].

$$\alpha_1 = \left(\frac{\text{new BS antenna height (m)}}{30.48\text{m}}\right)^2$$

$$\alpha_2 = \left(\frac{\text{new MS antenna height (m)}}{3 \text{ m}}\right)^\xi$$

$$\alpha_3 = \left(\frac{\text{new transmitter power}}{10 \text{ watts}}\right)^2$$

$$\alpha_4 = \frac{\text{new BS antenna gain with respect to } \lambda_c/2 \text{ dipole}}{4}$$

$$\alpha_5 = \text{different antenna-gain correction factor at the MS}$$

$$(2.195)$$

From these parameters, the correction factor α_o is

$$\alpha_o = \alpha_1 \cdot \alpha_2 \cdot \alpha_3 \cdot \alpha_4 \cdot \alpha_5 . \qquad (2.196)$$

The parameters β and $\mu_{\Omega_{r_o}}$ have been found from empirical measurements, and are listed in Table 2.3.

The value of n in (2.194) ranges between 2 and 3 with the exact value depending upon the carrier frequency and the geographic area. For $f_c < 450$ MHz in a suburban or open area, $n = 2$ is recommended. In an urban area with $f_c > 450$MHz, $n = 3$ is recommended. The value of ξ in (2.195) is also determined from empirical data

$$\xi = \begin{cases} 2 & \text{for a MS antenna height } > 10 \text{ m} \\ 3 & \text{for a MS antenna height } < 3 \text{ m} \end{cases} . \qquad (2.197)$$

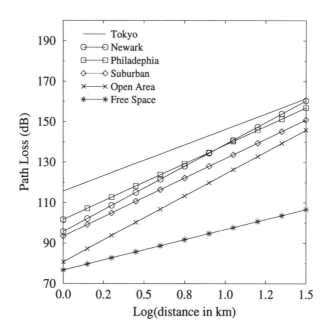

Figure 2.40 Path loss obtained by using Lee's method.

The path loss L_p is simply the difference between that transmitted and received field strengths, $L_p = \Omega_t - \mu_\Omega$. By using the above parameters for $\mu_{\Omega_{d_o}}$ and β, the following path losses (in dB) can be obtained

$$
L_p = \begin{cases}
76.71 + 20.0 \log_{10} d + n \log_{10}(f/900) - \alpha_o & \text{Free Space} \\
80.71 + 43.5 \log_{10} d + n \log_{10}(f/900) - \alpha_o & \text{Open Area} \\
93.41 + 38.4 \log_{10} d + n \log_{10}(f/900) - \alpha_o & \text{Suburban} \\
101.71 + 36.8 \log_{10} d + n \log_{10}(f/900) - \alpha_o & \text{Philadelphia} \\
95.71 + 43.1 \log_{10} d + n \log_{10}(f/900) - \alpha_o & \text{Newark} \\
115.71 + 30.5 \log_{10} d + n \log_{10}(f/900) - \alpha_o & \text{Tokyo}
\end{cases}
$$

(2.198)

These typical values from Lee's model are plotted in Fig. 2.40, and the values from Okumura and Hata's (large city) model are plotted in Fig. 2.41, for a BS height of 70 m, a MS antenna height of 1.5 m, and a carrier frequency of 900 MHz.

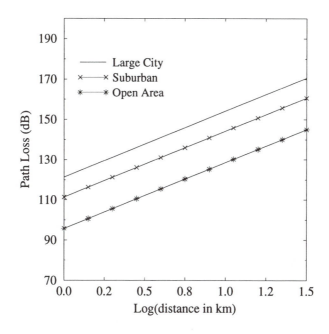

Figure 2.41 Path loss obtained by using Okumura and Hata's method.

2.5.2 Path Loss in Outdoor Microcells

Most of the future PCS microcellular systems are expected to operate in 1800-2000 MHz frequency bands. Some studies have suggested that the path losses experienced at 1845 MHz are about 10 dB larger than those experienced at 955 MHz when all other parameters are kept constant [55]. The COST231 study [56] has resulted in two models for urban microcellular propagation, the COST231-Hata model and the COST231-Walfish-Ikegami model.

COST231-Hata Model

The COST231-Hata model is based on the proposal by Mogensen [213] *et. al.* to extend Okumura and Hata's model for use in the 1500-2000 MHz frequency range where it is known that Okumura and Hata's model under estimates the path loss. The COST231-Hata model is expressed in terms of the carrier frequency $1500 \leq f_c \leq 2000$ (MHz), BS antenna height $30 \leq h_b \leq 200$ (m), MS antenna height $1 \leq h_m \leq 10$ (m), and distance $1 \leq d \leq 20$ (km). In particular, the path loss (in dB) with the COST231-Hata model is

$$L_p = A + B\log_{10}(d) + C \qquad (2.199)$$

where

$$A = 46.3 + 33.9\log_{10}(f_c) - 13.82\log_{10}(h_b) - a(h_m)$$
$$B = 44.9 - 6.55\log_{10}(h_b)$$
$$C = \begin{cases} 0 & \text{medium city and suburban areas} \\ & \text{with moderate tree density} \\ 3 & \text{for metropolitan centers} \end{cases}$$

Although both the Okumura and Hata and the COST231-Hata models are limited to BS antenna heights greater than 30 m, they can be used for lower BS antenna heights provided that the surrounding buildings are well below the BS antennas. They should not be used to predict path loss in urban canyons. The COST231-Hata model is good down to a path length of 1 km. It should not be used for smaller ranges, where path loss becomes highly dependent upon the local topography.

COST231-Walfish-Ikegami Model

The COST231-Walfish-Ikegami model is applicable to cases where the BS antennas are either above or below the roof tops. However, the model is not very accurate when the BS antennas are about the same height as the roof tops. For LOS propagation in a street canyon, the path loss (in dB) is

$$L_p = 42.6 + 26\log_{10}(d) + 20\log_{10}(f_c), \quad d \ge 20 \ m \qquad (2.200)$$

where the first constant is chosen so that L_p is equal to the free-space path loss at a distance of 20 m. The model parameters are the distance d (km) and carrier frequency f_c (MHz). As defined in Fig. 2.42, the path loss for non line-of-sight (NLOS) propagation is expressed in terms of the building heights, h_{Roof}, street widths, w, building separation, b, and road orientation with respect to the direct radio path, ϕ. The path loss is composed of three terms, viz.,

$$L_p = \begin{cases} L_o + L_{\text{rts}} + L_{\text{msd}} & \text{for } L_{\text{rts}} + L_{\text{msd}} \ge 0 \\ L_o & \text{for } L_{\text{rts}} + L_{\text{msd}} < 0 \end{cases} \qquad (2.201)$$

where L_o is the free-space loss, L_{rts} is the roof-to-street diffraction and scatter loss, and L_{msd} is the multi-screen diffraction loss. The free-space loss is

$$L_o = 32.4 + 20\log_{10}(d) + 20\log_{10}(f_c) \ . \qquad (2.202)$$

The roof-top-to-street diffraction and scatter loss is

$$L_{\text{rts}} = -16.9 - 10\log_{10}(w) + 10\log_{10}(f_c) + 20\log_{10}\Delta h_m + L_{\text{ori}} \qquad (2.203)$$

where

$$L_{\text{ori}} = \begin{cases} -10 + 0.354(\phi) , & 0 \le \phi \le 35° \\ 2.5 + 0.075(\phi - 35) , & 35 \le \phi \le 55° \\ 4.0 - 0.114(\phi - 55) , & 55 \le \phi \le 90° \end{cases} \qquad (2.204)$$

and

$$\Delta h_m = h_{\text{Roof}} - h_m . \qquad (2.205)$$

The multi-screen diffraction loss is

$$L_{\text{msd}} = L_{\text{bsh}} + k_a + k_d \log_{10}(d) + k_f \log_{10}(f_c) - 9\log_{10}(b) \qquad (2.206)$$

where

$$L_{\text{bsh}} = \begin{cases} -18\log_{10}(1 + \Delta h_b) & h_b > h_{\text{Roof}} \\ 0 & h_b \le h_{\text{Roof}} \end{cases} \qquad (2.207)$$

$$k_a = \begin{cases} 54 , & h_b > h_{\text{Roof}} \\ 54 - 0.8\Delta h_b , & d \ge 0.5\text{km and } h_b \le h_{\text{Roof}} \\ 54 - 0.8\Delta h_b d/0.5 , & d < 0.5\text{km and } h_b \le h_{\text{Roof}} \end{cases} \qquad (2.208)$$

$$k_d = \begin{cases} 18 , & h_b > h_{\text{Roof}} \\ 18 - 15\Delta h_b/h_{\text{Roof}} , & h_b \le h_{\text{Roof}} \end{cases} \qquad (2.209)$$

$$k_f = -4 + \begin{cases} 0.7(f_c/925 - 1) , & \text{medium city and suburban} \\ 1.5(f_c/925 - 1) , & \text{metropolitan area} \end{cases} \qquad (2.210)$$

and

$$\Delta h_b = h_b - h_{\text{Roof}} . \qquad (2.211)$$

The term k_a is the increase in path loss for BS antennas below the roof tops of adjacent buildings. The terms k_d and k_f control the dependency of the multi-screen diffraction loss on the distance and frequency, respectively. The model is valid for the following ranges of parameters, $800 \le f_c \le 2000$ (MHz), $4 \le h_b \le 50$ (m), $1 \le h_m \le 3$ (m), and $0.02 \le d \le 5$ (km). If no data on the structure of the buildings and roads are available, the following default values are recommended, $b = 20 \ldots 50$ (m), $w = b/2$, $\phi = 90°$, and $h_{\text{Roof}} = 3 \times$ number of floors + roof (m), where roof = 3 (m) pitched and 0 (m) flat. The COST231-Walfish-Ikegami model works best for $h_b \gg h_{\text{Roof}}$. Large prediction errors can be expected for $h_b \approx h_{\text{Roof}}$. The model is poor for $h_b \ll h_{\text{Roof}}$ because the terms in (2.208) do not consider wave guiding in street canyons and diffraction at street corners.

Path Loss in Street Microcells

For ranges less than 500 m and antenna heights less than 20 m, some empirical measurements have shown that the received signal strength for LOS propagation along city streets can be accurately described by the two-slope model [140], [131], [155], [338], [254], [327]

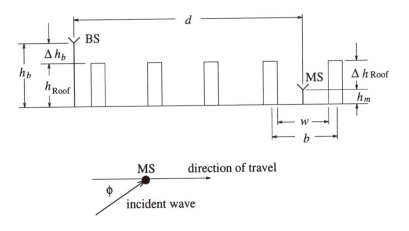

Figure 2.42   Definition of parameters used in the COST231-Walfish-Ikegami model.

$$\mu_\Omega = 10 \log_{10} \left(\frac{A}{d^a (1 + d/g)^b} \right) \quad \text{dBm} \tag{2.212}$$

where A is a constant and d (m) is the distance. Close into the BS, free space propagation will prevail so that $a = 2$. The parameter g is called the breakpoint and ranges from 150 to 300 m [140], [131], [155], [338]. At larger distances, an inverse-fourth to -eighth power law is experienced so that b ranges from 2 to 6. This is probably caused by increased shadowing at the greater distances [140]. The model parameters that were obtained by Harley [140] are listed in Tab. 2.4. Xia [342] has demonstrated that the breakpoint occurs where the Fresnel zone between the two antennas just touches the ground assuming a flat surface. This distance is

$$g = \frac{1}{\lambda_c} \sqrt{(\Sigma^2 - \Delta^2)^2 - 2(\Sigma^2 + \Delta^2)\left(\frac{\lambda_c}{2}\right)^2 + \left(\frac{\lambda_c}{2}\right)^4} \tag{2.213}$$

where $\Sigma = h_b + h_m$ and $\Delta = h_b - h_m$. For high frequencies this distance can be approximated as $g = 4 h_b h_m / \lambda_c$. Notice that the breakpoint is dependent on frequency, with the breakpoint at 1.9 GHz being about twice that for 900 MHz.

Street microcells may also exhibit NLOS propagation when a MS rounds a street corner as shown in Fig. 2.43. In this case, the average received signal strength can drop by 25-30 dB over distances as small as 10 m for low antenna heights in an area with multi-story buildings [40], [306], [196], [225], [272], and by 25-30 dB over distances of 45-50 m for low antenna heights in a region with

Base Antenna Height (m)	a	b	Break point g (m)
5	2.30	-0.28	148.6
9	1.48	0.54	151.8
15	0.40	2.10	143.9
19	-0.96	4.72	158.3

Table 2.4   Two-slope path loss parameters obtained by Harley, from [140].

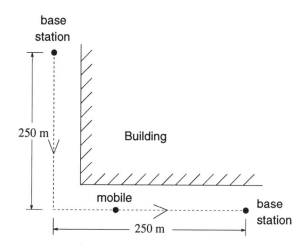

Figure 2.43   The corner effect in a street microcell environment.

only one- or two-story buildings [272]. This phenomenon is called the **corner effect**.

Grimlund and Gudmundson [131] have proposed an empirical street corner path loss model. Their model assumes LOS propagation until the MS reaches a street corner. The NLOS propagation after rounding a street corner is modeled by assuming LOS propagation from an imaginary transmitter that is located at the street corner having a transmit power equal to the received power at the street corner from the serving BS. That is, the received signal strength (in dBm) is given by

$$
\mu_\Omega = \begin{cases} 10\log_{10}\left(\frac{A}{d^a(1+d/g)^b}\right) & d \le d_c \\ 10\log_{10}\left(\frac{A}{d_c^a(1+d_c/g)^b} \cdot \frac{1}{(d-d_c)^a(1+(d-d_c)/g)^b}\right) & d > d_c \end{cases} \tag{2.214}
$$

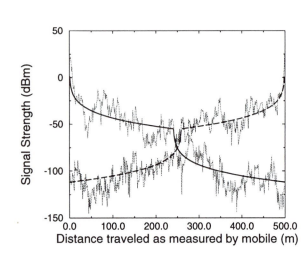

Figure 2.44   Average and instantaneous received signal strength for the street microcell environment in Fig. 2.43.   For the instantaneous received signal strength, $\sigma_\Omega = 6$ dB and $\phi_{\Omega_{(dB)}\Omega_{(dB)}}(d) = 0.1\sigma_\Omega^2$ at $d = 30$ m.

where d_c (m) is the distance between the serving BS and the corner. For the scenario depicted in Fig. 2.43, the received signal strength with this model is shown in Fig. 2.44. The heavy curves show the average received signal strength from the two BSs as the MS traverses the dashed path shown in Fig. 2.43. These curves were obtained by using $a = 2$, $b = 2$, $g = 150$ m, and $d_c = 250$ m in (2.214), and assuming that $\mu_\Omega = 1$ dBm at $d = 1$ m. The dotted curves superimposed on the heavy lines in Fig. 2.44 show the received signal strength with the combined effects of path loss, log-normal shadowing, and multipath-fading.

2.5.3   Path Loss in Indoor Microcells

Indoor microcellular systems are becoming very important for providing wire-less voice and data communications within the home and work-place. The characterization of in-building radio propagation is necessary for the effective deployment of these systems. In general, the path loss and shadowing charac-teristics vary greatly from one building to the next. Typical path loss exponents and shadow standard deviations are provided in Table 2.5 for several different types of buildings.

For multistory buildings, the RF attenuation between floors is important for frequency reuse on different floors of the same building.   Measurements have

Building	Frequency (MHz)	β	σ_Ω (dB)
Retail stores	914	2.2	8.7
Grocery stores	914	1.8	5.2
Office, hard partition	1500	3.0	7.0
Office, soft partition	900	2.4	9.6
Office, soft partition	1900	2.6	14.1

Table 2.5 Path loss exponents and shadow standard deviations for several different types of buildings, from [7].

indicated that the greatest floor loss occurs when the transmitter and receiver are separated by a single floor. Typically, the floor loss is 15 to 20 dB for one floor and an additional 6 to 10 dB per floor up to a separation of 4 floors. For 5 or more floors of separation, the overall floor loss will increase only a few dB for each additional floor. This effect is thought to be caused by signals diffracting up the sides of the building and signals scattering off the neighboring buildings. Also important for the deployment of indoor wireless systems is the building penetration loss. This loss depends on the frequency and height of the building. Turkmani *et. al.* [305] have shown that the building penetration losses decrease with increasing frequency, in particular they are 16.4, 11.6, and 7.6 dB at 441 MHz, 896.5 MHz, and 1400 MHz, respectively. In general the building penetration loss for signals propagating into a building tends to decrease with height, the reason being that a LOS path is more likely to exist at increased height. The building penetration loss decreases by about 2 dB per floor from ground level up to about 9 to 15 floors and then increases again [328]. Windows also have a significant effect on penetration loss. Plate glass provides an attenuation of about 6 dB, while lead lined glass provides an attenuation anywhere from 3 to 30 dB.

Appendix 2A
DERIVATION OF EQUATION (2.170)

This Appendix derives an expression for the second moment of a Ricean random variable in terms of its first moment. A Ricean random variable X has probability density function

$$p_X(x) = \frac{x}{\sigma^2} \exp\left\{-\frac{x^2 + s^2}{2\sigma^2}\right\} I_0\left(\frac{xs}{\sigma^2}\right) \quad x > 0 \qquad (2A.1)$$

and moments [257]

$$E[X^n] = (2\sigma^2)^{\frac{n}{2}} \exp\left\{-\frac{s^2}{2\sigma^2}\right\} \Gamma\left((2+n)/2\right) {}_1F_1\left(\frac{n+2}{2}, 1; \frac{s^2}{2\sigma^2}\right) \quad (2A.2)$$

where $\Gamma(\,\cdot\,)$ is the gamma function, and ${}_1F_1(a,\ b;\ x)$ is the confluent hypergeometric function. The first moment of X is

$$E[X] \equiv \Omega_v = (2\sigma^2)^{\frac{1}{2}} e^{-K} \frac{\sqrt{\pi}}{2} {}_1F_1(3/2, 1; K)\,, \quad (2A.3)$$

where $K = s^2/2\sigma^2$ is the Rice factor. The second moment of X is

$$\begin{aligned} E[X^2] \equiv \Omega_p &= 2\sigma^2 e^{-K} {}_1F_1(2, 1; K) \\ &= 2\sigma^2(K+1)\,. \end{aligned} \quad (2A.4)$$

Substituting $2\sigma^2$ from (2A.3) gives

$$\Omega_p = \frac{4e^{2K}(K+1)}{\pi {}_1F_1^2(3/2, 1; K)}\,\Omega_v^2 = C(K)\,\Omega_v^2\,. \quad (2A.5)$$

Note that $C(0) = 4/\pi$, $C(\infty) = 1$, and $4/\pi \le C(K) \le 1$ for $0 \le K \le \infty$.

Appendix 2B
DERIVATION OF EQUATION (2.187)

From (2.187), the composite distribution for the squared envelope $s = z_c^2$ is

$$\begin{aligned} p_{z_c^2}(x) &= \int_0^\infty \left(\frac{m}{w}\right)^m \frac{x^{m-1}}{\Gamma(m)} \exp\left\{-\frac{mx}{w}\right\} \\ &\quad \times \frac{\xi}{\sqrt{2\pi}\sigma_\Omega w} \exp\left\{-\frac{(10\log_{10}w - \mu_{\Omega_p})^2}{2\sigma_\Omega^2}\right\} dw\,. \end{aligned} \quad (2B.1)$$

The mean of the approximate log-normal distribution is

$$\begin{aligned} \mu_{(\text{new})} &= E[10\log_{10}(s)] \\ &= \int_0^\infty \int_0^\infty 10\log_{10}(x) \left(\frac{m}{w}\right)^m \frac{x^{m-1}}{\Gamma(m)} \exp\left\{-\frac{mx}{w}\right\} \\ &\quad \times \frac{\xi}{\sqrt{2\pi}\sigma_\Omega w} \exp\left\{-\frac{(10\log_{10}w - \mu_{\Omega_p})^2}{2\sigma_\Omega^2}\right\} dw\,dx \\ &= \frac{10\xi m^m}{\sqrt{2\pi}\sigma_\Omega \Gamma(m)} \int_0^\infty \frac{1}{w^{m+1}} \exp\left\{-\frac{(10\log_{10}w - \mu_{\Omega_p})^2}{2\sigma_\Omega^2}\right\} \\ &\quad \times \int_0^\infty \log_{10}(x)\, x^{m-1} e^{-\frac{mx}{w}}\, dx\,dw\,. \end{aligned} \quad (2B.2)$$

Assuming that m is an integer, the inner integral becomes [129, 4.352.2]

$$\int_0^\infty \log_{10}(x) x^{m-1} e^{-\frac{mx}{w}} dx = \frac{\Gamma(m)w^m}{m^m \ln 10} [\psi(m) - \ln(m/w)] \quad . \tag{2B.3}$$

Then by using the change of variables $x = 10\log_{10}(w)$ we obtain

$$\mu_{(\text{new})} = \xi[\psi(m) - \ln(m)] + \mu \tag{2B.4}$$

where $\psi(\,\cdot\,)$ is the Euler psi function, and

$$\psi(m) = -C + \sum_{k=1}^{m-1} \frac{1}{k} \tag{2B.5}$$

and $C \simeq 0.5772$ is Euler's constant. Likewise, the second moment of the approximate log-normal distribution is

$$E[(10\log_{10}(s))^2]$$
$$= \int_0^\infty \int_0^\infty [10\log_{10}(s)]^2 \left(\frac{m}{w}\right)^m \frac{x^{m-1}}{\Gamma(m)} \exp\left\{-\frac{mx}{w}\right\} \frac{\xi}{\sqrt{2\pi}\sigma_\Omega w}$$
$$\times \exp\left\{-\frac{(10\log_{10}w - \mu_{\Omega_p})^2}{2\sigma_\Omega^2}\right\} dw dx$$
$$= \frac{\xi m^m}{\sqrt{2\pi}\Gamma(m)} \int_0^\infty \frac{1}{w^{m+1}} \exp\left\{-\frac{(10\log_{10}w - \mu_{\Omega_p})^2}{2\sigma_\Omega^2}\right\}$$
$$\times \int_0^\infty [10\log_{10}(x)]^2 x^{m-1} e^{-\frac{mx}{w}} dx dw \quad . \tag{2B.6}$$

Assuming again that m is an integer, the inner integral is [129, 4.358.2]

$$\int_0^\infty [10\log_{10}(x)]^2 x^{m-1} e^{-\frac{mx}{w}} dx = \frac{(m-1)! w^m}{m^m \ln 10} \left([\psi(m) - \ln(m/w)]^2 + \zeta(2, m)\right) \tag{2B.7}$$

leading to

$$E[(10\log_{10}(s))^2] = \alpha^2 \left([\psi(m) - \ln(m)]^2 + \zeta(2, m)\right)$$
$$+ 2\alpha[\psi(m) - \ln(m)]\mu + \sigma^2 + \mu^2 \quad . \tag{2B.8}$$

where $\zeta(\,\cdot\,,\,\cdot\,)$ is Reimann's zeta function, and

$$\zeta(2, m) = \sum_{k=0}^\infty \frac{1}{(m+k)^2} \quad . \tag{2B.9}$$

Finally, the variance of the approximate log-normal distribution is

$$\sigma_{(\text{new})}^2 = E[(10\log_{10}(s))^2] - E^2[10\log_{10}(s)]$$
$$= \xi^2 \zeta(2, m) + \sigma_\Omega^2 \quad . \tag{2B.10}$$

Problems

2.1. Assume that $r(t) = r_I(t) + jr_Q(t)$ is a stationary random processes. Show that if $x(t)$ is a stationary random process, then the autocorrelation of $x(t)$ is

$$
\begin{aligned}
\mathrm{E}[x(t)x(t+\tau)] &= \mathrm{E}[r_I(t)r_I(t+\tau)]\cos 2\pi f_c\tau \\
&\quad -\mathrm{E}[r_Q(t)r_I(t+\tau)]\sin 2\pi f_c\tau \ .
\end{aligned}
$$

2.2. Suppose that an unmodulated carrier is transmitted over a channel characterized by 2-D isotropic scattering. Show that the psd of the received envelope $z(t) = |r(t)|$ is given by (2.58).

2.3. Show that for wide sense stationary (WSS) channels

$$
\begin{aligned}
\phi_H(f,m;\nu,\mu) &= \psi_H(f,m;\nu)\delta(\nu-\mu) \\
\phi_S(\tau,\eta;\nu,\mu) &= \psi_S(\tau,\eta;\nu)\delta(\nu-\mu) \ .
\end{aligned}
$$

2.4. Show that for uncorrelated scattering (US) channels

$$
\begin{aligned}
\phi_c(t,s;\tau,\eta) &= \psi_c(t,s;\tau)\delta(\eta-\tau) \\
\phi_S(\tau,\eta;\nu,\mu) &= \psi_S(\tau;\nu,\mu)\delta(\eta-\tau) \ .
\end{aligned}
$$

2.5. Consider the fading simulator in Fig. 2.24 with the low-pass filter described by (2.139). Derive the discrete autocorrelation function in (2.140).

2.6. Consider Jakes' fading simulator. With the choice that $\alpha = 0$ and $\beta_n = \pi n/(M+1)$ show that

$$
\begin{aligned}
< r_I(t)r_Q(t) > &= 0 \\
< r_Q^2(t) > &= M+1 \\
< r_I^2(t) > &= M
\end{aligned}
$$

2.7. Consider a zero-mean complex Gaussian random process $r(t)$ having the autocorrelation function $\phi_{rr}(\tau)$.

a) Determine the autocorrelation function of $z^2(t) = |r(t)|^2$.

b) Repeat part a) when the Gaussian random process has non-zero mean.

c) Consider the transmission of an unmodulated carrier over a channel characterized by the scattering environment shown in Fig. 2.5. Derive an expression for the autocorrelation of the squared-envelope.

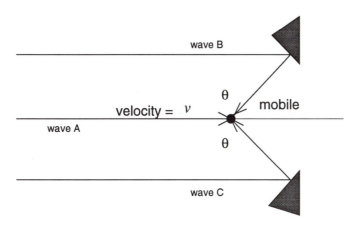

Figure 2.45 Scattering environment for Problem 2.9

2.8. Compute the average delay, the rms delay spread, and the approximate value of W_{50} for the power delay profiles that are defined in Tab. 2.34. Assuming a TDMA cellular system, what would you interpret as the difference between a "typical urban channel" and a "bad urban channel" from these values.

2.9. Consider the situation in Fig. 2.45 where a MS is moving in the vicinity of two local scatters that each have a reflection coefficient of 0.5. The reflection coefficient is the ratio of the amplitudes of the reflected and incident electric fields.

a) Derive an expression for the resultant signal from paths A, B, and C?

b) Plot the amplitude variation due to multipath-fading and calculate the Doppler frequency.

2.10. Consider a situation where the received signal is Rayleigh faded ($K = 0$), and suppose that the psd of the scatter component of the faded envelope $\hat{S}_{xx}(f)$ is not symmetrical about the carrier frequency, i.e., a form of non-isotropic scattering. Show that the envelope level crossing rate is given by

$$L_R = \sqrt{\frac{b_2}{b_0} - \frac{b_1^2}{b_0^2}} \cdot \frac{\rho}{\sqrt{\pi}} e^{-\rho^2}$$

where

$$\rho = \frac{R}{\sqrt{\Omega_p}} = \frac{R}{\sqrt{2b_0}} \quad .$$

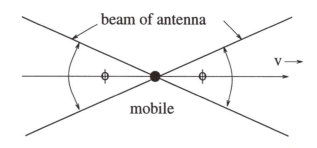

Figure 2.46 Mobile with directional antenna for Problem 2.12.

2.11. Suppose that a vertical monopole antenna is used and the probability density function of arriving plane waves is given by (2.36). Find the psd of the band-pass waveform $x(t)$.

2.12. Consider the situation in the Fig. 2.46, where the MS employs a directional antenna with a beam width of $\phi°$. Assume an isotropic scattering environment.

a) In receiving a radio transmission at 850 MHz, a Doppler frequency of 20 to 60 Hz is observed. What is the beam width of the MS antenna, and how fast is the MS traveling?

b) Suppose that the MS antenna has a beam width of 13°. What is the level-crossing rate with respect to the *rms* envelope level, assuming that the MS is traveling at a speed of 30 km/h?

2.13. The delay Doppler-spread function $S(\tau, \nu)$ for a multipath fading channel is nonzero for the range of values $0 \leq \tau \leq 1$ μs and $-40 \leq \lambda \leq 40$ Hz. Furthermore, $S(\tau, \nu)$ is uniform in the two variables τ and ν.

a) Find numerical values for the following parameters;

1. the average delay, μ_τ, and *rms* delay spread, σ_τ.

2. the Doppler spread, B_d

3. the approximate coherence time, T_c

4. the approximate coherence bandwidth, B_c

b) Given the answers in part a), what does it mean when the channel is

1. frequency-nonselective

2. slowly fading

3. frequency-selective

2.14. Consider a scattering environment where it is known that no plane waves arrive from either directly ahead or directly behind the direction of motion. We are interested in constructing a fading simulator similar to Jakes' method to account for this fact.

a) How might you modify Jakes' method to account for the above situation, assuming that you only need to generate one faded signal?

b) Assume that the received signal is of the form

$$r(t) = \sum_{n=1}^{N} e^{-j(\hat{\phi}_n + 2\pi f_m t \cos(\theta_n))}$$

where f_m is the maximum Doppler frequency, $\hat{\phi}_n$ is the random phase of the nth component, and $\theta_n = 2\pi(n - .5)/N$ is the angle of arrival for the nth copy of the signal. Following the method used for deriving Jakes' fading simulator and assuming that $N/4$ is even, show that $r(t)$ can be written in the form

$$r(t) = K \sum_{n=1}^{M} [\cos(\beta_n) + j \sin(\beta_n)] \cos(2\pi f_m \cos(\theta_n)t + \gamma_n)$$

where $M = N/4$.

1. What are the values of β_n and γ_n in terms of $\hat{\phi}_n$?
2. Determine K so that $E[z(t)^2] = 1$.
3. Assuming that $\beta_n = \pi n/M$, what is the correlation between the real and imaginary parts of $r(t)$? Is this a desirable result for the simulator?

2.15. Recall that Jakes' fading simulator is able to generate M faded envelopes $r_k(t)$, $k = 1, \ldots, M$, by using

$$r_k(t) = K \left\{ \sum_{n=1}^{M} [\cos(\beta_n) + j \sin(\beta_n)] \cos(2\pi f_m \cos(\theta_n)t + \gamma_{nk}) \right\}$$

where K is a normalization constant, $\theta_n = 2\pi n/N$, $\beta_n = \pi n/(M+1)$ and $\gamma_{nk} = \beta_n + 2\pi(k-1)/(M+1)$.

a) What are some of the problems with this technique?

b) It is claimed that this method can generate faded envelopes $r_j(t)$ and $r_k(t)$ that are almost uncorrelated for arbitrary j and k provided that

$$\gamma_{nj} - \gamma_{nk} = i\pi + \pi/2$$

for some integer i; otherwise, the correlation between certain pairs of faded envelopes may be significant. Is this claim true or false? Why?

2.16. Consider the following modification of Jakes' fading simulator

$$r_k(t) = \sqrt{\frac{2}{M}} \sum_{n=1}^{M} A_k(n) \left[\cos(\beta_n) + j\sin(\beta_n)\right] \cos(2\pi f_m \cos(\theta_n)t + \gamma_{nk})$$

where $M = 2^k$ for some integer k, $\beta_n = \pi n/M$, $\gamma_{nk} = \beta_n + 2\pi(k - 1)/(M + 1)$, and $A_k(n)$ denotes the nth co-ordinate of a Walsh-Hadamard codeword with $A_k(n) \in \{-1, +1\}$. It has been claimed that this method can yield uncorrelated faded envelopes, due to the orthogonality between the Walsh-Hadamard codewords.

a) By using simulation and/or analytic methods determine whether or not this method can yield faded envelopes having a crosscorrelation of zero at a delay of zero, i.e., $E[r_j(t)r_k^*(t)] = 0$, for $j \neq k$. For simulation purposes, choose $M = 8$ or 16.

Note: The Walsh-Hadamard codewords can be obtained from a Walsh-Hadamard matrix $\mathbf{H}_M$, where the kth row of $\mathbf{H}_M$ is the codeword $A_k(n)$. For $M = 2$

$$\mathbf{H}_2 = \begin{bmatrix} 1 & 1 \\ 1 & -1 \end{bmatrix} \,.$$

$\mathbf{H}_{2M}$ can be formed by

$$\mathbf{H}_{2M} = \begin{bmatrix} \mathbf{H}_M & \mathbf{H}_M \\ \mathbf{H}_M & -\mathbf{H}_M \end{bmatrix} \,.$$

2.17. Consider Fig. 2.47 and the following data

- The symbol transmission rate is 24300 symbols/s with 2 bits/symbol

- The channel bandwidth is 30 kHz

- The propagation environment is characterized by an *rms* delay spread of 1 ns

A MS is moving from base station A (BSA) to base station B (BSB). Base station C (BSC) is a co-channel BS with BSA.

a) Explain how you would construct a computer simulator to model the received signal power at the MS from (BSA) and (BSC), as the MS moves from BSA to BSB. Clearly state your assumptions and explain the relationship between the propagation characteristics in your model.

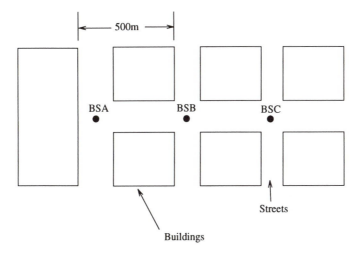

Figure 2.47   Base station and street layout for Problem 2.17.

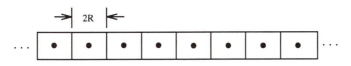

Figure 2.48   Proposed highway microcell system for Problem 2.18.

2.18. You are asked to design a highway microcell system as shown in Fig. 2.48. Each cell has length $2R$.

a) A BS with an omnidirectional antenna is placed at the center of each cell. Ignoring shadowing and envelope fading, determine the minimum reuse factor needed so that the worst case carrier-to-interference ratio, Λ, is at least 17 dB. State whatever assumptions you make.

b) Now suppose that directional antennas are used to divide each cell into two sectors with boundaries perpendicular to the highway. Repeat part a).

c) Consider again the sectored cell arrangement in part b). If shadowing is present with a standard deviation of σ_Ω dB, what is the probability of CCI on a cell boundary?

CO-CHANNEL INTERFERENCE

For cellular radio systems the radio link performance is usually limited by interference rather than noise and, therefore, the **probability of co-channel interference (CCI)** , is of primary concern. The definition of the probability of CCI depends on the the assumptions made about the radio receiver and propagation environment. At higher velocities, the radio receiver can usually average over the fast envelope variations by using coding and interleaving techniques. In this case, the transmission quality will be acceptable provided that the *average received carrier-to-interference ratio*, Λ, exceeds a receiver threshold Λ_{th}. The receiver threshold Λ_{th} is determined by the performance of the radio link in the presence of envelope fading. Once Λ_{th} has been determined, the variations in Λ due to path loss and shadowing will determine the probability of CCI. At lower velocities, the radio receiver cannot average over the fast envelope variations due to the delay constraints imposed by voice traffic. In this case, the transmission quality will be acceptable provided that the *instantaneous received carrier-to-interference ratio*, λ, exceeds another receiver threshold λ_{th}[1]. Once λ_{th} has been specified, variations in λ due to path loss, shadowing, and envelope fading, will determine the probability of CCI.

The effect of co-channel interference on the radio link performance depends on the ability of the radio receiver to reject co-channel interference. Some of the more advanced receivers incorporate sophisticated signal processing methods for the rejection or cancellation of co-channel interference, e.g., equalization and interference cancellation techniques. In this case, the radio receiver is more tolerant to co-channel interference and the receiver thresholds Λ_{th} and λ_{th} are reduced. This will have the effect of reducing the probability of CCI. Never-

---

[1]Note that Λ_{th} and λ_{th} are not the same.

theless, the effects of co-channel interference will not be completely eliminated so that the probability of CCI is still of interest.

The analysis of the probability of CCI for the log-normally shadowed signals that are typical in cellular frequency reuse systems requires the probability distribution of the interference power that is accumulated from several log-normal signals. Although there is no known exact expression for the probability distribution for the sum of log-normally random variables, several approximations have been derived by various authors. All of their approaches approximate the sum by another log-normal random variable. A method that matches the first two moments of the approximation has been developed by Fenton [98]. Sometimes this method is referred to as Wilkinson's method, as in [278]; here we called it the Fenton-Wilkinson method. Schwartz and Yeh developed another log-normal approximation that uses the exact first two moments for the sum of two log-normal random variables [278]. Schwartz and Yeh's method generally provides a more accurate approximation than the Fenton-Wilkinson method but it is more difficult to use. Prasad has corrected some errors in Schwartz and Yeh's paper in [249]. Another log-normal approximation is the cumulants matching approach suggested by Schleher [277]. With this approach different log-normal approximations are applied over different ranges of the composite distribution. A good comparison of the methods of Fenton-Wilkinson, Schwartz and Yeh, Farley, and Schleher has been undertaken by Beaulieu, Abu-Dayya, and McLane [22].

The above log-normal approximations have been extensively applied to the calculation of the probability of CCI in cellular systems. For example, Fenton's approach has been applied by Nagata and Akaiwa [227], Cox [60], Muammar and Gupta [222], and Daikoku and Ohdate [63]. Likewise, Schwartz and Yeh's approach has been applied by Yeh and Schwartz [346], Prasad and Arnbak [249], and Prasad, Kegel, and Arnbak [251].

Current literature also provides a thorough treatment of the probability of CCI when the signals are affected by fading only, including the work of Yao and Sheikh [344], Muammar [221], and Prasad and Kegel [250]. Section 3.3 shows that the probability of CCI is sensitive to the Rice factor of the desired signal, but it is insensitive to the number of interferers provided that the total interfering power remains constant. Calculations of the probability of CCI for signals with composite log-normal shadowing and fading have considered the cases of Rayleigh fading by Linnartz [192], Nakagami fading by Ho and Stüber [142], and Ricean fading by Austin and Stüber [20]. Sections 3.4 and 3.5 show that the shadowing has a more significant effect on the probability of CCI than fading. Furthermore, the probability of CCI is dominated by fading of the desired signal rather than fading of the interfering signals, e.g., with Ricean fading, the probability of CCI is sensitive to the Rice factor of the desired signal but is insensitive to the Rice factor of interfering signals. Finally, all of the above references assume a channel characterized by frequency non-selective (flat) fading. If the channel exhibits frequency selective fading, then the same general methodology can be used but an appropriate expression must be used for λ. This expression depends on the type of receiver that is employed, e.g., a maximum likelihood sequence estimation (MSLE) receiver for TDMA systems.

Most of the literature dealing with the probability of CCI assumes that the co-channel interfering signals add noncoherently. The probability of CCI has also been evaluated for the case of coherently added Rayleigh faded interferers and a Ricean faded desired signal by Prasad and Kegel [252], [250]. Coherent interferers are assumed to arrive at the receiver antenna with the same carrier phase. However, as discussed by Prasad and Kegel [253] and Linnartz [192], it is more realistic to assume noncoherent addition of interfering signals in mobile radio systems because of the scattering environment. The assumption of coherently added interferers generally leads to pessimistic predictions of the probability of CCI.

This chapter examines the probability of CCI in terms of both Λ and λ, under the assumption of noncoherent interferers. Section 3.1 treats the probability of CCI in terms of Λ, where various approximations are employed for the power sum of multiple log-normally shadowed interferers. The various approximations are compared in terms of their accuracy. The probability of CCI in terms of λ is treated in Section 3.3 for multiple Rayleigh or Ricean faded interferers without shadowing, in Section 3.4 for multiple log-normally shadowed Nakagami faded interferers, and in Section 3.5 for multiple log-normally shadowed Ricean faded interferers.

3.1 MULTIPLE LOG-NORMAL INTERFERERS

The noncoherent power sum of N_I log-normally shadowed interferers is

$$L = \sum_{k=1}^{N_I} L_k = \sum_{k=1}^{N_I} 10^{\Omega_{k(\text{dB})}/10} \tag{3.1}$$

where the $\Omega_{k\ (\text{dB})}$ are Gaussian random variables with means μ_{Ω_k} and variances $\sigma_{\Omega_k}^2$, and the $L_k = 10^{\Omega_{k(\text{dB})}/10}$ are log-normal random variables. Unfortunately, there is no known closed form expression for the probability density function of the sum of multiple ($N_I \geq 2$) log-normal random variables. However, there is a general consensus that the sum of independent log-normal random variables can be approximated by another log-normal random variable with appropriately chosen parameters. That is,

$$L = \sum_{k=1}^{N_I} 10^{\Omega_k(\text{dB})/10} \approx 10^{Z_{(\text{dB})}/10} = \tilde{L} \tag{3.2}$$

where $Z_{(\text{dB})}$ is a Gaussian random variable with mean μ_Z and variance σ_Z^2. Several methods have been suggested in the literature for this approximation by Fenton [98], Schwartz and Yeh [278], Farley [278], and others. Each of these methods provides varying degrees of accuracy over specified ranges of the shadow standard deviation σ_Ω, the power sum L, and the number of interferers N_I.

3.1.1 Fenton-Wilkinson Method

In the Fenton-Wilkinson method, the mean μ_Z and variance σ_Z^2 of $Z_{(\text{dB})}$ are obtained by matching the first two moments of the power sum L with the first two moments of the approximation $\tilde{L}$. To derive the appropriate moments, it is more convenient to use natural logarithms. Hence, we write

$$L_k = 10^{\Omega_k \ (\text{dB})/10} = e^{\xi \Omega_k \ (\text{dB})} = e^{\hat{\Omega}_k} \tag{3.3}$$

where $\xi = (\ln 10)/10 = 0.23026$ and $\hat{\Omega}_k = \xi \Omega_k \ (\text{dB})$. Note that $\mu_{\hat{\Omega}_k} = \xi \mu_{\Omega_k}$ and $\sigma_{\hat{\Omega}_k}^2 = \xi^2 \sigma_{\Omega_k}^2$. The rth moment of the log-normal random variable L_k can be obtained from the moment generating function of the Gaussian random variable $\hat{\Omega}_k$ as

$$\text{E}[L_k^r] = \text{E}[e^{r\hat{\Omega}_k}] = e^{r\mu_{\hat{\Omega}_k} + (1/2)r^2\sigma_{\hat{\Omega}_k}^2} \ . \tag{3.4}$$

To find the appropriate moments for the log-normal approximation we can use (3.4) and equate the first two moments on both sides of $L \approx e^{\hat{Z}} = \hat{L}$, where $\hat{Z} = \xi Z_{(\text{dB})}$. For example, suppose that $\hat{\Omega}_1, \ldots, \hat{\Omega}_{N_I}$ are independent with means $\mu_{\hat{\Omega}_1}, \ldots, \mu_{\hat{\Omega}_{N_I}}$ and identical variances $\sigma_{\hat{\Omega}}^2$. Identical variances are often assumed because the standard deviation of log-normal shadowing is largely independent of the radio path length [174], [176]. Then

$$\mu_L = \left(\sum_{k=1}^{N_I} e^{\mu_{\hat{\Omega}_k}} \right) e^{(1/2)\sigma_{\hat{\Omega}}^2} = e^{\mu_{\hat{Z}} + (1/2)\sigma_{\hat{Z}}^2} \tag{3.5}$$

$$\sigma_L^2 = \left(\sum_{k=1}^{N_I} e^{2\mu_{\hat\Omega_k}}\right) e^{\sigma_{\hat\Omega}^2}(e^{\sigma_{\hat\Omega}^2}-1) = e^{2\mu_{\hat z}}e^{\sigma_{\hat z}^2}(e^{\sigma_{\hat z}^2}-1) \ . \qquad (3.6)$$

By squaring the first equation and dividing by the second we can first solve for $\sigma_{\hat Z}^2$ in terms of the known values of $\mu_{\hat\Omega_1}, \ \ldots, \ \mu_{\hat\Omega_{N_I}}$ and $\sigma_{\hat\Omega}^2$. Then, $\mu_{\hat Z}$ can be obtained from the first equation. This procedure yields the following solution:

$$\mu_{\hat Z} = \frac{\sigma_{\hat\Omega}^2-\sigma_{\hat Z}^2}{2} + \ln\left(\sum_{k=1}^{N_I} e^{\mu_{\hat\Omega_k}}\right) \qquad (3.7)$$

$$\sigma_{\hat Z}^2 = \ln\left((e^{\sigma_{\hat\Omega}^2}-1)\frac{\sum_{k=1}^{N_I} e^{2\mu_{\hat\Omega_k}}}{\left(\sum_{k=1}^{N_I} e^{\mu_{\hat\Omega_k}}\right)^2}+1\right) \ . \qquad (3.8)$$

The accuracy of this log-normal approximation can be measured in terms of how accurately the first two moments of $10\log_{10}L$ are estimated, and how well the cumulative distribution function (cdf) of $10\log_{10}L$ is described by a Gaussian cdf. It has been reported in [278] that the Fenton-Wilkinson method breaks down for $\sigma_\Omega > 4$ dB. Unfortunately, for cellular radio applications the standard deviation of log-normal shadowing typically ranges from 6 to 12 dB. However, as pointed out in [22], the Fenton-Wilkinson method only breaks down if one considers the application of the Fenton-Wilkinson method for the prediction of the first and second moments of the power sum $10\log_{10}L$. Moreover, in problems relating to the probability of CCI in cellular radio systems, we are usually interested in the tails of both the complementary distribution function (cdfc) $F_L^c(x) = \Pr(L \geq x)$ and the cdf $F_L(x) = 1 - F_L^c(x) = \Pr(L < x)$. In this case, we are interested in the accuracy of the approximation

$$F_L(x) \approx \Pr(e^{\hat Z} \geq x) = Q\left(\frac{\ln x - \mu_{\hat Z}}{\sigma_{\hat Z}}\right) \qquad (3.9)$$

especially for large and small values of x. It will be shown later that the Fenton-Wilkinson method, despite its simplicity, can approximate the tails of the cdf and cdfc functions with reasonable accuracy.

3.1.2 Schwartz and Yeh's Method

Schwartz and Yeh's method [278] is also based upon the assumption that the power sum is log-normally distributed. However, unlike the Fenton-Wilkinson method, Schwartz and Yeh's method uses exact expressions for the first two moments of the sum of two log-normal random variables. Nesting and recursion techniques are then used to find exact values for the first two moments for the sum of N_I log-normal random variables. For example, suppose that $L = L_1 + L_2 + L_3$. The exact first two moments of $\ln(L_1 + L_2)$ are computed. We then define $Z_2 = \ln(L_1 + L_2)$ as a new Gaussian random variable, let $L = e^{Z_2} + L_3$, and again compute the exact first two moments of $\ln L$. Since the procedure is recursive we only need to detail Schwartz and Yeh's method for the case when $N_I = 2$, i.e.,

$$L = e^{\hat{\Omega}_1} + e^{\hat{\Omega}_2} \approx e^{\hat{Z}} = \hat{L} \tag{3.10}$$

or

$$\hat{Z} \approx \ln\left(e^{\hat{\Omega}_1} + e^{\hat{\Omega}_2}\right) \tag{3.11}$$

where the Gaussian random variables $\hat{\Omega}_1$ and $\hat{\Omega}_2$ have means $\mu_{\hat{\Omega}_1}$ and $\mu_{\hat{\Omega}_2}$, and variances $\sigma^2_{\hat{\Omega}_1}$ and $\sigma^2_{\hat{\Omega}_2}$, respectively.

Define the Gaussian random variable $\hat{\Omega}_d \overset{\Delta}{=} \hat{\Omega}_2 - \hat{\Omega}_1$ so that

$$\mu_{\hat{\Omega}_d} = \mu_{\hat{\Omega}_2} - \mu_{\hat{\Omega}_1} \tag{3.12}$$

$$\sigma^2_{\hat{\Omega}_d} = \sigma^2_{\hat{\Omega}_1} + \sigma^2_{\hat{\Omega}_2} . \tag{3.13}$$

Taking the expectation of both sides of (3.11) and assuming that the approximation holds with equality gives

$$\begin{aligned}
\mu_{\hat{Z}} &= \mathrm{E}\left[\ln\left(e^{\hat{\Omega}_2} + e^{\hat{\Omega}_1}\right)\right] \\
&= \mathrm{E}\left[\ln\left(e^{\hat{\Omega}_1}\left(1 + e^{\hat{\Omega}_2 - \hat{\Omega}_1}\right)\right)\right] \\
&= \mathrm{E}\left[\hat{\Omega}_1\right] + \mathrm{E}\left[\ln\left(1 + e^{\hat{\Omega}_d}\right)\right] .
\end{aligned} \tag{3.14}$$

The second term in (3.14) is

$$\mathrm{E}\left[\ln\left(1 + e^{\hat{\Omega}_d}\right)\right] = \int_{-\infty}^{\infty} [\ln(1 + e^x)] \, p_{\hat{\Omega}_d}(x) dx . \tag{3.15}$$

We now use the power series expansion

$$\ln(1 + x) = \sum_{k=1}^{\infty} c_k x^k, \qquad c_k = \frac{(-1)^{k+1}}{k} \tag{3.16}$$

where $|x| < 1$. To ensure convergence of the power series and the resulting series of integrals, the integration in (3.15) is broken into ranges as follows:

$$\int_{-\infty}^{\infty} [\ln(1+e^x)] \, p_{\hat{\Omega}_d}(x) dx \;=\; \int_{-\infty}^{0} [\ln(1+e^x)] \, p_{\hat{\Omega}_d}(x) dx$$

$$+ \int_{0}^{\infty} [\ln(1+e^{-x}) + x] \, p_{\hat{\Omega}_d}(x) dx \;. \quad (3.17)$$

The second integral is obtained by using the identity

$$\begin{aligned} \ln(1+e^x) &= \ln[(e^{-x}+1)e^x] \\ &= \ln(1+e^{-x}) + \ln(e^x) \\ &= \ln(1+e^{-x}) + x \;. \end{aligned} \quad (3.18)$$

After a very long derivation that is detailed in [278],

$$\mu_{\hat{Z}} = \mu_{\hat{\Omega}_1} + G_1 \quad (3.19)$$

where

$$\begin{aligned} G_1 &= \mu_{\hat{\Omega}_d} \Phi\left(\frac{\mu_{\hat{\Omega}_d}}{\sigma_{\hat{\Omega}_d}}\right) + \frac{\sigma_{\hat{\Omega}_d}}{\sqrt{2\pi}} e^{-\mu_{\hat{\Omega}_d}^2/2\sigma_{\hat{\Omega}_d}^2} \\ &\quad + \sum_{k=1}^{\infty} c_k e^{k^2\sigma_{\hat{\Omega}_d}^2/2} \left[e^{k\mu_{\hat{\Omega}_d}} \Phi\left(\frac{-\mu_{\hat{\Omega}_d} - k\sigma_{\hat{\Omega}_d}^2}{\sigma_{\hat{\Omega}_d}}\right) + T_1 \right] \end{aligned} \quad (3.20)$$

with

$$T_1 = e^{-k\mu_{\hat{\Omega}_d}} \Phi\left(\frac{\mu_{\hat{\Omega}_d} - k\sigma_{\hat{\Omega}_d}^2}{\sigma_{\hat{\Omega}_d}}\right) \;. \quad (3.21)$$

The variance can be computed in a similar fashion, resulting in the expression [278]

$$\sigma_{\hat{Z}}^2 = \sigma_{\hat{\Omega}_1}^2 - G_1^2 - 2\sigma_{\hat{\Omega}_1}^2 G_3 + G_2 \quad (3.22)$$

where

$$\begin{aligned} G_2 &= \sum_{k=1}^{\infty} b_k T_2 + \left[1 - \Phi\left(-\frac{\mu_{\hat{\Omega}_d}}{\sigma_{\hat{\Omega}_d}}\right)\right] (\mu_{\hat{\Omega}_d}^2 + \sigma_{\hat{\Omega}_d}^2) \\ &\quad + \frac{\mu_{\hat{\Omega}_d} \sigma_{\hat{\Omega}_d}}{\sqrt{2\pi}} e^{-\mu_{\hat{\Omega}_d}^2/(2\sigma_{\hat{\Omega}_d}^2)} \\ &\quad + \sum_{k=1}^{\infty} b_k e^{-(k+1)\mu_{\hat{\Omega}_d} + (k+1)^2\sigma_{\hat{\Omega}_d}^2/2} \Phi\left(\frac{\mu_{\hat{\Omega}_d} - \sigma_{\hat{\Omega}_d}^2(k+1)}{\sigma_{\hat{\Omega}_d}}\right) \\ &\quad - 2\sum_{k=1}^{\infty} c_k e^{-\mu_{\hat{\Omega}_d} k + k^2\sigma_{\hat{\Omega}_d}^2/2} \left[\mu_{\hat{\Omega}_k} \Phi\left(-\frac{\mu_{\hat{\Omega}_k}}{\sigma_{\hat{\Omega}_d}}\right) - \frac{\sigma_{\hat{\Omega}_d}}{\sqrt{2\pi}} e^{-\mu_{\hat{\Omega}_k}^2/(2\sigma_{\hat{\Omega}_d}^2)}\right] \end{aligned} \quad (3.23)$$

$$G_3 = \sum_{k=0}^{\infty} (-1)^k e^{k^2 \sigma_{\hat{\Omega}_d}^2 / 2} T_1 + \sum_{k=0}^{\infty} (-1)^k T_2 \tag{3.24}$$

with

$$T_2 = e^{\mu_{\hat{\Omega}_d}(k+1) + (k+1)^2 \sigma_{\hat{\Omega}_d}^2 / 2} \Phi \left(\frac{-\mu_{\hat{\Omega}_d} - (k+1)\sigma_{\hat{\Omega}_d}^2}{\sigma_{\hat{\Omega}_d}} \right) \tag{3.25}$$

and

$$b_k = \frac{2(-1)^{k+1}}{k+1} \sum_{j=1}^{k} j^{-1} \tag{3.26}$$

$$\mu_{\hat{\Omega}_k} = -\mu_{\hat{\Omega}_d} + k\sigma_{\hat{\Omega}_d}^2 . \tag{3.27}$$

It has been reported in [278] that approximately 40 terms are required in the infinite summations to achieve four significant digits of accuracy in the moments. On the next step of the recursion it is important that we let $\sigma_{\hat{\Omega}_1}^2 = \sigma_{\hat{Z}}^2$ and $\mu_{\hat{\Omega}_1} = \mu_{\hat{Z}}$; otherwise, the procedure fails to converge.

3.1.3   Farley's Method

Consider N_I normal random variables $\hat{\Omega}_k$ each with mean $\mu_{\hat{\Omega}}$ and variance $\sigma_{\hat{\Omega}}^2$. Farley approximated the cdfc of the power sum

$$L = \sum_{k=1}^{N_I} = e^{\hat{\Omega}_k} \tag{3.28}$$

as

$$\mathrm{P_r}(L > x) \approx 1 - \left[1 - Q \left(\frac{\ln x - \mu_{\hat{\Omega}}}{\sigma_{\hat{\Omega}}} \right) \right]^{N_I} . \tag{3.29}$$

As shown in [22], Farley's approximation is actually a strict lower bound on the cdfc. To obtain this result let

$$F_L^c(x) = \mathrm{P_r} \left(L_1 + L_2 + \cdots + L_{N_I} > x \right) \tag{3.30}$$

and define the two events

$$A = \{ \text{at least one } L_i > x \}$$
$$B = A^c, \text{ the complement of event } A . \tag{3.31}$$

Events A and B are mutually exclusive and partition the sample space. Therefore,

$$\begin{aligned} \mathrm{P_r}(L > x) &= \mathrm{P_r}(L > x, \, A) + \mathrm{P_r}(L > x, \, B) \\ &= \mathrm{P_r}(A) + \mathrm{P_r}(L > x, \, B) . \end{aligned} \tag{3.32}$$

The second term in (3.32) is positive for continuous density functions such as the log-normal density. For example, the event

$$C = \{x/N_I < L_i < x, \forall\, i\} \tag{3.33}$$

is a subset of the event B. Under the assumption that the $\hat{\Omega}_k$ are independent and identically distributed, the probability of event C is

$$\mathrm{P_r}(C) = \left[Q\left(\frac{\ln(x/N_I) - \mu_{\hat{\Omega}}}{\sigma_{\hat{\Omega}}}\right) - Q\left(\frac{\ln x - \mu_{\hat{\Omega}}}{\sigma_{\hat{\Omega}}}\right)\right]^{N_I} > 0\; . \tag{3.34}$$

Therefore, $\mathrm{P_r}(L > x) > \mathrm{P_r}(A)$. Since the L_i are independent and identically distributed

$$\begin{aligned}
\mathrm{P_r}(A) &= 1 - \prod_{i=1}^{N_I} \mathrm{P_r}(L_i \le x) \\
&= 1 - \left[1 - Q\left(\frac{\ln x - \mu_{\hat{\Omega}}}{\sigma_{\hat{\Omega}}}\right)\right]^{N_I}\; .
\end{aligned} \tag{3.35}$$

Finally, we have the lower bound on the cdfc

$$\mathrm{P_r}(L > x) > 1 - \left[1 - Q\left(\frac{\ln x - \mu_{\hat{\Omega}}}{\sigma_{\hat{\Omega}}}\right)\right]^{N_I} \tag{3.36}$$

or, equivalently, the upper bound on the cdf

$$\mathrm{P_r}(L \le x) > \left[1 - Q\left(\frac{\ln x - \mu_{\hat{\Omega}}}{\sigma_{\hat{\Omega}}}\right)\right]^{N_I}\; . \tag{3.37}$$

3.1.4 Numerical Comparisons

Fig. 3.1 compares the cdf for $N_I = 2$ and $N_I = 6$ log-normal random variables with the various log-normal approximations. Likewise, Figs. 3.2 - 3.4 provide comparisons of the various log-normal approximations for the cdfc. Exact results are also shown that have been obtained by computer simulation. Observe that the cdfc is approximated quite well for all the methods, but the best approximation depends on the number of interferers, shadow standard deviation, and range of distribution. The cdf is approximated less accurately, especially for $N_I = 6$ log-normal random variables.

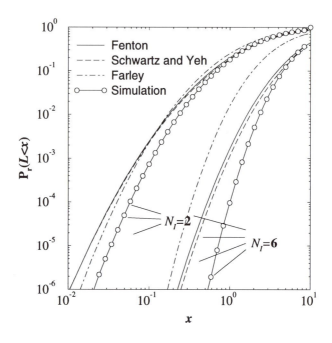

Figure 3.1 Comparison of the cdf for the sum of two and six log-normal random variables with various approximations; $\sigma_\Omega = 6$ dB.

3.2 PROBABILITY OF CCI

Consider the situation shown in Fig. 1.14, where a mobile station (MS) is at distance d_0 from the desired base station (BS) and at distances $d_k, k = 1, 2, \cdots, N_I$ from the first tier of N_I co-channel BSs. For convenience, let define the vector $\mathbf{d} = (d_0, \ d_1, \cdots, \ d_{N_I})$ as the set of distances for a particular MS location. The average received carrier-to-interference ratio as a function of the vector $\mathbf{d}$ is

$$\Lambda_{(\mathrm{dB})}(\mathbf{d}) = \Omega_{(\mathrm{dB})}(d_0) - 10\log_{10}\sum_{k=1}^{N_I} 10^{\Omega_{(\mathrm{dB})}(d_k)/10} \ . \tag{3.38}$$

and the probability of CCI is given by

$$O(\mathbf{d}) = \mathrm{P_r}\left(\Lambda_{(\mathrm{dB})}(\mathbf{d}) < \Lambda_{\mathrm{th}(\mathrm{dB})}\right) \ . \tag{3.39}$$

For the case of a single interferer $(N_I = 1)$ the sum on the right side of (3.38) only has one term. Therefore, $\Lambda_{(\mathrm{dB})}(\mathbf{d})$ has the Gaussian density

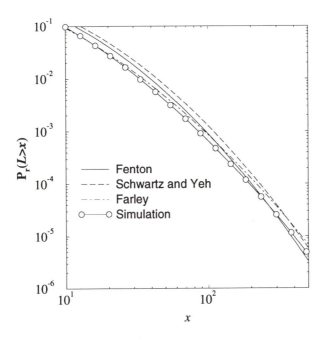

Figure 3.2 Comparison of the cdfc for the sum of two log-normal random variables with various approximations; $\sigma_\Omega = 6$ dB.

$$p_{\Lambda_{(dB)}(d)}(x) = \frac{1}{\sqrt{4\pi}\sigma_\Omega} \exp\left\{-\frac{(x - \mu_{\Lambda(d)})^2}{4\sigma_\Omega^2}\right\} \tag{3.40}$$

where

$$\mu_{\Lambda(d)} = \mu_{\Omega(d_0)} - \mu_{\Omega(d_1)} \tag{3.41}$$

and the probability of CCI is

$$
\begin{aligned}
O(d) &= \Pr(\Lambda_{(dB)}(d) < \Lambda_{th(dB)}) \\
&= \int_{-\infty}^{\Lambda_{th(dB)}} \frac{1}{\sqrt{4\pi}\sigma_\Omega} \exp\left\{-\frac{(x - \mu_{\Lambda(d)})^2}{4\sigma_\Omega^2}\right\} dx \\
&= Q\left(\frac{\mu_{\Lambda(d)} - \Lambda_{th(dB)}}{\sqrt{2}\sigma_\Omega}\right).
\end{aligned}
\tag{3.42}
$$

Now consider the worst case situation shown in Fig. 3.5, where a MS is located on a cell fringe at distance $D - R$ from a single co-channel BS.[2] Although Fig. 3.5 depicts the forward channel, we note that the worst case interference on the reverse channel from a single co-channel MS is the same. In either case

---

[2] Here we approximate the cell radius as the distance from the center to an edge of a cell.

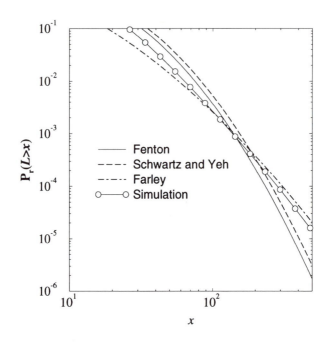

Figure 3.3 Comparison of the cdfc for the sum of six log-normal random variables with various approximations; $\sigma_\Omega = 6$ dB.

$$\mu_{\Lambda(\mathbf{d})} = \mu_{\Omega(R)} - \mu_{\Omega(D-R)} \tag{3.43}$$

and the probability of CCI is

$$O(R) = Q\left(\frac{\mu_{\Omega(R)} - \mu_{\Omega(D-R)} - \Lambda_{\text{th(dB)}}}{\sqrt{2}\sigma_\Omega}\right) . \tag{3.44}$$

A relationship can be derived between the probability of CCI on a cell fringe and the area averaged probability of CCI. To do so, we assume a uniform spatial distribution of MSs and use the approximation $d_1 \approx D - d_0$. Then using (3.44) along with the assumption that $\Lambda(\mathbf{d}) \gg 1$ and $\Lambda_{\text{th}} \gg 1$ yields

$$\begin{aligned} O &= \frac{1}{\pi R^2} \int_0^R O(r)2\pi r \; dr \\ &= Q(\hat{X}) - \exp\left\{\hat{X}\hat{Y} + \hat{Y}^2/2\right\} Q(\hat{X} + \hat{Y}) \end{aligned} \tag{3.45}$$

where

$$\hat{X} = \frac{\mu_{\Lambda(R)} - \Lambda_{\text{th(dB)}}}{\sqrt{2}\sigma_\Omega}, \qquad \hat{Y} = \frac{2\sqrt{2}\sigma_\Omega}{\beta\xi} \tag{3.46}$$

where $xi = 10/\ln 10$. The first term of (3.45) is the probability of CCI on the

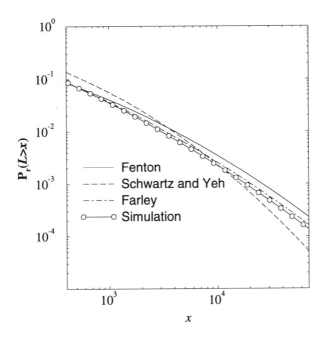

Figure 3.4 Comparison of the cdfc for the sum of six log-normal random variables with various approximations; $\sigma_\Omega = 12$ dB.

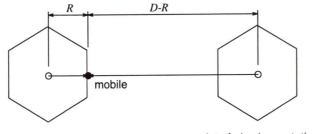

Figure 3.5 Worst case interference from a single co-channel BS.

cell fringe, and the second term is a correction factor.

Using (3.44) along with the simple path loss model in (1.6) gives

$$O(R) = Q\left(\frac{10\log_{10}\left(\frac{D}{R} - 1\right)^{\beta} - \Lambda_{\text{th(dB)}}}{\sqrt{2}\sigma_\Omega}\right) \ . \tag{3.47}$$

The quantity

$$M_\Lambda = 10 \log_{10} \left(\frac{D}{R} - 1 \right)^\beta - \Lambda_{\text{th(dB)}} \tag{3.48}$$

is the carrier-to-interference ratio margin on the cell fringe required to sustain a probability of CCI equal to $O(R)$. For hexagonal cells, a geometric argument (see Problem 1.2) can be used to show that the reuse cluster size, N, is related to the co-channel reuse factor, D/R, by

$$\frac{D}{R} = \sqrt{3N} \ . \tag{3.49}$$

Substituting (3.49) into (3.48) gives

$$N = \frac{1}{3} \left[10^{\frac{M_\Lambda + \Lambda_{\text{th(dB)}}}{10\beta}} + 1 \right]^2 \ . \tag{3.50}$$

In order to achieve high spectral efficiency, it is desirable to make N or, equivalently, D/R as small as possible. Since M_Λ is determined by the acceptable probability of CCI and is therefore fixed, a reduction in N can only be achieved by making Λ_{th} small. Recall that Λ_{th} depends on the radio link design and the multipath-fading characteristics of the channel. Therefore, radio links must be designed that will operate over multipath-fading channels with the minimum possible threshold Λ_{th}.

3.2.1   Multiple Log-normal Interferers

For the case of multiple interferers, we first obtain the appropriate values for the mean $\mu_{\hat{Z}}$ and the variance $\sigma_{\hat{Z}}^2$ of the log-normal approximation. Then the probability of CCI can be readily obtained from (3.39). The mean and variance of $Z_{\text{(dB)}} = \hat{Z}/\xi$ are

$$\mu_Z = \mu_{\hat{Z}}/\xi \tag{3.51}$$
$$\sigma_Z^2 = \sigma_{\hat{Z}}^2/\xi^2 \ . \tag{3.52}$$

Then

$$\Lambda_{\text{(dB)}}(\mathbf{d}) = \Omega_{\text{(dB)}}(d_0) - Z_{\text{(dB)}}(d_1, \, d_2, \, \ldots, \, d_{N_I}) \tag{3.53}$$

where we have again shown the dependency of the cochannel interference on the set of distances. Note that $\Lambda_{\text{(dB)}}(\mathbf{d})$ has mean and variance

$$\mu_{\Lambda(\mathbf{d})} = \mu_{\Omega(d_0)} - \mu_Z \tag{3.54}$$
$$\sigma_{\Lambda(\mathbf{d})}^2 = \sigma_\Omega^2 + \sigma_Z^2 \ . \tag{3.55}$$

Therefore, the probability of CCI is

$$O(\mathbf{d}) = Q\left(\frac{\mu_{\Omega(d_0)} - \mu_Z - \Lambda_{\text{th(dB)}}}{\sqrt{\sigma_\Omega^2 + \sigma_Z^2}}\right) . \tag{3.56}$$

For the forward channel, the worst case probability of CCI can be obtained by assuming that the MS is located at the corner of the cell and using the corresponding **d** vector. The area averaged probability of CCI is then obtained by averaging over the random location of the MS, i.e.,

$$O = \int_{R^N} O(\mathbf{d}) p_{\mathbf{d}}(\mathbf{d}) d\mathbf{d} \tag{3.57}$$

where $p_{\mathbf{d}}(\mathbf{d})$ is the probability density function of the random vector **d** and R^N is the region of integration. Usually, $p_{\mathbf{d}}(\mathbf{d})$ is impossible to evaluate in closed form so that Monte Carlo integration becomes necessary. For the reverse channel, the same procedure applies, but we must now average over the random locations of all the co-channel MSs.

3.3 MULTIPLE RICEAN/RAYLEIGH INTERFERERS

In microcellular environments, the received signal often consists of a direct line of sight (LOS) component, or perhaps a specular component, accompanied by a diffuse component. In this case, the envelope of the received signal experiences Ricean fading. In the same environment, the co-channel signals are often assumed to be Rayleigh faded, because a direct LOS between the co-channel cells is not likely to exist and the propagation path lengths are much longer. In this section, we calculate the probability of CCI for the case of fading only. The combined effect of shadowing and fading is deferred until the next section. Let the instantaneous power in the desired signal and the N_I interfering signals be denoted by s_0 and s_k, $k = 1, \cdots, N_I$, respectively. Note that $s_i = z_i^2$ and z_i^2 is the squared-envelope. For a specified receiver threshold λ_{th}, the probability of CCI is

$$O = \text{Pr}\left(\lambda < \lambda_{\text{th}}\right) \equiv \text{Pr}\left(s_0 < \lambda_{\text{th}} \sum_{k=1}^{N_I} s_k\right) \tag{3.58}$$

where $\lambda = s_0 / \sum_{k=1}^{N_I} s_k$. The instantaneous received signal power, s_0, has the non-central chi-square (Ricean fading) distribution in (2.44), while the instantaneous power of each interferer, s_k, has the exponential distribution (Rayleigh fading) in (2.39).

For the case of a single interferer, the probability of CCI reduces to the simple closed form [344]

$$O = \frac{\lambda_{\text{th}}}{\lambda_{\text{th}} + b_1} \exp\left\{ -\frac{K b_1}{\lambda_{\text{th}} + b_1} \right\} \tag{3.59}$$

where K is the Rice factor of the desired signal, $b_1 = \Omega_0/(K+1)\Omega_1$, and $\Omega_k = \text{E}[s_k]$. If the desired signal is Rayleigh faded, then the probability of CCI can be obtained by setting $K = 0$ in (3.59). For the case of multiple interferers, each with mean power Ω_k, the probability of CCI has the closed form [344]

$$O = 1 - \sum_{k=1}^{N_I} \left[1 - \frac{\lambda_{\text{th}}}{\lambda_{\text{th}} + b_k} \exp\left\{ -\frac{K b_k}{\lambda_{\text{th}} + b_k} \right\} \right] \prod_{\substack{j=1 \\ j \neq k}}^{N_I} \frac{b_j}{b_j - b_k} \tag{3.60}$$

where $b_k = \Omega_0/(K+1)\Omega_k$. This expression is only valid if $\Omega_i \neq \Omega_j$ when $i \neq j$, i.e., the different interferers have different mean power. If some of the interferers have the same mean power, then an appropriate expression for the probability of CCI can be derived in straight forward manner. If all the interferers have the same mean power, then the total interference power $s_M = \sum_{k=1}^{N_I} s_k$ has the Gamma density function

$$p_{s_M}(x) = \frac{x^{N_I - 1}}{\Omega_1^{N_I}(N_I - 1)!} \exp\left\{ -\frac{x}{\Omega_1} \right\} . \tag{3.61}$$

The probability of CCI can be derived as [344]

$$\begin{aligned} O &= \frac{\lambda_{\text{th}}}{\lambda_{\text{th}} + b_1} \exp\left\{ -\frac{K b_1}{\lambda_{\text{th}} + b_1} \right\} \\ &\times \sum_{k=0}^{N_I - 1} \left(\frac{b_1}{\lambda_{\text{th}} + b_1} \right)^k \sum_{m=0}^{k} \binom{k}{m} \frac{1}{m!} \left(\frac{K \lambda_{\text{th}}}{\lambda_{\text{th}} + b_1} \right)^m . \end{aligned} \tag{3.62}$$

Again, if the desired signal is Ricean faded, then the probability of CCI with multiple Rayleigh faded interferers can be obtained by setting $K = 0$ in either (3.60) or (3.62), which ever is appropriate. In Fig. 3.6, the CCI outage probability is plotted as a function of the carrier-to-interference ratio

$$\Lambda = \frac{\Omega_0}{N_I \Omega_1} \tag{3.63}$$

for various Rice factors and a single interferer. Observe that the Rice factor of the desired signal has a significant effect on the probability of CCI. Fig. 3.7 plots the probability of CCI for $K = 0$ and 7 and varying numbers of interferers. Observe that the number of interferers does not affect the probability of CCI as much as the Rice factor, provided that the total interfering power remains constant.

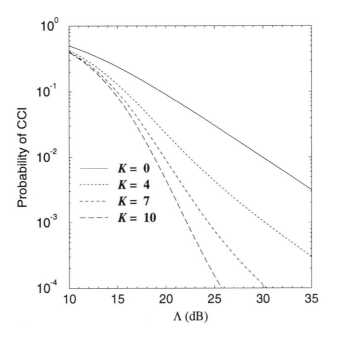

Figure 3.6 Probability of CCI with a single interferer. The desired signal is Ricean faded with various Rice factors, while the interfering signal is Rayleigh faded; $\lambda_{\text{th}} = 10.0$ dB.

3.4 MULTIPLE LOG-NORMAL NAKAGAMI INTERFERERS

The probability of CCI has been evaluated in the literature for a single Nakagami interferer [338] and multiple Nakagami interferers [2], [345], in the absence of shadowing. Here we analytically formulate the probability of CCI with multiple log-normal Nakagami interferers. For the case when the interfering signals have the same shadowing and fading statistics, we derive an exact mathematical expression for the probability of CCI. Let the instantaneous power in the desired signal and the N_I interfering signals be denoted by s_0 and s_k, $k = 1, \cdots, N_I$, respectively. Again, for a specified receiver threshold λ_{th}, the probability of CCI is

$$O = \text{P}_\text{r}\left(\lambda < \lambda_{\text{th}}\right) \equiv \text{P}_\text{r}\left(s_0 < \lambda_{\text{th}} \sum_{k=1}^{N_I} s_k\right) \qquad (3.64)$$

where $\lambda = s_0 / \sum_{k=1}^{N_I} s_k$. Since the kth signal is affected by log-normal shadowing and Nakagami fading, s_k has the composite density function

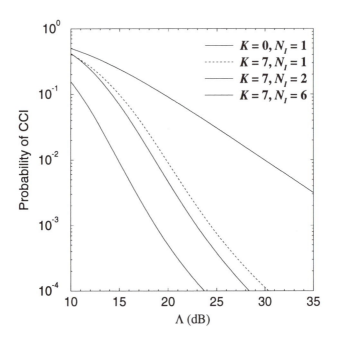

Figure 3.7   Probability of CCI with multiple interferers. The desired signal is Ricean faded with various Rice factors, while the interfering signals are Rayleigh faded and of equal power; $\lambda_{\text{th}} = 10.0$ dB.

$$p_{s_k}(x) = \int_0^\infty \left(\frac{m_k}{\Omega_k}\right)^{m_k} \frac{x^{m_k-1}}{\Gamma(m_k)} \exp\left\{-\frac{m_k x}{\Omega_k}\right\} \tag{3.65}$$
$$\times \frac{\xi}{\sqrt{2\pi}\sigma_{\Omega_k}\Omega_k} \exp\left\{-\frac{(10\log_{10}\Omega_k - \mu_{\Omega_k})^2}{2\sigma_{\Omega_k}^2}\right\} d\Omega_k \quad .$$

Let $W = \sum_{k=1}^{N_I} s_k$ be the total power from the N_I interfering signals, $X = s_0/W$, and $Y = W$. Then the joint probability density function of X and Y is $p_{XY}(xy) = y p_{s_0,W}(xy, y)$ and

$$p_X(x) = \int_0^\infty y p_{s_0}(xy) p_W(y) dy \quad . \tag{3.66}$$

It follows that the probability of CCI is

$$O = \Pr\left(\frac{s_0}{W} < \lambda_{\text{th}}\right)$$
$$= 1 - \int_{\lambda_{\text{th}}}^\infty dx \int_0^\infty y p_{s_0}(xy) p_W(y) dy \tag{3.67}$$

Substituting the Nakagami density for s_0 and integrating with respect to x gives the conditional probability [128]

$$\Pr\left(\frac{s_0}{W} < \lambda_{\text{th}} \mid \Omega_0\right) = 1 - \sum_{h=0}^{m_0-1} \left(\frac{m_0\lambda_{\text{th}}}{\Omega_0}\right)^h \frac{1}{h!} \int_0^\infty \exp\left\{-\frac{m_0\lambda_{\text{th}}y}{\Omega_0}\right\} y^h p_W(y)dy$$

(3.68)

3.4.1   Statistically Identical Interferers

Here we assume statistically identical co-channel interferers so that $\sigma_{\Omega_k} = \sigma_\Omega$ and $m_k = m_I$, $i = 1, \ldots, N_I$. Following Linnartz [192], the integral in (3.68) can be obtained by using Laplace transform techniques. The Laplace transform of the probability density function $p_W(y)$ is

$$\mathbf{L}_W(s) = \int_0^\infty e^{-sy} p_W(y)dy$$

(3.69)

The integral in (3.68) is then equal to the hth derivative of $\mathbf{L}_W(s)$ with respect to s evaluated at the point $s = (m_0\lambda_{\text{th}})/\Omega_0$. That is,

$$\int_0^\infty e^{-sy} y^h p_W(y)dy = (-1)^h \mathbf{L}_W^{(h)}(s)$$

(3.70)

$$= (-1)^h \frac{d^h}{ds^h}\left\{\prod_{k=1}^{N_I} \int_0^\infty e^{-sy_k} p_{s_k}(y_k)dy_k\right\}$$

where the last line follows under the assumption of statistically independent interferers. By using the composite distribution in (3.65) with $m_k = m_I$

$$\int_0^\infty e^{-sy_k} p_{s_k}(y_k)dy_k = \frac{1}{\sqrt{\pi}} \int_{-\infty}^\infty \frac{m_I^{m_I} e^{-x^2}}{(10^{(\mu_{\Omega_k}+\sqrt{2}\sigma_\Omega x)/10}s + m_I)^{m_I}} dx .$$

(3.71)

Averaging over the log-normal shadowing distribution of the desired signal gives the final result

$$O = 1 - \int_0^\infty \left\{\sum_{h=0}^{m_0-1} \left(-\frac{m_0\lambda_{\text{th}}}{\Omega_0}\right)^h \frac{1}{h!}\right. $$

(3.72)

$$\times \frac{d^h}{ds^h}\left\{\prod_{k=1}^{N_I} \int_{-\infty}^\infty \frac{m_I^{m_I} e^{-x^2} dx}{\sqrt{\pi}(10^{(\mu_{\Omega_k}+\sqrt{2}\sigma_\Omega)x/10}s + m_I)^{m_I}}\right\}\Bigg|_{s=\frac{m_0\lambda_{\text{th}}}{\Omega_0}}\right\}$$

$$\times \frac{\xi}{\sqrt{2\pi}\sigma_\Omega\Omega_0} \exp\left\{-\frac{(10\log_{10}\Omega_0 - \mu_{\Omega_0})^2}{2\sigma_\Omega^2}\right\} d\Omega_0 .$$

Equation (3.72) is an exact expression for shadowed Nakagami fading channels. When $m_d = m_I = 1$, it reduces to the simple expression obtained by Linnartz [192] for shadowed Rayleigh fading channels. If the path loss associated with each interferer is the same, then $\Omega_k = \Omega_I$ and the product in (3.72) reduces to taking the N_Ith power. Let

$$F(s) = \int_{-\infty}^{\infty} \frac{e^{-x^2}}{(10^{(\mu\Omega_I + \sqrt{2}\sigma_\Omega x)/10}s + m_I)^{m_I}}\,dx \qquad (3.73)$$

and use the identity [128]

$$
\begin{aligned}
G(s) &= \frac{d^h}{ds^h}[F(s)]^{N_I} \\
&= N_I\binom{h-N_I}{N_I}\sum_{i=1}^{h}(-1)^i\binom{h}{i}\frac{[F(s)]^{N_I-i}}{N_I-i}\frac{d^h}{ds^h}[F(s)]^i \ . \qquad (3.74)
\end{aligned}
$$

Observe that $G(s)$ is just a function of the derivatives of $F(s)$, and

$$
\begin{aligned}
\frac{d^w}{dw}F(s) &= \frac{d^w}{ds^w}\left\{\int_{-\infty}^{\infty}\frac{e^{-x^2}}{(10^{(\mu\Omega_I+\sqrt{2}\sigma_\Omega x)/10}s + m_I)^{m_I}}\,dx\right\} \qquad (3.75) \\
&= (-1)^w\frac{(m_I+w-1)!}{(m_I-1)!}\int_{-\infty}^{\infty}\frac{(10^{(\mu\Omega_I+\sqrt{2}\sigma_\Omega x)/10})^w e^{-x^2}}{(10^{(\mu\Omega_I+\sqrt{2}\sigma_\Omega x)/10}s + m_I)^{m_I+w}}\,dx \ .
\end{aligned}
$$

We can obtain $G(s)$ from (3.74) and (3.75), and substitute it into (3.72). Then by using a change of variables the probability of CCI in (3.72) becomes

$$
\begin{aligned}
O &= 1 - \sum_{h=0}^{m_0-1}\left(-m_0\lambda_{\text{th}}10^{-\mu\Omega_0/10}\right)^h\frac{m_I^{m_I N_I}}{\sqrt{\pi}^{N_I+1}h!} \qquad (3.76) \\
&\quad \times \int_{-\infty}^{\infty}10^{-\sqrt{2}\sigma_\Omega x h/10}e^{-x^2}G\left(m_0\lambda_{\text{th}}10^{-(\mu\Omega_0+\sqrt{2}\sigma_\Omega x h)/10}\right)dx \ .
\end{aligned}
$$

The integrals in (3.75) and (3.76) can be efficiently computed by using Hermite-Gauss quadrature integration. Applying the Hermite-Gauss quadrature formula to (3.75) gives

$$F^{(w)}(s) = (-1)^w\frac{(m_I+w-1)!}{(m_I-1)!}\sum_{t=1}^{N_p}H_{x_t}\frac{10^{(\mu\Omega_I+\sqrt{2}\sigma_\Omega x_t)w/10}}{(10^{(\mu\Omega_I+\sqrt{2}\sigma_\Omega x_t)/10}s + m_I)^{m_I+w}}$$
$$(3.77)$$

where H_{x_t} are weight factors, x_t are the zeros of the Hermite polynomial $H_p(x)$, and N_p is the order of the Hermite polynomial. By using this result and the values in Table 3.1 (listed for convenience) we have

Zeros x_i	Weight Factors H_{x_i}
± 0.27348104613815	$5.079294790166 \times 10^{-1}$
± 0.82295144914466	$2.806474585285 \times 10^{-1}$
± 1.38025853919888	$8.381004139899 \times 10^{-2}$
± 1.95178799091625	$1.288031153551 \times 10^{-2}$
± 2.54620215784748	$9.322840086242 \times 10^{-4}$
± 3.17699916197996	$2.711860092538 \times 10^{-5}$
± 3.86944790486012	$2.320980844865 \times 10^{-7}$
± 4.68873893930582	$2.654807474011 \times 10^{-10}$

Table 3.1   Zeros and weight factors of 16 order Hermite polynomials [1].

$$
O = 1 - \sum_{h=0}^{m_0-1} \left(-m_0 \lambda_{\text{th}} 10^{\mu_{\Omega_0}/10} \right)^h \frac{m_I^{m_I N_I}}{\sqrt{\pi}^{N_I+1} h!} \tag{3.78}
$$

$$
\times \sum_{\ell=1}^{p} H_{x_\ell} 10^{-\sqrt{2}\sigma_\Omega x_\ell h/10} G \left(m_0 \lambda_{\text{th}} 10^{-(\mu_{\Omega_0} + \sqrt{2}\sigma_\Omega x h)/10} \right) dx \ .
$$

Fig. 3.8 shows the probability of CCI as a function of the carrier-to-interference ratio

$$
\Lambda = \mu_{\Omega_0} - \mu_{\Omega_I} - 10 \log_{10}(N_I) \quad \text{dB} \ . \tag{3.79}
$$

Results are plotted for six interfering signals and varying degrees of fading on the desired and interfering signals. Observe that the probability of CCI is insensitive to changes in the m values for interfering signals. This phenomenon demonstrates that co-channel interference is dominated by the fading of the desired signal rather than fading of the interfering signals. Fig. 3.9 shows the probability of CCI for different values of the shadow standard deviation σ_Ω. We can conclude that the number of interferers and the shadow standard deviation have the most significant effect on the probability of CCI.

3.5   MULTIPLE LOG-NORMAL RICEAN/RAYLEIGH INTERFERERS

This section presents an exact method for evaluating the probability of co-channel interference for Ricean/Rayleigh faded channels with log-normal shadowing. The results can be applied for a Ricean faded desired signal and a single

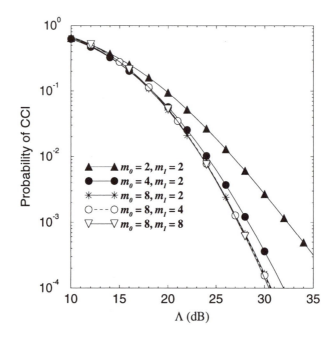

Figure 3.8 Probability of CCI when the desired and interfering signals are Nakagami faded. Results are shown for various fading distribution parameters; $\sigma_\Omega = 6$ dB, $\lambda_{\mathrm{th}} = 10.0$ dB.

Rayleigh faded interferer, or vice versa. It can also be applied for a Rayleigh faded desired signal with multiple Ricean or Rayleigh faded interfering signals. Once again, let the instantaneous power in the desired signal and the N_I interfering signals be denoted by s_0 and $s_k, k = 1, \cdots, N_I$, respectively. For a specified receiver threshold λ_{th}, the probability of CCI is, again,

$$O = \mathrm{P_r}\left(\lambda < \lambda_{\mathrm{th}}\right) \equiv \mathrm{P_r}\left(s_0 < \lambda_{\mathrm{th}} \sum_{k=1}^{N_I} s_k\right) \tag{3.80}$$

where $\lambda = s_0/\sum_{k=1}^{N_I} s_k$ and each s_k has either a composite log-normal exponential (Rayleigh fading) distribution or a composite log-normal non-central chi-square (Ricean fading) distribution. The $s_k, k = 0, \ldots, N_I$ in (3.80) can be reordered such that $\left(s_0 < \sum_{k=1}^{N_I} \lambda_{\mathrm{th}} s_k\right) = \left(\tilde{s}_0 < \sum_{k=1}^{N_I} \delta_k \tilde{s}_k\right)$, where $\tilde{s}_0$ is exponentially distributed, the $\tilde{s}_k, k = 1, \ldots, N_I$ are either exponentially or non-central chi-square distributed. When the desired signal is Rayleigh faded $\tilde{s}_0 = s_o$ and $\tilde{s}_k = s_k$ and $\delta_i = \lambda_{\mathrm{th}}$. Otherwise, when the desired signal is Ricean faded and a single Rayleigh interferer is present, we observe that $\mathrm{P_r}(s_0 < \lambda_{\mathrm{th}} s_1)$ $= \mathrm{P_r}(s_1/(s_0)/\delta) > 1)$. Therefore, using $\tilde{s}_0 = s_1$, $\tilde{s}_1 = s_0$, and $\delta_1 = 1/\lambda_{\mathrm{th}}$, the

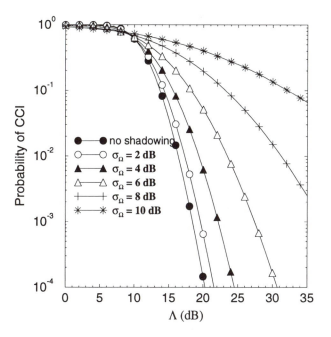

Figure 3.9 Probability of CCI when the desired and interfering signals are Nakagami faded. Results are shown for various shadow standard deviations; $m_0 = 8$, $m_I = 2$, $\lambda_{\text{th}} = 10.0$ dB.

probability of CCI is $1 - \text{P}_\text{r}(s_1/s_0/\lambda_{\text{th}} < 1) = 1 - \text{P}_\text{r}(\tilde{s}_0/\delta\tilde{s}_1 < 1)$. Thus, let $X = \tilde{s}_0/W$ and $Y = W$, where $W = \sum_{k=1}^{N_I} \delta_k \tilde{s}_k$. The joint probability density function of X and Y is $p_{XY}(x,y) = y\, p_{\tilde{s}_0,W}(xy, y)$ and

$$p_X(x) = \int_0^\infty y p_{\tilde{s}_0}(xy)p_W(y)dy \ . \qquad (3.81)$$

Therefore,

$$O = \text{P}_\text{r}\left(\frac{\tilde{s}_0^2}{W} < 1\right) = 1 - \int_1^\infty dx \int_0^\infty y p_{\tilde{s}_0}(xy)p_W(y)dy \ . \qquad (3.82)$$

Substituting the exponential density for $p_{\tilde{s}_0}(xy)$ and integrating with respect to x gives the conditional probability

$$\text{P}_\text{r}\left(\frac{\tilde{s}_0^2}{W} < 1 \Big| \Omega_0\right) = 1 - \int_0^\infty \exp\left\{-\frac{y}{\Omega_0}\right\} p_W(y)dy \qquad (3.83)$$

where $\Omega_0 = \text{E}[\tilde{s}_0]$. Following Linnartz [192], the integral in (3.83) can be simplified by using Laplace transform techniques. Since the $\tilde{s}_k$, $k = 1, \ldots, N_I$ are all independent random variables, $p_W(y)$ is the convolution of the densities of the $\delta_k \tilde{s}_k$. Hence, (3.83) becomes

$$
\Pr \left(\frac{\tilde{s}_0^2}{W} < 1 \, | \Omega_0 \right) = 1 - \prod_{k=1}^{N_I} F_{\tilde{s}_k}(s) \Bigg|_{s = \delta_k / \Omega_0} \tag{3.84}
$$

where $F_{\tilde{s}_k}(s)$ is the Laplace transform of probability density function $p_{s_k}(x)$, given by

$$
\begin{aligned}
F_{\tilde{s}_k}(s) &= \int_0^\infty e^{-sx} \int_0^\infty \frac{K_k + 1}{\Omega_k} \exp \left\{ -K_k - \frac{(K_k + 1)x}{\Omega_k} \right\} \\
&\quad \times I_0 \left(2 \sqrt{\frac{K_k(K_k + 1)x}{\Omega_k}} \right) p_{\Omega_k}(\Omega_k) d\Omega_k \, dx \\
&= \int_0^\infty \frac{K_k + 1}{K_k + 1 + \delta_k s \Omega_k} \exp \left\{ -\frac{\delta_k s \Omega_k K_k}{K_k + 1 + \delta_k v \Omega_k} \right\} \\
&\quad \times \frac{\xi}{\sqrt{2\pi} \sigma_{\Omega_k} \Omega_k} \exp \left\{ -\frac{(10\log_{10}\Omega_k - \mu_{\Omega_k})^2}{2\sigma_{\Omega_k}^2} \right\} d\Omega_k
\end{aligned} \tag{3.85}
$$

where K_k is the Rice factor of the kth signal. Averaging over shadowing distribution for the desired signal yields the final result

$$
\begin{aligned}
O &= 1 - \int_0^\infty \left(\prod_{k=1}^{N_I} F_{\tilde{s}_k}(\delta_k / \Omega_k) \right) \\
&\quad \times \frac{\xi}{\sqrt{2\pi} \sigma_{\Omega_0} \Omega_0} \exp \left\{ -\frac{(10\log_{10}\Omega_0 - \mu_{\Omega_0})^2}{2\sigma_{\Omega_0}^2} \right\} d\Omega_0
\end{aligned} \tag{3.86}
$$

The integrals in (3.85) and (3.86) can be efficiently computed using Hermite-Gauss quadrature integration, as explained earlier. Corresponding expressions for Rayleigh fading can be obtained by setting the $K_k = 0$ in (3.85).

3.5.1 Single Interferer

For a Rayleigh faded desired signal and a Ricean faded interferer, (3.86) can be used directly with $K = 0$, and $\delta_1 = \lambda_{\text{th}}$. If we assume the simple path loss model in (1.6), and define the normalized reuse distance as d_1 / d_0, where d_0 and d_1 are the radio path lengths of the desired and interfering signals, respectively, then the average carrier-to-interference ratio is

$$
\Lambda = \mu_{\Omega_0} - \mu_{\Omega_1} = 10\beta \log_{10}(d_1 / d_0) \quad \text{dB} . \tag{3.87}
$$

The probability of CCI is plotted against the normalized reuse distance in Fig. 3.10, where it is shown to be insensitive to the Rice factor of the interferer.

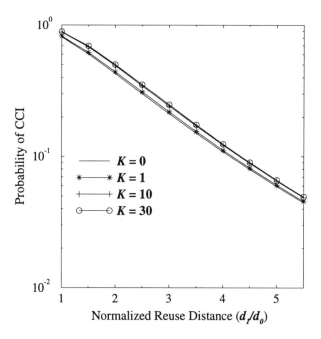

Figure 3.10 Probability of CCI outage against the normalized reuse distance for a Rayleigh faded desired signal and one Ricean faded interferer. The Rice factors of the interfering signal varied; $\lambda_{th} = 10$ dB, $\sigma_{\Omega_0} = \sigma_{\Omega_1} = 6$ dB, $\beta = 4$.

Likewise, Fig. 3.11 plots the probability of CCI against the normalized reuse distance when the desired signal is Ricean faded and there is a single Rayleigh faded interferer. Observe the strong dependency of the probability of CCI on the Rice factor of the desired signal.

3.5.2 Multiple Interferers

For a Rayleigh faded desired signal, (3.86) can be used directly leading to the same (not shown) result as the single interferer case, the probability of CCI is insensitive to the Rice factors of the interferers. For a Ricean faded desired signal with multiple Ricean/Rayleigh interferers, a more detailed analysis is required. For example, with two Rayleigh faded interferers and a Ricean faded desired signal

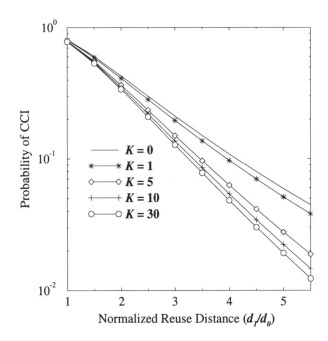

Figure 3.11 Probability of CCI outage against the normalized reuse distance for a Ricean faded desired signal and one Rayleigh faded interferer. The Rice factors of the desired signal are varied; $\delta = 10$ dB, $\sigma_{\Omega_0} = \sigma_{\Omega_1} = 6$ dB, $\beta = 4$.

$$
\begin{aligned}
\mathrm{P_r}\left(s_0 < \lambda_{\mathrm{th}} s_2\right) &= \mathrm{P_r}\left(\frac{s_1}{s_0/\lambda_{\mathrm{th}} - s_2} > 1 \,|s_0/\lambda_{\mathrm{th}} - s_2 > 0\right) \\
&\quad \times \mathrm{P_r}\left(s_0/\lambda_{\mathrm{th}} - s_2 > 0\right) \\
&+ \mathrm{P_r}\left(-\frac{s_1}{s_2 - s_0/\lambda_{\mathrm{th}}} < 1 \,|s_0/\lambda_{\mathrm{th}} - s_2 < 0\right) \\
&\quad \times \mathrm{P_r}\left(s_0/\lambda_{\mathrm{th}} - s_2 < 0\right) \ .
\end{aligned}
$$

(3.88)

Since $s_0/\lambda_{\mathrm{th}} - s_2$ can be negative, the conditional density $p_X(x|W > 0)$ must be used in the above calculation. Unfortunately, this does not result in a simple multiplication of Laplace transforms as before and, hence, alternative methods for finding the exact pdf must be employed. This is an open research problem. If approximations are employed, then one possibility is to use the results of Section 3.4 to approximate the Rice distribution with a Nakagami distribution.

Problems

3.1. Suppose that a MS is affected by six co-channel interferers as shown in Fig. 1.14. Each of the interfering signals is independently shadowed with a shadow standard deviation of σ_Ω dB. The path loss is described by the model in (1.6). It is assumed that the receiver can average over the fast envelope fading. By using the Fenton-Wilkinson method, calculate the worst case probability of CCI for a 7-cell reuse pattern. Plot your results as a function of σ_Ω and β.

3.2. This problem uses computer simulation to verify the usefulness of Schwartz and Yeh's approximation and the Fenton-Wilkinson approximation for the sum of two log-normal random variables.

Consider the sum of two log-normal random variables

$$L = e^{\hat{\Omega}_1} + e^{\hat{\Omega}_2}$$

where the Gaussian random variables $\hat{\Omega}_1$ and $\hat{\Omega}_2$ are independent and identically distributed with zero mean and variance σ_Ω^2. By using Schwartz and Yeh's method plot the values of $\mu_{\hat{Z}}$ and $\sigma_{\hat{Z}}^2$ (in dB), as a function of the variance σ_Ω^2. Repeat for the Fenton-Wilkinson method. Now obtain the same results by using computer simulation and compare the analytical results.

3.3. Show that the area averaged probability of CCI with a single co-channel interferer is given by (3.45).

3.4. Derive equation (3.59).

3.5. Derive equation (3.62).

3.6. Consider a microcellular environment where a Ricean faded signal is affected by a single Rayleigh faded interferer. Neglect the effect of path loss and shadowing. Suppose that the transmission quality is deemed acceptable if both the instantaneous carrier-to-noise ratio and the instantaneous carrier-to-interference ratio exceed the thresholds, γ_{th} and λ_{th}, respectively. Analogous to (3.62) derive an expression for the probability of CCI.

4

MODULATED SIGNALS AND THEIR POWER SPECTRAL DENSITIES

Modulation is a process where the message information is embedded into the radio carrier. Message information can be transmitted in the amplitude, frequency, or phase of the carrier, or a combination of these, in either analog or digital form. For mobile radio applications it is desirable to use bandwidth and power resources most efficiently. The primary objective of bandwidth and power efficient modulation is to maximize the bandwidth efficiency, measured in bits/s/Hz, while achieving a prescribed bit error probability with a minimum expenditure of power resources. Good bit error rate performance must be achieved in the presence of a variety of channel impairments including fading, Doppler spread, intersymbol interference, adjacent and co-channel interference, and thermal noise. Furthermore, portable and mobile radio transmitters normally use power efficient nonlinear amplifiers to conserve battery resources. Because of the amplifier nonlinearities, modulation techniques with a relatively constant envelope are often used. All first generation cellular systems used analog FM. However, the pressing need for greater bandwidth efficiency has lead to the use of digital modulation techniques in second generation digital cellular standards. The North American IS-54 and Japanese PDC systems use $\pi/4$-DQPSK, the European GSM system uses Gaussian minimum shift keying (GSMK), and the Motorola Integrated Radio System (MIRS) uses orthogonal frequency division multiplexing (OFDM).

Section 4.1 begins this chapter with a brief treatment of analog FM. Although all second generation cellular systems use digital modulation, analog FM still finds widespread application. For example, in North America, the second generation common air interfaces require dual mode radios; the IS-54 system uses analog FM with $\pi/4$-DQPSK and the IS-95 system uses analog FM with CDMA. The treatment of analog FM is followed by a basic generic description of digitally modulated signals in Section 4.2. Sections 4.3 through 4.9 then provide a detailed discussion of various linear and nonlinear digital modulations techniques that are suitable for mobile radio applications, including $\pi/4$-DQPSK, GMSK, OFDM, and others. Since bandwidth efficiency is of great concern in mobile radio systems, the chapter concludes with a treatment of the power spectral densities (psds) of digitally modulated signals in Section 4.10.

4.1 ANALOG FREQUENCY MODULATION

Frequency modulation (FM) is an angle modulation technique where the instantaneous frequency of the carrier, $f_i(t)$, is varied linearly with the baseband modulating waveform, $m(t)$, i.e.,

$$f_i(t) = f_c + k_f m(t) \tag{4.1}$$

where f_c is the carrier frequency and k_f (Hz/v) is the frequency sensitivity of the modulator. The instantaneous carrier phase is

$$\phi_i(t) = \int_0^t f_i(t)dt = 2\pi f_c t + 2\pi k_f \int_0^t m(t)\ dt\ . \tag{4.2}$$

The complex low-pass FM signal is

$$v(t) = A \exp\left\{ 2\pi k_f \int_0^t m(t)\ dt \right\} \tag{4.3}$$

and the band-pass FM signal is

$$
\begin{aligned}
s(t) &= \text{Re}\left\{ v(t)e^{j2\pi f_c t} \right\} \\
&= A\cos\left(2\pi f_c t + 2\pi k_f \int_0^t m(t)\ dt \right)\ .
\end{aligned} \tag{4.4}
$$

where $\text{Re}\{z\}$ is the real part of z. Note that FM is a constant envelope modulation technique making it suitable for nonlinear amplification.

If the modulating waveform is the pure sinusoid $m(t) = A_m \cos(2\pi f_m t)$, then the instantaneous frequency deviation is

$$
\begin{aligned}
f_i(t) &= f_c + k_f A_m \cos(2\pi f_m t) \\
&= f_c + \Delta_f \cos(2\pi f_m t)
\end{aligned} \tag{4.5}
$$

where $\Delta_f = k_f A_m$ is called the **peak frequency deviation**. The corresponding instantaneous carrier phase is

$$\begin{aligned} \phi_i(t) &= 2\pi \int_0^t f_i(t)\, dt \\ &= 2\pi f_c t + \frac{\Delta_f}{f_m} \sin(2\pi f_m t) \\ &= 2\pi f_c t + \beta \sin(2\pi f_m t) \ . \end{aligned} \tag{4.6}$$

The dimensionless ratio $\beta = \Delta_f / f_m$ is called the **modulation index**. The corresponding band-pass signal is

$$s(t) = A \cos\left(2\pi f_c t + \beta \sin(2\pi f_m t)\right) \ . \tag{4.7}$$

An FM signal is a nonlinear function of the modulating waveform $m(t)$ and, therefore, the spectral characteristics of $s(t)$ cannot be obtained directly from the spectral characteristics of $m(t)$. However, the bandwidth of $s(t)$ depends on β. If $\beta < 1$, then narrow-band FM is generated, where the spectral widths of $s(t)$ and $m(t)$ are about the same, i.e., $2f_m$. If $\beta \gg 1$, then wide-band FM is generated, where the spectral width of $s(t)$ is slightly greater than $2\Delta_f$. For an arbitrary modulation index, the approximate bandwidth of the FM signal is

$$W \approx 2\Delta_f + 2f_m = 2\Delta_f \left(1 + \frac{1}{\beta}\right) \tag{4.8}$$

a relation known as **Carson's rule**. Unfortunately, typical cellular radio systems use a modulation index in the range $1 \lesssim \beta \lesssim 3$ where Carson's rule is not very accurate. Furthermore, the message waveform $m(t)$ is not a pure sinusoid so that Carson's rule does not directly apply. To obtain a more accurate estimate of the bandwidth requirements for an arbitrary waveform $m(t)$, first determine the ratio of the frequency deviation Δ_f corresponding to the maximum amplitude of the modulating waveform $m(t)$ to the highest frequency component B that is present in $m(t)$. The resulting ratio, denoted by D, is called the **deviation ratio**. Then replace β by D and f_m by B in Carson's rule, giving

$$W \approx 2\Delta_f + 2B = 2\Delta_f \left(1 + \frac{1}{D}\right) \ . \tag{4.9}$$

This approximation will over estimate the bandwidth requirements. A more accurate estimate of the bandwidth requirements must be obtained from simulation or measurements.

4.1.1 Demodulation of Analog FM Signals

Two basic devices can be used for demodulation of analog FM signals, the frequency discriminator and the phase-locked loop (PLL). We first discuss FM detection with a discriminator detector. However, a PLL detector can offer better performance in a fading environment due to an extended threshold region [111]. Therefore, a discussion of PLL detection is also provided.

Discriminator Detection

A frequency discriminator consists of a differentiator followed by an envelope detector. The complex low-pass representation for the transfer function of a differentiator is

$$C(f) = \begin{cases} j2\pi a \left(f + \frac{B}{2}\right), & \frac{-B}{2} \le f \le \frac{B}{2} \\ 0, & \text{elsewhere} \end{cases} \quad (4.10)$$

where B is the signal bandwidth and $2\pi a$ is the slope of the transfer function. The equivalent band-pass filter has transfer function

$$H(f) = C(f - f_c) + C^*(-f - f_c) \ . \quad (4.11)$$

When the complex low-pass signal

$$v(t) = A \exp\left\{2\pi k_f \int_0^t m(t)dt\right\} \quad (4.12)$$

is applied to the differentiator, the Fourier transform of the low-pass output signal is

$$W(f) = C(f)V(f) \ . \quad (4.13)$$

Invoking properties of the Fourier transform it follows that the time domain output is

$$\begin{aligned} w(t) &= a\left(\frac{dv(t)}{dt} + j\pi B v(t)\right) \\ &= j\pi B a A \left(1 + \frac{2k_f}{B}m(t)\right) \exp\left\{j2\pi k_f \int_0^t m(t)dt\right\} \ . \end{aligned} \quad (4.14)$$

The envelope at the output of the differentiator is given by

$$|w(t)| = \pi B a A \left(1 + \frac{2k_f}{B}m(t)\right) \quad (4.15)$$

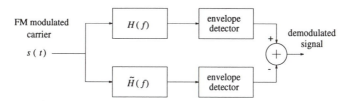

Figure 4.1 Discriminator detector for analog FM.

provided that

$$\left|\frac{2k_f}{B}m(t)\right| < 1 \ . \tag{4.16}$$

To remove the bias of πBaA, a second differentiator can be used having the complex low-pass transfer function

$$\tilde{C}(f) = C(-f) \ . \tag{4.17}$$

The filter $\tilde{C}(f)$ has a slope opposite to $C(f)$, and for the input $v(t)$ it yields an output $\tilde{w}(t)$ having the envelope

$$|\tilde{w}(t)| = \pi BaA\left(1 - \frac{2k_f}{B}m(t)\right) \ . \tag{4.18}$$

The difference between the two envelopes

$$|w(t)| - |\tilde{w}(t)| = 4\pi k_f aAm(t) \tag{4.19}$$

yields an output that is proportional to $m(t)$ and free of bias. A functional diagram of the discriminator detector is shown in Fig. 4.1.

Phase-Locked Loop Detection

A PLL consists of three major components, a loop multiplier, a VCO, and a loop filter, connected in the arrangement shown in Fig. 4.2. There are many excellent references that detail the behavior of PLLs, such as those by Gardner [111] and Viterbi [320]. The VCO is initially adjusted so that the following two conditions occur when the control voltage is zero: i) the frequency of the VCO is equal to the carrier frequency f_c, and ii) the VCO output has a 90° phase shift with respect to the unmodulated carrier waveform. Referring to (4.4), the input to the PLL is

$$s(t) = A\cos\left(2\pi f_c t + \phi_1(t)\right) \tag{4.20}$$

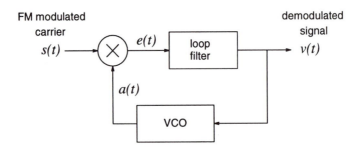

Figure 4.2   Phase-locked loop.

where

$$\phi_1(t) = 2\pi k_f \int_0^t m(t)dt \ . \tag{4.21}$$

For and input signal $v(t)$, the output of the VCO is

$$a(t) = A_v \sin\left(2\pi f_c t + \phi_2(t)\right) \tag{4.22}$$

where

$$\phi_2(t) = 2\pi k_v \int_0^t v(t)dt \tag{4.23}$$

and k_v is the frequency sensitivity of the VCO in Hz/v. The output of the multiplier (or mixer) in Fig. 4.2 is

$$e(t) = -\frac{1}{2}\left\{ k_m A A_v \sin(\phi_e(t)) + k_m A A_v \sin(4\pi f_c t + \phi_1(t) + \phi_2(t))\right\} \tag{4.24}$$

where

$$\phi_e(t) = \phi_1(t) - \phi_2(t) \tag{4.25}$$

is the phase error and k_m is the multiplier gain. The output of the loop filter is

$$v(t) = \int_{-\infty}^{\infty} e(\tau)h(t - \tau)d\tau \tag{4.26}$$

where $h(t)$ is the impulse response of the loop filter. The loop filter is a low-pass filter that eliminates the double frequency component in (4.24). Combining (4.23) to (4.26) yields the following nonlinear integro-differential equation that describes the dynamic behavior of the PLL:

$$\dot\phi_e(t) = \dot\phi_1(t) - \pi k_m k_v A A_v \int_{-\infty}^{\infty} \sin[\phi_e(\tau)]h(t - \tau)d\tau \ . \tag{4.27}$$

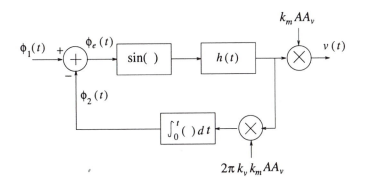

Figure 4.3   Nonlinear phase-locked loop model.

The above equation suggests the nonlinear negative feedback system shown in Fig. 4.3. Unfortunately, the analysis of this nonlinear system is very difficult over the entire range of phase errors. However, when the phase error is close to zero this difficulty can be remedied by making the approximation $\sin(\phi_e(t)) \approx \phi_e(t)$, resulting in the *linear* model shown in Fig. 4.4. When $\phi_e(t) = 0$ the PLL is said to be in phase lock. Under this condition $\phi_2(t) = \phi_1(t)$, and (4.23) and (4.21) imply

$$
\begin{aligned}
v(t) &= \frac{1}{2\pi k_v}\dot\phi_1(t) \\
&= \frac{k_f}{k_v}m(t) \ .
\end{aligned}
\tag{4.28}
$$

In other words, the output of the phase locked loop is the same as the message waveform $m(t)$, except for a scale factor.

Taking Laplace transforms gives the frequency domain representation in Fig. 4.5, from which we obtain the relations

$$
\Phi_e(s) = \Phi_1(s) - \Phi_2(s)
\tag{4.29}
$$

$$
\Phi_2(s) = \frac{2\pi k_v k_m A A_v H(s)\Phi_e(s)}{s}
\tag{4.30}
$$

$$
\frac{\Phi_2(s)}{\Phi_1(s)} = \frac{2\pi k_v k_m A A_v H(s)}{s + 2\pi k_v k_m A A_v H(s)} \ .
\tag{4.31}
$$

From the above equations, two quantities can be obtained; one is the loop phase error

$$
\frac{\Phi_e(s)}{\Phi_1(s)} = \frac{\Phi_1(s) - \Phi_2(s)}{\Phi_1(s)} = 1 - \frac{\Phi_2(s)}{\Phi_1(s)}
$$

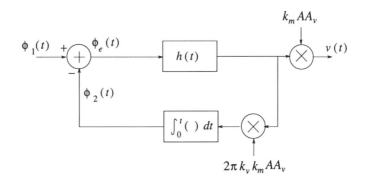

Figure 4.4   Linear phase-locked loop model for small phase errors.

$$= \frac{1}{1 + G(s)} \tag{4.32}$$

where

$$G(s) = \frac{2\pi k_v k_m A A_v H(s)}{s} \tag{4.33}$$

is the open loop transfer function. The other quantity is the closed loop transfer function

$$\frac{V(s)}{\Phi_1(s)} = \frac{k_m A A_v s H(s)}{s + 2\pi k_v k_m A A_v H(s)} . \tag{4.34}$$

From (4.32) it is apparent that if $|G(s)|$ is large over all frequencies of interest, then $\Phi_e(s)$ approaches zero and phase lock is achieved, Moreover, under this condition the output $V(s)$ in (4.34) is

$$V(s) \approx \frac{s\Phi_1(s)}{2\pi k_v} . \tag{4.35}$$

In the time domain the above equation implies that

$$v(t) = \frac{1}{2\pi k_v}\dot{\phi}_1(t) \tag{4.36}$$

Using (4.21) gives

$$v(t) = \frac{k_f}{k_v}m(t) \tag{4.37}$$

which is the exact same result we derived earlier in (4.28).

The performance and complexity of the PLL is determined by the transfer function of the loop filter $H(s)$. The simplest form is when $H(s) = 1$; that is,

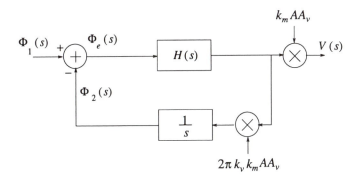

Figure 4.5 Frequency domain representation of linear phase-locked loop model.

no loop filter is used at all. The resulting structure is called a first order PLL. Sometimes, the transfer function is chosen to be a first order filter of the form

$$H(s) = \frac{1 + \tau_1 s}{1 + \tau_2 s} \ . \tag{4.38}$$

This results in a second order PLL.

First Order Phase-Locked Loop:

When $H(s) = 1$, the loop phase error in (4.32) is

$$\frac{\Phi_e(s)}{\Phi_1(s)} = \frac{s}{s + 2\pi k_v k_m A A_v} \tag{4.39}$$

and the closed loop transfer function in (4.34) is

$$\frac{V(s)}{\Phi_1(s)} = \frac{k_m A A_v s}{s + 2\pi k_v k_m A A_v} \ . \tag{4.40}$$

To examine the behavior of the loop, suppose that the message waveform is an impulse, i.e., $m(t) = \delta(t)$. Then $\phi_1(t) = 2\pi k_f u(t)$ and $\Phi_1(s) = 2\pi k_f / s$. From (4.40), The output of the PLL is

$$V(s) = \frac{2\pi k_f k_m A A_v}{s + 2\pi k_v k_m A A_v} \ . \tag{4.41}$$

Since this is indeed the impulse response, it is apparent that the PLL behaves like a first order system and, hence, it is called a first order loop. The 3 dB corner frequency of the output is $k_v k_m A A_v$. Hence, to avoid excessive high frequency roll-off the bandwidth of the message waveform must satisfy $B < k_v k_m A A_v$.

Now suppose that the message waveform is a unit step, i.e., $m(t) = \alpha u(t)$. Then $\phi_1(t) = 2\pi k_f \alpha t u(t)$ and $\Phi_1(s) = 2\pi \alpha k_f / s^2$. The output is

$$V(s) = \frac{2\pi \alpha k_f k_m A A_v}{s(s + 2\pi k_v k_m A A_v)} \tag{4.42}$$

or in the time domain

$$v(t) = \frac{\alpha k_f}{k_v} \left(1 - \exp\left\{-2\pi k_v k_m A A_v t\right\}\right) u(t) \tag{4.43}$$

After all the transients have died out, the desired voltage $v(t) = (\alpha k_f / k_v) u(t)$ is obtained. The phase error for this input is, from (4.32),

$$\Phi_e(s) = \frac{2\pi \alpha k_f}{s(s + 2\pi k_v k_m A A_v)} \tag{4.44}$$

or in the time domain

$$\phi_e(t) = \frac{\alpha k_f}{k_v k_m A A_v} \left(1 - \exp\left\{-2\pi k_v k_m A A_v t\right\}\right) u(t) \ . \tag{4.45}$$

The steady state phase error is

$$\phi_e(\infty) = \frac{\alpha k_f}{k_v k_m A A_v} \ . \tag{4.46}$$

To achieve a small steady state phase error the open loop gain $k_v k_m A A_v$ must be large. More insight can be obtained by using (4.27). At steady state, $\dot{\phi}_e(t) = 0$ so that

$$\sin \phi_e(t) = \frac{2 k_f \alpha}{k_m k_v A A_v} \ . \tag{4.47}$$

Since $|\sin \phi_e(t)| \le 1$, the magnitude of the input step must satisfy the condition $\alpha < k_m k_v A A_v / (2 k_f)$. If this condition is not satisfied the loop falls out of lock.

Finally, when using a PLL demodulator many circuit designers insist on using a limiter before the phase locked loop to remove any amplitude variations that may be present in the received signal. However, Gardner [111] has shown that a limiter is not necessary and has no practical effect on the performance since information is not transmitted in the amplitude of the carrier anyway. Nevertheless, the use of a limiter simplifies the performance analysis of analog FM.

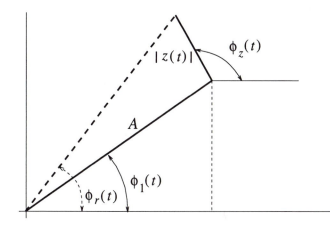

Figure 4.6   Phasor representation of an FM signal received in AWGN.

4.1.2   Performance of Analog FM

Nonfading Channel with AWGN

Consider the transmission of the low-pass FM waveform

$$v(t) = Ae^{\phi_1(t)} \tag{4.48}$$

in the presence of complex additive white Gaussian noise (AWGN)

$$z(t) = z^I(t) + jz^Q(t) = |z(t)|e^{j\phi_z(t)} \tag{4.49}$$

where

$$\phi_z(t) = \text{Tan}^{-1}\left(\frac{z^Q(t)}{z^I(t)}\right) . \tag{4.50}$$

The received low-pass waveform is

$$r(t) = Ae^{\phi_1(t)} + |z(t)|e^{j\phi_z(t)} \tag{4.51}$$

which has the phasor representation shown in Fig. 4.6.

The phase of the received signal is

$$\phi_r(t) = \phi_1(t) + \text{Tan}^{-1}\left\{\frac{|z(t)|\sin[\phi_z(t) - \phi_1(t)]}{A + |z(t)|\cos[\phi_z(t) - \phi_1(t)]}\right\} \tag{4.52}$$

The amplitude of the received signal is of no consequence because information is not transmitted in the amplitude. In order to obtain insightful results, consider the region of high signal-to-noise ratio where $A \gg |z(t)|$. Under this condition

$$\phi_r(t) \approx \phi_1(t) + \frac{|z(t)|}{A} \sin[\phi_z(t) - \phi_1(t)] \; . \tag{4.53}$$

If a PLL detector is used, then the output (c.f. (4.28)) is

$$
\begin{aligned}
v(t) &= \frac{1}{2\pi k_v} \dot{\phi}(t) \\
&= \frac{k_f}{k_v} m(t) + \tilde{z}(t)
\end{aligned}
\tag{4.54}
$$

where the noise term $\tilde{z}(t)$ is

$$\tilde{z}(t) = \frac{1}{2\pi k_v A} \frac{d}{dt} \left\{ |z(t)| \sin[\phi_z(t) - \phi_1(t)] \right\} \; . \tag{4.55}$$

The noise at the output of the PLL can be considered to be independent of the modulating signal. Under this condition the noise term in (4.55) can be written as

$$
\begin{aligned}
\tilde{z}(t) &= \frac{1}{2\pi k_v A} \frac{d}{dt} \left\{ |z(t)| \sin[\phi_z(t)] \right\} \\
&= \frac{1}{2\pi k_v A} \dot{z}^Q(t) \; .
\end{aligned}
\tag{4.56}
$$

Hence, the noise at the PLL output depends on the carrier amplitude A and the quadrature component of the noise $z^Q(t)$. The power spectral density corresponding to (4.56) is

$$S_{\tilde{z}\tilde{z}}(f) = \frac{f^2}{k_v^2 A^2} S_{z^Q z^Q}(f) \tag{4.57}$$

where $S_{z^Q z^Q}(f) = N_o$ is the power spectral density of the white Gaussian noise process $z^Q(t)$. If the output of the PLL is processed by an ideal low-pass filter having a bandwidth equal to the message bandwidth B, then the power spectral density of the noise at the output of the filter is

$$S_{\tilde{z}\tilde{z}}(f) = \begin{cases} \frac{N_o f^2}{k_v^2 A^2} & |f| \le B \\ 0 & \text{otherwise} \end{cases} \; . \tag{4.58}$$

The total noise power is

$$
\begin{aligned}
P_n &= \frac{N_o}{k_v^2 A^2} \int_{-B}^{B} f^2 df \\
&= \frac{2 N_o B^3}{3 k_v^2 A^2} \; .
\end{aligned}
\tag{4.59}
$$

The average signal power at the output of the PLL is

$$P_s = \left(\frac{k_f}{k_v}\right)^2 < m^2(t) > .$$ (4.60)

Therefore, the output signal-to-noise ratio (SNR) is

$$\rho_o = \frac{P_s}{P_n} = \frac{3k_f^2 A^2 < m^2(t) >}{2N_o B^3} .$$ (4.61)

The input carrier-to-noise ratio (CNR) can be defined as

$$\rho_i = \frac{A^2}{2W N_o}$$ (4.62)

where $W N_o$ is the average noise power in the transmission bandwidth of the FM signal.[1] Therefore,

$$\frac{\rho_o}{\rho_i} = \frac{3k_f^2 < m^2(t) > W}{B^3} .$$ (4.63)

Note that the average noise power in (4.59) is inversely proportional to the carrier power, a phenomenon known as the **noise quieting** effect. Also, ρ_o is proportional to the square of the frequency sensitivity of the FM modulator. Since the peak frequency deviation is proportional to the frequency sensitivity and the deviation ratio is equal to the peak frequency deviation divided by the message bandwidth, it follows that ρ_o is proportional to the square of the deviation ratio.

For the particular case of a sinusoidal modulating waveform

$$m(t) = A_m \cos 2\pi f_m t .$$ (4.64)

The peak frequency deviation is $\Delta_f = A_m k_f$ and (4.63) becomes

$$\frac{\rho_o}{\rho_i} = \frac{3\beta^2 W}{2f_m} .$$ (4.65)

Notice that ρ_o is proportional to the square of the modulation index β.

---

[1] The FM demodulator is preceded by an IF filter of bandwidth W to reject the out-of-band noise.

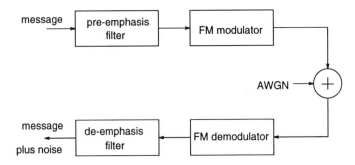

Figure 4.7   Pre-emphasis and de-emphasis filters for noise reduction in FM systems.

Pre-emphasis and De-emphasis

It was shown in the last section that the average noise power at the PLL output depends on the square of the frequency, as shown in (4.59). Unfortunately, the psd of a typical speech waveform $m(t)$ diminishes rapidly at the band edges $\pm B$. Therefore, a significant amount of noise is introduced into the received message waveform to accommodate the relatively weak high frequency content of $m(t)$. One solution is to use pre-emphasis in the transmitter and de-emphasis in the receiver as illustrated in Fig. 4.7. The pre-emphasis filter amplifies the high frequency component of the message waveform prior to FM modulation. After FM demodulation the de-emphasis filter attenuates the high frequency components. The pre-emphasis and de-emphasis filters are chosen so that the message waveform is undistorted. However, the noise power is reduced by the de-emphasis filter.

Let $H_p(f)$ and $H_d(f)$ denote the transfer functions of the pre-emphasis and de-emphasis filters, respectively. Then

$$H_d(f) = \frac{1}{H_p(f)} \ . \tag{4.66}$$

The psd of the noise at the output of the de-emphasis filter is

$$S_{dd}(f) = S_{\tilde{z}\tilde{z}}(f)|H_d(f)|^2 \tag{4.67}$$

where $S_{\tilde{z}\tilde{z}}(f)$ is given by (4.58). Hence the average noise power at the output of the de-emphasis filter is

$$P_{n,d} = \frac{N_o}{k_v^2 A^2} \int_{-B}^{B} f^2 |H_d(f)|^2 df \ . \tag{4.68}$$

Since the message waveform is unaffected, the improvement in the output SNR is the ratio of the noise power without pre-emphasis to the noise power with pre-emphasis

$$\frac{\text{SNR without pre-emphasis}}{\text{SNR with pre-emphasis}} = \frac{2B^3}{3\int_{-B}^{B} f^2 |H_d(f)|^2 \, df} \ . \tag{4.69}$$

Rayleigh Fading and the Threshold Effect

The expression for ρ_o in (4.61) is valid only if ρ_i in (4.62) is large. However, in the region of small ρ_i, the approximation in (4.53) no longer applies. In this case, Rice [268] has derived the output SNR as

$$\rho_o = \frac{3\rho_i(W/2B)^3}{1 + 8\sqrt{3}\rho_i(W/2B)^2 Q(\sqrt{2\rho_i})} \ . \tag{4.70}$$

Rice derived this expression under the following assumptions: i) the modulating waveform is a sinusoid that produces a frequency deviation equal to $\Delta_f = W/2$, ii) the average output signal power is measured in the absence of noise, and iii) the average output noise power is calculated in the presence of an unmodulated carrier. For large ρ_i, ρ_o is approximately $3\rho_i(W/2B)^3$. This is the same value obtained from (4.65) when the frequency deviation is $\Delta_f = W/2$. Fig. 4.8 plots ρ_o in (4.70), clearly showing a **thresholding effect**. Note that ρ_o deviates significantly from a linear function of ρ_i, when $\rho_i < 10$ dB.

If the channel exhibits flat Rayleigh fading, then the received signal amplitude is Rayleigh distributed, so that the input CNR has the exponential distribution

$$p(\rho_i) = \frac{1}{\bar{\rho}_i} e^{-\rho_i/\bar{\rho}_i} \ . \tag{4.71}$$

Using (4.70) and (4.71) gives the average output SNR

$$\bar{\rho}_o = \int_0^\infty \frac{3\rho_i(W/2B)^3}{1 + 8\sqrt{3}\rho_i(W/2B)^2 Q(\sqrt{2\rho_i})} \frac{1}{\bar{\rho}_i} e^{-\rho_i/\bar{\rho}_i} d\rho_i \ . \tag{4.72}$$

This expression is plotted in Fig. 4.9. Note that fading has significantly deteriorated or softened the receiver threshold.

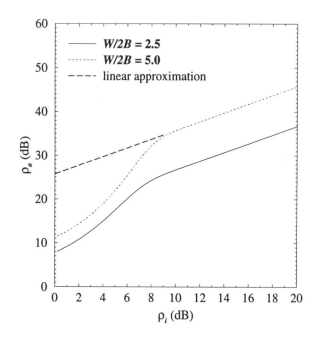

Figure 4.8 Output SNR of an FM demodulator against the input CNR for an AWGN channel.

4.2 BASIC DESCRIPTION OF DIGITALLY MODULATED SIGNALS

With any modulation technique, the band-pass waveform can be expressed in the form

$$s(t) = \text{Re}\left\{v(t)e^{j2\pi f_c t}\right\} \tag{4.73}$$

where $v(t)$ is the **complex envelope**, f_c is the carrier frequency. For any digital modulation scheme, $v(t)$ can be written in the *standard form*

$$v(t) = A\sum_k b(t - kT, \mathbf{x}_k) \tag{4.74}$$

where A is the carrier amplitude, $\mathbf{x}_k = (x_k,\ x_{k-1},\ \ldots,\ x_{k-K})$ is the source symbol sequence, T is the symbol or baud duration, and $b(t, \mathbf{x}_i)$ is an **equivalent shaping function** of duration T. The exact form of $b(t, \mathbf{x}_k)$ and the memory length K depends on the type of modulation that is employed. Several examples are provided later in this chapter where information is transmitted in the amplitude, phase, and frequency of the carrier. In each case, the modulated

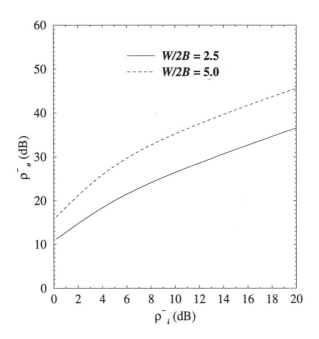

Figure 4.9 Output SNR of an FM demodulator against the input CNR for a Rayleigh fading channel.

signal will be represented in the standard form in (4.74) so as to simplify the task of finding its psd.

By expanding (4.73), the band-pass waveform can also be expressed in the **quadrature form**

$$s(t) = v^I(t)\cos(2\pi f_c t) - v^Q(t)\sin(2\pi f_c t) \tag{4.75}$$

where

$$v(t) = v^I(t) + jv^Q(t) \ . \tag{4.76}$$

The waveforms $v^I(t)$ and $v^Q(t)$ are known as the in-phase and quadrature components, respectively, of $v(t)$. Finally (4.73) can be expressed in the **envelope-phase form**

$$s(t) = a(t)\cos(2\pi f_c t + \phi(t)) \tag{4.77}$$

where

$$a(t) = \sqrt{(v^I(t))^2 + (v^Q(t))^2} \tag{4.78}$$

$$\phi(t) \;\; = \;\; \mathrm{Tan}^{-1} \left[\frac{x^Q(t)}{x^I(t)} \right] \; . \tag{4.79}$$

All three of the above representations will be used interchangeably.

4.2.1   Vector-space Representation

For digital modulation schemes, the complex envelope $v(t)$ during the kth baud $kT \le t \le (k+1)T$ belongs to a finite set of waveforms $\{v_i(t)\}_{i=1}^M$. Consequently, the waveforms $\{v_i(t)\}_{i=1}^M$ can be represented in terms of a set complex of orthonormal basis functions $\{\varphi_n(t)\}_{n=1}^N$ that are defined on the interval $[0, T]^2$. The parameter N is the dimensionality of the vector-space that is needed to represent the signal set and will be defined shortly. The basis functions $\{\varphi_n(t)\}_{n=1}^N$ are orthonormal meaning that

$$\int_0^T \varphi_i(t) \varphi_j^*(t) dt = \delta_{ij} \tag{4.80}$$

where δ_{ij} is the dirac delta function. Expressed in terms of the basis functions

$$v_i(t) = \sum_{m=1}^N v_{im} \varphi_m(t), \;\; i = 1, \, \dots, \, M \tag{4.81}$$

where

$$v_{im} = \int_0^T v_i(t) \varphi_m^*(t) dt \; . \tag{4.82}$$

It follows that the waveforms $\{v_i(t)\}_{i=1}^M$ have the vector representation

$$\mathbf{v}_i = (v_{i1}, \, v_{i2}, \, \dots, \, v_{iN}), \;\; i = 1, \, \dots, M \; . \tag{4.83}$$

We now present one method for obtaining the appropriate set of orthonormal basis functions.

4.2.2   Gram-Schmidt Orthonormalization Procedure

The Gram-Schmidt orthonormalization procedure has two basic steps.

---

[2]The band-pass waveforms $\{s_i(t)\}_{i=1}^M$ can also be represented in terms of a set of *real* orthonormal basis functions that are defined on the interval $[0, T]$.

Step 1: Determine if the set of waveforms $\{v_i(t)\}_{i=1}^M$ is linearly dependent. If so, then there exists a set of complex co-efficients $\{a_i\}_{i=1}^M$, not all zero, such that

$$a_1 v_1(t) + a_2 v_2(t) + \cdots + a_M v_M(t) = 0 \ . \tag{4.84}$$

Suppose, without loss of generality, that $a_M \neq 0^3$. Then

$$v_M(t) = -\left(\frac{a_1}{a_M} v_1(t) + \frac{a_2}{a_M} v_2(t) + \cdots + \frac{a_{M-1}}{a_M} v_M(t) \right) \ . \tag{4.85}$$

Next consider the reduced signal set $\{v_i(t)\}_{i=1}^{M-1}$. If this set of waveforms is linearly dependent, then there exists another set of complex co-efficients $\{b_i\}_{i=1}^{M-1}$, not all zero, such that

$$b_1 v_1(t) + b_2 v_2(t) + \cdots + b_{M-1} v_{M-1}(t) = 0 \ . \tag{4.86}$$

Suppose that $b_{M-1} \neq 0$. Then

$$v_{M-1}(t) = -\left(\frac{b_1}{b_{M-1}} v_1(t) + \frac{b_2}{b_{M-1}} v_2(t) + \cdots + \frac{b_{M-2}}{b_{M-1}} v_M(t) \right) \ . \tag{4.87}$$

We continue in this fashion until a set $\{v_i(t)\}_{i=1}^N$ of linearly independent waveforms is obtained. The value of N defines the dimension of the vector-space, and $N \leq M$ with equality if and only if the original set of waveforms $\{v_i(t)\}_{i=1}^M$ is linearly independent. If $N < M$, then the set of linearly independent waveforms $\{v_i(t)\}_{i=1}^N$ is not unique, but any one will do.

Step 2: From the set $\{v_i(t)\}_{i=1}^N$ construct the set of N orthonormal basis functions $\{\varphi_i(t)\}_{i=1}^N$ as follows. First, let

$$\varphi_1(t) = \frac{v_1(t)}{2E_1} \tag{4.88}$$

where E_1 is the energy in the band-pass waveform $s_1(t)$, given by

$$\begin{aligned} E_1 &= \int_0^T s_1^2(t)dt \\ &= \frac{1}{2} \int_0^T |v_1(t)|^2 dt \ . \end{aligned} \tag{4.89}$$

Then

$$v_1(t) = \sqrt{2E_1}\,\varphi_1(t) = v_{11}\varphi_1(t) \tag{4.90}$$

---

^{3}If $a_M = 0$, then the signal set can be permuted so that $a_M \neq 0$.

where

$$v_{11} = \sqrt{2E_1} \ . \tag{4.91}$$

Next, by using the waveform $v_2(t)$ we obtain

$$v_{21} = \int_0^T v_2(t)\varphi_1^*(t)dt \tag{4.92}$$

along with the intermediate function

$$y_2(t) = v_2(t) - v_{21}\varphi_1(t) \ . \tag{4.93}$$

Note that $y_2(t)$ is orthogonal to $\varphi_1(t)$ over the interval $[0, T]$. Hence, the second basis function is

$$
\begin{aligned}
\varphi_2(t) &= \frac{y_2(t)}{\sqrt{\int_0^T |y_2(t)|^2 dt}} \tag{4.94} \\
&= \frac{v_2(t) - v_{21}\varphi_1(t)}{\sqrt{2E_2 - |v_{21}|^2}} \ . \tag{4.95}
\end{aligned}
$$

Continuing in the above fashion, we define the ith intermediate function

$$y_i(t) = v_i(t) - \sum_{j=1}^{i-1} v_{ij}\varphi_j(t) \tag{4.96}$$

where

$$v_{ij} = \int_0^T v_i(t)\varphi_j^*(t)dt \ . \tag{4.97}$$

The set of functions

$$\varphi_i(t) = \frac{y_i(t)}{\sqrt{\int_0^T |y_i(t)|^2}} \qquad i = 1, \ 2, \ ,\ldots, \ N \tag{4.98}$$

is the required set of orthonormal basis functions.

4.2.3 Signal Energy and Correlations

Consider the set of band-pass waveforms

$$s_m(t) = \text{Re}\left\{ v_m(t)e^{j2\pi f_c t} \right\} \ , \quad m = 1, 2, \ldots, M \tag{4.99}$$

defined on the interval $[0, T]$ and their vector representations $\{\mathbf{s}_m\}_{m=1}^M$. By using the band-pass equivalent of (4.81), the energy in the waveform $s_m(t)$ is

$$
\begin{aligned}
E_m &= \int_0^T s_m^2(t)dt \\
&= \int_0^T \left(\sum_{i=1}^N s_{mi}\varphi_i(t) \right)^2 dt \\
&= \sum_{i=1}^N s_{mi}^2 = \|\mathbf{s}_m\|^2 .
\end{aligned}
\tag{4.100}
$$

Using the identity $\text{Re}\{z\} = (z + z^*)/2$, the energy in the waveform $s_m(t)$ is also equal to

$$
\begin{aligned}
E_m &= \frac{1}{2}\int_0^T \|v_m(t)\|^2 dt \\
&\quad + \frac{1}{4}\int_0^T \|v_m(t)\|^2 \cos(4\pi f_c t + 2\phi(t))dt \\
&\approx \frac{1}{2}\int_0^T \|v_m(t)\|^2 dt = \frac{1}{2}\|\mathbf{v}_m\|^2 .
\end{aligned}
\tag{4.101}
$$

The above approximation is only valid when $f_c T \gg 1$, i.e., the bandwidth of the complex envelope is much less than the carrier frequency so that the double frequency term can be neglected. Stated another way, the approximation is valid when the symbol duration T contains many cycles of the carrier.

The correlation between the waveforms $s_m(t)$ and $s_k(t)$ is

$$
\begin{aligned}
\rho_{km} &= \frac{1}{\sqrt{E_k E_m}} \int_0^T s_m(t)s_k(t)dt \\
&= \frac{1}{\sqrt{E_k E_m}} \int_0^T \sum_{n=1}^N s_{mn}\varphi_n(t) \sum_{i=1}^N s_{ki}\varphi_i(t)dt \\
&= \frac{1}{\sqrt{E_k E_m}} \sum_{n=1}^N s_{mn}s_{kn} \\
&= \frac{\mathbf{s}_m \cdot \mathbf{s}_k}{\|\mathbf{s}_m\|\|\mathbf{s}_k\|} .
\end{aligned}
\tag{4.102}
$$

Again, if $f_c T \gg 1$, then the correlation can be expressed in terms of the low-pass signal vectors as

$$
\rho_{km} = \text{Re}\left\{ \frac{\mathbf{v}_m \cdot \mathbf{v}_k^*}{\|\mathbf{v}_m\|\|\mathbf{v}_k\|} \right\} .
\tag{4.103}
$$

Finally, the squared Euclidean distance between $s_k(t)$ and $s_m(t)$ is

$$
\begin{aligned}
d_{km}^2 &= \|\mathbf{s}_m - \mathbf{s}_k\|^2 \\
&= \|\mathbf{s}_m\|^2 - \|\mathbf{s}_k\|^2 - 2\mathbf{s}_m \cdot \mathbf{s}_k \\
&= E_m + E_k - 2\sqrt{E_m E_k}\rho_{mk} \ .
\end{aligned}
\tag{4.104}
$$

If all the $s_i(t)$ have energy equal to E, then the squared Euclidean distance is

$$
d_{km}^2 = 2E\left(1 - \rho_{km}\right) \ .
\tag{4.105}
$$

4.3   AMPLITUDE SHIFT KEYING (ASK)

Amplitude shift keying (ASK) is a linear modulation technique with nonconstant envelope. ASK has the advantage of being more spectrally efficient than other modulation schemes having constant envelope. However, amplifier nonlinearities will degrade the performance of ASK.

With ASK, the complex envelope is

$$
v(t) = A \sum_n x_n h_a(t - nT) \ .
\tag{4.106}
$$

where $\{x_n\} = \{x_n^I + jx_n^Q\}$ is source symbol sequence, and $h_a(t)$ is an **amplitude shaping pulse**. Expressed in standard from, the complex envelope is

$$
v(t) = A \sum_k b(t - kT, \mathbf{x}_k)
\tag{4.107}
$$

where

$$
b(t, \mathbf{x}_k) = x_k h_a(t) \ .
\tag{4.108}
$$

The ASK band-pass signal has the quadrature representation

$$
s(t) = A \sum_n \left\{ x_n^I h_a(t - nT) \cos 2\pi f_c t - x_n^Q h_a(t - nT) \sin 2\pi f_c t \right\}
\tag{4.109}
$$

and the envelope-phase representation

$$
s(t) = A \sum_n |x_n| h_a(t - nT) \cos(2\pi f_c t + \arg(x_n))
\tag{4.110}
$$

where

$$|x_n| = \sqrt{(x_n^I)^2 + j(x_n^Q)^2} \qquad (4.111)$$

$$\arg(x_n) = \mathrm{Tan}^{-1}\left(\frac{x_n^Q}{x_n^I}\right) . \qquad (4.112)$$

It is apparent that both the amplitude and the phase of an ASK signal depend on the complex symbol.

Signal Space Representation:

The appropriate choice of basis functions for the real ASK band-pass signals can be obtained from (4.109) as

$$\varphi_1(t) = \sqrt{\frac{2}{T}}h_a(t)\cos(2\pi f_c t) \qquad (4.113)$$

$$\varphi_2(t) = -\sqrt{\frac{2}{T}}h_a(t)\sin(2\pi f_c t) . \qquad (4.114)$$

Using these basis functions

$$s_i(t) = \sqrt{E_A}x_i^I\varphi_1(t) + \sqrt{E_A}x_i^Q\varphi_2(t) , \qquad 0 \le t \le T \qquad (4.115)$$

where $E_A = \frac{A^2 T}{2}$ and $x_i^I, x_i^Q \in \{\pm 1, \pm 3, \ldots\}$. In terms of E_A, the energy of symbol x_m is $E_m = E_A|x_m|^2$ and the average symbol energy is $E_{av} = E_A \mathrm{E}[|x_m|^2]$. A very popular form of ASK is M-ary quadrature amplitude modulation (M-QAM), where the source symbols are chosen from an M-ary constellation such that $x_i = x_i^I + jx_i^Q$ with $x_i^I, x_i^Q \in \{\pm 1, \pm 3, \ldots, \pm(N-1)\}$ and $N = \sqrt{M}$. The signal-space diagram for 16-QAM is shown in Fig. 4.10.

It is also possible to construct a signal-space diagram for the complex low-pass ASK signals. An appropriate basis function is

$$\varphi_1(t) = \sqrt{\frac{1}{T}}h_a(t) . \qquad (4.116)$$

Using this basis function

$$v_i(t) = \sqrt{2E_A}x_i\varphi_1(t) \qquad 0 \le t \le T . \qquad (4.117)$$

Note that only one basis function is needed, because each basis function defines a 2-D complex plane. The signal-space diagram for 16-QAM is shown in Fig. 4.11 and is identical to the one shown in Fig. 4.10 except that the signal points are $\sqrt{2}$ further apart.

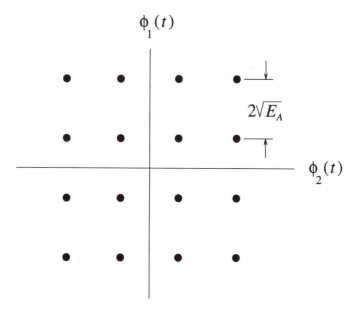

Figure 4.10   Real signal-space diagram for 16-QAM signals.

4.4   ORTHOGONAL FREQUENCY
DIVISION MULTIPLEXING (OFDM)

Orthogonal frequency division multiplexing (OFDM) is a modulation technique that has been recently suggested for use in cellular radio [29], digital audio broadcasting [99], and digital video broadcasting. OFDM is to block modulation scheme where source symbols are transmitted in parallel by employing a (large) number of orthogonal sub-carriers. A block of N serial source symbols, each of duration T_s, is converted into a block of N parallel modulated symbols, each with duration $T = NT_s$. The block length N is chosen so that $NT_s \gg \sigma_\tau$, where σ_τ is the rms delay spread of the channel. Since the symbol rate on each sub-carrier is much less than the serial source rate, the effects of delay spread are greatly reduced. This has practical advantages, because it may reduce or even eliminate the need for equalization. Although the block length N is chosen so that $NT_s \gg \sigma_\tau$, the channel dispersion will still cause consecutive blocks to overlap. This results in some residual ISI that will degrade the performance. This residual ISI can be eliminated at the expense of channel capacity by using **guard intervals** between the blocks that are at least as long as the maximum expected channel impulse response.

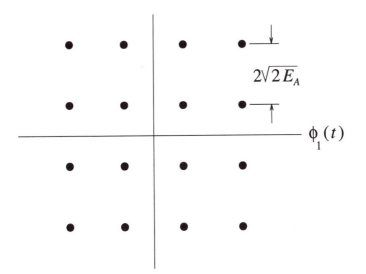

Figure 4.11 Complex signal-space diagram for 16-QAM signals.

The complex envelope of an OFDM signal is described by

$$v(t) = A \sum_{k} \sum_{n=0}^{N-1} x_{k,n} \phi_n(t - kT) \tag{4.118}$$

where the orthogonal waveforms $\{\phi_n(t)\}$ are chosen as

$$\phi_n(t) = h_a(t) \exp\left\{ j \frac{2\pi \left(n - \frac{N-1}{2}\right) t}{T} \right\}, \quad n = 1, \ldots, N-1 . \tag{4.119}$$

If a rectangular amplitude shaping pulse $h_a(t) = u_T(t)$ is chosen, then the frequency separation of the sub-carriers, $1/T$, ensures that the sub-carriers are orthogonal and phase continuity is maintained from one symbol to the next, but is twice the minimum required for orthogonality with coherent detection [256]. At epoch k, N source symbols are transmitted by using the N distinct pulses. The source symbols $x_{k,n}$ are often chosen from a QAM constellation.

The OFDM signal can also be expressed in the standard form

$$v(t) = A \sum_{k} b(t - kT, \mathbf{x}_k) \tag{4.120}$$

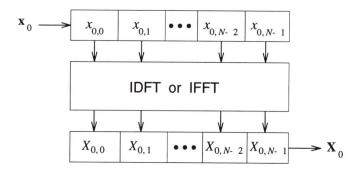

Figure 4.12   Block diagram of OFDM transmitter using IDFT or IFFT.

where

$$b(t, \mathbf{x}_k) = h_a(t) \sum_{n=0}^{N-1} x_{k,n} \exp\left\{ j \frac{2\pi \left(n - \frac{N-1}{2}\right) t}{T} \right\} . \qquad (4.121)$$

and $\mathbf{x}_k = \{x_{k,0}, \ x_{k,1}, \ \ldots, \ x_{k,N-1}\}$ is the source symbol block at epoch k.

A key advantage of using OFDM is that the modulation can be achieved in the discrete-domain by using an inverse discrete Fourier transform (IDFT), or the more computationally efficient inverse fast Fourier transform (IFFT). Considering the source symbol block at epoch $k = 0$, ignoring the frequency offset $\exp\left\{-j\frac{2\pi(N-1)t}{2T}\right\}$, and assuming $h_a(t) = U_T(t)$, the complex envelope has the form

$$v(t) = A \sum_{n=0}^{N-1} x_{0,n} \exp\left\{ \frac{j2\pi nt}{NT_s} \right\} , \quad 0 \le t \le T . \qquad (4.122)$$

The complex envelope is sampled at epochs $t = kT_s$ to yield the sequence

$$X_{0,k} = v(kT_s) = A \sum_{n=0}^{N-1} x_{0,n} \exp\left\{ \frac{j2\pi nk}{N} \right\} , \quad k = 0, 1, \ldots, N-1 . \qquad (4.123)$$

The samples are then passed through a D/A converter and carrier modulated. Note that the sampled envelope has duration NT_s and the samples $\mathbf{X}_0 = \{X_{0,i}\}_{i=0}^{N-1}$ are the just the IFFT of the block $A\mathbf{x}_0$. This leads to the basic OFDM transmitter shown in Fig. 4.12.

To mitigate the effects of the channel delay spread, a guard interval consisting of a cyclic suffix of length G can be added to the sequence $\mathbf{X}_0$. The transmitted sequence with guard interval is

$$X_{0,k}^g = A \sum_{n=0}^{N-1} x_{0,n} \exp\left\{\frac{j2\pi nk}{N}\right\} , \quad k = 0, 1, \ldots, N+G-1 . \quad (4.124)$$

To avoid a reduction in data rate, the source symbol duration with guard interval is $T_s^g = T_s/(1 + \frac{G}{N})$.

4.5 PHASE SHIFT KEYING (PSK)

The M-PSK complex envelope can be written in the standard form

$$v(t) = A \sum_k b(t - kT, \mathbf{x}_k) \quad (4.125)$$

where

$$b(t, \mathbf{x}_k) = h_a(t) \exp\left\{j\frac{\pi}{M}x_k h_p(t)\right\} \quad (4.126)$$

$h_p(t)$ is the **phase shaping pulse**, $h_a(t)$ is the amplitude shaping pulse, and M is the alphabet size. The source symbols are chosen from the set $x_n \in \{\pm1, \pm3, \ldots, \pm(M-1)\}$. The signal-space diagram for 8-PSK is shown in Fig. 4.13, where $E = A^2T/2$ is the symbol energy.

The phase shaping pulse is normally chosen to be the rectangular pulse $h_p(t) = u_T(t)$. The amplitude shaping pulse is very often chosen to be a **square root raised cosine** pulse, where the Fourier transform of $h_a(t)$ is

$$H_a(f) = \begin{cases} \sqrt{T} & 0 \leq |f| \leq (1-\beta)/2T \\ \sqrt{\frac{T}{2}\left[1 - \sin\frac{\pi T}{\beta}\left(f - \frac{1}{2T}\right)\right]} & (1-\beta)/2T \leq |f| \leq (1+\beta)/2T \end{cases}$$

$$(4.127)$$

If receiver implements the same filter, then the overall pulse has the raised cosine spectrum $|H_a(f)|^2$. If the channel is affected by flat fading and additive white Gaussian noise, then an equal partition of the filtering operations between the transmitter and receiver optimizes the SNR at the output of the receiver filter at the sampling instants [355]. The **roll-off factor** β usually lies between 0 and 1, and defines the **excess bandwidth** $100\beta\%$. Using a smaller β results in a more compact power density spectrum, but the link performance becomes more sensitive to errors in symbol timing.

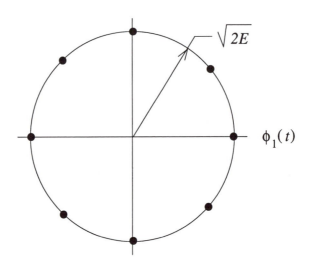

Figure 4.13 Complex signal-space diagram for 8-PSK signals.

The time domain pulse $h_a(t)$, obtained by taking the inverse Fourier transform of $H_a(f)$, is

$$h_a(t) = 4\beta \, \frac{\cos[(1+\beta)\pi t/T] + \sin[(1-\beta)\pi t/T](4\beta t/T)^{-1}}{\pi\sqrt{T}[1 - 16\beta^2 t^2/T^2]} \, . \qquad (4.128)$$

A typical square root raised cosine pulse with a roll-off factor of $\beta = 0.5$ is shown in Fig. 4.14. Strictly speaking the pulse in (4.128) is noncausal, but in practice a truncated time domain pulse can be used. For example, in Fig. 4.14 the pulse is truncated to $6T$ and time shifted by $3T$ to yield a causal pulse.

4.6 $\pi/4$-DQPSK

Unlike QPSK that uses 4 carrier phases, $\pi/4$-DQPSK uses 8 carrier phases and information is transmitted in the differential carrier phase. Let θ_n be the absolute carrier phase for the nth symbol, and let $\Delta\theta_n = \theta_n - \theta_{n-1}$ be the differential carrier phase. With $\pi/4$-DQPSK, the differential phase is related to the quaternary source sequence $\{x_n\}$, $x_n \in \{\pm 1, \pm 3\}$ through the mapping

$$\Delta\theta_n = \begin{cases} -3\pi/4 & , \quad x_n = -3 \\ -\pi/4 & , \quad x_n = -1 \\ +\pi/4 & , \quad x_n = +1 \\ +3\pi/4 & , \quad x_n = +3 \end{cases} \qquad (4.129)$$

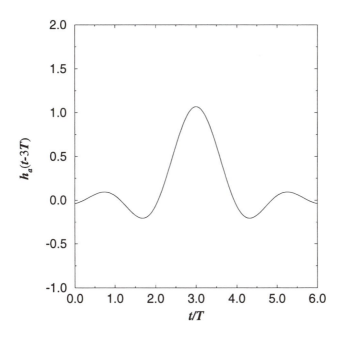

Figure 4.14 Square root raised cosine pulse with roll-off factor $\beta = 0.5$.

Since the source sequence $\{x_n\}$ is random the above mapping is arbitrary, except that the phase differences must be $\pm\pi/4$ and $\pm 3\pi/4$. The differential phase with the mapping in (4.129) can be expressed in the convenient algebraic form

$$\Delta\theta_n = x_n \frac{\pi}{4} \ . \tag{4.130}$$

This algebraic expression allows write the complex envelope of the $\pi/4$-DQPSK signal in the standard form

$$v(t) = A \sum_k b(t - kT, \mathbf{x}_k) \tag{4.131}$$

where

$$
\begin{aligned}
b(t, \mathbf{x}_k) &= h_a(t) \exp\left\{ j \left(\theta_{k-1} + x_k \frac{\pi}{4} \right) \right\} \\
&= h_a(t) \exp\left\{ j \frac{\pi}{4} \left(\sum_{n=-\infty}^{k-1} x_n + x_k \right) \right\} \ .
\end{aligned}
\tag{4.132}
$$

The summation in the exponent represents the accumulated carrier phase, while the last term is the phase change due to the kth symbol.

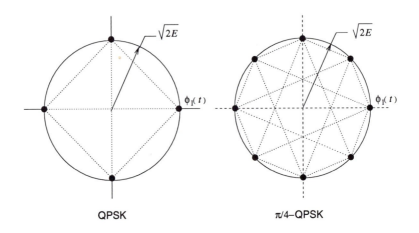

QPSK   π/4–QPSK

Figure 4.15   Complex signal-space diagram QPSK and π/4-DQPSK signals.

The signal-space diagrams for QPSK and $\pi/4$-DQPSK are shown in Fig. 4.15, where $E = A^2T/2$ is the symbol energy. The dotted lines in Fig. 4.15 show the allowable phase transitions. The exact phase trajectories depend on the shaping functions. The phase shaping pulse with $\pi/4$-DQPSK is $h_p(t) = u_T(t)$ and the amplitude shaping pulse $h_a(t)$ is often chosen to be the square root raised cosine pulse in (4.128). The phase diagram for $\pi/4$-DQPSK with square root raised cosine amplitude pulse shaping is shown in Fig. 4.16. Note that the phase trajectories do not pass through the origin. This property reduces the envelope fluctuations of the signal, making it less susceptible to amplifier nonlinearities. It also reduces the dynamic range that is required of the power amplifier.   Finally, we observe that the carrier phase of $\pi/4$-DQPSK changes by $\pm\pi/4$ or $\pm 3\pi/4$ radians during *every* symbol period. This property makes symbol synchronization is easier with $\pi/4$-DQPSK as compared to QPSK.

4.7   CONTINUOUS PHASE MODULATION (CPM)

Some of the modulation techniques used in mobile radio systems are special cases of the broad class of frequency modulation techniques called continuous phase modulation (CPM), where the carrier phase varies in a continuous manner. A comprehensive treatment of CPM is provided in Anderson *et. al.* [8]. CPM schemes are attractive because they have constant envelope and excellent

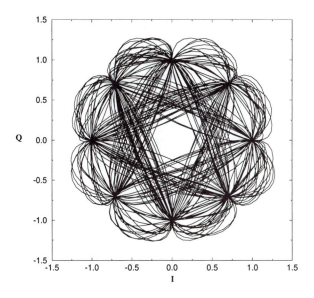

Figure 4.16 Phase diagram for $\pi/4$-DQPSK with square root raised cosine amplitude pulse shaping; $\beta = 0.5$.

spectral characteristics. The complex envelope of any CPM signal has the form

$$
\begin{aligned}
v(t) &= A \exp\left\{ j2\pi k_f \int_{-\infty}^{t} \sum_{n} x_n h_f(\tau - nT)d\tau \right\} \\
&\equiv A \exp\{j\phi(t)\}
\end{aligned}
\tag{4.133}
$$

where A is the amplitude, k_f is the **peak frequency deviation**, $h_f(t)$ is the **frequency shaping pulse**, $\{x_n\}$ is the source symbol sequence, and T is the symbol duration. The elements of the symbol sequence are chosen from the alphabet $\mathcal{X} = \{\pm 1, \pm 3, \cdots, \pm(M-1)\}$, where M is the alphabet size. The instantaneous frequency deviation from the carrier is

$$
f_{\text{dev}}(t) = \frac{1}{2\pi}\dot{\phi}(t) = k_f \sum_{n} x_n h_f(t - nT) \ .
\tag{4.134}
$$

During the time interval $kT \leq t \leq (k+1)T$, the phase $\phi(t)$ is

$$
\phi(t) = 2\pi k_f \int_{-\infty}^{kT} \sum_{n=-\infty}^{k-1} x_n h_f(\tau - nT)d\tau + 2\pi k_f \int_{kT}^{t} x_k h_f(\tau - kT)d\tau \ .
\tag{4.135}
$$

Sometimes $\phi(t)$ is called the **excess phase** . Note that the excess phase is continuous as long as the shaping function $h_f(t)$ does not contain impulses.

The frequency shaping pulse $h_f(t)$ can assume a variety of forms. A **full response** shaping function has a duration equal to the symbol period T, while a **partial response** shaping function has a duration greater than T. This section considers full response shaping functions, while the discussion of partial response shaping function is deferred to Section 4.9. Some possible full response frequency shaping pulses are

$$
h_f(t) = \begin{cases}
u_T(t) & \text{non-return-to-zero (NRZ)} \\[2mm]
\sin(\pi t/T)u_T(t) & \text{half sinusoid (HS)} \\[2mm]
\frac{1}{2}\left[1 - \cos\left(\frac{2\pi t}{T}\right)\right] u_T(t) & \text{raised cosine (RC)} \\[2mm]
1 - \frac{|t - T/2|}{T/2} & \text{triangular (T)}
\end{cases}
\tag{4.136}
$$

Since (4.135) represents the excess phase for the interval $kT \leq t \leq (k+1)T$, all such intervals can be combined together to write $v(t)$ in the standard form

$$
v(t) = A \sum_k b(t - kT, \mathbf{x}_k)
\tag{4.137}
$$

where

$$
b(t, \mathbf{x}_k) = \exp\left\{ j\left(\beta(T) \sum_{n=-\infty}^{k-1} x_n + x_k \beta(t)\right)\right\} u_T(t)
\tag{4.138}
$$

and

$$
\beta(t) = \begin{cases}
0 & , \ t < 0 \\
2\pi k_f \int_0^t h_f(\tau)d\tau & , \ 0 \leq t \leq T \\
\beta(T) & , \ t \geq T
\end{cases}
\tag{4.139}
$$

The first term in the exponent of (4.138) represents the accumulated excess phase, while the second term is the excess phase trajectory for the current source symbol. The **average frequency deviation** of a full response CPM waveform is

$$
\overline{k_f} = k_f \frac{1}{T} \int_0^T h_f(t)dt
\tag{4.140}
$$

and the **modulation index** is

$$
\begin{aligned}
h &= \frac{\beta(T)}{\pi} \\
&= 2k_f T \cdot \frac{1}{T} \int_0^T h_f(t) dt \\
&= 2\overline{k_f} T \ .
\end{aligned}
\tag{4.141}
$$

Observe that an infinite variety of CPM signals can be generated by choosing different frequency shaping pulses $h_f(t)$, modulation indices h, and alphabet sizes M. For example, **continuous phase frequency shift keying** (CPFSK) is a special type of CPM that is obtained by using the rectangular shaping function $h_f(t) = u_T(t)$. For CPFSK

$$
\overline{k_f} = k_f
\tag{4.142}
$$

$$
h = 2k_f T
\tag{4.143}
$$

$$
\beta(t) = \begin{cases} 0 & , \ t < 0 \\ 2\pi k_f t = \pi h t / T & , \ 0 \le t \le T \\ 2\pi k_f T = \pi h & , \ t \ge T \end{cases} \ .
\tag{4.144}
$$

CPM signals can be conveniently described by sketching the excess phase

$$
\phi(t) = \beta(T) \sum_{n=-\infty}^{k-1} x_n + x_k \beta(t - kT)
\tag{4.145}
$$

for all possible symbol sequences $\{x_k\}$. This plot is called a **phase tree**, and a typical phase tree is shown in Fig. 4.17 for binary CPFSK. Since the shaping function is rectangular, the phase trajectories of CPFSK are linear but with a discontinuous first derivative. In general, a more compact power density spectrum is obtained by using shaping functions that create a phase trajectory with continuous higher order derivatives.

The modulation index is often restricted to be a rational number, $h = m/p$, where m and p are integers that have no common factors. This constraint ensures that the number of **phase states** is finite which has practical implications for some receiver designs. If m is even, then a full response CPM signal has the phase states

$$
\Theta_s = \left\{ 0, \ \frac{\pi m}{p}, \ \frac{2\pi m}{p}, \ \cdots, \ \frac{(p-1)\pi m}{p} \right\}
\tag{4.146}
$$

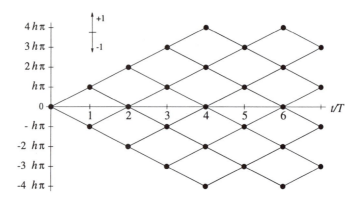

Figure 4.17 Phase tree of binary CPFSK with an arbitrary modulation index. CPFSK is characterized by linear phase trajectories.

while if m is odd the phase states are

$$\Theta_s = \left\{ 0, \ \frac{\pi m}{p}, \ \frac{2\pi m}{p}, \ \cdots, \ \frac{(2p-1)\pi m}{p} \right\} \ . \tag{4.147}$$

Hence, there are p phase states for even m, while there are $2p$ phase states for odd m. For example, if $h = 1/4$, then the CPM signal has the phase states

$$\Theta_s = \left\{ 0, \ \frac{\pi}{4}, \ \frac{\pi}{2}, \ \frac{3\pi}{4}, \ \pi, \ \frac{5\pi}{4}, \ \frac{3\pi}{2}, \ \frac{7\pi}{4} \right\} \ . \tag{4.148}$$

CPM signals cannot be represented as discrete points in a signal-space, because the phase is continuous and time-varying. Nevertheless, a signal-space diagram can still be constructed. Given the CPM complex envelope in (4.133), it follows that an appropriate choice for the basis function is

$$\varphi_1(t) = \sqrt{\frac{1}{T}} u_T(t)$$

so that

$$v(t) = \sqrt{2\mathcal{E}} \exp\left\{ j\phi(t) \right\} \varphi_1(t) \ , \quad kT \le t \le (k+1)T \tag{4.149}$$

where $\mathcal{E} = A^2 T/2$. Fig. 4.18 shows the signal-space diagram for binary CPM with $h = 1/4$. Since binary modulation is used, the phase trajectories are only to adjacent phase states as shown by the dotted lines in Fig. 4.18.

Multi-h CPM is another special type of CPM, where the modulation index for each symbol interval is cyclically selected from the set $\{h_1, h_2, \ldots, h_H\}$ of H

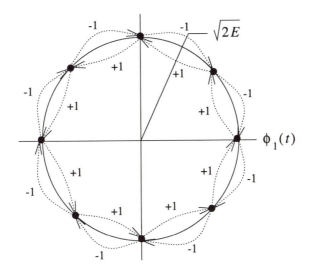

Figure 4.18   Complex signal-space diagram for binary CPM signals with $h = 1/4$.

rational numbers. That is

$$h_i = h_{i+H} \; . \tag{4.150}$$

Multi-h CPM waveforms are attractive because they often have better spectral properties than single h CPM, i.e., a narrower main lobe and faster roll-off of side lobes.

Example 4.1_____

Fig. 4.19 shows the phase trellis for binary multi-h CPM with the rectangular shaping function $h_s(t) = u_T(t)$ and modulation indices $h_n = 1/4$ and $h_{n+1} = 3/4$.

_____

4.8   MINIMUM SHIFT KEYING (MSK)

Minimum shift keying (MSK) is a special case of binary CPFSK, with modulation index $h = 1/2$. For MSK, (4.144) becomes

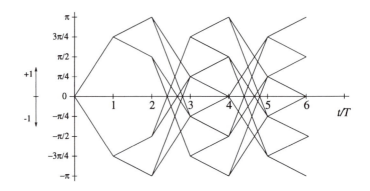

Figure 4.19   Phase trellis of multi-h CPFSK.

$$\beta(t) = \begin{cases} 0 & , \ t < 0 \\ 2\pi k_f t = \pi t/2T & , \ 0 \le t \le T \\ 2\pi k_f T = \pi/2 & , \ t \ge T \end{cases} \qquad (4.151)$$

The *carrier* phase, $\phi_i(t) = 2\pi f_c t + \phi(t)$, during the time interval $kT \le t \le (k+1)T$ is

$$\begin{aligned} \phi_i(t) &= 2\pi f_c t + \frac{\pi}{2} \sum_{n=-\infty}^{k-1} x_n + \frac{\pi}{2} x_k \frac{t - kT}{T} \\ &= 2\pi \left(f_c + \frac{\pi x_k}{2T} \right) t + \frac{\pi}{2} \sum_{n=-\infty}^{k-1} x_n - \frac{\pi k}{2} x_k \ . \end{aligned} \qquad (4.152)$$

The excess phase of an MSK signal can be described in terms of the phase trellis and signal-space diagrams shown in Figs. 4.20 and 4.21, respectively. In Fig. 4.20, we assume $\phi(0) = 0$.

The MSK band-pass waveform is

$$s(t) = A \cos \left(\left(2\pi f_c + \frac{\pi x_k}{2T} \right) t + \frac{\pi}{2} \sum_{n=-\infty}^{k-1} x_n - \frac{\pi k}{2} x_k \right) \ , \quad kT \le t \le (k+1)T$$

$$(4.153)$$

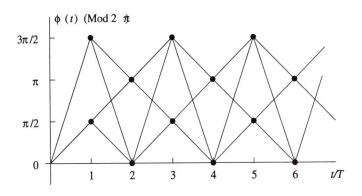

Figure 4.20   Phase-trellis diagram for MSK.

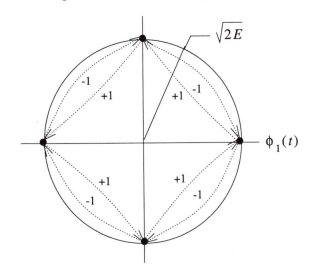

Figure 4.21   Complex signal-space diagram for MSK signals.

From (4.153, we note that an MSK signal has one of two possible frequencies in the time interval $kT \leq t \leq (k + 1)T$. These frequencies are

$$f_L = f_c - \frac{1}{4T} \qquad \text{and} \qquad f_U = f_c + \frac{1}{4T} \ . \qquad (4.154)$$

The difference between these frequencies is $\Delta f = f_U - f_L = 1/(2T)$. This is the minimum frequency separation to ensure orthogonality between two sinusoids of duration T with coherent demodulation [256], hence the name *minimum* shift keying. By applying various trigonometric identities to (4.153) we can write

$$s(t) = A \left[x_k^R h_a(t - k2T) \cos(2\pi f_c t) \right. \tag{4.155}$$
$$\left. - x_k^I h_a(t - k2T - T) \sin(2\pi f_c t) \right], \ kT \leq t \leq (k+1)T$$

where

$$x_k^R = -x_{k-1}^I x_{2k-1} \tag{4.156}$$

$$x_k^I = x_k^R x_{2k} \tag{4.157}$$

$$h_a(t) = \cos\left(\frac{\pi t}{2T}\right), \quad -T \leq t \leq T . \tag{4.158}$$

Note that $\{x_k^R\}$ and $\{x_k^I\}$ are independent binary symbol sequences that take on elements from the set $x_k^R, x_k^I \in \{-1, +1\}$, and the half sinusoid amplitude shaping pulse $h_a(t)$ has duration $2T$ and $h_a(t - T) = \sin(\pi t/(2T)), 0 \leq t \leq 2T$. Therefore, MSK is mathematically equivalent to offset quadrature amplitude shift keying (OQASK) with a half-sinusoid amplitude shaping pulse.

4.9 PARTIAL RESPONSE CPM

Partial response CPM signals are a broad class of signals characterized by a frequency shaping pulse $h_f(t)$ of duration greater than T[4]. For example, if $h_f(t)$ has duration KT, then

$$h_f(t) = h_f(t) u_{KT}(t) \tag{4.159}$$

where

$$u_{KT}(t) = \sum_{k=0}^{K-1} u_T(t - kT) . \tag{4.160}$$

The advantage of using partial response CPM is an improvement in the spectral characteristics of the modulated signals by providing both a narrower main lobe and faster roll-off of side lobes.

The partial response shaping function can be written in the form

$$h_f(t) = \sum_{k=0}^{K-1} h_f(t) u_T(t - kT)$$

---

[4] Linear partial response schemes can be defined similarly by using an amplitude shaping pulse $h_a(t)$ of duration greater than T.

$$= \sum_{k=0}^{K-1} h_{f,k}(t - kT) \qquad (4.161)$$

where

$$h_{f,k}(t) = h_f(t + kT)u_T(t) . \qquad (4.162)$$

An equivalent shaping function of duration T can be derived by noting that the baseband modulating signal has the form

$$
\begin{aligned}
x(t) &= \sum_n x_n h_f(t - nT) = \sum_n \sum_{k=0}^{K-1} x_n h_{f,k}(t - (n+k)T) \\
&= \sum_m \sum_{k=0}^{K-1} x_{m-k} h_{f,k}(t - mT) .
\end{aligned}
\qquad (4.163)
$$

It follows that

$$x(t) = \sum_m h_f(t - mT, \mathbf{x}_m) \qquad (4.164)$$

where

$$h_f(t - mT, \mathbf{x}_m) = \sum_{k=0}^{K-1} x_{m-k} h_{f,k}(t - mT) \qquad (4.165)$$

and

$$\mathbf{x}_m = (x_m, x_{m-1}, \ldots, x_{m-K+1}) . \qquad (4.166)$$

Therefore, the partial response shaping function, $h_f(t, x_n)$, of duration KT can been replaced by an equivalent shaping function $h_f(t, \mathbf{x}_m)$ of duration T that depends on the current and $K - 1$ past source symbols.

Example 4.2 KREC Shaping────────────────────────

For KREC signaling , $h_f(t) = u_{KT}(t)$. Hence,

$$h_f(t, \mathbf{x}_n) = x_n h_{f,0}(t) + x_{n-1} h_{f,1}(t) + \cdots + x_{n-K+1} h_{f,K-1}$$

where

$$h_{f,0}(t) = h_{f,1}(t) = \cdots = h_{f,K-1} = u_T(t) .$$

Therefore,

$$h_f(t, \mathbf{x}_n) = (x_n + x_{n-1} + \cdots + x_{n-K+1})u_T(t) .$$

Example 4.3 KRC Shaping————————————————————————————

For KRC signaling ,

$$h_f(t) = \left[1 - \cos\left(\frac{2\pi t}{KT} \right) \right] u_{KT}(t) \ .$$

Hence,

$$h_f(t, \mathbf{x}_n) = x_n h_{f,0}(t) + x_{n-1} h_{f,1}(t) + \cdots + x_{n-K} h_{f,K-1}(t)$$

where

$$h_{f,k}(t) = \left[1 - \cos\left(\frac{2\pi(t + KT)}{KT} \right) \right] u_T(t) \ .$$

——

It follows from the above development that a partial response CPM signal can be written in the standard form

$$v(t) = A \sum_k b(t - kT, \mathbf{x}_k) \tag{4.167}$$

where

$$b(t, \mathbf{x}_k) = \exp\left\{ j \left(\sum_{i=0}^{k-1} \beta(T, \mathbf{x}_i) + \beta(t, \mathbf{x}_k) \right) \right\} u_T(t) \tag{4.168}$$

and

$$\beta(t, \mathbf{x}_k) = \begin{cases} 0 & , \ t < 0 \\ 2\pi k_f \int_0^t h_f(\tau, \mathbf{x}_k) d\tau & , \ 0 \le t \le T \\ \beta(T, \mathbf{x}_k) & , \ t \ge T \end{cases} \tag{4.169}$$

4.9.1   Gaussian Minimum Shift Keying (GMSK)

It will be shown in Section 4.10 that MSK has all the desirable attributes for mobile radio, except for a compact power density spectrum. This can be alleviated by low-pass filtering the modulating signal

$$x(t) = \sum_n x_n u_T(t - nT) \tag{4.170}$$

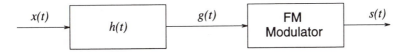

Figure 4.22 Pre-modulation filtered MSK. The modulating signal is low-pass filtered to remove the high frequency components prior to modulation.

prior to modulation, as shown in Fig. 4.22. Such filtering removes the higher frequency components in $x(t)$ and results in a more compact spectrum. In general, the low-pass premodulation filter is chosen to have the following three properties:

- narrow bandwidth and a sharp transition band.

- low overshoot impulse response.

- preservation of the output pulse area to ensure a phase shift of $\pi/2$.

GMSK is a special type of partial response CPM that uses a low-pass premodulation filter having the transfer function

$$H(f) = \exp\left\{-\left(\frac{f}{B}\right)^2 \frac{\ln 2}{2}\right\} \qquad (4.171)$$

where B is the bandwidth of the filter. It is apparent that $H(f)$ is bell shaped about $f = 0$, hence the name Gaussian MSK. A rectangular pulse $\text{rect}(t/T) = u_T(t + T/2)$ transmitted through this filter yields the frequency shaping pulse

$$h_f(t) = \sqrt{\frac{2\pi}{\ln 2}}(BT) \int_{t-T/2}^{t+T/2} \exp\left\{-\frac{2\pi^2(BT)^2 x^2}{\ln 2}\right\} dx \ . \qquad (4.172)$$

The excess phase change over the time interval from $-T/2$ to $T/2$ is

$$\phi(T/2) - \phi(-T/2) = x_0\beta_0(T) + \sum_{\substack{n=-\infty \\ n \neq 0}}^{\infty} x_n\beta_n(T) \qquad (4.173)$$

where

$$\beta_n(T) = \frac{\pi h}{\int_{-\infty}^{\infty} h_f(\nu) d\nu} \int_{-T/2-nT}^{T/2-nT} h_f(\nu) \, d\nu \ . \qquad (4.174)$$

The first term in (4.173) is the desired term, and the second term is the intersymbol interference (ISI) introduced by the premodulation filter. For the

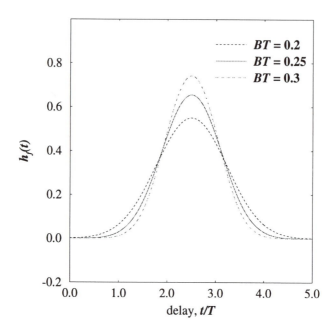

Figure 4.23 GMSK frequency shaping pulse for various normalized filter bandwidths BT.

version of GMSK used in the GSM system, $x_n \in \{-1, +1\}$ and $h = 1/2$ is chosen so that a total phase shift of $\pi/2$ is maintained.

Fig. 4.23 plots a GMSK frequency shaping pulse (truncated to $5T$ and time shifted by $2.5T$) for various normalized filter bandwidths BT. Because the frequency shaping pulse has a duration greater than T, ISI is introduced. As BT decreases, the ISI increases. Thus, while a smaller value of BT results in a more compact power density spectrum, the induced ISI will degrade the bit error rate performance. This leads to a tradeoff in the choice of BT. Some studies have indicated that $BT = 0.25$ provides a good tradeoff for cellular radio systems [226].

4.9.2   Tamed Frequency Modulation (TFM)

Tamed frequency modulation (TFM) is a special type of partial response binary CPM that was introduced by Jager and Dekker [153]. To define TFM signals, recall that the excess carrier phase for MSK obeys the difference equation

$$\phi(nT + T) - \phi(nT) = x_n \frac{\pi}{2} \ . \tag{4.175}$$

For TFM, the excess phase trajectory is smoothed by using the partial response condition

$$\phi(nT + T) - \phi(nT) = \frac{\pi}{2} \left(\frac{x_{n-1}}{4} + \frac{x_n}{2} + \frac{x_{n+1}}{4} \right) \ . \tag{4.176}$$

The maximum excess phase change over any bit interval is equal to $\pi/2$. To complete the definition of the TFM signal, an appropriate premodulation filter must be defined. If the premodulation filter has impulse response $h_f(t)$, then the excess phase can be written as

$$\phi(t) = \sum_{m=-\infty}^{\infty} x_m \beta(t - mT) + \phi_o \tag{4.177}$$

where

$$\beta(t) = 2\pi k_f \int_{-\infty}^{t} h_f(t)dt \tag{4.178}$$

and ϕ_o is an arbitrary constant. The excess phase change over the time interval $(nT, (n + 1)T)$ is

$$\begin{aligned}
\phi(nT + T) - \phi(nT) &= \sum_{m=-\infty}^{\infty} x_m \left(\beta(nT + T - mT) - \beta(nT - mT) \right) \\
&= \sum_{\ell=-\infty}^{\infty} x_{n-\ell} \left(\beta(\ell T + T) - \beta(\ell T) \right) \ . \tag{4.179}
\end{aligned}$$

Expanding (4.176) in more detail gives

$$\phi(nT+T)-\phi(nT) = \frac{\pi}{2} \left(\ldots + x_{n-2} \cdot 0 + \frac{x_{n-1}}{4} + \frac{x_n}{2} + \frac{x_{n+1}}{4} + x_{n+2} \cdot 0 + \ldots \right) . \tag{4.180}$$

Comparing (4.179) and (4.180) gives the condition

$$\beta(\ell T + T) - \beta(\ell T) = \begin{cases} \frac{\pi}{8} & , \ \ell = 1 \\ \frac{\pi}{4} & , \ \ell = 0 \\ \frac{\pi}{8} & , \ \ell = -1 \\ 0 & , \ \text{otherwise} \end{cases} \ . \tag{4.181}$$

From the definition of $\beta(t)$ in (4.178) the above equation leads to

$$\int_{\ell T}^{(\ell+1)T} h_f(t)dt = \begin{cases} \frac{\pi}{8} & , \ |\ell| = 1 \\ \frac{\pi}{4} & , \ \ell = 0 \\ 0 & , \ \text{otherwise} \end{cases} \ . \tag{4.182}$$

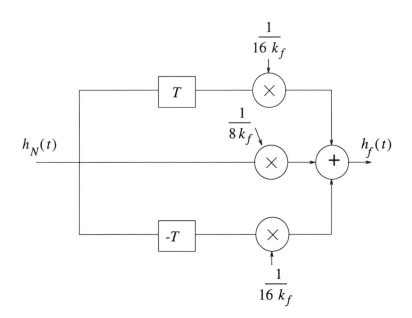

Figure 4.24 Filter to generate a TFM frequency shaping pulse.

One possibility is to use a single pulse $h_N(t)$ that satisfies the third Nyquist criterion [234], [245]

$$\int_{(2\ell-1)T/2}^{(2\ell+1)T/2} h_N(t)dt = \left\{ \begin{array}{ll} 1 & , \quad \ell = 0 \\ 0 & , \quad \ell \neq 0 \end{array} \right. \tag{4.183}$$

and generate $h_f(t)$ by using scaling and delay operations through the filter shown in Fig. 4.24. The transfer function of this filter is

$$\begin{aligned} S(f) &= \frac{1}{16k_f}e^{-j2\pi fT} + \frac{1}{8k_f} + \frac{1}{16k_f}e^{j2\pi fT} \\ &= \frac{1}{4k_f}\cos^2(\pi fT) \ . \end{aligned} \tag{4.184}$$

The overall pulse $h_f(t)$ has the form

$$\begin{aligned} H_f(f) &= H_N(f)S(f) \\ &= H_N(f)\frac{1}{4k_f}\cos^2(\pi fT) \ . \end{aligned} \tag{4.185}$$

The filter $S(f)$ ensures that the phase constraint in (4.176) is satisfied. However, $H_N(f)$ determines the shape of the phase trajectories and, hence, can influence the TFM power density spectrum. In general, $H_N(f)$ has the form

$$H_N(f) = \frac{\pi f T}{\sin(\pi f T)} N_1(f) \qquad (4.186)$$

where $N_1(f)$ is the Fourier transform of a pulse that satisfies the first Nyquist criterion [234], [245]. One example is the raised cosine pulse $|H_a(f)|^2$ where $H_a(f)$ is defined in (4.127). Consider, for example, the raised cosine pulse with $\beta = 0$, i.e.,

$$N_1(f) = \begin{cases} 1 & , \quad 0 \leq |f| \leq 1/2T \\ 0 & , \quad \text{otherwise} \end{cases} \qquad (4.187)$$

Using (4.185)–(4.187) gives

$$H_f(f) = \frac{1}{4k_f} \frac{\pi f T}{\sin(\pi f T)} \cos^2(\pi f T) \ . \qquad (4.188)$$

The corresponding frequency shaping pulse $\pi k_f h_f(t)$ is plotted in Fig. 4.25.

Generalized tamed frequency modulation (GTFM) is an extension of TFM where the phase difference has the form

$$\phi_i(nT + T) - \phi_i(nT) = \frac{\pi}{2} \left(a x_{n-1} + b x_n + a x_{n+1} \right) \ . \qquad (4.189)$$

The constants a and b satisfy the condition $2a + b = 1$ so that the maximum change in $\phi_i(t)$ during one symbol period is restricted to $\pm \pi/2$. A large class of signals can be constructed by varying the value of b and by varying the pulse response $\tilde{\beta}(t)$. TFM is a special case of GTFM where $b = 0.5$.

4.10 POWER SPECTRAL DENSITIES OF DIGITALLY MODULATED SIGNALS

A digitally modulated signal can be written in the generic form

$$
\begin{aligned}
s(t) &= \text{Re} \left\{ v(t) e^{j(2\pi f_c t + \phi_T)} \right\} \\
&= \frac{1}{2} \left\{ v(t) e^{j(2\pi f_c t = \phi_T)} + v^*(t) e^{-j(2\pi f_c t = \phi_T)} \right\}
\end{aligned} \qquad (4.190)
$$

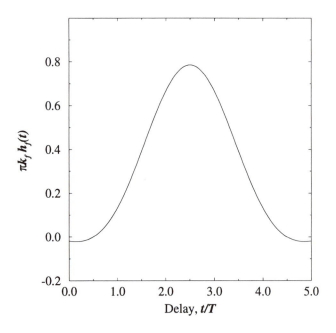

Figure 4.25   TFM frequency shaping pulse.

where ϕ_T is a random phase uniformly distributed over $(-\pi, \pi)$ The autocorrelation function of $s(t)$ is

$$
\begin{aligned}
\phi_{ss}(\tau) &= \mathrm{E}\left[s(t+\tau)s(t)\right] \\
&= \frac{1}{4}\mathrm{E}\left[\left(v(t+\tau)e^{j(2\pi f_c t + 2\pi f_c \tau + \phi_T)} + v^*(t+\tau)e^{-j(2\pi f_c t + 2\pi f_c \tau + \phi_T)}\right)\right. \\
&\qquad\left. \times \left(v(t)e^{j(2\pi f_c t + \phi_T)} + v^*(t)e^{-j(2\pi f_c t + \phi_T)}\right)\right] \\
&= \frac{1}{4}\mathrm{E}\left[v(t+\tau)v(t)e^{j(4\pi f_c t + 2\pi f_c \tau + 2\phi_T)} + v^*(t+\tau)v(t)e^{-j2\pi f_c \tau}\right. \\
&\qquad\left. + v(t+\tau)v^*(t)e^{j2\pi f_c \tau} + v^*(t+\tau)v^*(t)e^{-j(4\pi f_c t + 2\pi f_c \tau + 2\phi_T)}\right] \ .
\end{aligned}
$$

$$(4.191)$$

To proceed further, we note that

$$
\mathrm{E}\left[e^{\pm j(4\pi f_c t + 2\pi f_c \tau + 2\phi_T)}\right] = 0 \ .
$$

$$(4.192)$$

where E[·] denotes the ensemble average. Using this result,

$$\phi_{ss}(\tau) = \frac{1}{2}\phi_{vv}^*(\tau)e^{-j2\pi f_c\tau} + \frac{1}{2}\phi_{vv}(\tau)e^{j2\pi f_c\tau} \quad . \tag{4.193}$$

Finally, the power density spectrum is the Fourier transform of $\phi_{ss}(\tau)$, i.e.,

$$S_{ss}(f) = \frac{1}{2}[S_{vv}(f - f_c) + S_{vv}^*(-f - f_c)] \tag{4.194}$$

where $S_{vv}(f)$ is the power density spectrum of the complex low-pass signal. However, $S_{vv}(f)$ is real, even though $v(t)$ and $\phi_{vv}(\tau)$ are complex; this property follows from the fact that $\phi_{vv}(\tau) = \phi_{vv}^*(-\tau)$ as shown in Appendix A. Therefore,

$$S_{ss}(f) = \frac{1}{2}[S_{vv}(f - f_c) + S_{vv}(f + f_c)] \quad . \tag{4.195}$$

From the above expression, it is apparent that the psd of the band-pass waveform $s(t)$ is completely determined by the psd of its complex envelope $v(t)$.

4.10.1 Psd of a Complex Envelope

We have seen that the complex envelope of any digitally modulated signal can be expressed in the form

$$v(t) = A\sum_k b(t - kT, \mathbf{x}_k) \quad . \tag{4.196}$$

The autocorrelation of $v(t)$ is

$$\begin{aligned}
\phi_{vv}(t + \tau, t) &= \frac{1}{2}\mathrm{E}\left[v(t + \tau)v^*(t)\right] \tag{4.197}\\
&= \frac{A^2}{2}\sum_i\sum_k \mathrm{E}\left[b(t + \tau - iT, \mathbf{x}_i)b^*(t - kT, \mathbf{x}_k)\right] \quad .
\end{aligned}$$

Observe that $v(t)$ is a cyclostationary random process, meaning that the autocorrelation function $\phi_{vv}(t + \tau, t)$ is periodic in t with period T. To see this property, first note that

$$\begin{aligned}
\phi_{vv}(t + T + \tau, t + T) &= \frac{A^2}{2}\sum_i\sum_k \mathrm{E}\left[b(t + T + \tau - iT, \mathbf{x}_i)b^*(t + T - kT, \mathbf{x}_k)\right]\\
&= \frac{A^2}{2}\sum_{i'}\sum_{k'} \mathrm{E}\left[b(t + \tau - i'T, \mathbf{x}_{i'+1})b^*(t - k'T, \mathbf{x}_{k'+1})\right] \quad .
\end{aligned}$$

$$\tag{4.198}$$

Under the assumption that the source sequence is a stationary random process we can write

$$\phi_{vv}(t+T+\tau, t+T) = \frac{A^2}{2} \sum_{i'} \sum_{k'} \mathrm{E}\left[b(t+\tau-i'T, \mathbf{x}_{i'})b^*(t-k'T, \mathbf{x}_{k'})\right]$$

$$= \phi_{vv}(t+\tau, t) . \tag{4.199}$$

Since $v(t)$ is cyclostationary, the autocorrelation $\phi_{vv}(\tau)$ can be obtained by taking the time average of $\phi_{vv}(t+\tau, t)$, given by

$$\phi_{vv}(\tau) = \, <\phi_{vv}(t+\tau, t)>$$

$$= \frac{A^2}{2} \sum_{i} \sum_{k} \frac{1}{T} \int_{0}^{T} \mathrm{E}\left[b(t+\tau-iT, \mathbf{x}_i)b^*(t-kT, \mathbf{x}_k)\right] dt$$

$$= \frac{A^2}{2T} \sum_{i} \sum_{k} \int_{-kT}^{-kT+T} \mathrm{E}\left[b(z+\tau-(i-k)T, \mathbf{x}_i)b^*(z, \mathbf{x}_k)\right] dz$$

$$= \cdot \frac{A^2}{2T} \sum_{m} \sum_{k} \int_{-kT}^{-kT+T} \mathrm{E}\left[b(z+\tau-mT, \mathbf{x}_{m+k})b^*(z, \mathbf{x}_k)\right] dz$$

$$= \frac{A^2}{2T} \sum_{m} \sum_{k} \int_{-kT}^{-kT+T} \mathrm{E}\left[b(z+\tau-mT, \mathbf{x}_{m})b^*(z, \mathbf{x}_0)\right] dz$$

$$= \frac{A^2}{2T} \sum_{m} \int_{-\infty}^{\infty} \mathrm{E}\left[b(z+\tau-mT, \mathbf{x}_{m})b^*(z, \mathbf{x}_0)\right] dz . \tag{4.200}$$

where $\langle \cdot \rangle$ denotes time averaging. The psd of $v(t)$ is

$$S_{vv}(f) = \mathrm{E}\left[\frac{A^2}{2T} \sum_{m} \int_{-\infty}^{\infty} \int_{-\infty}^{\infty} b(z+\tau-mT, \mathbf{x}_{m})b^*(z, \mathbf{x}_0)dze^{-j2\pi f\tau}d\tau\right]$$

$$= \mathrm{E}\left[\frac{A^2}{2T} \sum_{m} \int_{-\infty}^{\infty} b(z+\tau-mT, \mathbf{x}_{m})e^{-j2\pi f(z+\tau-mT)}d\tau \right.$$

$$\left. \times \int_{-\infty}^{\infty} b^*(z, \mathbf{x}_0)e^{j2\pi fz}dze^{-j2\pi fmT}\right]$$

$$= \mathrm{E}\left[\frac{A^2}{2T} \sum_{m} \int_{-\infty}^{\infty} b(\tau', \mathbf{x}_{m})e^{-j2\pi f\tau'}d\tau' \right.$$

$$\left. \times \int_{-\infty}^{\infty} b^*(z, \mathbf{x}_0)e^{j2\pi fz}dze^{-j2\pi fmT}\right]$$

$$= \frac{A^2}{2T} \sum_{m} \mathrm{E}\left[B(f, \mathbf{x}_{m})B^*(f, \mathbf{x}_0)\right] e^{-j2\pi fmT} \tag{4.201}$$

where $B(f, \mathbf{x}_m)$ is the Fourier transform of $b(t, \mathbf{x}_m)$. To express the power density spectrum in a more convenient form, let

$$S_{b,m}(f) = \frac{1}{2} \mathrm{E}\left[B(f, \mathbf{x}_m)B^*(f, \mathbf{x}_0)\right] \ . \tag{4.202}$$

Then

$$S_{vv}(f) = \frac{A^2}{T} \sum_m S_{b,m}(f) e^{-j2\pi f mT} \ . \tag{4.203}$$

Note that the psd in (4.203) depends on the correlation properties of the source sequence $\mathbf{x}$ and the form of the equivalent pulse shaping function $b(t, \mathbf{x}_m)$. Now suppose that the source characteristics are such that $\mathbf{x}_m$ and $\mathbf{x}_0$ are uncorrelated for $|m| \geq K$. Then

$$S_{b,m}(f) = S_{b,K}(f) \tag{4.204}$$

where

$$
\begin{aligned}
S_{b,K}(f) &= \frac{1}{2} \mathrm{E}\left[B(f, \mathbf{x}_m)\right] \mathrm{E}\left[B^*(f, \mathbf{x}_0)\right] & |m| > K \\
&= \frac{1}{2} \mathrm{E}\left[B(f, \mathbf{x}_0)\right] \mathrm{E}\left[B^*(f, \mathbf{x}_0)\right] & |m| > K \\
&= \frac{1}{2} \left|\mathrm{E}\left[B(f, \mathbf{x}_0)\right]\right|^2 \ , & |m| \geq K \ .
\end{aligned}
\tag{4.205}
$$

It follows that

$$S_{vv}(f) = S_{vv}^c(f) + S_{vv}^d(f) \tag{4.206}$$

where

$$
\begin{aligned}
S_{vv}^c(f) &= \frac{A^2}{T} \sum_{|m|<K} \left(S_{b,m}(f) - S_{b,K}(f)\right) e^{-j2\pi f mT} \\
S_{vv}^d(f) &= \frac{A^2}{T} S_{b,K}(f) \sum_m e^{-j2\pi f mT} \ .
\end{aligned}
\tag{4.207}
$$

The terms $S_{vv}^c(f)$ and $S_{vv}^d(f)$ represent the continuous and discrete portions of the psd. To see more clearly that $S_{vv}^d(f)$ represents the discrete portion, we use the identity

$$T \sum_m e^{-j2\pi f mT} = \sum_n \delta\left(f - \frac{n}{T}\right) \tag{4.208}$$

to write

$$S_{vv_d}(f) = \left(\frac{A}{T}\right)^2 S_{b,K}(f) \sum_n \delta\left(f - \frac{n}{T}\right) \ . \tag{4.209}$$

Finally, by using the property $S_{b,-m}(f) = S_{b,m}^*(f)$, the continuous portion of the psd can be written as

$$
\begin{aligned}
S_{vv_c}(f) &= \frac{A^2}{T}\left(S_{b,0}(f) - S_{b,K}(f)\right) \\
&\quad + \frac{A^2}{T}\sum_{m=1}^{K}\left\{\left(S_{b,m}(f) - S_{b,K}(f)\right)e^{-j2\pi fmT}\right. \\
&\qquad\qquad \left. + \left(S_{b,m}^*(f) - S_{b,K}(f)\right)e^{j2\pi fmT}\right\} \\
&= \frac{A^2}{T}\left(S_{b,0}(f) - S_{b,K}(f)\right) \\
&\quad + \frac{A^2}{T}2\mathrm{Re}\left\{\sum_{m=1}^{K}\left(S_{b,m}(f) - S_{b,K}(f)\right)e^{-j2\pi fmT}\right\} \;. \quad (4.210)
\end{aligned}
$$

Note that the ensemble average and Fourier transform are interchangeable linear operators. Therefore, if the complex envelope $v(t)$ has zero mean, i.e., $E[b(t, \mathbf{x}_0)] = 0$, then $E[B(f, \mathbf{x}_0)] = 0$. Under this condition

$$
S_{b,K}(f) = \frac{1}{2}\left|E[B(f, \mathbf{x}_0)]\right|^2 = 0 \;. \quad (4.211)
$$

Hence, if $b(t, \mathbf{x}_0)$ has zero mean, then $S_{vv}(f)$ contains no discrete components and $S_{vv}(f) = S_{vv}^c(f)$. Conversely, if $b(t, \mathbf{x}_0)$ has nonzero mean, then $S_{vv}(f)$ will contain discrete components.

Alternative Method

An alternative method of computing the psd is as follows. From the first line in (4.201)

$$
\begin{aligned}
S_{vv}(f) &= E\left[\frac{A^2}{2T}\sum_m \int_{-\infty}^{\infty}\int_{-\infty}^{\infty} b(z + \tau - mT, \mathbf{x}_m)b^*(z, \mathbf{x}_0)dz\,e^{-j2\pi f\tau}d\tau\right] \\
&= \frac{A^2}{2T}\sum_m \int_{-\infty}^{\infty}\int_{-\infty}^{\infty} E\left[b(\tau', \mathbf{x}_m)b^*(z, \mathbf{x}_0)\right] \\
&\qquad\qquad \times e^{-j2\pi f(\tau' - z)}dz\,d\tau'\,e^{-j2\pi fmT} \quad (4.212)
\end{aligned}
$$

Therefore, $S_{b,m}(f)$ is given by the double Fourier transform

$$
S_{b,m}(f) = \int_{-\infty}^{\infty}\int_{-\infty}^{\infty} \phi_{b,m}(\tau', z)e^{-j2\pi f(\tau' - z)}dz\,d\tau' \;. \quad (4.213)
$$

where

$$\phi_{b,m}(\tau', z) = \frac{1}{2} \mathrm{E}\left[b(\tau', \mathbf{x}_m)b^*(z, \mathbf{x}_0)\right] \ . \tag{4.214}$$

Uncorrelated Source Symbols

For the special case of uncorrelated source symbols

$$b(t, \mathbf{x}_m) = b(t, x_m) \tag{4.215}$$

so that $b(t, \mathbf{x}_m)$ depends on one symbol only. In this case,

$$S_{b,0}(f) = \frac{1}{2} \mathrm{E}\left[|B(f, x_0)|^2\right] \tag{4.216}$$

$$S_{b,m}(f) = \frac{1}{2} |\mathrm{E}\left[B(f, x_0)\right]|^2 \qquad |m| \geq 1 \ . \tag{4.217}$$

Hence, $S_{vv}(f)$ is given by (4.206), where

$$S_{vv_d}(f) = \frac{A^2}{T^2} S_{b,1}(f) \sum_n \delta\left(f - \frac{n}{T}\right) \tag{4.218}$$

$$S_{vv_c}(f) = \frac{A^2}{T}\left(S_{b,0}(f) - S_{b,1}(f)\right) \ . \tag{4.219}$$

Once again, if $b(t, x_m)$ has zero mean, then $S_{b,1}(f) = 0$ (no discrete spectral components) and

$$S_{vv}(f) = \frac{A^2}{T} S_{b,0}(f) \ . \tag{4.220}$$

Linear Full Response Modulation

For linear full response modulation schemes, $b(t, \mathbf{x}_k) = x_k h_a(t)$ and $B(f, \mathbf{x}_k) = x_k H_a(f)$. From (4.202)

$$S_{b,m}(f) = \phi_{xx}(m) |H_a(f)|^2 \ . \tag{4.221}$$

Hence, from (4.203) the psd of the complex envelope is

$$S_{vv}(f) = \frac{A^2}{T} |H_a(f)|^2 S_{xx}(f) \tag{4.222}$$

where

$$S_{xx}(f) = \sum_m \phi_{xx}(n)e^{-j2\pi fmT} \ . \tag{4.223}$$

With uncorrelated source symbols

$$S_{b,0}(f) = \sigma_x^2 |H_a(f)|^2 \tag{4.224}$$

$$S_{b,m}(f) = \frac{1}{2} |\mu_x|^2 |H_a(f)|^2 , \quad |m| \geq 1 . \tag{4.225}$$

where $\mu_x = \mathrm{E}[x_m]$. The psd $S_{vv}(f)$ is then given by (4.206), (4.218), and (4.219). If $\mu_x = 0$, then $S_{b,1}(f) = 0$ and

$$S_{vv}(f) = \frac{A^2}{T} \sigma_x^2 |H_a(f)|^2 . \tag{4.226}$$

Linear Partial Response Modulation

For linear partial response modulation schemes

$$b(t, \mathbf{x}_m) = h_a(t, \mathbf{x}_m)$$
$$= \sum_{k=0}^{K-1} x_{m-k} h_{a,k}(t) \tag{4.227}$$

and

$$B(f, \mathbf{x}_m) = \sum_{k=0}^{K-1} x_{m-k} H_{a,k}(f) . \tag{4.228}$$

From (4.202),

$$S_{b,m}(f) = \frac{1}{2} \mathrm{E} \left[\sum_{k=0}^{K-1} x_{m-k} H_{a,k}(f) \sum_{\ell=0}^{K-1} x_{-\ell}^* H_{a,\ell}^*(f) \right]$$
$$= \sum_{k=0}^{K-1} \sum_{\ell=0}^{K-1} \frac{1}{2} \phi_{xx}(m - k + \ell) H_{a,k}(f) H_{a,\ell}^*(f) . \tag{4.229}$$

With uncorrelated source symbols, $\phi_{xx}(m - k + \ell) = \sigma_x^2 \delta(m - k + \ell)$. Hence,

$$S_{b,m}(f) = \sigma_x^2 \sum_{\ell=0}^{K-1} H_{a,m+\ell}(f) H_{a,\ell}^*(f) . \tag{4.230}$$

Example 4.4 Duobinary Signaling———————————————————

For duobinary signaling, $K = 2$ and $h_{a,0}(t) = h_{a,1}(t) = \text{Sa}(\pi t/T)$ and $H_{a,0}(f) = H_{a,1}(f) = T\text{rect}(fT)$. With uncorrelated source symbols

$$S_{b,m}(f) = \frac{1}{2}\text{E}\left[\left(x_0^* H_{a,0}^*(f) + x_{-1}^* H_{a,1}^*(f)\right)\left(x_m H_{a,0}(f) + x_{m-1} H_{a,1}(f)\right)\right]$$

$$= \begin{cases} 2\sigma_x^2 T^2 \text{rect}(fT) & , \quad m = 0 \\ \sigma_x^2 T^2 \text{rect}(fT) & , \quad m = -1, 1 \\ 0 & , \quad \text{elsewhere} \end{cases}$$

and from (4.203)

$$S_{vv}(f) = 4A^2 T\sigma_x^2 \cos^2(\pi fT)\text{rect}(fT) \ . \tag{4.231}$$

Example 4.5 Modified Duobinary Signaling———————————————————

For modified duobinary signaling, $K = 3$ and $h_{a,0}(t) = h_{a,2}(t) = \text{Sa}(\pi t/T)$ and $h_{a,1}(t) = 0$. With uncorrelated source symbols,

$$S_{b,m}(f) = \begin{cases} 2\sigma_x^2 T^2 \text{rect}(fT) & , \quad m = 0 \\ -\sigma_x^2 T^2 \text{rect}(fT) & , \quad m = -2, 2 \\ 0 & , \quad \text{elsewhere} \end{cases}$$

and from (4.203)

$$S_{vv}(f) = 4A^2 T\sigma_x^2 \sin^2(2\pi fT)\text{rect}(fT) \ .$$

4.10.2   Psd of ASK

The psd of ASK with uncorrelated zero-mean source symbols is given by (4.226). With square root raised cosine pulse shaping, $H_a(f)$ has the form defined in (4.127) with $h_a(t)$ in (4.128). In practice, the pulse $h_a(t)$ is truncated to length τ yielding the new pulse $\tilde{h}_a(t) = h_a(t)\text{rect}(t/\tau)$. The Fourier transform of the pulse $\tilde{h}_a(t)$ is $\tilde{H}_a(f) = H_a(f) * \tau\text{sinc}(f\tau)$, where $*$ denotes the operation of convolution. The psd of ASK with the pulse $\tilde{h}_a(t)$ is again obtained from (4.226). As shown in Fig. 4.26, pulse truncation leads to side lobe regeneration.

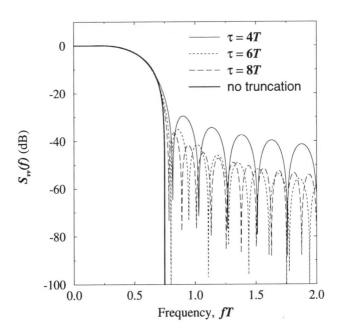

Figure 4.26 Psd of ASK with a truncated square root raised cosine pulse with various truncation lengths; $\beta = 0.5$.

4.10.3   Psd of OFDM

The psd of an OFDM signal can be obtained by treating OFDM as independent modulation on orthogonal sub-carriers that are separated in frequency by $1/T$. For a signal constellation with zero mean and the amplitude shaping pulse $h_a(t)$, the psd of the complex envelope is

$$S_{vv}(f) = \frac{A^2}{T}\sigma_x^2 \sum_{n=0}^{N-1} \left| H_a\left(f - \frac{1}{T}\left(n - \frac{N-1}{2} \right) \right) \right|^2 \qquad (4.232)$$

where, $\sigma_x^2 = \frac{1}{2}E[|x_{k,n}|^2]$.

As an example, the psd of OFDM with the rectangular amplitude shaping pulse $h_a(t) = u_T(t)$ $(H_a(f) = \text{Sa}(\pi fT))$ is shown for $N = 4$ and $N = 32$ in Figs. 4.27 and 4.28, respectively. For large values of N the psd becomes more flat in the $N/T = 1/T_s$ bandwidth containing containing the sub-carriers, and only the sub-carriers near the band edge contribute to out of band power. Therefore, as the block length N becomes large, the spectral efficiency approaches that of

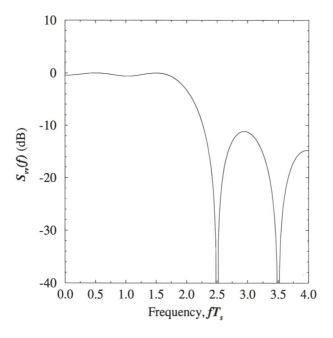

Figure 4.27 Psd of OFDM with $N = 4$.

single carrier modulation with ideal Nyquist filtering. This is a significant advantage offered by OFDM. Further improvement in the psd can be obtained by using a square root raised cosine pulse shaping, as shown in Fig. 4.29. However, square root raised cosine pulse shaping destroys the sub-carrier orthogonality which leads to a degraded error rate performance.

4.10.4   Psd of PSK

For PSK signals with the uncorrelated source symbols and the equivalent shaping function in (4.126), the psd is given by (4.206) and (4.216)–(4.219). To evaluate the psd, define

$$m_b(t) = \mathrm{E}[b(t, \mathbf{x}_k)] \ . \tag{4.233}$$

Then

$$S_{b,1}(f) = \frac{1}{2} |M_b(f)|^2 \ . \tag{4.234}$$

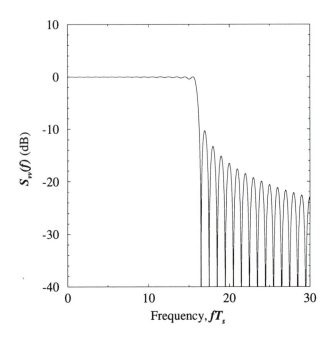

Figure 4.28   Psd of OFDM with $N = 32$.

To compute $m_b(t)$ we assume that the source symbols are equiprobable and defined by the set

$$x_k \in \{2i - 1 - M : i = 1, \, 2, \, \cdots, \, M\} \ . \tag{4.235}$$

Then by taking the ensemble average over x_k it can be shown that

$$
\begin{aligned}
m_b(t) &= \mathrm{E}\left[\exp\left\{j\frac{\pi}{M}x_k h_p(t)\right\}\right] h_a(t) \\
&= \operatorname{sinf}\left(\frac{\pi}{M}h_p(t)\right) h_a(t)
\end{aligned}
\tag{4.236}
$$

where $\operatorname{sinf}(x)$ is defined by

$$\operatorname{sinf}(x) \triangleq \frac{\sin Mx}{M \sin x} \ . \tag{4.237}$$

To compute $S_{b,0}(f)$, we use (4.213) to obtain

$$S_{b,0}(f) = \int_{-\infty}^{\infty} \int_{-\infty}^{\infty} \phi_{b,0}(\tau', z) e^{-j2\pi f(\tau' - z)} dz d\tau' \tag{4.238}$$

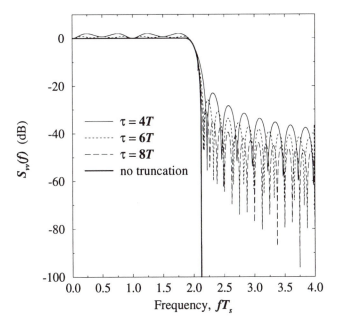

Figure 4.29 Psd of OFDM with $N = 4$ and truncated square root raised cosine pulse shaping; $\beta = 0.25$.

where

$$
\begin{aligned}
\phi_{b,0}(\tau', z) &= \frac{1}{2}\mathrm{E}\left[b(\tau', x_0)b^*(z, x_0)\right] \\
&= \frac{1}{2}\mathrm{E}\left[e^{jk_p(h_p(\tau')-h_p(z))x_0}\right]h_a(\tau')h_a(z) \\
&= \frac{1}{2}\mathrm{sinf}\left(k_p(h_p(\tau') - h_p(z))\right)h_a(\tau')h_a(z) \ . \quad (4.239)
\end{aligned}
$$

Example 4.6 NRZ Phase Shaping Function————————————

If the phase shaping function is $h_p(t) = u_T(t)$, then

$$
m_b(t) = \mathrm{sinf}(k_p)h_a(t) = \mathrm{sinf}(\pi/M)h_a(t) = 0 \ .
$$

Also,

$$\phi_{b,0}(\tau',z) \;=\; \frac{1}{2}\frac{\sin \pi\left(u_T(\tau') - u_T(z)\right)}{M\sin \frac{\pi(u_T(\tau') - u_T(z))}{M}}h_a(\tau')h_a(z)$$

$$=\; \frac{1}{2}\frac{\pi\cos\pi\left(u_T(\tau') - u_T(z)\right)}{\pi\cos\frac{\pi(u_T(\tau') - u_T(z))}{M}}h_a(\tau')h_a(z)$$

$$=\; \frac{1}{2}h_a(\tau')h_a(z) \;.$$

where the second step follows from L'Hôpital's rule. Hence,

$$S_{b,0}(f) \;=\; \frac{1}{2}\int_0^\infty \int_0^\infty h_a(\tau')h_a(z)e^{-j2\pi f(\tau'-z)}d\tau'dz$$

$$=\; \frac{1}{2}\int_0^\infty h_a(\tau')e^{-j2\pi f\tau'}d\tau'\int_0^\infty h_a(z)e^{j2\pi fz}dz$$

$$=\; \frac{1}{2}\left|H_a(f)\right|^2 \;.$$

If $h_a(t) = u_T(t)$, then

$$S_{vv}(f) = \frac{A^2 T}{2}\left(\frac{\sin \pi fT}{\pi fT}\right)^2 \;. \tag{4.240}$$

To fairly compare bandwidth efficiencies with different M, the frequency variable should be normalized by the bit interval T_b. For M-ary signaling $T = T_b\log_2 M$. Hence,

$$S_{vv}(f) = \frac{A^2 T_b \log_2 M}{2}\left(\frac{\sin \pi fT_b\log_2 M}{\pi fT_b\log_2 M}\right)^2 \;. \tag{4.241}$$

Likewise, if the amplitude shaping pulse is a square root raised cosine pulse, then the psd is

$$S_{vv}(f) = \frac{A^2}{T}\left|H_a(f)\right|^2 \tag{4.242}$$

where $H_a(f)$ is defined in (4.127). The psd for this case is shown in Fig. 4.26.

4.10.5 Psd of $\pi/4$-DQPSK

To find the power density spectrum of $\pi/4$-DQPSK we first compute the autocorrelation

$$\phi_{b,m}(\tau',z) = \frac{1}{2}\mathrm{E}\left[b(\tau',\mathbf{x}_m)b^*(z,\mathbf{x}_0)\right] \;. \tag{4.243}$$

For $m > 0$,

$$
\begin{aligned}
\phi_{b,m}(\tau', z) &= \frac{1}{2} \mathrm{E} \left[h_a(\tau') \exp \left\{ j\frac{\pi}{4} \sum_{n=1}^{m} x_n \right\} h_a(z) \exp \left\{ j\frac{\pi}{4} x_0 \right\} \right] \\
&= \frac{1}{2} \mathrm{E} \left[\exp \left\{ j\frac{\pi}{4} \sum_{n=1}^{m} x_m \right\} \right] h_a(\tau') h_a(z) \\
&= 0 .
\end{aligned}
\tag{4.244}
$$

For $m = 0$,

$$
\begin{aligned}
\phi_{b,m}(\tau', z) &= \frac{1}{2} \mathrm{E} \left[h_a(\tau') \exp \left\{ j\frac{\pi}{4} x_0 \right\} h_a(z) \exp \left\{ -j\frac{\pi}{4} x_0 \right\} \right] \\
&= \frac{1}{2} h_a(\tau') h_a(z) .
\end{aligned}
\tag{4.245}
$$

Taking the double Fourier transform gives

$$
\begin{aligned}
S_{b,0}(f) &= \int_{-\infty}^{\infty} \int_{-\infty}^{\infty} \phi_{b,m}(\tau', z) e^{-j2\pi f(\tau' - z)} dz d\tau' \\
&= \frac{1}{2} |H_a(f)|^2 .
\end{aligned}
\tag{4.246}
$$

With a square root raised cosine amplitude shaping pulse the psd is

$$
S_{vv}(f) = \frac{A^2}{T} |H_a(f)|^2
\tag{4.247}
$$

where $H_a(f)$ is defined in (4.127). Note that $\pi/4$-DQPSK has the same power density spectrum as QPSK.

4.10.6   Psd of Full Response CPM

Recall that the equivalent shaping function for CPM is given by (4.138). To compute the psd, we define the auxiliary function

$$
r(t, x_k) \triangleq e^{jx_k\beta(t)} u_T(t)
\tag{4.248}
$$

and calculate its mean and autocorrelation function. If M-ary signaling is used with the values of x_k defined in (4.235), then

$$
\begin{aligned}
m_r(t) &= \mathrm{E}[r(t, x_k)] \\
&= \frac{1}{M} \sum_{i=1}^{M} e^{j(2i-1-M)\beta(t)} u_T(t) \\
&= \mathrm{sinf}(\beta(t)) u_T(t) .
\end{aligned}
\tag{4.249}
$$

where $\text{sinf}(x)$ is defined in (4.237). Also

$$\phi_{r,m}(t,t') = \frac{1}{2}\mathrm{E}\left[r(t,x_m)r^*(t',x_0)\right] \ . \tag{4.250}$$

Evaluating the above expression for $m = 0$ gives the following result which will be used later

$$
\begin{aligned}
\phi_{r,0}(t,t') &= \frac{1}{2}\mathrm{E}\left[r(t,x_0)r^*(t',x_0)\right] \\
&= \frac{1}{2}\mathrm{E}\left[e^{jx_0\beta(t)}e^{-jx_0\beta(t')}\right]u_T(t)u_T(t') \\
&= \frac{1}{2}\mathrm{E}\left[e^{j(x_0\beta(t)-x_0\beta(t'))}\right]u_T(t)u_T(t') \\
&= \frac{1}{2}\text{sinf}\left(\beta(t)-\beta(t')\right)u_T(t)u_T(t') \ .
\end{aligned}
\tag{4.251}
$$

To evaluate the psd, it is necessary to compute the autocorrelation of $b(t,\mathbf{x}_m)$. This can be done as follows

$$
\begin{aligned}
\phi_{b,m}(t,t') &= \frac{1}{2}\mathrm{E}\left[b(t,\mathbf{x}_m)b^*(t',\mathbf{x}_0)\right] \\
&= \frac{1}{2}\mathrm{E}\left[\exp\left\{j\beta(T)\sum_{k=0}^{m-1}x_k\right\}r(t,x_m)r^*(t',x_0)\right] \\
&= \frac{1}{2}\mathrm{E}\left[\left(\prod_{k=0}^{m-1}r(T,x_k)\right)r(t,x_m)r^*(t',x_0)\right] \\
&= \frac{1}{2}\mathrm{E}\left[\left(\prod_{k=1}^{m-1}r(T,x_k)\right)r(t,x_m)r(T,x_0)r^*(t',x_0)\right] \ .
\end{aligned}
\tag{4.252}
$$

Now suppose that the source sequence is uncorrelated. Then for $m > 0$

$$
\begin{aligned}
\phi_{b,m}(t,t') &= \frac{1}{2}\left[m_r(T)\right]^{m-1}m_r(t)\phi_{r,0}(T,t') \\
&= \frac{1}{2}\left[\text{sinf}\beta(T)\right]^{m-1}\left[\text{sinf}\beta(t)\right]\left[\text{sinf}\left(\beta(T)-\beta(t')\right)\right]u_T(t)u_T(t')
\end{aligned}
\tag{4.253}
$$

where we have used (4.251). Likewise, for $m = 0$

$$\begin{aligned}
\phi_{b,0}(t,t') &= \frac{1}{2}\mathrm{E}\left[b(t,\mathbf{x}_0)b^*(t',\mathbf{x}_0)\right]\\
&= \frac{1}{2}\mathrm{E}\left[e^{j(x_0\beta(t)-jx_0\beta(t'))}\right]u_T(t)u_T(t')\\
&= \frac{1}{2}\mathrm{sinf}\left(\beta(t)-\beta(t')\right)u_T(t)u_T(t')\\
&= \phi_{r,0}(t,t') .
\end{aligned} \tag{4.254}$$

Finally, the psd is obtained by using (4.253) and (4.254) along with (4.203) and (4.213).

Alternative Method

There is an alternate method for obtaining the psd that provides more insight. Similar to the way that (4.210) was derived, we use (4.203) along with the property $S_{b,-m}(f) = S_{b,m}^*(f)$ to obtain

$$S_{vv}(f) = \frac{A^2}{T}\left(S_{b,0}(f) + 2\mathrm{Re}\left\{\sum_{m=1}^{\infty}S_{b,m}(f)e^{-j2\pi fmT}\right\}\right) . \tag{4.255}$$

Taking the double Fourier transforms of (4.253) and (4.254) gives

$$S_{b,m}(f) = \begin{cases} S_{r,0}(f) & m=0\\ m_r^{m-1}(T)M_r(f)\hat{M}_r^*(f) & m>0 \end{cases} \tag{4.256}$$

where

$$\begin{aligned}
m_r^{m-1}(T) &= \mathrm{sinf}^{m-1}\beta(T)\\
M_r(f) &= \mathcal{F}[m_r(t)u_T(t)] = \mathcal{F}[\mathrm{sinf}\beta(t)u_T(t)]\\
\hat{M}_r^*(f) &= \frac{1}{2}\mathrm{E}\left[r(T,x_0)R^*(f,x_0)\right] = \frac{1}{2}\mathrm{E}\left[e^{j2\pi\overline{k_f}Tx_0}R^*(f,x_0)\right]
\end{aligned}$$

$\mathcal{F}[\,\cdot\,]$ denotes the Fourier transform and

$$R^*(f,x_0) = \mathcal{F}[r^*(t,x_0)] = \mathcal{F}\left[e^{-jx_0\beta(t)}u_T(t)\right] . \tag{4.257}$$

Then,

$$
\begin{aligned}
S_{vv}(f) &= \frac{A^2}{T}\left(S_{r,0}(f) + 2\mathrm{Re}\left\{M_r(f)\hat{M}_r^*(f)\right.\right. \\
&\qquad\left.\left. \times \sum_{m=1}^{\infty} m_r^{m-1}(T)e^{-j2\pi fmT}\right\}\right) \\
&= \frac{A^2}{T}\left(S_{r,0}(f) + 2\mathrm{Re}\left\{M_r(f)\hat{M}_r^*(f)\right.\right. \\
&\qquad\left.\left. \times \sum_{n=0}^{\infty}\left[m_r(T)e^{-j2\pi fT}\right]^n e^{-j2\pi fT}\right\}\right) \quad .
\end{aligned}
$$

$$(4.258)$$

Observe that

$$|r(t, x_k)| = \left|e^{jx_k\beta(t)}u_T(t)\right| = 1 \tag{4.259}$$

so that

$$\left|m_r(T)e^{-j2\pi fT}\right| = |m_r(T)| \le 1 \ . \tag{4.260}$$

The implication of equation (4.260) is that two separate cases must be considered when evaluating the psd.

Case 1: $|m_r(T)| < 1$

In this case the sum in (4.258) converges so that

$$S_{vv}(f) = \frac{A^2}{T}\left(S_{r,0}(f) + 2\mathrm{Re}\left\{\frac{M_r(f)\hat{M}_r^*(f)}{\exp\{j2\pi fT\} - m_r(T)}\right\}\right) \quad . \tag{4.261}$$

and the psd has no discrete components.

Case 2: $|m_r(T)| = 1$

This case is possible only if

$$|m_r(T)| = \left|\mathrm{E}\left[e^{jx_k\beta(t)}\right]\right| = 1 \ . \tag{4.262}$$

For this condition to be true we must have

$$e^{jx_k\beta(T)} = e^{jc} \ , \quad \forall \ k \tag{4.263}$$

where c is a constant. However, $\beta(T) = 2\pi\overline{k_f}T$ so that $x_k\beta(T) = 2\pi\overline{k_f}T$ mod(2π) for all i. Then $r(T, x_0) = \exp\{j2\pi\overline{k_f}T\}$ is a constant so that

$$m_r(T) = \mathrm{E}[r(T, x_0)] = e^{j2\pi \overline{k_f} T} \tag{4.264}$$

and

$$\hat{M}_r^*(f) = M_r^*(f) e^{j2\pi \overline{k_f} T} \; . \tag{4.265}$$

Hence, the psd is

$$
\begin{aligned}
S_{vv}(f) &= \frac{A^2}{T} \left(S_{r,0}(f) + |M_r(f)|^2 2\mathrm{Re}\left\{ \sum_{m=1}^{\infty} e^{j2\pi(f-\overline{k_f})mT} \right\} \right) \\
&= \frac{A^2}{T} \Big(S_{r,0}(f) - |M_r(f)|^2 + |M_r(f)|^2 \\
&\quad \times \sum_{m=-\infty}^{\infty} e^{-j2\pi(f-\overline{k_f})mT} \Big) \\
&= \frac{A^2}{T} \left(S_{r,0}(f) - |M_r(f)|^2 + \frac{1}{T}|M_r(f)|^2 \sum_{n=-\infty}^{\infty} \delta\left(f - \overline{k_f} - \frac{n}{T} \right) \right) \\
&= \frac{A^2}{T} \left(S_{r,0}(f) - |M_r(f)|^2 \right) \\
&\quad + \left(\frac{A}{T} \right)^2 \sum_{m=-\infty}^{\infty} \left| M_r(\overline{k_f} + n/T) \right|^2 \delta\left(f - \overline{k_f} - \frac{n}{T} \right) \; .
\end{aligned}
\tag{4.266}
$$

Clearly, the second term in the above expression is a discrete spectral component. Finally, if x_k assumes the values defined in (4.235), then

$$x_k 2\pi \overline{k_f} T = 2\pi \overline{k_f} T \qquad \mathrm{mod}\ (2\pi) \; . \tag{4.267}$$

However, $h = \beta(T)/\pi = 2\overline{k_f}T$. Therefore,

$$x_k h\pi = h\pi \qquad \mathrm{mod}\ (2\pi) \; . \tag{4.268}$$

Hence, h must be an integer for there to be a discrete spectral component.

Psd of CPFSK

Suppose that h is a noninteger so that the psd has no discrete components. Then

$$
\begin{aligned}
R(f, x_0) &= \int_0^T e^{j2\pi k_f t x_0} \cdot e^{-j2\pi f t} dt \\
&= T e^{-j\pi(f - x_0 k_f)T} \mathrm{Sa}\left(\pi(f - k_f x_0)T \right)
\end{aligned}
\tag{4.269}
$$

where

$$M_r(f) = \mathrm{E}\left[R(f, x_0)\right]$$

$$= \frac{T}{M} \sum_{i=1}^{M} e^{-j\pi(f-(2i-1-M)k_f)T}$$

$$\times \mathrm{Sa}\left(\pi\left(f - (2i - 1 - M)k_f\right)T\right) \tag{4.270}$$

$$S_{r,0}(f) = \frac{1}{2}\mathrm{E}\left[|R(f, x_0)|^2\right]$$

$$= \frac{T^2}{M} \sum_{i=1}^{M} \mathrm{Sa}^2\left(\pi\left(f - (2i - 1 - M)k_f\right)T\right) . \tag{4.271}$$

Also,

$$\hat{M}_r^*(f) = \frac{T}{2M} \sum_{i=1}^{M} e^{j\pi(f+(2i-1-M)k_f)T}$$

$$\times \mathrm{Sa}\left(\pi\left(f - (2i - 1 - M)k_f\right)T\right) \tag{4.272}$$

These expressions are used in (4.261) to obtain the psd. If h is an integer, then the psd will have a discrete component and

$$M_r(f) = \frac{T}{M} e^{-j2\pi fT} \sin(\pi fT) \sum_{i=1}^{M} e^{-j\pi(f-(2i-1-M)k_f)T} . \tag{4.273}$$

This expression can be used in (4.265) to obtain the psd.

If a binary modulation ($M = 2$) is used, the above expressions simplify even more. For the case when h is not an integer

$$\Phi_{r,0}(f) = \frac{T^2}{4}\left\{\mathrm{Sa}^2\left(\pi(f - k_f)T\right) + \mathrm{Sa}^2\left(\pi(f + k_f)T\right)\right\} \tag{4.274}$$

$$M_r(f) = \left\{\frac{T}{2} e^{-j\pi(f+k_f)T} \mathrm{Sa}\left(\pi(f + k_f)T\right)\right.$$

$$\left. e^{-j\pi(f-k_f)T} \mathrm{Sa}\left(\pi(f - k_f)T\right)\right\} \tag{4.275}$$

$$\hat{M}_r^*(f) = \frac{T}{4}\left\{ e^{j\pi(f-k_f)T} \mathrm{Sa}\left(\pi(f + k_f)T\right)\right.$$

$$\left. + e^{j\pi(f+k_f)T} \mathrm{Sa}\left(\pi(f - k_f)T\right)\right\} \tag{4.276}$$

$$m_r(T) = \mathrm{sinf}(2\pi k_f T) = \mathrm{sinf}(h\pi) . \tag{4.277}$$

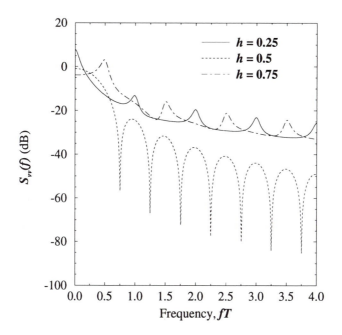

Figure 4.30a   Psd of binary CPFSK for various modulation indices.

When h is an integer

$$\Phi_{r,0}(f) = \frac{T^2 \sin^2 \pi f T}{2} \left\{ \left(\frac{1}{\pi(f - k_f)T} \right)^2 + \left(\frac{1}{\pi(f + k_f)T} \right)^2 \right\} \quad (4.278)$$

$$M_r(f) = \frac{T e^{-j\pi f T} \sin \pi f T}{2} \left\{ \frac{1}{\pi(f - k_f)T} + \frac{1}{\pi(f + k_f)T} \right\} . \quad (4.279)$$

Figs. 4.30a and 4.30b plot the psd $4S_v(f)/A^2 T$ against the frequency fT. The psd of MSK corresponds to $h = 0.5$. Note that modulation indices other than $h = 0.5$ result in a narrower main lobe than MSK, but larger sidelobes. Fig. 4.30b demonstrates the appearance of discrete components as $h \to 1$.

4.10.7   Psd of MSK

An alternative method for computing the psd of MSK starts by recognizing that MSK is equivalent to OQASK with a half-sinusoid amplitude shaping pulse. It follows from (4.155) that the MSK baseband signal has the form

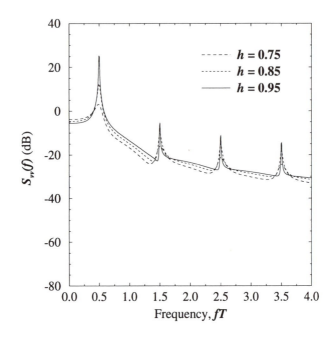

Figure 4.30b   Psd of binary CPFSK for various modulation indices.

$$v(t) = A \sum_k b(t - 2kT, x_k) \tag{4.280}$$

where

$$b(t, x_k) = x_k^I h_a(t) + j x_k^Q h_a(t - T) \tag{4.281}$$

and

$$h_a(t) = \cos\left(\frac{\pi t}{2T}\right) \quad . \tag{4.282}$$

The Fourier transform of (4.281) is

$$B(f, x_k) = \left(x_k^I + j x_k^Q \exp\{-j 2\pi f T\}\right) H_a(f) \quad . \tag{4.283}$$

Since the source sequence is uncorrelated, the psd can be computed from (4.216) and (4.217). Since the source sequence has zero mean, $S_{b,1}(f) = 0$. Also,

$$
\begin{aligned}
S_{b,0}(f) &= \frac{1}{2} \mathrm{E}\left[(x_k^I)^2 + (x_k^Q)^2\right] |H_a(f)|^2 \\
&= \mathrm{E}[\,(x_k^I)^2\,] |H_a(f)|^2 \\
&= |H_a(f)|^2 \quad .
\end{aligned}
\tag{4.284}
$$

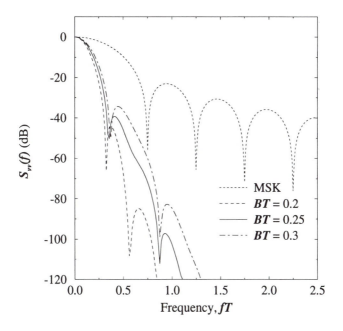

Figure 4.31   Psd of GMSK with various normalized filter bandwidths BT.

From (4.220)

$$S_{vv}(f) = \frac{32A^2T}{\pi^2}\left[\frac{\cos 2\pi fT}{1 - 16f^2T^2}\right]^2 . \tag{4.285}$$

Once again, the psd of MSK is plotted in Fig. 4.30a. Comparing with Fig. 4.26, we observe MSK has fairly large sidelobes as compared to $\pi/4$-DQPSK with truncated square root raised cosine pulse shaping.

4.10.8   Psd of GMSK and TFM

GSMK and TFM are special cases of partial response CPM. In general, the psd of partial response CPM is difficult to obtain except for a rectangular shaping function. One approach suggested by Garrison [112] is to approximate the modulating pulses by using a large number of rectangular subpulses with quantized amplitudes.

Fig. 4.31 plots the psd of GMSK with various normalized filter bandwidths BT. Note that a smaller BT results in a more compact psd. Likewise, Fig. 4.32 plots

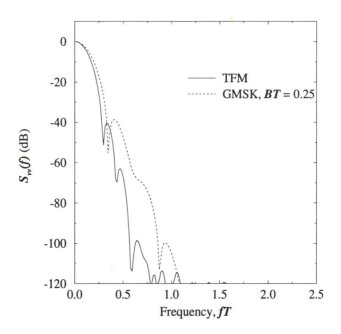

Figure 4.32   Psd of TFM and GMSK with $BT = 0.25$.

the psd of TFM and GMSK with $BT = 0.25$. Observe that the psd of TFM compares well with that of GMSK. This is not surprising since their frequency pulses are quite similar, see Figs. 4.23 and 4.25.

Problems

4.1. The input to an FM receiver consists of an unmodulated carrier along with an interfering sine wave. The interference is 20 dB below the carrier level, and the frequency separation between them is 15 kHz. Assuming that the receiver uses a VCO with a sensitivity of $k_v = 0.2$ kHz/v, determine the receiver output voltage.

4.2. Suppose that the transfer functions of the pre-emphasis and de-emphasis filters of an FM system are scaled as follows

$$H_p(f) = k \left(1 + \frac{jf}{f_o} \right)$$

and

$$H_d(f) = \frac{1}{k}\left(\frac{1}{1 + jf/f_o}\right) .$$

The scaling factor k selected so that the average power of the emphasized message signal is the same as the average power of the message signal $m(t)$.

a) Find the value of k that satisfies this requirement for the case when the power spectral density of the message signal $m(t)$ is

$$\Phi_m(f) = \begin{cases} \frac{S_o}{1+(f/f_o)^2}, & -W \le f \le W \\ 0, & \text{elsewhere} \end{cases} .$$

b) What is the corresponding value of the improvement in the output SNR by using this pair of pre-emphasis and de-emphasis filters?

4.3. A CPM signal is generated from a baseband signal with a half-sinusoid shaping function.

a) Find the peak frequency deviation k_f so that $\beta(T) = \pi/2$.

b) Sketch the phase tree and phase trellis if the source symbol sequence is

$$\mathbf{x} = \{+3, -1, +1, +3, -3, +1, -1\} .$$

4.4. Sketch the phase-tree, the phase trellis, and the state diagram for partial response CPM with $h = 1/2$ and

$$h_f(t) = u_{2T_b}(t) .$$

4.5. a) Generate a shaping function of duration $3T$ by convolving a NRZ shaping function of duration T with a NRZ shaping function of duration $2T$.

b) Define and sketch the three segments of the shaping function, $h_{f,k}(t)$, $k = 0, 1, 2$.

c) Sketch the baseband signal if the symbol sequence is

$$\mathbf{x} = \{+1, -1, +1, -1, -1\} .$$

4.6. Determine the number of phase states in the phase trellis diagram for the cases

a) full response binary CPFSK with either $h = 2/3$ or $h = 3/4$.

b) partial response $K = 3$ binary CPFSK with either $h = 2/3$ aro 3/4.

4.7. Consider a multi-h CPM waveform with the h sequence $\{h_1, h_2\} = \{\frac{1}{4}, \frac{1}{2}\}$ and the frequency shaping pulse $h_f(t) = U_T(t)$. In the ith signaling interval the excess phase changes by $\pm\pi/4$ radians if $h_1 = \frac{1}{4}$ is used and by $\pm\pi/2$ radians if $h_2 = \frac{1}{2}$ is used. Plot the phase-trellis assuming that the initial phase at time $t = 0$ is 0 radians. Indicate the phase trajectory for the symbol sequence $\mathbf{x} = \{1, -1, -1, 1, 1, 1, -1\}$.

4.8. Show that 16-QAM can be represented as a superposition of two four-phase constant envelope signals where each component is amplified separately before summing, i.e.,

$$s(t) = [A_n \cos 2\pi f_c t + B_n \sin 2\pi f_c t] + [C_n \cos 2\pi f_c t + D_n \sin 2\pi f_c t]$$

where $\{A_n\}$, $\{b_n\}$, $\{C_n\}$, and $\{D_n\}$ are statistically independent binary sequences with elements from the set $\{-1, +1\}$. Thus, show that the resulting signal is equivalent to

$$s(t) = I_n \cos 2\pi f_c t + Q_n \sin 2\pi f_c t$$

and determine I_n and Q_n in terms of A_n, B_n, C_n, and D_n.

4.9 Consider TFM with the frequency shaping pulse

$$H_f(f) = \frac{\pi}{4k_f} \frac{\pi fT}{\sin(\pi fT)} \cos^2(\pi fT) \ .$$

Suppose that this pulse is obtained by exciting a filter $\tilde{h}(t)$ with a gate function rect(t/T). Find and sketch the impulse response of the filter $\tilde{h}(t)$.

4.10. Consider the case of uncorrelated source symbols.

a) Show that if the symbols are equiprobable, then

$$\mathrm{E}\left[|B(f, d_0)|^2\right] - |\mathrm{E}\left[B(f, d_0)\right]|^2 = \frac{1}{2M^2} \sum_{i=1}^{M} \sum_{k=1}^{M} |B(f, d(i)) - B(f, d(k))|^2$$

for $d_0 \in \{d(1), d(2), \ldots, d(M)\}$.

b) Compute the value of part a) for $M = 2$.

4.11 Consider the complex low-pass binary modulated signal

$$v(t) = A \sum_n x_n h_a(t - nT)$$

where $x_n \in \{-1, +1\}$. The data sequence $\{x_n\}$ is correlated such that

$$\phi_{xx}(n) = \frac{1}{2} E[x_k x_{k+n}^*] = \rho^{|n|} .$$

Compute the power density spectrum of $v(t)$.

4.12 Compute the power density spectrum of 4-PSK with the half sinusoid phase shaping pulse

$$h_p(t) = \sin\left(\frac{\pi t}{T}\right) u_T(t) .$$

5

DIGITAL SIGNALING ON FLAT FADING CHANNELS

The performance of a digital modulation scheme is degraded by many transmission impairments including fading, delay spread, Doppler spread, co-channel and adjacent channel interference, and noise. Fading causes a very low instantaneous received signal-to-noise ratio (SNR) or carrier-to-noise ratio (CNR) when the channel exhibits a deep fade, delay spread causes intersymbol interference (ISI) between the transmitted symbols, and a large Doppler spread is indicative of rapid channel variation and necessitates a receiver with a fast convergent algorithm. Co-channel interference, adjacent channel interference, and noise, are all additive distortions that degrade the bit error rate performance by reducing the CNR or SNR.

This chapter derives the bit error rate performance of digital signaling on frequency non-selective (flat) fading channels with AWGN. Flat fading channel models are appropriate for narrow-band land mobile radio systems or mobile satellite systems. Flat fading channels affect all frequency components of a narrow-band signal in exactly the same way and, therefore, do not introduce amplitude or phase distortion into the received signal. Frequency selective channels distort the transmitted signal and will be the subject of Chapter 6. Flat fading channel will be shown to significantly degrade the bit error rate performance unless appropriate countermeasures are taken. Diversity and coding techniques are well known methods for combating fading. The basic idea of diversity systems is to provide the receiver with multiple replicas of the same information bearing signal, where the replicas are affected by uncorrelated fading. Coding techniques introduce a form of time diversity into the transmitted signal which can be exploited to mitigate the effects of fading.

215

The remainder of this chapter is organized as follows. Section 5.1 introduces a vector representation for digital signaling on flat fading channels with additive white Gaussian noise (AWGN). Section 5.2 derives the structure of the optimum coherent receiver for the detection of known signals in AWGN. It will be shown that the optimum coherent receiver generally requires knowledge of the complex fading gain (amplitude and phase), although for some types of modulated signals such as PSK only the phase is required. Section 5.3 evaluates the performance of the various digital signaling techniques that were introduced in Chapter 4 on flat fading channels. Section 5.4 considers a simplified receiver structure for PSK and $\pi/4$-QPSK signaling that uses differential detection. A differential detector is a suboptimum receiver that detects phase changes between successive signaling intervals. Section 5.5 discusses various types of diversity and diversity combining techniques that can be used to combat the effects of fading. Finally, Section 5.5.7 introduces the concept of macroscopic diversity that is useful for mitigating the effects of shadow variations.

5.1   VECTOR REPRESENTATION OF RECEIVED SIGNALS

Suppose that one the M complex low-pass signals $\{v_k(t)\}_{k=1}^M$, say $v_i(t)$, is transmitted. For a flat fading channel, the received complex envelope is

$$w(t) = g(t)v_i(t) + z(t), \quad 0 \le t \le T \tag{5.1}$$

where $g(t) = \alpha(t)e^{j\phi(t)}$ is the complex fading gain introduced by the channel, and $z(t)$ is zero-mean complex AWGN with a power spectral density (psd) of N_o watts/Hz. The basic detection problem at the receiver is to determine which message waveform $v_k(t)$ was transmitted from the observation of received signal $w(t)$.

If the channel changes very slowly with respect to the symbol duration, i.e., $f_m T \ll 1$, then $g(t)$ will effectively remain constant over the symbol duration[1]. Under this condition, the explicit time dependency of $g(t)$ can be removed so the received signal becomes

$$w(t) \;\; = \;\; gv_i(t) + z(t), \quad 0 \le t \le T \tag{5.2}$$

where the fading gain $g = \alpha e^{j\phi}$ is a complex Gaussian random variable. If the Gaussian random process has zero (non-zero) mean then the magnitude α

---

[1] For land mobile radio applications $f_m T \ll 1$ is a reasonable assumption.

is Rayleigh (Ricean) distributed and the phase ϕ is uniformly distributed over $[-\pi, \pi]$.

To facilitate the derivation of the optimum receiver, and the analysis of the probability of bit error, it is useful to introduce a signal-space representation for the received signals. If the channel is affected by AWGN, then the required basis functions can be obtained by using the Gram-Schmidt orthonormalization procedure described in Section 4.2.2. However, if the channel is affected by colored noise, then the Gram-Schmidt orthonormalization procedure fails. In this case, the more general Karhunen-Loéve expansion can be used to obtain the basis functions, as described in [256]. By using the Gram-Schmidt orthonormalization procedure, the received signal can be expressed as

$$w(t) = \sum_{n=1}^{\infty} w_n \phi_n(t) \tag{5.3}$$

where

$$w_n = \int_{-\infty}^{\infty} w(t)\phi_n^*(t)dt \ . \tag{5.4}$$

Although the signal set $\{v_i(t)\}_{i=1}^{M}$ can be represented as points in an N-D signal-space, a signal-space of infinite dimension is required to represent the noise waveform $z(t)$. It follows that the received signal can be written as

$$
\begin{aligned}
w(t) &= \sum_{n=1}^{\infty} w_n \varphi_n(t) \\
&= \sum_{n=1}^{N} (gv_{in} + z_n) \varphi_n(t) + \sum_{n=N+1}^{\infty} z_n \varphi_n(t) \ .
\end{aligned}
\tag{5.5}
$$

We now show that the remainder process

$$\tilde{z}(t) - \sum_{n=1}^{N} z_n \varphi_n(t) \tag{5.6}$$

is irrelevant in the receiver decision process because

$$E[\tilde{z}(t)w_j^*] = 0 \ , \quad j = 1, \ldots, N \ . \tag{5.7}$$

This result, known as Wozencraft's irrelevance theorem [339]. It implies that sufficient statistics are provided by the N-dimensional vector

$$\mathbf{w} = g\mathbf{v}_m + \mathbf{z} \tag{5.8}$$

where

$$w_n = gv_{mn} + z_n, \quad n = 1, 2, \ldots, N \; . \tag{5.9}$$

Theorem 5.1 _____

If $z(t)$ is additive white Gaussian noise, then

$$E[\tilde{z}(t)w_j^*] = 0 \; , \quad j = 1, \; \ldots, N \; . \tag{5.10}$$

Proof

$$
\begin{aligned}
E[\tilde{z}(t)r_j^*] &= E\left[\tilde{z}(t)\int_0^T w^*(s)\varphi_j(s)ds\right] \\
&= E\left[\left(z(t) - \sum_{n=1}^N z_n\varphi_n(t)\right)\int_0^T w^*(s)\varphi_j(s)ds\right] \\
&= E\left[\int_0^T (gv_m^*(s) + z^*(s))\left(z(t) - \sum_{n=1}^N z_n\varphi_n(t)\right)\varphi_j(s)ds\right] \\
&= \int_0^T E\left[z^*(s)z(t) - z^*(s)\sum_{n=1}^N z_n\varphi_n(t)\right]\varphi_j(s)ds \\
&= \int_0^T \left(2\phi_{zz}(t-s) - \sum_{n=1}^N E[z^*(s)z_n]\varphi_n(t)\right)\varphi_j(s)ds \; .
\end{aligned}
$$

However,

$$
\begin{aligned}
E[z^*(s)z_n] &= E\left[\int_0^T z^*(s)z(v)\varphi_n^*(v)dv\right] \\
&= \int_0^T 2\phi_{zz}(v-s)\varphi_n^*(v)dv \; .
\end{aligned}
$$

Therefore follows that

$$E[\tilde{z}(t)w_j^*] = \int_0^T 2\phi_{zz}(t-s)\varphi_j(s)ds$$
$$- \int_0^T \sum_{n=1}^N \left(\int_0^T 2\phi_{zz}(v-s)\varphi_n^*(v)dv \right) \varphi_n(t)\varphi_j(s)ds .$$

If the channel is affected by AWGN with a psd of N_o watts/Hz, then the autocorrelation of the noise process $z(t)$ is

$$\phi_{zz}(t-s) = N_o\delta(t-s) .$$

Therefore

$$\begin{aligned}
E[\tilde{z}(t)w_j^*] &= 2N_o\varphi_j(t) - 2\int_0^T \sum_{n=1}^N N_o\varphi_n^*(s)\varphi_j(s)\varphi_n(t)ds \\
&= 2N_o\varphi_j(t) - 2N_o \sum_{n=1}^N \delta_{nj}\varphi_n(t)ds \\
&= 2N_o\varphi_j(t) - 2N_o\varphi_j(t) \\
&= 0 .
\end{aligned}$$

5.2 DETECTION OF KNOWN SIGNALS IN ADDITIVE WHITE GAUSSIAN NOISE

The maximum *a posteriori* probability (MAP) receiver observes the received vector $\mathbf{w}$ and decides in favor of the message vector $\mathbf{v}_i$ that maximizes the *a posteriori* probability $P(g\mathbf{v}_i|\mathbf{w})$. The conditional probability of decision error, given that the message vector $\mathbf{v}_m$ was transmitted is

$$P_{e|m} = 1 - P_r(g\mathbf{v}_m|\mathbf{w}) \tag{5.11}$$

and the average probability of decision error is

$$P_e = \sum_{m=1}^M P_{e|m}P_m \tag{5.12}$$

where P_m is the *a priori* probability of $\mathbf{v}_m$. The MAP receiver clearly minimizes the probability of decision error.

By using Bayes' theorem, the *a posteriori* probability $P_r(g\mathbf{v}_m|\mathbf{w})$ can be expressed in the form

$$P_r(g\mathbf{v}_m|\mathbf{w}) = \frac{p(\mathbf{w}|g\mathbf{v}_m)P_m}{p(\mathbf{w})} \ , \quad m = 1, \cdots, M \ . \tag{5.13}$$

Since the probability density function (pdf) of the received vector $p(\mathbf{w})$ is independent of the transmitted message, the MAP decision rule has the simpler form

$$\text{choose } \mathbf{v}_m \text{ if } \quad p(\mathbf{w}|g\mathbf{v}_m)P_m \geq p(\mathbf{w}|g\mathbf{v}_{\hat{m}})P_{\hat{m}} \quad \forall \ \hat{m} \neq m \ . \tag{5.14}$$

The MAP receiver requires knowledge of the complex channel gain g, implying that the receiver must employ an adaptive channel estimator.

A receiver that maximizes $p(\mathbf{w}|g\mathbf{v}_m)$ regardless of the *a priori* messages probabilities is called a maximum likelihood (ML) receiver . The ML decision rule is

$$\text{choose } \mathbf{v}m \text{ if } \quad p(\mathbf{w}|g\mathbf{v}_m) \geq p(\mathbf{w}|g\mathbf{v}_{\hat{m}}) \quad \forall \ \hat{m} \neq m \ . \tag{5.15}$$

If the messages are equally likely, i.e., $P_k = 1/M$, then selection of the signal vector that maximizes $p(\mathbf{w}|g\mathbf{v}_m)$ also maximizes $p(g\mathbf{v}_m|\mathbf{w})$. Under this condition the ML receiver also minimizes the probability of error. The *a priori* probabilities will be equal when the source coding is good. In practice, an ML receiver is quite often implemented regardless of the *a priori* message probabilities, because they may not be known.

To proceed further we need to specify the joint conditional pdf $p(\mathbf{w}|g\mathbf{v}_m)$. For an AWGN channel, the components of the noise vector $\mathbf{z}$ are zero-mean complex Gaussian random variables with the covariance function

$$\phi_{zz}(k - j) = \frac{1}{2}\mathrm{E}[z_k z_j^*] = N_o \delta_{kj} \tag{5.16}$$

where N_o is the one-sided noise psd. Therefore, the joint conditional pdf $p(\mathbf{w}|g\mathbf{v}_m)$ has the multivariate Gaussian density

$$
\begin{aligned}
p(\mathbf{w}|g\mathbf{v}_m) &= \prod_{i=1}^{N} \frac{1}{\pi N_o} \exp\left\{ -\frac{1}{N_o}|w_i - gv_{mi}|^2 \right\} \\
&= \frac{1}{(\pi N_o)^N} \exp\left\{ -\frac{1}{N_o}\|\mathbf{w} - g\mathbf{v}_m\|^2 \right\} \ .
\end{aligned} \tag{5.17}
$$

By using this conditional pdf in (5.15), it is apparent that the ML receiver chooses the $\mathbf{v}_m$ that maximizes the metric (distance measure)

$$
\begin{aligned}
\mu(\mathbf{v}_m) &= -\|\mathbf{w} - g\mathbf{v}_m\|^2 \\
&= -\|\mathbf{w}\|^2 + 2\mathrm{Re}\left\{ \mathbf{w} \cdot g^* \mathbf{v}_m^* \right\} - |g|^2\|\mathbf{v}_m\|^2 \ .
\end{aligned} \tag{5.18}
$$

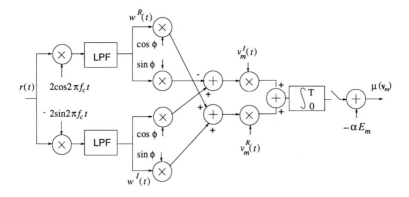

Figure 5.1 Correlation receiver for detection of known signals in additive Gaussian noise.

In other words, the maximum likelihood receiver decides in favor of the scaled message vector $g\mathbf{v}_m$ that is closest in Euclidean distance to the received vector $\mathbf{w}$. Since $\|\mathbf{w}\|^2$ is independent of $\mathbf{v}_m$ and $\|\mathbf{v}_m\|^2 = 2E_m$, the receiver just needs to maximize the metric

$$
\begin{aligned}
\mu(\mathbf{v}_m) &= \text{Re}\left\{\mathbf{w}\cdot g^*\mathbf{v}_m^*\right\} - |g|^2 E_m \\
&= \text{Re}\left\{\int_0^T w(t)g^*v_m^*(t)dt\right\} - |g|^2 E_m \\
&\equiv \text{Re}\left\{\int_0^T w(t)e^{-j\phi}v_m^*(t)dt\right\} - \alpha E_m \ .
\end{aligned}
\tag{5.19}
$$

The last line in (5.19) follows because all the $\mu(\mathbf{v}_m)$ can be divided by α without altering the decision process. One possible receiver structure based on (5.19) is shown in Fig. 5.1 where $r(t)$ is the received bandpass signal. The low-pass filters (LPFs) are just used to reject the double frequency term after demodulation. This receiver structure is called a **correlation receiver** .

Various simplifications can be made depending on the type of modulation used. If the message waveforms have equal energy, then αE_m is the same for all m and the metric in (5.19) simplifies to

$$
\mu(\mathbf{v}_m) = \text{Re}\left\{\mathbf{w}\cdot e^{-j\phi}\mathbf{v}_m^*\right\}
\tag{5.20}
$$

$$
= \text{Re}\left\{\int_0^T w(t)e^{-j\phi}v_m^*(t)dt\right\}
\tag{5.21}
$$

In this case, the receiver does not require the complete complex channel gain $g = \alpha e^{j\phi}$, but only the random carrier phase ϕ. The random carrier phase can be obtained in practice by using a phase locked loop.

5.3 ERROR PROBABILITY WITH COHERENT DETECTION

Suppose that the receiver employs coherent detection, and the complex channel gain g is known exactly at the receiver. The ML receiver decides in favor of the message vector $\mathbf{v}_m$ that minimizes the Euclidean distance $\|\mathbf{w} - g\mathbf{v}_m\|^2$. Consider two message vectors $\mathbf{v}_j$ and $\mathbf{v}_k$ that are separated *at the receiver* by the squared Euclidean distance $|g|^2\|\mathbf{v}_j - \mathbf{v}_k\|^2 = \alpha^2\|\mathbf{v}_j - \mathbf{v}_k\|^2$. The Euclidean distance between the vectors $g\mathbf{v}_j$ and $g\mathbf{v}_k$ is invariant to their translation and rotation. We therefore choose a rotation $e^{j\phi}$ and translation $\hat{\mathbf{v}}$ such that the vector through the points $e^{j\phi}g(\mathbf{v}_j - \hat{\mathbf{v}}$ and $e^{j\phi}g(\mathbf{v}_k - \hat{\mathbf{v}}$ is coincident with one of the basis functions that define the N-D signal-space. The Gaussian noise along this vector has variance N_o. A decision boundary can be established at the midpoint between points as shown in Fig. 5.2. Then the **pairwise error probability** associated with the message vectors $\mathbf{v}_j$ and $\mathbf{v}_k$ is simply

$$\mathrm{P_r}(\mathbf{v}_j, \mathbf{v}_k) = Q\left(\sqrt{\frac{\alpha^2 d_{jk}^2}{4N_o}}\right) . \tag{5.22}$$

where $d_{jk}^2 = \|\mathbf{v}_j - \mathbf{v}_k\|^2$ is the squared Euclidean distance between $\mathbf{v}_j$ and $\mathbf{v}_k$. This concept can be readily extended to compute error probabilities for arbitrary sets of signal vectors. The idea is to first a define convex decision region R_m around each of the signal points $g\mathbf{v}_m$ in the N-D signal space. For an ML receiver, the decision regions are defined by

$$R_m = \left\{ \mathbf{w} : \|\mathbf{w} - g\mathbf{v}_m\|^2 \leq \|\mathbf{w} - g\mathbf{v}_{\hat{m}}\|^2 , \ \forall \hat{m} \neq m \right\} . \tag{5.23}$$

The conditional error probability associated with $\mathbf{v}_m$ is

$$P_{e|m} = \mathrm{P_r}(\mathbf{w} \notin R_m) \tag{5.24}$$

and the average symbol error probability is given by (5.12).

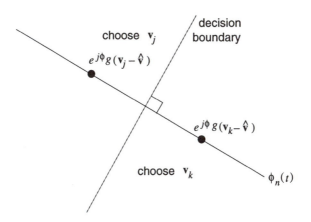

Figure 5.2 Two received signal points in an N-D signal-space.

5.3.1 Error Probability of MSK

Since the error probability is invariant to rotation, the error probability can be computed directly from the signal-space diagram in Fig. 4.21, from which we observe that the message vectors are separated by the squared Euclidean distance $d_{12} = 2\sqrt{2E}$ at the decision instants. From (5.22), the bit error probability is

$$P_b(\gamma_b) = Q(\sqrt{2\gamma_b}) \qquad (5.25)$$

where

$$\gamma_b = \alpha^2 E/N_o \qquad (5.26)$$

is the received bit energy-to-noise ratio. We shall see that MSK has the same bit error rate performance as BPSK, QPSK, 2-QAM, and 4-QAM, if ideal coherent detection is used.

If the channel is Rayleigh faded, then α in (5.26) is a Rayleigh random variable. By using a transformation of random variables, γ_b has the exponential pdf

$$p_{\gamma_b}(x) = \frac{1}{\bar{\gamma}_b} e^{-x/\bar{\gamma}_b} \qquad (5.27)$$

where $\bar{\gamma}_b$ is the average received bit-energy-to-noise ratio. The average probability of bit error is

$$
\begin{aligned}
P_b &= \int_0^\infty P_b(x) p_{\gamma_b}(x) dx \\
&= \frac{1}{2}\left[1 - \sqrt{\frac{\bar{\gamma}_b}{1+\bar{\gamma}_b}}\right] \approx \frac{1}{4\bar{\gamma}_b} \qquad (5.28)
\end{aligned}
$$

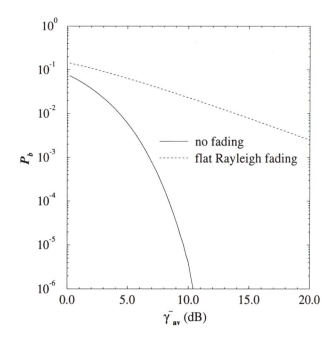

Figure 5.3 Bit error probability for MSK with coherent detection on an AWGN channel and a flat Rayleigh fading channel with AWGN.

The MSK bit error probability is plotted in Fig. 5.3 for an AWGN channel and a Rayleigh fading channel with AWGN. Observe that Rayleigh fading converts an exponential dependency of the bit error probability on the bit energy-to-noise ratio into an inverse linear one. This behavior is typical for any uncoded modulation scheme in Rayleigh fading, and results in a huge loss in performance unless appropriate countermeasures are taken.

An easy method for deriving the structure of the coherent detector for MSK begins with the fact that MSK is equivalent to OQASK with a half-sinusoid shaping function. From (4.155), the MSK band-pass signal is

$$s(t) = \sqrt{E}x_k^R\varphi_1(t) - \sqrt{E}x_k^I\varphi_2(t) , \quad kT \le t \le (k+1)T \qquad (5.29)$$

where

$$\varphi_1(t) \;=\; \frac{2}{\sqrt{T}}\cos\left(\frac{\pi t}{2T}\right)\cos(2\pi f_c t) , \quad 0 \le t \le 2T \qquad (5.30)$$

$$\varphi_2(t) \;=\; \frac{2}{\sqrt{T}}\sin\left(\frac{\pi t}{2T}\right)\sin(2\pi f_c t) , \quad 0 \le t \le 2T \qquad (5.31)$$

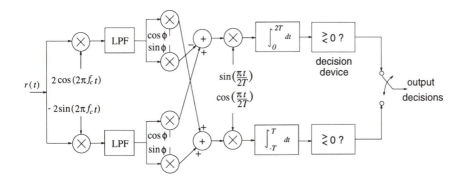

Figure 5.4 Coherent detector for MSK signals.

are the orthonormal basis functions. Since the source symbol sequences $\{x_n^R\}$ and $\{x_n^I\}$ provide independent modulation on the in-phase and quadrature carrier components, a suitable coherent detector for MSK is shown in Fig. 5.4. Note that the source symbols on the in-phase and quadrature carrier components must be detected over intervals of length $2T$, the duration of the amplitude shaping pulse, and bit decisions are made every T seconds.

5.3.2 Error Probability of M-QAM

Consider an M-QAM system having a constellation size $M = 2^{2m}$ for some integer m. To derive the error probability for M-QAM consider, for example, the 16-QAM signal constellation in Fig. 5.5. The dotted lines in Fig. 5.5 define the decision regions. The M-QAM system can be viewed as two pulse amplitude modulated (PAM) systems in quadrature, each having $\sqrt{M} = 2^m$ constellation points and one-half the power of the QAM system. For example, the 16-QAM system can be treated as two 4-PAM systems. The error probability in the $\sqrt{M}$-PAM system is (see Problem 5.1)

$$P_{\sqrt{M}} = 2 \left(1 - \frac{1}{\sqrt{M}} \right) Q \left(\sqrt{\frac{6}{M-1} \frac{\gamma_s}{2}} \right) \tag{5.32}$$

where γ_s is the average received symbol energy-to-noise ratio for the M-QAM signal constellation. The probability of correct symbol reception in the M-QAM system is

$$P_c = (1 - P_{\sqrt{M}})^2 \tag{5.33}$$

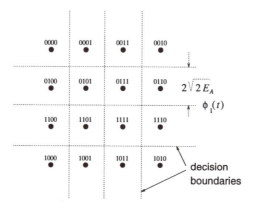

Figure 5.5 Complex signal-space diagram for 16-QAM constellation with Gray coding. Adjacent symbols differ in only one bit position. The dotted lines define decision regions.

and the probability of symbol error is

$$P_M = 1 - P_c = 1 - (1 - P_{\sqrt{M}})^2 \ . \tag{5.34}$$

If the channel is Rayleigh faded, then γ_s has the exponential pdf

$$p_{\gamma_s}(x) = \frac{1}{\bar{\gamma}_s} e^{-x/\bar{\gamma}_s} \tag{5.35}$$

where $\bar{\gamma}_s$ is the received symbol energy-to-noise ratio averaged over the signal constellation and the fade distribution. It follows that the average symbol error probability is

$$P_M = \int_0^\infty P_M(x) p_{\gamma_s}(x) dx \tag{5.36}$$

which, unfortunately, does not exist in closed form.

Calculation of the probability of bit error for M-QAM requires a mapping between the symbols and source bits. Uncoded QAM systems usually employ **Gray coding,** where the source bits are mapped onto symbols such that adjacent symbols differ in only one bit position. A Gray code mapping for 16-QAM is included in Fig. 5.5. At high $\bar{\gamma}_s$, errors are made to adjacent symbols with high probability and, therefore, the bit error probability is closely approximated by $P_b \approx P_M/\log_2 M$. This approximation (which is actually a lower bound) becomes loose at low $\bar{\gamma}_s$.

Fig. 5.6 plots the (approximate) bit error probability against the average received *bit* energy-to-noise ratio, $\bar{\gamma}_b = \bar{\gamma}_s/\log_2 M$, for several values of M. No-

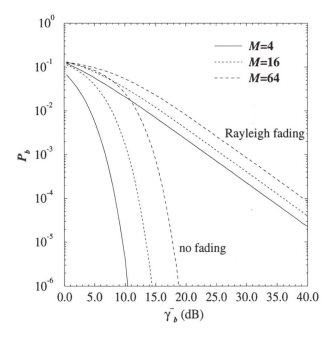

Figure 5.6 Bit error probability for M-QAM on an AWGN channel and a Rayleigh fading channel with AWGN.

tice that the required $\bar{\gamma}_b$ increases with the alphabet size M. However, the bandwidth efficiency also increases with M, since the baud rate is $\log_2 M$ bits/symbol.

5.3.3   Error Probability of OFDM

For an AWGN channel, the error probability of OFDM can be calculated by taking advantage of the property that the OFDM sub-carriers are orthogonal. The optimum receiver for OFDM on an AWGN channel consists of a bank of correlators, each operating on one sub-carrier. The overall symbol error probability is equal to the symbol error probability associated with one of the sub-carriers.

A key advantage of using OFDM is that the correlation receiver can be implemented in discrete time by using a discrete Fourier transform (DFT) or a fast Fourier transform (FFT). Suppose that the discrete-time sequence

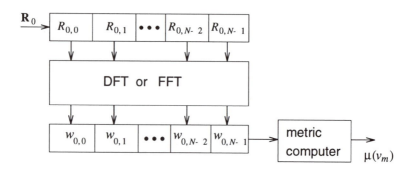

Figure 5.7 Block diagram of OFDM receiver implemented by using a DFT or an FFT.

$\mathbf{X}_0^g = \{X_{0,i}^g\}_{i=0}^{N+G-1}$ is transmitted over a flat fading channel with complex gain g. The corresponding received sequence is $\mathbf{R}_0^g = \{R_{0,i}^g\}_{i=0}^{N+G-1}$, where $R_{0,k}^g = gX_k^g + z_k$. The receiver first removes the guard interval according to $R_{0,k} = R_{0,G+(k-G)_N}^g$, $0 \le k \le N-1$, where $(k)_N$ is the residue of k modulo N. Demodulation is then performed by computing the FFT on the block $\mathbf{R}_0$ to yield the N decision variables

$$w_{0,i} = \frac{1}{N} \sum_{k=0}^{N-1} R_{0,k} e^{-\frac{j2\pi ik}{N}} , \quad i = 0, \ldots, N-1 . \tag{5.37}$$

For *each* of the $w_{0,i}$, the ML receiver decides in favor of the *signal point* x_m that maximizes the metric

$$\mu(v_m) = -\|w_{0,i} - gv_m\|^2 \quad i = 0, \ldots, N-1 \tag{5.38}$$

where $v_m = \sqrt{2E_A}x_m$. A block diagram of an OFDM receiver is shown in Fig. 5.7.

A more interesting issue is the effect of Doppler on the receiver performance. In our analysis, we will assume that the guard interval is always longer than the maximum channel delay, so that the effects of ISI are removed and the blocks can be analyzed independently. However, variations in the complex channel gain over the duration of a block causes a loss of subchannel orthogonality that results in the introduction of **interchannel interference** (ICI) . Suppose that the channel is characterized by flat fading channel and the channel gain g changes over a block of N symbols. Ignoring AWGN, the discrete-time received sequence after removal of the guard interval is

$$R_{0,k} = g_{G+(k-G)_N} X_{0,k} . \tag{5.39}$$

The output of the FFT demodulator becomes

$$w_i = \sqrt{2E_A} \sum_{n=0}^{N-1} x_{0,n} H(n-i) \tag{5.40}$$

where

$$H(n-i) = \sum_{k=0}^{N-1} g_{G+(k-G)_N} e^{j\frac{2\pi}{N}(n-i)k} , \quad 0 \le i \le N-1 . \tag{5.41}$$

To highlight the mechanism of the channel induced distortion, (5.40) can be rewritten as

$$w_i = \sqrt{2E_A} H(0) x_{0,i} + c_i \tag{5.42}$$

where

$$c_i = \sqrt{2E_A} \sum_{\substack{n=1 \\ n \ne i}}^{N-1} x_{0,n} H(n-i) . \tag{5.43}$$

Note that $H(0)$ is a multiplicative noise term and c_i is the additive ICI term. If the channel is time-invariant, then $g_k = g$ and $w_i = g\sqrt{2E_A} x_{0,i}$; otherwise w_i is a function of all the data symbols within a block and, hence, interchannel interference (ICI) is introduced.

For N sufficiently large, the central limit theorem can be invoked and the ICI modeled as a Gaussian random process. Since $x_{0,n}$ and $H(n-i)$ are independent random variables and $E[x_{0,n}] = 0$, it follows that $E[c_l] = 0$. Since $E_A E[x_{0,n} x_{0,m}^*] = E_{av} \delta_{nm}$, where E_{av} is the average symbol energy, the autocorrelation of c_l is

$$\phi_{cc}(r) = \frac{1}{2} E[c_i c_{i+r}^*] = E_{av} \sum_{n \ne i, i+r} E[H(n-i) H^*(n-i-r)] . \tag{5.44}$$

If we further assume $E[|g_k|^2] = 1$ and isotropic scattering, the autocorrelation becomes

$$\phi_{cc}(r) = E_{av} \delta_r - \frac{E_{av}}{N^2} \sum_{k=0}^{N-1} \sum_{k'=0}^{N-1} J_0(2\pi f_m T_s^g (k-k'))$$
$$\cdot \left[\exp\left(j\frac{2\pi k' r}{N} \right) + (1 - \delta_r) \exp\left(j\frac{2\pi k r}{N} \right) \right] \tag{5.45}$$

where f_m is the maximum Doppler frequency. Note that the autocorrelation is not influenced by the positioning of the guard interval, due to the symmetry of the summations over k.

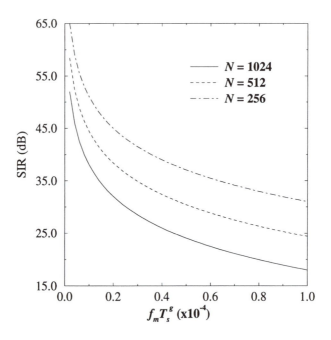

Figure 5.8 Signal-to-interference ratio due to ICI.

For symbol-by-symbol detection, it is sufficient to examine the variance of the ICI, which can be simplified as

$$\phi_{cc}(0) = E_{\mathrm{av}} - \frac{E_{\mathrm{av}}}{N^2}\left\{ N + 2\sum_{i=1}^{N-1}(N-i)J_0(2\pi f_m T_s^g i) \right\}, \tag{5.46}$$

where the fact that $J_0(\ \cdot\ )$ is an even function has been used. Note that variance of the ICI terms is only a function of E_{av}, N, T_s^g, and f_m, but is independent of the signal constellation. Fig. 5.8 plots the signal-to-interference ratio

$$\mathrm{SIR} = \frac{E_{\mathrm{av}}}{\phi_{cc}(0)} \tag{5.47}$$

as a function of $f_m T_s^g$ for several values of N.

Suppose that the symbols $x_{k,n}$ are chosen from a 16-QAM alphabet. From Section 5.3.2, the symbol error probability for 16-QAM is

$$P_M = 3Q\left(\sqrt{\frac{1}{5}\gamma_s}\right)\left[1 - \frac{3}{4}Q\left(\sqrt{\frac{1}{5}\gamma_s}\right)\right] \tag{5.48}$$

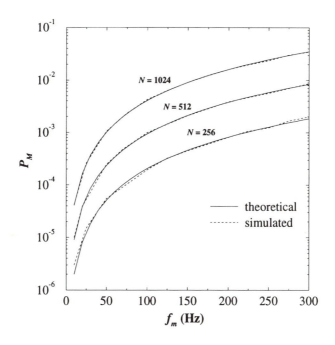

Figure 5.9 Error floor due to ICI with 16-QAM.

where γ_s is the average received symbol energy-to-noise ratio. With Rayleigh fading, the symbol error probability is obtained by averaging (5.48) over the pdf in (5.35). The error floor due to ICI can be obtained by substituting the SIR in (5.47) for $\bar{\gamma}_s$ yielding the result shown in Fig. 5.9. Simulation results are also shown in Fig. 5.9 to verify the the Gaussian approximation for the ICI. Fig. 5.10 shows the performance of OFDM with $N = 512$ subcarriers, a 16-QAM signal constellation, and a 20 Mbps bit rate for various Doppler frequencies. At low $\bar{\gamma}_b$, additive noise dominates the performance so that Doppler has little effect. However, at $\bar{\gamma}_b$ ICI dominates the performance and causes an error floor.

5.3.4   Error Probability of M-PSK

To derive the error probability of M-PSK consider, for example, the 8-PSK signal constellation and associated decision regions shown in Fig. 5.11. Suppose that the message vector $v_0 = \sqrt{2E}$ is transmitted. The received signal vector is

$$w = \alpha e^{j\phi} v_0 + z \ . \tag{5.49}$$

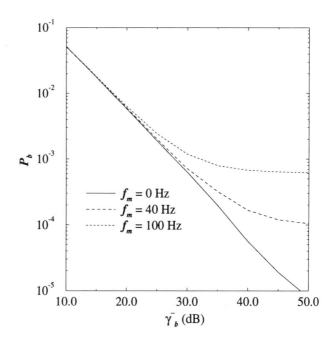

Figure 5.10   Bit error probability for 16-QAM OFDM on a Rayleigh fading channel with various Doppler frequencies.

Since the error probability is invariant to rotation of the signal set, we can arbitrarily set $\phi = 0$ so that

$$
\begin{aligned}
w &= \alpha v_0 + z \\
&= \alpha \sqrt{2E} + z
\end{aligned}
\tag{5.50}
$$

It follows that $w = w^R + jw^I$ is a complex Gaussian random variable with pdf

$$
p_w(w) = \frac{1}{2\pi N_o} \exp\left\{ -\frac{1}{2N_o} \left| w - \alpha\sqrt{2E} \right|^2 \right\} .
\tag{5.51}
$$

Using the bivariate transformation

$$
R = \sqrt{(w^R)^2 + (w^I)^2} , \qquad \Theta = \mathrm{Tan}^{-1}\left(\frac{w^I}{w^R} \right)
\tag{5.52}
$$

gives

$$
p_{R,\Theta}(r, \theta) = \frac{r}{2\pi N_o} \exp\left\{ -\frac{1}{2N_o} \left(r^2 - 2\sqrt{2}r\alpha E \cos\theta + 2\alpha^2 E^2 \right) \right\} .
\tag{5.53}
$$

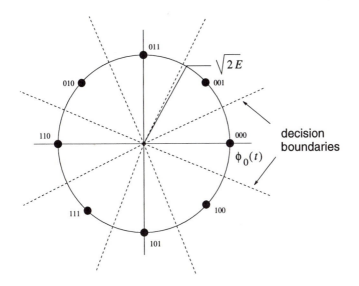

Figure 5.11 Complex signal-space diagram for 8-PSK along and the associated decision regions.

with M-PSK, only the phase conveys information and, therefore, we extract the marginal pdf

$$p_\Theta(\theta) = \int_0^\infty p_{R,\Theta}(r,\theta)dr \qquad (5.54)$$

$$= \frac{1}{2\pi}e^{-\gamma_s}\left(1 + \sqrt{4\pi\gamma_s}\cos\theta e^{\gamma_s\cos^2\theta}\frac{1}{\sqrt{2\pi}}\int_{-\sqrt{2\gamma_s}\cos\theta}^\infty e^{-x^2/2}dx\right)$$

where $\gamma_s = \alpha^2 E/N_o$ is the received symbol energy-to-noise ratio. The probability of symbol error is just the probability that Θ lies outside the region $-\pi/M \leq \Theta \leq \pi/M$. Thus

$$P_M(\gamma_s) = 1 - \int_{-\pi/M}^{\pi/M} p(\theta)d\theta \ . \qquad (5.55)$$

For $M = 2$ or $M = 4$, an exact expression can be obtained for the symbol error probability and the bit error probability. In particular

$$P_b = Q(\sqrt{2\gamma_b}) \ . \qquad (5.56)$$

where $\gamma_b = \gamma_s/\log_2 M$ is the received bit energy-to-noise ratio.

If the channel is Rayleigh faded, then γ has the exponential pdf in (refqamsnr). The average symbol error probability is

$$P_M = \int_0^\infty P_M(x) p_{\gamma_s}(x) dx \ . \tag{5.57}$$

For $M = 2$ or $M = 4$

$$P_b = \frac{1}{2} \left[1 - \sqrt{\frac{\bar{\gamma}_b}{1 + \bar{\gamma}_b}} \right] \ . \tag{5.58}$$

Differential Encoding:

Coherent receivers extract the carrier phase from the received signal. However, random data creates a phase ambiguity in the absolute carrier phase. For BPSK, a 180° phase ambiguity is present, while for QPSK a 90° phase ambiguity is present. This phase ambiguity can be eliminated by using **differential encoding** , where information is transmitted in the phase difference between successive symbols. Sometimes this is called differentially encoded PSK or DPSK. Coherent demodulation of DPSK begins as described above. After making symbol decisions, a differential decoder is used to determine the phase difference between two successive symbol decisions. The bit error probability is higher than that with absolute phase encoding, being roughly two times as great. However, this translates into only a very small performance loss.

5.4   DIFFERENTIAL DETECTION

DPSK can also be detected noncoherently by using differentially coherent detection, where the receiver compares the phase of the received signal between two successive signaling intervals. If the carrier phase changes slowly with respect to the symbol duration, then the phase difference between waveforms received in two successive signaling intervals will be independent of the absolute carrier phase. However for fading channels, the carrier phase changes very rapidly when the channel exhibits a deep fade. This leads to an error floor that increases with the Doppler frequency.

Suppose that binary DPSK is used. Let θ_n denote the absolute carrier phase for the nth symbol, and $\Delta\theta_n = \theta_n - \theta_{n-1}$ denote the differential carrier phase. Several mappings between the differential carrier phase and the source symbols.

Here we consider the mapping

$$\Delta\theta_n = \begin{cases} 0 & , \quad x_n = +1 \\ \pi & , \quad x_n = -1 \end{cases} \tag{5.59}$$

The complex envelope of the transmitted signal is

$$v(t) = A \sum_n h_a(t - nT)e^{j\theta_n} \tag{5.60}$$

and the complex envelope of the received signal is

$$w(t) = \alpha A \sum_n h_a(t - nT)e^{j\theta_n + \phi} + z(t) \tag{5.61}$$

where $g = \alpha e^{j\phi}$ is the complex channel gain.

A block diagram of a differentially coherent receiver for DPSK is shown in Fig. 5.12. The values of X_k, X_{kd}, Y_k and Y_{kd} in Fig. 5.12 are

$$\begin{aligned} X_k &= 2\alpha E \cos(\theta_n + \phi) + z^R \\ X_{kd} &= 2\alpha E \cos(\theta_{n-1} + \phi) + z_d^R \\ Y_k &= 2\alpha E \sin(\theta_n + \phi) + z^I \\ Y_{kd} &= 2\alpha E \sin(\theta_{n-1} + \phi) + z_d^I \end{aligned} \tag{5.62}$$

where

$$E = \frac{A^2}{2} \int_0^T h_a^2(t)dt \tag{5.63}$$

is the bit energy, and the noise terms are

$$\begin{aligned} z^R &= A \int_{kT}^{(k+1)T} z^R(t)h_a(t)dt \\ z_d^R &= A \int_{(k-1)T}^{kT} z^R(t)h_a(t)dt \\ z^I &= A \int_{kT}^{(k+1)T} z^I(t)h_a(t)dt \\ z_d^I &= A \int_{(k-1)T}^{kT} z^I(t)h_a(t) \end{aligned} \tag{5.64}$$

Note that z^R, z_d^R, z^I, and z_d^I are independent Gaussian random variables with variance $2EN_o$.

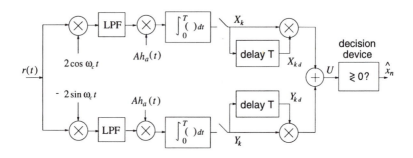

Figure 5.12   Differentially coherent receiver for DPSK signals.

In the absence of noise, it is easy to verify that $U = 4\alpha^2 E^2 x_n$. To determine the pdf of the decision variable U, it is convenient to express U as

$$U = \text{Re}\{Z_k Z_{kd}^*\} = \frac{1}{2}\left(Z_k Z_{kd}^* + Z_k^* Z_{kd}\right) \qquad (5.65)$$

where

$$Z_k = X_k + jY_k \qquad (5.66)$$
$$Z_{kd} = X_{kd} + jY_{kd} \ . \qquad (5.67)$$

It can be shown by using characteristic functions that $U = W - Y$, where W and Y are non-central and central chi-square random variables with densities [292]

$$f_W(w) = \frac{1}{2EN_o} e^{-\frac{(wx_n + 4\alpha^2 E^2)}{2EN_o}} I_0\left(\frac{2\sqrt{wx_n\alpha^2 E^2}}{EN_o}\right), \ wx_n \geq 0 \quad (5.68)$$

$$f_Y(y) = \frac{1}{2EN_o} e^{-\frac{yx_n}{2EN_o}}, \ yx_n \geq 0 \ . \qquad (5.69)$$

The pdf of U is

$$f_U(u) = \int_{R_{UV}} f_W(v) f_Y(v - u) dv$$

$$= \begin{cases} \frac{1}{4EN_o} \exp\left\{\frac{x_n u - 2\alpha^2 E^2}{2EN_o}\right\}, & -\infty < x_n u < 0 \\ \frac{1}{4EN_o} \exp\left\{\frac{x_n u - 2\alpha^2 E^2}{2EN_o}\right\} Q\left(\sqrt{\frac{2\alpha^2\mathcal{E}}{N_o}}, \sqrt{\frac{2x_n u}{EN_o}}\right), & 0 < x_n u < \infty \end{cases}$$
$$(5.70)$$

where $Q(a, b)$ is the Marcum Q function. From (5.70), the bit error probability of DPSK with differential detection is

$$P_b = \int_0^\infty \frac{1}{4EN_o} \exp\left\{-\frac{u + 2\alpha^2 E^2}{2EN_o}\right\} du = \frac{1}{2} e^{-\gamma_b} \qquad (5.71)$$

where $\gamma_b = \alpha^2 E/N_o$ is the received bit energy-to-noise ratio. For a Rayleigh fading channel, α is Rayleigh distributed so the received bit energy-to-noise ratio has the exponential pdf in (5.27). It follows that the bit error probability with Rayleigh fading is

$$P_b = \frac{1}{2(1 + \bar{\gamma}_b)} \; . \tag{5.72}$$

5.4.1 Differential Detection of $\pi/4$-DQPSK

The above results can be extended to differential detection of $\pi/4$-DQPSK. Once again the complex envelopes of the transmitted and received signals are given by (5.60) and (5.61), respectively. However, with $\pi/4$-DQPSK, $\Delta\theta_n = 3\pi x_n/4$ where $x_n \in \{\pm 1, \pm 3\}$. A block diagram of the differentially coherent receiver for $\pi/4$-DQPSK is shown in Fig. 5.13. The values of X_k, X_{kd}, Y_k and Y_{kd} are again given by (5.62). The detector outputs are

$$U = \text{Re}\{Z_k Z_{kd}^*\} = \frac{1}{2}(Z_k Z_{kd}^* + Z_k^* Z_{kd}) \tag{5.73}$$

$$V = \text{Im}\{Z_k Z_{kd}^*\} = \frac{1}{j2}(Z_k Z_{kd}^* - Z_k^* Z_{kd}) \tag{5.74}$$

where Z_k and Z_{kd} are defined in (5.66) and (5.67), respectively. In the absence of noise, it can be verified that the detector outputs are

$$\begin{aligned}
U &= -a, & V &= -a, & x_n &= -3 \\
U &= a, & V &= -a, & x_n &= -1 \\
U &= a, & V &= a, & x_n &= +1 \\
U &= -a, & V &= a, & x_n &= +3
\end{aligned} \tag{5.75}$$

where $a = 2\sqrt{2}\alpha^2 E^2$. The bit error probability for $\pi/4$-DQPSK with Gray coding is quite complicated to derive, but can be expressed in terms of well known functions [256]

$$P_b = Q(a, b) - \frac{1}{2}I_0(ab)\exp\left\{-\frac{1}{2}\left(a^2 + b^2\right)\right\} \tag{5.76}$$

where

$$a = \sqrt{2\gamma_b\left(1 - \frac{1}{\sqrt{2}}\right)} \tag{5.77}$$

$$b = \sqrt{2\gamma_b\left(1 + \frac{1}{\sqrt{2}}\right)} \tag{5.78}$$

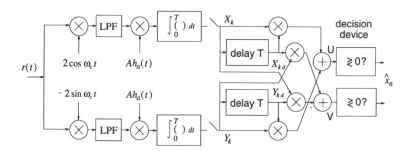

Figure 5.13   Differentially coherent receiver for $\pi/4$-QPSK signals.

and γ_b is the bit energy-to-noise ratio. Once again, if the channel is faded, then the bit error probability can be obtained by averaging over the fade distribution.

5.5   DIVERSITY TECHNIQUES

Rayleigh fading has been shown to convert an exponential dependency of the bit error probability on the signal-to-noise ratio into an inverse linear one, thereby resulting in a very large signal-to-noise ratio penalty. Diversity is one very effective solution that exploits the principle of providing the receiver with multiple faded replicas of the same information bearing signal. Assume independent diversity branches, and let p denote the probability that the instantaneous signal-to-noise ratio is below some critical threshold on a particular diversity branch. Then p^L is the probability that the instantaneous signal-to-noise ratio is below the same critical value on L diversity branches.

The methods by which diversity can be achieved generally fall into seven categories: i) space, ii) angle, iii) polarization, iv) field v) frequency, vi) multipath, and vii) time. Space diversity is achieved by using multiple receiver antennas. The spatial separation between the multiple antennas is chosen so that the diversity branches experience uncorrelated fading. Chapter 2 showed that a spatial separation of about a half-wavelength will suffice with 2-D isotropic scattering. Angle diversity or direction diversity requires a number of directional antennas. Each antenna selects plane waves arriving from a narrow range of angles, so that uncorrelated branches are achieved. Polarization diversity exploits the property that a scattering environment tends to depolarize a signal. Receiver antennas having different polarizations can be used to obtain diversity. The combination of vertical monopole and patch antennas are commonly used in cellular subscriber units for this purpose. Field diversity exploits the fact that the electric and magnetic field components at any point are uncorrelated. It suffers from the disadvantage that the receiver front end needs to be duplicated. Frequency diversity is obtained by using multiple channels that are separated by at least the coherence bandwidth of the channel. In many cases, this can be several hundred kilohertz. Since multiple frequencies are needed, frequency diversity is not a bandwidth efficient solution for FDMA or TDMA systems. Frequency hop spread spectrum can exploit frequency diversity through the principle of fast frequency hopping, where each symbol is transmitted sequentially on multiple hops (or carriers) that experience uncorrelated fading. Multipath diversity is obtained by resolving multipath components at different delays by using direct sequence spread spectrum signaling along with a RAKE receiver. Spread spectrum concepts will be discussed in detail in Chapter 8. Time diversity is obtained by using multiple time slots that are separated by at least the coherence time of the channel. Error correction coding can be viewed as an efficient method of time diversity. Unfortunately, the coherence time of the channel depends on the velocity of the mobile station (MS). Slow moving MSs have channels will have a large coherence time. Under this condition, it may not be possible to obtain time diversity without introducing unacceptable delay into the transmitted signals.

5.5.1 Diversity Combining

Diversity combining refers to the method by which the signals from the diversity branches are combined. There are several ways of categorizing diversity combining methods. Diversity combining that takes place before detection is called **predetection combining**, while diversity combining that takes place after detection is called **postdetection combining**. With ideal coherent detection, there is no difference in performance between predetection and postdetection combining. For differentially coherent detection there is a slight difference in performance. In any case, there are many methods of combining the diversity branches including maximal ratio, equal gain, selective, and switched combining.

The received complex envelopes for the diversity branches are

$$w_k(t) = g_k v_m(t) + z_k(t) \quad , \qquad k = 1, \ldots, L$$
$$, \qquad m = 1, \ldots, M \qquad (5.79)$$

where $g_k = \alpha_k e^{-j\phi_k}$ is the complex fading gain associated with the kth branch. Note that the noise processes $z_k(t)$ are assumed to be independent from branch to branch. The corresponding low-pass received signal vectors are

$$\mathbf{w}_k = g_k \mathbf{v}_m + \mathbf{z}_k \qquad \begin{aligned} k &= 1, \ldots, L \\ m &= 1, \ldots, M \end{aligned} \qquad (5.80)$$

where

$$w_{k,i} = g_k v_{m,i} + z_{k,i} \; . \qquad (5.81)$$

The fading gains of the various diversity branches typically have some degree of correlation, the degree of which depends on the type of diversity being used and the propagation environment. For analytical purposes, the assumption is often made that the diversity branches are uncorrelated. However, branch correlation reduces the achievable diversity gain and, therefore, the uncorrelated branch assumption gives optimistic results. Nevertheless, we will evaluate the performance of the various diversity combining techniques under the assumption of uncorrelated branches.

Finally, the fade distribution will affect the diversity gain. In general, the *relative* advantage of diversity is greater for Rayleigh fading than Ricean fading, because as the Rice factor K increases there is less difference between the instantaneous received signal-to-noise ratios on the various diversity branches. However, the performance will always be better with Ricean fading than with Rayleigh fading, for a given average received signal-to-noise ratio and diversity order. For our purpose, we will detail the performance with Rayleigh fading.

5.5.2 Predetection Selective Combining

Suppose that ideal selective combining (SC) is used, so that branch giving the highest signal-to-noise ratio is selected at any instant. For radio links that use continuous transmission, i.e., FMDA systems, SC is not very practical, because it requires continuous monitoring of all the diversity branches. If such monitoring is performed, then it is probably better to use maximal ratio combining, as discussed in the next section, since the implementation is no more complex and the performance is better. However, in TDMA systems, a form of SC can sometimes be implemented where the diversity branch is selected prior to the transmission of a TDMA burst. The selected branch is then used for the duration of the entire burst. For example, in the base-to-mobile link specified in the IS-54 common air interface [78] signals are transmitted in all the slots of a frame regardless of whether or not they carry voice traffic. Therefore, branch monitoring is easy to implement at the MS. In this section, however, we evaluate selection diversity under the assumption of continuous branch selection.

With Rayleigh fading, the instantaneous received bit energy-to-noise ratio on the kth diversity branch has the exponential pdf

$$p_{\gamma_k}(x) = \frac{1}{\bar{\gamma}_c} e^{-x/\bar{\gamma}_c} \tag{5.82}$$

where $\bar{\gamma}_c$ is the average received bit energy-to-noise ratio for each diversity branch. With SC the branch with the largest bit energy-to-noise ratio is always selected so that the effective instantaneous bit energy-to-noise ratio is

$$\gamma_b^s = \max\{\gamma_1, \gamma_2, \cdots, \gamma_L\} \tag{5.83}$$

where L is the number of diversity branches. If the diversity branches are independently faded then order statistics gives the cumulative distribution function (cdf)

$$F_{\gamma_b^s}(x) = \Pr\left[\gamma_1 \le x \bigcap \gamma_2 \le x \bigcap \cdots \bigcap \gamma_L \le x\right] = \left[1 - e^{-x/\bar{\gamma}_c}\right]^L . \tag{5.84}$$

Differentiating the above expression gives the pdf

$$p_{\gamma_b^s}(x) = \frac{L}{\bar{\gamma}_c}\left[1 - e^{-x/\bar{\gamma}_c}\right]^{L-1} e^{-x/\bar{\gamma}_c} . \tag{5.85}$$

The average bit energy-to-noise ratio with SC is

$$
\begin{aligned}
\bar{\gamma}_b^s &= \int_0^\infty x p_{\gamma_b^s}(x)\,dx \\
&= \int_0^\infty \frac{Lx}{\bar{\gamma}_c}\left[1 - e^{-x/\bar{\gamma}_c}\right]^{L-1} e^{-x/\bar{\gamma}_c}\,dx \\
&= \bar{\gamma}_c \sum_{k=1}^L \frac{1}{k} .
\end{aligned}
\tag{5.86}
$$

Fig. 5.14 plots the cdf $F_{\gamma_b^s}(x)$. Note that the largest diversity gain is obtained by using 2-branch diversity and diminishing returns are obtained with increasing L. This is typical for all diversity techniques.

To compute the bit error probability, slow fading is assumed where the bit error probability can be obtained by averaging over the pdf of γ_b^s. For example, DPSK with differential detection has the bit error probability

$$P_b(\gamma_b^s) = \frac{1}{2} e^{-\gamma_b^s} . \tag{5.87}$$

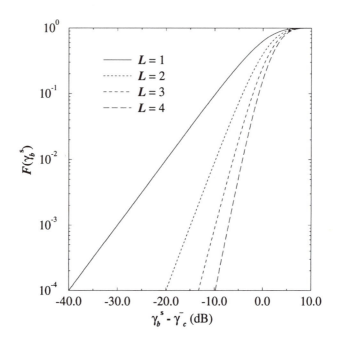

Figure 5.14 Cdf of γ_b^s for predetection selective combining; $\bar{\gamma}_c$ is the average branch bit energy-to-noise ratio.

Hence, with SC

$$
\begin{aligned}
P_b &= \int_0^\infty P_b(x) p_{\gamma_b^s}(x)\,dx \\
&= \int_0^\infty \frac{L}{2\bar{\gamma}_c} \exp\left\{-\left(1+\frac{1}{\bar{\gamma}_c}\right)x\right\}\left(1-\exp\left\{-\frac{x}{\bar{\gamma}_c}\right\}\right)^{L-1} dx \\
&= \frac{L}{2\bar{\gamma}_c}\sum_{n=0}^{L-1}\binom{L-1}{n}(-1)^n \int_0^\infty \exp\left\{-\left(1+\frac{1}{/\bar{\gamma}_c}+\frac{n}{\bar{\gamma}_c}\right)x\right\}dx \\
&= \frac{L}{2}\sum_{n=0}^{L-1}\frac{\binom{L-1}{n}(-1)^n}{1+n+\bar{\gamma}_c}
\end{aligned}
\tag{5.88}
$$

where we have used the binomial expansion

$$
(1-x)^{L-1} = \sum_{n=0}^{L-1}\binom{L-1}{n}(-1)^n x^n \ .
\tag{5.89}
$$

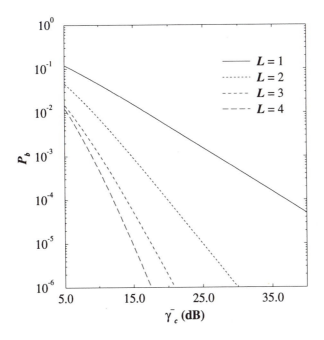

Figure 5.15 Bit error probability for predetection selective combining and differentially detected DPSK.

The error probability in (5.88) is plotted in Fig. 5.15. Notice that diversity offers a very large improvement in the performance. Equation (5.88) shows that at large bit energy-to-noise ratios, the bit error probability is proportional to $\bar{\gamma}_c^{-L}$. Again, the largest diversity gain is achieved with 2-branch diversity and diminishing returns are realized with increasing L.

5.5.3 Predetection and Postdetection Maximal Ratio Combining

With maximal ratio combining (MRC), the diversity branches must be weighted by their respective complex fading gains and combined. MRC results in an ML receiver as we will now show and, therefore, gives the best possible performance among the diversity combining techniques.

Referring to (5.80), the vector

$$\mathbf{w} \overset{\triangle}{=} (\mathbf{w}_1, \ \mathbf{w}_2, \ \cdots, \ \mathbf{w}_L) \tag{5.90}$$

has the multivariate Gaussian distribution

$$
\begin{aligned}
p(\mathbf{w}|\mathbf{g}, \mathbf{v}_m) &= \prod_{k=1}^{L} \prod_{i=1}^{N} \frac{1}{\pi N_o} \exp\left\{ -\frac{1}{N_o} |w_{k,i} - g_k v_{m,i}|^2 \right\} \\
&= \frac{1}{(\pi N_o)^{LN}} \exp\left\{ -\frac{1}{N_o} \sum_{k=1}^{L} \|\mathbf{w}_k - g_k \mathbf{v}_m\|^2 \right\} \tag{5.91}
\end{aligned}
$$

where $\mathbf{g} = (g_1, g_2, \ldots, g_L)$ is the channel vector. From this expression, the ML receiver chooses the message vector $\mathbf{v}_m$ that maximizes the metric

$$
\begin{aligned}
\mu(\mathbf{v}_m) &= -\sum_{k=1}^{L} \|\mathbf{w}_k - g_k \mathbf{v}_m\|^2 \\
&= -\sum_{k=1}^{L} \left\{ \|\mathbf{w}_k\|^2 - 2\mathrm{Re}\left\{\mathbf{w}_k \cdot g_k^* \mathbf{v}_m^*\right\} + |g_k|^2 \|\mathbf{v}_m\|^2 \right\} \ . \tag{5.92}
\end{aligned}
$$

Since $\sum_{k=1}^{L} \|\mathbf{w}_k\|^2$ is independent of $\mathbf{v}_m$ and $\|\mathbf{v}_m\|^2 = 2\mathcal{E}_m$, the receiver just needs to maximize the metric

$$
\begin{aligned}
\mu(\mathbf{v}_m) &= \sum_{k=1}^{L} \mathrm{Re}\left\{\mathbf{w}_k \cdot g_k^* \mathbf{v}_m^*\right\} - \mathcal{E}_m \sum_{k=1}^{L} |g_k|^2 \\
&= \sum_{k=1}^{L} \mathrm{Re}\left\{ g_k^* \int_0^T w_k(t) v_m^*(t) dt \right\} - \sum_{k=1}^{L} |g_k|^2 \mathcal{E}_m \tag{5.93}
\end{aligned}
$$

If signals have equal energy then the last term can be neglected, since it is the same for all message vectors, resulting in

$$
\begin{aligned}
\mu(\mathbf{v}_m) &= \sum_{k=1}^{L} \mathrm{Re}\left\{\mathbf{w}_k \cdot g_k^* \mathbf{v}_m^*\right\} \\
&= \sum_{k=1}^{L} \mathrm{Re}\left\{ g_k^* \int_0^T w_k(t) v_m^*(t) dt \right\} \tag{5.94}
\end{aligned}
$$

From the above development, an ML receiver can be constructed. The portion of the receiver that generates $\mu(\mathbf{v}_m)$ for arbitrary m is shown in Fig. 5.16 (assuming equal energy signals). It is apparent that the ML receiver must have complete knowledge of the channel vector $\{\mathbf{g}\}$.

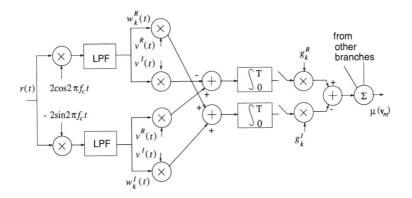

Figure 5.16 Receiver for maximal ratio postdetection diversity combining. Equal energy signals are assumed.

The receiver in Fig. 5.16 uses postdetection combining, where weighting and combining is performed after integration. However, the ML receiver can also be implemented by using predetection combining. To obtain this receiver structure we simply rewrite the metric in the form

$$
\mu(\mathbf{v}_m) = \mathrm{Re}\left\{\left(\sum_{k=1}^{L} g_k^* \mathbf{w}_k\right) \cdot \mathbf{v}_m^*\right\}
$$

$$
= \int_0^T \mathrm{Re}\left\{\left(\sum_{k=1}^{L} g_k^* w_k(t)\right) v_m^*(t)\right\} dt \qquad (5.95)
$$

The resulting structure can be obtained from inspection of this equation and is shown in Fig. 5.17. Note that the weighting and combining is performed before integration.

The envelope of the composite signal is

$$
\alpha_M = A \sum_{k=1}^{L} \alpha_k^2 \qquad (5.96)
$$

and the sum of the branch noise powers is

$$
P_{n,\mathrm{tot}} = P_n \sum_{k=1}^{L} \alpha_k^2 \qquad (5.97)
$$

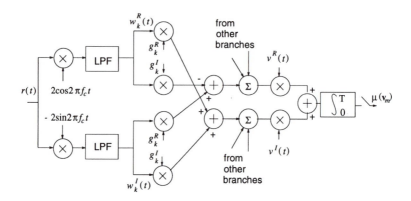

Figure 5.17 Receiver for maximal ratio predetection diversity combining. Equal energy signals are assumed.

where P_n is the noise power in each diversity branch. Hence, the bit energy-to-noise ratio with MRC is

$$\gamma_b^{mr} = \frac{\alpha_M^2}{2P_{n,tot}} = \sum_{k=1}^{L} \frac{A^2\alpha_k^2}{2P_n} = \sum_{k=1}^{L} \gamma_k \qquad (5.98)$$

which is the sum of the bit energy-to-noise ratios of the diversity branches. If all the diversity branches provide the same average power (which is a reasonable assumption for antenna diversity) and the branches are uncorrelated, then γ_b^{mr} has a chi-square distribution with $2L$ degrees of freedom. That is,

$$p_{\gamma_b^{mr}}(x) = \frac{1}{(L-1)!(\bar{\gamma}_c)^L} x^{L-1} e^{-x/\bar{\gamma}_c} \qquad (5.99)$$

where

$$\bar{\gamma}_c = \mathrm{E}[\gamma_k] \qquad k = 1, \ldots, L . \qquad (5.100)$$

The cdf of γ_b^{mr} is

$$F_{\gamma_b^{mr}}(x) = 1 - e^{-x/\bar{\gamma}_c} \sum_{k=1}^{L-1} \frac{1}{k!} \left(\frac{x}{\bar{\gamma}_c}\right)^k . \qquad (5.101)$$

It follows from (5.98) that the average bit energy-to-noise ratio with MRC is

$$\bar{\gamma}_b^{mr} = \sum_{k=1}^{L} \bar{\gamma}_k = \sum_{k=1}^{L} \bar{\gamma}_c = L\bar{\gamma}_c . \qquad (5.102)$$

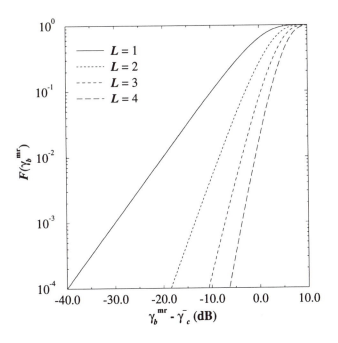

Figure 5.18 Cdf of γ_b^{mr} for maximal ratio combining; $\bar{\gamma}_c$ is the average branch bit energy-to-noise ratio.

Fig. 5.14 plots the cdf $F_{\gamma_b^{mr}}(x)$.

Since MRC is a coherent detection technique we must limit our attention to coherent signaling techniques, e.g., BPSK and M-QAM. For example, if BPSK is used then resulting error probability is:

$$
\begin{aligned}
P_b &= \int_0^\infty P_b(x) p_{\gamma_b^{mr}}(x) dx \\
&= \int_0^\infty Q\left(\sqrt{2x}\right) \frac{1}{(L-1)!(\bar{\gamma}_c)^L} x^{L-1} e^{-x/\bar{\gamma}_c} dx \\
&= \left(\frac{1-\mu}{2}\right)^L \sum_{k=0}^{L-1} \binom{L-1+k}{k} \left(\frac{1+\mu}{2}\right)^k
\end{aligned}
\tag{5.103}
$$

where

$$
\mu = \sqrt{\frac{\bar{\gamma}_c}{1+\bar{\gamma}_c}}
\tag{5.104}
$$

The last step follows after some algebra. The expression in (5.103) is plotted in Fig. 5.19. Once again, diversity significantly improves the performance.

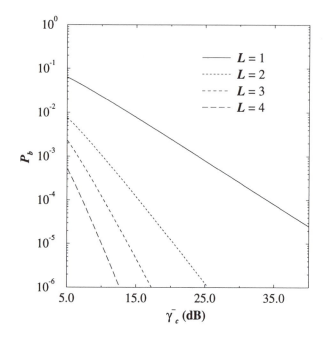

Figure 5.19 Bit error probability for predetection maximal ratio combining and coherent BPSK signaling.

5.5.4 Predetection and Postdetection Coherent Equal Gain Combining

Predetection and postdetection coherent equal gain combining (EGC) is similar to MRC because the diversity branches are co-phased, but different from MRC because the diversity branches are not weighted. In practice, such a scheme is useful for modulation techniques having equal energy symbols, e.g., M-PSK. With signals of unequal energy, the complete channel vector $\mathbf{g} = (g_1, g_2, \ldots, g_L)$ is required anyway and MRC should be used. With EGC, the receiver maximizes the metric

$$
\begin{aligned}
\mu(\mathbf{v}_m) &= \sum_{k=1}^{L} \mathrm{Re} \left\{ \mathbf{w}_k \cdot e^{-j\phi_k} \mathbf{v}_m^* \right\} \\
&= \sum_{k=1}^{L} \mathrm{Re} \left\{ e^{-j\phi_k} \int_0^T w_k(t) v_m^*(t) dt \right\} .
\end{aligned}
\tag{5.105}
$$

This expression leads to a coherent postdection equal gain receiver. Once again, the receiver can be implemented by using predetection combining. To obtain this receiver structure we rewrite (5.105) as

$$
\mu(\mathbf{v}_m) = \mathrm{Re}\left\{ \left(\sum_{k=1}^{L} e^{-j\phi_k} \mathbf{w}_k \right) \cdot \mathbf{v}_m^* \right\}
$$

$$
= \int_0^T \mathrm{Re}\left\{ \left(\sum_{k=1}^{L} e^{-j\phi_k} w_k(t) \right) v_m^*(t) \right\} dt \ . \tag{5.106}
$$

The postdetection and predetection receiver structures are identical to those shown in Figs. 5.16 and 5.17, except that we use $g_k = e^{j\phi_k}$ rather than $g_k = \alpha_k^{j\phi_k}$, i.e., the branches are co-phased but not weighted.

The envelope of the composite signal is

$$
\alpha_E = A \sum_{k=1}^{L} \alpha_k \tag{5.107}
$$

and the sum of the branch noise powers is LP_n. The resulting bit energy-to-noise ratio is

$$
\gamma_b^{\mathrm{eg}} = \frac{\alpha_E^2}{2LP_n} \tag{5.108}
$$

The cdf and pdf for γ_b^{eg} does not exist in closed form for $L > 2$. However, for $L = 2$ the cdf is equal to

$$
F_{\gamma_b^{\mathrm{eg}}}(x) = 1 - e^{-2x/\bar{\gamma}_c} - \sqrt{\pi \frac{x}{\bar{\gamma}_c}} e^{-x/\bar{\gamma}_c} \left(1 - 2Q\left(\sqrt{2\frac{x}{\bar{\gamma}_c}} \right) \right) \ . \tag{5.109}
$$

Differentiating the above expression yields the pdf

$$
p_{\gamma_b^{\mathrm{eg}}}(x) = \frac{1}{\bar{\gamma}_c} e^{-2x/\bar{\gamma}_c} + \sqrt{\pi} e^{-x/\bar{\gamma}_c} \left(\frac{1}{2\sqrt{x\bar{\gamma}_c}} - \frac{1}{\bar{\gamma}_c}\sqrt{\frac{x}{\bar{\gamma}_c}} \right)
$$

$$
\times \left(1 + 2Q\left(\sqrt{2\frac{x}{\bar{\gamma}_c}} \right) \right) \ . \tag{5.110}
$$

The mean value of the output bit energy-to-noise ratio is

$$
\bar{\gamma}_b^{\mathrm{eg}} = \frac{1}{2LP_n} \mathrm{E}\left[\left(A^2 \sum_{k=1}^{L} \alpha_k \right)^2 \right]
$$

$$
= \frac{A^2}{2LP_n} \sum_{j=1}^{L} \sum_{\ell=1}^{L} \mathrm{E}\left[\alpha_k \alpha_\ell \right] \ . \tag{5.111}
$$

Using the result from Chapter 2 that $E[\alpha_k^2] = 2\sigma^2$, and $E[\alpha_k] = \sqrt{\pi/2}\,\sigma$, along with the fact that the fading on the branches is uncorrelated, i.e., $E[\alpha_k\alpha_\ell] = E[\alpha_k]E[\alpha_\ell]$ for $k \neq \ell$, gives

$$
\begin{aligned}
\bar{\gamma}_b^{\text{eg}} &= \frac{A^2}{2LP_n}\left(2L\sigma^2 + L(L-1)\frac{\pi\sigma}{2}\right) \\
&= \frac{A^2\sigma^2}{P_n}\left(1 + (L-1)\frac{\pi}{4}\right) \\
&= \bar{\gamma}_c\left(1 + (L-1)\frac{\pi}{4}\right) \ .
\end{aligned}
\tag{5.112}
$$

The bit error probability with 2-branch combining can be using the pdf in (5.110). For example, with coherent BPSK signaling the bit error probability is (see Problem 5.15)

$$
\begin{aligned}
P_b &= \int_0^\infty P_b(x)p_{\gamma_b^{\text{mr}}}(x)dx \\
&= \frac{1}{2}\left(1 - \sqrt{1 - \mu^2}\right)
\end{aligned}
\tag{5.113}
$$

where

$$
\mu = \frac{1}{1 + \bar{\gamma}_c} \ .
\tag{5.114}
$$

5.5.5   Predetection Switched Combining

A switched combiner scans through the diversity branches until it finds one that has a signal-to-noise ratio exceeding a specified threshold. This diversity branch is used until the signal-to-noise ratio again drops below the threshold. When this happens another diversity branch is chosen that has a signal-to-noise ratio exceeding the threshold. There are several variations of switched diversity. Here, we analyze two-branch **switch and stay** combining (SSC). With SSC, the receiver switches to, and stays with, the alternate branch when the bit energy-to-noise ratio drops below a specified threshold. It does this regardless of whether or not the bit energy-to-noise ratio with the alternate branch is above or below the threshold.

Let the bit energy-to-noise ratios associated with the two branches be denoted by γ_1 and γ_2, and let the switching threshold be denoted by T. By using (5.82), the probability that γ_i is less than T is

$$
\begin{aligned}
q &= \Pr[\gamma_i < T] \\
&= 1 - e^{-T/\bar{\gamma}_c} \ .
\end{aligned}
\tag{5.115}
$$

Likewise, the probability that γ_i is less than S is:

$$p = 1 - e^{-S/\bar{\gamma}_c} \tag{5.116}$$

Let γ_b^{sw} denote the bit energy-to-noise ratio at the output of the switched combiner. Then

$$\Pr[\gamma_b^{\mathrm{sw}} \leq S] = \Pr\left[\{\gamma_b^{\mathrm{sw}} \leq S | \gamma_b^{\mathrm{sw}} = \gamma_1\} \bigcup \{\gamma_b^{\mathrm{sw}} \leq S | \gamma_b^{\mathrm{sw}} = \gamma_2\}\right] \tag{5.117}$$

Note that γ_1 is statistically identical to γ_2. Therefore, we can assume that branch 1 is currently in use. It follows that

$$\Pr[\gamma_b^{\mathrm{sw}} \leq S] = \begin{cases} \Pr\left[\gamma_1 \leq T \bigcap \gamma_2 \leq S\right], & S < T \\ \Pr\left[(T \leq \gamma_1 \leq S) \bigcup (\gamma_1 \leq T \bigcap \gamma_2 \leq S)\right], & S \geq T \end{cases} \tag{5.118}$$

The region $S < T$ corresponds to the case where γ_1 has dropped below the threshold T and a switch to branch 2 is initiated, but $\gamma_2 < T$ so that the switch does not result in a γ_{sw} greater than T. On the other hand, the region $S \geq T$ corresponds to the case when either γ_1 is between T and S or when γ_1 has dropped below the threshold T so that a switch to branch 2 occurs, and $T \leq \gamma_2 \leq S$. Since γ_1 and γ_2 are independent, the above probabilities are

$$\Pr\left[\gamma_1 \leq T \bigcap \gamma_2 \leq S\right] = qp \tag{5.119}$$

$$\Pr\left[(T \leq \gamma_1 \leq S) \bigcup \left(\gamma_1 \leq T \bigcap \gamma_2 \leq S\right)\right] = p - q + qp \tag{5.120}$$

Therefore,

$$\Pr[\gamma_b^{\mathrm{sw}} \leq S] = \begin{cases} qp & S < T \\ p - q + qp & S \geq T \end{cases}. \tag{5.121}$$

Fig. 5.20 plots the cdf $F_{\gamma_b^{\mathrm{sw}}}(x)$ for several values of the normalized threshold $R = 10\log_{10}(T/\bar{\gamma}_c)$ (dB). Observe that SSC always performs worse than SC except at the switching threshold, where the performance is the same. Since SSC offers the most improvement just above the threshold level, the threshold level should be chosen as γ_{th}, the minimum acceptable instantaneous bit energy-to-noise ratio that the radio system can tolerate and still provide an acceptable bit error probability. Finally, the optimum threshold, $T = R\bar{\gamma}_c$, depends on $\bar{\gamma}_c$. Since $\bar{\gamma}_c$ varies due to path loss and shadowing, the threshold must be adaptive.

The probability of bit error can be also be computed for SSC. The pdf for γ_b^{sw} is

$$p_{\gamma_b^{\mathrm{sw}}}(x) = \begin{cases} q\frac{1}{\bar{\gamma}_c}e^{-x/\bar{\gamma}_c} & , \quad x < T \\ (1 + q)\frac{1}{\bar{\gamma}_c}e^{-x/\bar{\gamma}_c} & , \quad x \geq T \end{cases} \tag{5.122}$$

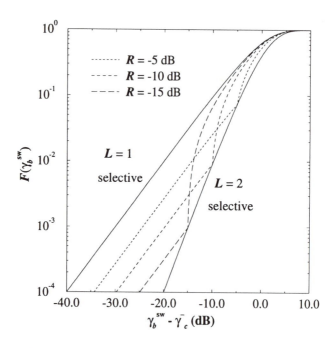

Figure 5.20   Cdf of γ_{sw} for 2-branch switched diversity for several values of the normalized threshold $R = 10\log_{10}(T/\bar{\gamma}_c)$; $\bar{\gamma}_c$ is the average branch bit energy-to-noise ratio.

If DPSK is used, then the probability of error is

$$P_b = \int_0^{\infty} P_b(x)p_{\gamma_b^{\mathrm{sw}}}(x)dx$$

$$= \frac{1}{2(1+\bar{\gamma}_c)}\left(q + (1-q)e^{-T}\right) \ . \tag{5.123}$$

The above expression is plotted in Fig. 5.21 for several values of the threshold T. The performance with a threshold level of $T = 0$ is the same as using no diversity, because no switching occurs. The performance changes little for $T > 6$. As T increases, the probability of switching q also increases, as shown in Fig. 5.22. It is desirable to keep q as small as possible to minimize the number of switches.

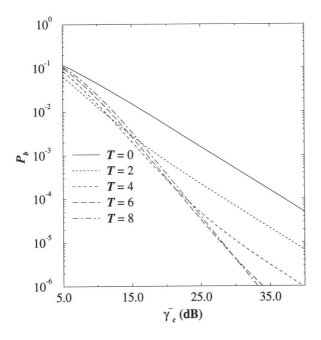

Figure 5.21 Bit error probability for 2-branch predetection switched combining and differentially detected DPSK.

5.5.6 Postdetection Equal Gain Combining with Differential Detection

Postdetection equal gain combining has a simple implementation and very good performance when used in conjunction with differential detection. Differential detection circumvents the need to co-phase and weight the diversity branches. The overall receiver structure is shown in Fig. 5.23. The structure of the individual differential detectors depends on the type of modulation that is being used. For DPSK, the detector is shown in Fig. 5.12, while for $\pi/4$-QPSK the detector is shown in Fig. 5.13. In the latter case, the U and V branches are combined separately.

For DPSK the decision variable at the output of the combiner is, from (5.65),

$$U = \sum_{k=1}^{L} U_i = \frac{1}{2} \sum_{k=1}^{L} (Z_k Z_{kd}^* + Z_k^* Z_{kd}) \ . \tag{5.124}$$

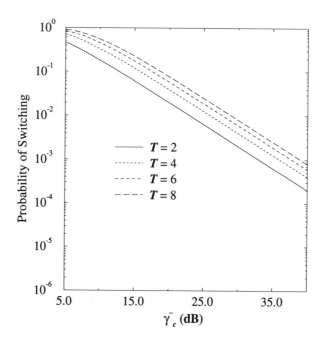

Figure 5.22 Probability of switching for two-branch predetection switched combining.

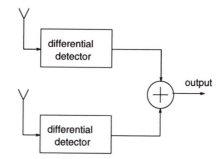

Figure 5.23 Differential detection with postdetection equal gain combining.

Once again, by using characteristic functions it can be shown that $U = W - Y$, where W and Y are non-central and central chi-square random variables with $2L$ degrees of freedom, respectively, and having the densities

$$f_W(w) = \frac{1}{2EN_o} \left(\frac{w}{s^2}\right)^{\frac{L-1}{2}} \exp\left\{-\frac{(s^2 + w)}{2EN_o}\right\} I_{L-1}\left(\sqrt{w}\frac{s}{EN_o}\right) \quad (5.125)$$

$$f_Y(y) = \left(\frac{1}{2EN_o}\right)^L \frac{1}{(L-1)!} y^{L-1} \exp\left\{-\frac{y}{2EN_o}\right\} . \quad (5.126)$$

where

$$s^2 = 4E \sum_{k=1}^{L} \alpha_k^2 \quad (5.127)$$

is the non-centrality parameter. After some algebraic detail, the probability of error can be expressed in the closed form [256]

$$P_b = \frac{1}{2^{2L-1}} e^{-\gamma_t} \sum_{k=0}^{L-1} b_k \gamma_t^k \quad (5.128)$$

where

$$b_k = \frac{1}{k!} \sum_{n=0}^{L-1-k} \binom{2L-1}{n} \quad (5.129)$$

and

$$\gamma_t = \sum_{k=1}^{L} \gamma_k . \quad (5.130)$$

Averaging over the pdf in (5.99) gives the result

$$P_b = \frac{1}{2^{2L-1}(L-1)!(1+\bar{\gamma}_c)^L} \sum_{k=1}^{L-1} b_k(L-1+k)! \left(\frac{\bar{\gamma}_c}{1+\bar{\gamma}_c}\right)^k . \quad (5.131)$$

This can be manipulated in the same form as (5.103) with

$$\mu = \frac{\bar{\gamma}_c}{1+\bar{\gamma}_c} . \quad (5.132)$$

The various diversity combining techniques are compared in Fig. 5.24 for differentially detected DPSK signals. It is apparent that SSC results in the worst performance, followed by SC, and postdetection EGC. If differential detection is used, predetection and postdetection EGC have the same performance. Finally, it does not make sense to use MRC with differential detection. Since MRC requires knowledge of the channel vector **g**, we might as well use coherent detection and achieve better performance. Therefore, a curve for MRC is not included in Fig. 5.24.

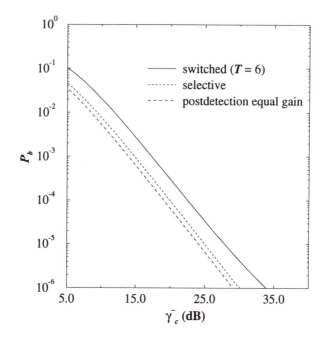

Figure 5.24 Comparison of 2-branch diversity combining techniques for differentially detected DPSK.

5.5.7 Macroscopic Diversity

Diversity techniques that are used to mitigate the effects of envelope fading are sometimes called microscopic diversity. Another type of diversity that can be used in mobile radio applications is macroscopic diversity. This type of diversity is intended to combat the effects of shadow fading. With macroscopic diversity, signals that are received from a mobile station (MS) by two or more geographically separated base stations (BSs) are combined together. A diversity advantage is obtained if the signals experience some degree of uncorrelated shadowing. This is usually the case, because the large terrain features often vary with the direction from the MS. It is quite safe to assume that the signals on the different macroscopic diversity branches will be affected by independent envelope fading. As discussed in Chapter 2, shadowing is often modeled as being log-normally distributed meaning that the average received bit energy-to-noise ratio per diversity branch, $\bar{\gamma}_c$, (in dB) has the normal distribution

$$p(\bar{\gamma}_{c\ (\text{dB})}) = \frac{1}{\sqrt{2\pi}\sigma_{\bar{\gamma}_c}} \exp\left\{-\frac{\left(\bar{\gamma}_{c\ (\text{dB})} - \mu_{\bar{\gamma}_c}\right)^2}{2\sigma_{\bar{\gamma}_c}^2}\right\} \tag{5.133}$$

where $\mu_{\bar{\gamma}_c} = \text{E}[\bar{\gamma}_{c\ (\text{dB})}]$.

Note that the average received carrier-to-noise ratio $\Gamma = \Omega_p/N$ and the average received bit energy-to-noise ratio $\bar{\gamma}_c$ are closely related. The average received bit energy is $\bar{\mathcal{E}}_b = \Omega_p/R_b$ where $R_b = 1/T_b$ is the bit rate, and the noise psd is $N_o = N/B_w$ where N is the total noise power and B_w is the receiver noise bandwidth. It follows that

$$\bar{\gamma}_c = \frac{\bar{\mathcal{E}}_b}{N_o} = \Gamma\frac{B_w}{R_b} \ . \tag{5.134}$$

Several methods may be used for macroscopic diversity combining. The most effective and simplest to use is selective combining, where the BS that provides the largest average received bit energy-to-noise ratio is selected as the serving BS. Let $\bar{\gamma}_{\text{th}\ (\text{dB})}$ be a specified threshold level, and let $\bar{\gamma}_{ck\ (\text{dB})}$ be the average received bit energy-to-noise ratio from BS k. Then

$$\begin{aligned}
\text{Pr}(\bar{\gamma}_{ck\ (\text{dB})} \leq \bar{\gamma}_{\text{th}\ (\text{dB})}) &= \int_{-\infty}^{\bar{\gamma}_{\text{th}\ (\text{dB})}} \frac{1}{\sqrt{2\pi}\sigma_{\bar{\gamma}_c}} \exp\left\{-\frac{\left(\bar{\gamma}_{ck\ (\text{dB})} - \mu_{\bar{\gamma}_{ck}}\right)^2}{2\sigma_{\bar{\gamma}_c}^2}\right\} \\
&= Q\left(\frac{\mu_{\bar{\gamma}_{ck}} - \bar{\gamma}_{\text{th}\ (\text{dB})}}{\sigma_{\bar{\gamma}_c}}\right)
\end{aligned} \tag{5.135}$$

Let $\mathbf{d} = (d_0,\ d_1,\ \ldots,\ d_M)$ be the set of distances between the MS and M different BSs. The probability of thermal noise can be defined as the probability that the average received bit energy-to-noise ratio from all M BSs is below $\bar{\gamma}_{\text{th}}$, and is given by

$$O(\mathbf{d}) = \prod_{k=0}^{M-1} \text{Pr}(\bar{\gamma}_{ck\ (\text{dB})} \leq \bar{\gamma}_{\text{th}\ (\text{dB})}) = \prod_{k=0}^{M-1} Q\left(\frac{\mu_{\bar{\gamma}_{ck}} - \bar{\gamma}_{\text{th}\ (\text{dB})}}{\sigma_{\bar{\gamma}_c}}\right) \ . \tag{5.136}$$

The probability of thermal noise depends on the set of mean values $\{\mu_{\bar{\gamma}_{ck}}\}_{k=0}^{M-1}$ which in turn depend on the distances between the MS and the BSs and other factors such as antenna heights. For the purpose of illustration, assume that the $\mu_{\bar{\gamma}_{ck}}$ are all equal to $\mu_{\bar{\gamma}_c}$. Then

$$O = \left[Q\left(\frac{\mu_{\bar{\gamma}_c} - \bar{\gamma}_{\text{th}\ (\text{dB})}}{\sigma_{\bar{\gamma}_c}}\right)\right]^M \ . \tag{5.137}$$

Fig. 5.25 plots the probability of thermal noise O as a function of the thermal noise margin $M = \mu_{\bar{\gamma}_c} - \bar{\gamma}_{\text{th}\ (\text{dB})}$ for several values of M and $\sigma_{\bar{\gamma}_c} = 8$ dB. This

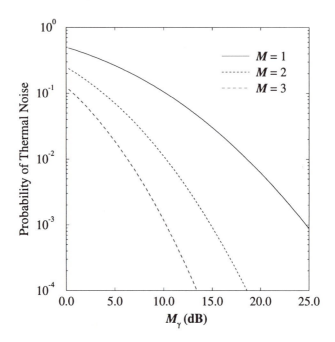

Figure 5.25 Thermal noise outage against the thermal noise margin $\mu_{\bar{\gamma}_c}$ − $\bar{\gamma}_{\text{th (dB)}}$ with BS diversity; $\sigma_{\bar{\gamma}_c} = 8$ dB.

figure shows that BS diversity can provide a very significant improvement in outage. It must be remembered, however, that the results in Fig. 5.25 show the effect of using macroscopic diversity under ideal conditions, where the multiple signals are independently shadowed and the $\mu_{\bar{\gamma}_{ck}}$ are the same for all diversity branches. Any deviation from these conditions will diminish the gains.

Problems

5.1. Derive the expression for the symbol error probability of $\sqrt{M}$-PAM in (5.32).

5.2. Show that the symbol error probability for coherent M-ary PSK is bounded by $p \leq P_M \leq 2p$, where

$$p = Q\left(\sqrt{2\gamma}\sin\frac{\pi}{M}\right)$$

and γ is the symbol energy-to-noise ratio.

5.3. Suppose that BPSK signaling is used with coherent detection. The channel is affected by flat Rayleigh fading and log-normal shadowing with a shadow standard deviation of σ_Ω dB. Plot the average probability of bit error against the average received bit energy-to-noise ratio $\bar{\Omega}_v$ under the assumption that the MS is stationary, i.e., use the Susuki distribution in (2.180). Plot several curves with different values of σ_Ω.

5.4. Consider the differentially coherent receiver shown in Fig. refdpsk. Show that the pdf of U is given by (5.70).

5.5. Consider binary CPFSK modulation with modulation index $h \leq 0.5$. Compute the minimum squared Euclidean distance between any pair of band-pass waveforms as given by

$$D^2_{\min} = \lim_{n \to \infty} \min_{i,j} \int_0^{nT} \left[s(t; \mathbf{x}^{(i)}) - s(t; \mathbf{x}^{(j)}) \right]^2 dt$$

where $s(t; \mathbf{x}^{(i)})$ and $s(t; \mathbf{x}^{(j)})$ are the two band-pass signals whose phase trajectories diverge at time $t = 0$ and remerge sometime later. What is the pairwise error probability between two such signals?

5.6. The squared Euclidean distance between a pair of CPM band-pass waveforms, $s(t; \mathbf{x}^{(i)})$ and $s(t; \mathbf{x}^{(j)})$, is

$$D^2 = \int_0^\infty \left[s(t; \mathbf{x}^{(i)}) - s(t; \mathbf{x}^{(j)}) \right]^2 dt$$

Show that

$$D^2 = 2(\log_2 M) E_b \frac{1}{T} \int_0^\infty [1 - \cos \Delta_\phi(t)] \, dt$$

where M is the symbol alphabet size, E_b is the energy per bit, and $\Delta_\phi(t)$ is the phase difference between the two signals.

5.7. Construct a differential detector for MSK signaling. Obtain an expression for the probability of bit error for differentially detected MSK on an AWGN channel.

5.8. Suppose that GMSK signaling is used. Unfortunately, a GMSK pulse is noncausal and, therefore, a truncated version of the pulse is usually employed in a practical system, i.e., the time domain pulse is

$$g(t) = \left[A \sqrt{\frac{2\pi}{\ln 2}} B \int_{t-T_b/2}^{t+T_b/2} \exp \left\{ -\frac{2\pi^2 B^2 x^2}{\ln 2} \right\} \, dx \right] u_{2L_T T_b}(t - L_T T_b) \ .$$

Compute the maximum value of the ISI term in (4.173) as a function of the normalized filter bandwidth BT_b when $L_T = 3$.

5.9. Suppose that two-branch predetection selective combining is used. However, the branches are mismatched such that $\bar{\gamma}_1 \neq \bar{\gamma}_2$ where the $\bar{\gamma}_i$, $i = 1,\ 2$ are the average received bit energy-to-noise ratios for the two branches. Plot the cdf of γ_b^s against the average normalized bit energy-to-noise ratio $10\log_{10}(\gamma_b^s/\bar{\gamma}_t)$, where $\bar{\gamma}_t = (\bar{\gamma}_1 + \bar{\gamma}_2)/2$. Show several curves while varying the ratio $\xi = \bar{\gamma}_1/\bar{\gamma}_2$.

5.10. Consider using predetection selective combining with coherent BPSK. For BPSK, the probability of bit error is $P_b(\gamma_b^s) = Q(\sqrt{2\gamma_b^s})$ The instantaneous bit energy-to-noise ratio is given by (5.85).

a) Derive an expression for the average bit error probability

$$P_b = \int_0^\infty P_b(x)p_{\gamma_b^s}(x)dx$$

b) Repeat part a) for two-branch switched diversity combining where the pdf of γ_b^{sw} is given by (5.122).

c) Plot and compare the results in parts a) and b) for two-branch diversity.

5.11. Suppose that binary DPSK signaling ($x_k \in \{-1, +1\}$) is used on a flat Rayleigh fading channel with 3-branch diversity. The diversity branches are assumed to experience uncorrelated fading. The signal that is received over each diversity branch is corrupted with AWGN having a one-sided psd of N_o watts/Hz. The noise processes that are associated with the diversity branches are mutually uncorrelated.

a) Suppose that a separate differential detector is used on each diversity branch, yielding three independent estimates of each transmitted bit, i.e., for x_k the receiver generates the three independent estimates $(\hat{x}_k^1, \hat{x}_k^2, \hat{x}_k^3)$. Majority logic combining is then used to combine the three estimates together to yield the final decision $\hat{x}_k$, i.e.,

$$\hat{x}_k = \begin{cases} -1 & \text{if two or more } \hat{x}_k^i = -1 \\ +1 & \text{if two or more } \hat{x}_k^i = +1 \end{cases}$$

Find an expression for the probability of bit error, P_b. Evaluate P_b for $\bar{\gamma}_c = 20$ dB, where $\bar{\gamma}_c$ is the average received branch bit energy-to-noise ratio.

b) Evaluate the probability of bit error for $\bar{\gamma}_c = 20$ dB if the receiver uses 3-branch diversity with postdetection equal gain combining. Compare with the result in part a).

c) Generalize the expression for the probability of bit error in part a) to L-branch diversity.

5.12. Derive (5.103) for BPSK and maximal ratio combining.

5.13. Derive (5.128) for DPSK and postdetection equal gain combining.

5.14. The bit error probability MSK signaling on a Rayleigh fading channel with additive white Gaussian noise is

$$P_b = \frac{1}{2}\left[1 - \sqrt{\frac{\bar{\gamma}_b}{1 + \bar{\gamma}_b}}\right]$$

a) Derive a Chernoff upper bound (see Appendix A) on the probability of bit error for the same channel and compare the Chernoff bound with the exact error probability.

b) Repeat part a) if the receiver employs L-branch diversity. Assume uncorrelated diversity branches with $\bar{\gamma}_1 = \bar{\gamma}_2 = \cdots \bar{\gamma}_L = \bar{\gamma}_c$.

5.15. Suppose that BPSK modulation is used with two-branch diversity and coherent equal gain combining. Assume uncorrelated diversity branches with $\bar{\gamma}_1 = \bar{\gamma}_2 = \bar{\gamma}_c$. Show that the probability of bit error for a Rayleigh fading channel is given by (5.113).

5.16. Consider a system that uses L-branch selection diversity. The instantaneous received signal power on each diversity branch, $s_{0,i}$, $i = 1, \ldots, L$, has the non-central chi-square (Ricean fading) distribution in (2.44). The instantaneous received signal power from each interferer on each diversity branch, $s_{k,i}$, $i = 1, \ldots, L$ has the exponential (Rayleigh fading) distribution in (2.39). Note that all the $s_{0,i}$ and $s_{k,i}$ are all independent. Let $\lambda_i = s_{0,i}/\sum_{k=1}^{N_I} s_{k,i}$, $i = 1, \ldots, L$ be the instantaneous carrier-to-interference ratio for each diversity branch and $\lambda_s = \max_i \lambda_i$. Derive an expression for the probability of CCI

$$O = \mathrm{P_r}(\lambda_s < \lambda_{\mathrm{th}}) \ .$$

Plot the probability of CCI against λ_{th} for various L.

5.17. Suppose that two-branch antenna diversity is used with selective combining. However, the branches have correlated fading so that the maximum diversity gain is not achieved. Let γ_1 and γ_2 be the joint pdf for the instantaneous bit energy-to-noise ratio for each diversity branch, and let $\bar{\gamma}_c = \mathrm{E}[\gamma_i]$. It is known that joint pdf of γ_1 and γ_2 is

$$p_{\gamma_1,\gamma_2}(x_1, x_2) = \frac{1}{\bar{\gamma}_c(1-\rho^2)} I_0 \left(\frac{2\rho\sqrt{x_1 x_2}}{\bar{\gamma}_c(1-\rho^2)} \right) \exp \left\{ -\frac{x_1 + x_2}{\bar{\gamma}_c(1-\rho^2)} \right\}$$

where ρ is magnitude of the covariance of the two complex, jointly Gaussian random processes that are associated with each diversity branch. Derive an expression for the cdf of the output of the selective combiner

$$\gamma_s = \max\{\gamma_1, \gamma_2\} \; .$$

Plot the cdf for various ρ. What conclusions can you make?

6

DIGITAL SIGNALING ON ISI CHANNELS

Land mobile radio channels are modeled as fading dispersive channels, because of the multipath propagation and the randomly changing medium characteristics. Many types of impairments are observed on these channels such as multipath spread (or delay spread), fading, Doppler spread, nonlinear distortion, frequency offset, phase jitter, impulse noise, thermal noise, and co-channel and adjacent channel interference arising from spectrum sharing. This chapter concentrates on the effects of delay spread, fading, Doppler spread, thermal noise, and co-channel interference. Delay spread causes interference between adjacent symbols, known as intersymbol interference (ISI), a large Doppler spread indicates rapid channel variations and necessitates a fast convergent algorithm when an adaptive receiver is employed, and fading results in a very low received signal-to-noise ratio or signal-to-interference ratio when the channel exhibits a deep fade.

An adaptive equalizer is an arrangement of adjustable filters at the receiver whose purpose is to mitigate the combined effect of ISI and noise [199], [259]. Two broad categories of equalizers have been documented extensively in the literature; symbol-by-symbol equalizers and sequence estimators. Symbol-by-symbol equalizers include a decision device to make symbol-by-symbol decisions on the received symbol sequence, while sequence estimators make decisions on sequences of received symbols. Many structures and adaptive algorithms have been proposed for each type of equalizer for different channel characteristics. Sequence estimators are generally more complex than symbol-by-symbol equalizers, but can potentially offer better performance.

This chapter begins with a brief survey of adaptive equalization techniques. This is followed by a discussion of ISI channel modeling in Section 6.2. The optimum receiver for digital signaling on an ISI channel is presented in Section 6.3. Section 6.4 provides a treatment of symbol-by-symbol equalizers and Section 6.5 provides a treatment of sequence estimators. Section 6.6 provides an analysis of the bit error rate performance of maximum likelihood sequence estimation (MLSE) on static ISI channels and multipath fading ISI channels. Finally, Section 6.7 analyzes the performance of fractionally-spaced MLSE receivers on ISI channels.

6.1   OVERVIEW OF EQUALIZATION TECHNIQUES

6.1.1   Symbol-by-symbol Equalizers

Lucky [197], [198] was the first to develop an adaptive (linear) equalizer for digital communication systems in the mid-1960s. This equalizer was based on the peak distortion criterion, where the equalizer forces the ISI to zero, and it is called a **zero-forcing** (ZF) equalizer. Soon after, Proakis and Miller [257], Lucky et. al. [199], and Gersho [115] developed the linear **LMS equalizer**, based on the least mean square (LMS) criterion. The LMS equalizer is more robust than the ZF equalizer, because the latter ignores the effects of noise. Thaper [301] examined the performance of trellis coded modulation for high speed data transmission on voiceband telephone channels, and proposed a simple receiver structure that used an adaptive linear equalizer. He reported that the performance was close to ideal, but his work did not include the more severely distorted multipath fading ISI channels.

Linear equalizers have the drawback of enhancing channel noise while trying to eliminate ISI, a characteristic known as **noise enhancement**. As a result, satisfactory performance is unattainable with linear equalizers for channels having severe amplitude distortion. In 1967, Austin [17] proposed the nonlinear **decision feedback equalizer** (DFE) to mitigate noise enhancement. Because only the precursor ISI is eliminated by the feedforward filter of the DFE, noise enhancement is greatly reduced. To eliminate the postcursor ISI, the estimated symbols are fed back through the feedback filter of the DFE. However, this introduces error propagation which can seriously degrade the performance of the DFE and complicate analysis of its performance. Belfiore and Park [23] proposed an equivalent DFE, called a **predictive DFE**, by using a linear predictor as the feedback filter. This structure is useful when a DFE is combined with a sequence estimator for equalization and decoding of trellis-coded modulation on an ISI channel [89].

Early adaptive equalizers were implemented by using a transversal filter with a tap-spacing equal to the signal interval, T, known as symbol-spaced equalizers. The performance of a symbol-spaced equalizer is very sensitive to the sampling instant and can be very poor with an improperly chosen sampling time [30], [311], [121]. Even with perfect timing and matched filtering, the symbol-spaced equalizer cannot realize the optimal linear receiver because of the finite tapped delay line structure. Brady [30], Monson [216], Ungerboeck [311], and Gitlin and Weinstein [121] solved this problem by proposing a **fractionally-spaced equalizer** (FSE), where the tap-spacing is less than T. If a symbol-spaced equalizer is preceded by a matched filter, then an FSE and a symbol-spaced equalizer are equivalent. However, the exact matched filter is difficult to obtain in practical applications because its structure depends on the unknown channel characteristics and, hence, an FSE is quite attractive. It can also be argued that the FSE can achieve an arbitrary linear filter with a finite-length fractionally-spaced tapped delay line. Hence, the FSE is expected to outperform a (finite-length) symbol-spaced equalizer even with ideal matched filtering and sampling.

In the 1980's, Gersho and Lim [116], Mueller and Salz [223], and Wesolowski [330] proposed an interesting decision-aided equalizer, known as an **ISI canceller**. Theoretically, ISI cancellers can eliminate ISI completely without any noise enhancement. However, a decision-aided mechanism is employed in the equalizer so that it suffers from error propagation, similar to a DFE.

Various adaptation algorithms have been proposed to adjust the equalizer coefficients. The **LMS algorithm**, proposed by Widrow and Hoff [333], and analyzed by Gitlin *et. al.* [120], Mazo [207], [208], Ungerboeck [309], and Widrow *et. al.* [334], is the most popular because of its simplicity and numerical stability. However, the LMS algorithm converges very slowly for channels with severe amplitude distortion. This slow convergence is intolerable for many practical applications. For example, Hsu *et. al.* [149] reported that the LMS algorithm is not suitable for an HF shortwave ionospheric channel, because the channel has severe amplitude distortion when a deep fade occurs and the channel characteristics change very rapidly.

A considerable research effort has been directed to finding a fast-convergent algorithms for adaptive equalizers. In 1974, Gordard [127] described a fast-convergent algorithm later known as the **recursive least square** (RLS) algorithm. This algorithm utilizes all available information from the beginning of processing, and converges much faster than the LMS algorithm. Unfortunately, the computational complexity is proportional to N^2, where N is the order of the equalizer, which is too high for many practical applications. To reduce the complexity, Falconer and Ljung [94], and Cioffi and Kailath [46] developed different fast RLS algorithms in 1978 and 1984, respectively. These algorithms have a complexity proportional to the equalizer order N. However, when the algorithms are implemented with finite precision arithmetic, they tend to become unstable. Examples of this numerical instability were reported by Mueller [224].

Another RLS algorithm, called the **recursive least square lattice** (RLSL) algorithm, was investigated by Morf *et. al.* [219], Satorius [275], [276], Friedlander [106], and Ling and Proakis [190]. The RLSL algorithm has a higher complexity that the fast RLS algorithms, but has better numerical stability. However, numerical instability of the RLSL algorithm was still reported by Perl [246].

Some applications of symbol-by-symbol equalization techniques to multipath fading channels were studied by Monson [216], [217], [218], Hsu *et. al.* [149], Ling and Proakis [191], and Eleftheriou and Falconer [82]. For rapidly time-varying channels, a reinitialization procedure might be needed for fast-convergent algorithms in order to avoid numerical instability [80]. Finally, Wong and McLane [337] examined the performance of trellis-coded modulation for HF radio channels having in-band spectral nulls. They considered both linear and non-linear equalization and proposed a modified DFE (MDFE).

6.1.2   Sequence Estimation

The Viterbi algorithm was originally devised by Viterbi for maximum likelihood decoding of convolutional codes [321], [322]. Forney recognized the analogy between an ISI channel and a convolutional encoder, and applied the Viterbi algorithm for the detection of digital signals corrupted by ISI and additive white Gaussian noise [100]. Because of the efficiency of the Viterbi algorithm, the implementation of optimum **maximum likelihood sequence estimation** (MLSE) for detecting ISI-corrupted signals is feasible.

After Forney's initial work [100], the MLSE receiver was modified and extended. Magee and Proakis [201] proposed an adaptive MLSE receiver that employed an adaptive channel estimator for estimating the channel impulse response. Ungerboeck [310] developed a simpler MLSE that also accounted for the effect of carrier phase errors and sampling time errors. Acampora [3] used

MLSE for combining convolutional decoding and equalization, and extended the application of MLSE to quadrature amplitude modulation (QAM) systems [4].

MLSE has a complexity that grows exponentially with the size of signal constellation and the length of channel impulse response. MLSE is impractical for systems having a large signal constellation and/or having a long channel impulse response. Considerable research has been undertaken to reduce the complexity of MLSE while retaining most of its performance. Early efforts concentrated on shaping the original channel impulse response into the one having a shorter length. Then a sequence estimator with a smaller number of states can be applied. In [260], Qureshi and Newhall employed a linear equalizer as the shaping filter. This method is quite successful if the original channel and the desired channel have a similar channel spectrum. Falconer and Magee [93], and Beare [21] adaptively optimized the linear equalizer and the desired channel response, by minimizing the mean square error between the output of the equalizer and the desired channel. This scheme has improved performance when the original channel is quite different from the desired one, but it has a higher complexity. As mentioned earlier, linear equalizers enhance the channel noise. Lee and Hill [178] proposed using a DFE to truncate the channel impulse response so as to reduce the system complexity and mitigate noise enhancement.

Another approach for reducing the complexity of MLSE lies in simplifying the Viterbi algorithm itself. By employing suitable decision regions, Vermuelen [316] and Foschini [102] observed that only a small number of likely paths need to be extended to obtain a near maximum likelihood performance. Wesolowski [331] employed a DFE to determine a small set of likely signal points, and then used the Viterbi algorithm to find the most likely sequence path through a reduced-state trellis. Clark also proposed some similar detection methods in [49], [48].

Recently, two novel reduced-state sequence estimation techniques have been proposed. Eyuboğlu and Qureshi [91] proposed **reduced-state sequence estimation** (RSSE), a technique that is especially useful for systems with large signal constellations. Duel-Hallen and Heegard [76], [75] proposed **delayed decision-feedback sequence estimation** (DDFSE), a technique that is useful for channels with long impulse responses (DDFSE can be applied on channels with an infinite impulse response). Chevillat and Eleftheriou [38] independently proposed the same algorithm, but for a finite length channel. Both RSSE and DDFSE use the Viterbi algorithm to search for the most likely path, and provide a good performance/complexity trade-off. In both schemes, a feedback mechanism must be introduced to compute the branch metrics, because of the reduction in the number of system states. This feedback introduces error propagation. However, the effect of the error propagation is much smaller than with a DFE [91], [76]. Eyuboğlu and Qureshi [91] also observed that for channels with a finite channel impulse response, DDFSE can be conveniently modeled as a special case of RSSE. Eyuboğlu and Qureshi [89] and Chevillat and Eleftheriou [38] also suggested using RSSE for systems employing trellis-coded modulation. Sheen and Stüber have obtained error probability upper bounds and approximations for RSSE and DDFSE for uncoded systems [282] and trellis-coded systems [283].

Eyuboğlu and Forney [90] proposed a combined precoding and coded modulation technique that achieves the best coding gain of any known trellis code. With their technique, equalization is achieved by using Tomlinson-Harashima precoding [90], which requires that the channel impulse response be known at the transmitter.

For decoding convolutional codes, a sequential decoding algorithm is a good alternative to the Viterbi algorithm, especially when the encoder has a long constraint length and the system has a moderate-to-high SNR [189]. It is apparent that **sequential sequence estimation** (SSE) can be applied for detecting ISI-corrupted signals. Long and Bush [195], [194], and Xiong *et. al.* [342] reported some results on this application. In [195], [194], the Fano algorithm [95] was employed as the detection algorithm, and a DFE was used to determine the path to be extended. If the DFE makes correct decisions most of the time, then the number of nodes visited by the Fano algorithm can be reduced. The multiple stack algorithm [37] was employed in [342] for avoiding the erasure or buffer overflow problem encountered with sequential detection algorithms. Systems with an infinite impulse response were also considered in [342].

Applications of sequence estimation techniques to multipath fading ISI channels were studied by D'aria and Zingarelli [64], D'avella *et. al.* [65], and Eleftheriou and Falconer [82]. MLSE was employed for equalizing UHF land mobile radio channels in [64, 65], and employed for equalizing HF shortwave ionospheric channels in [82]. Tight upper bound on the error probability of digital signaling on fading ISI channels with MLSE have been provided by Sheen and Stüber for uncoded systems [281] and trellis-coded systems [284]. Katz and Stüber [160] have applied SSE for the detection of trellis-coded signals on multipath fading ISI channels.

6.2 MODELING OF ISI CHANNELS

Chapter 4 showed that the complex envelope of any modulated signal can be expressed in the general form

$$v(t) = A \sum_k b(t - kT, \mathbf{x}_k) \ . \tag{6.1}$$

This chapter restricts attention to linear modulation schemes where

$$b(t, \mathbf{x}_k) = x_k h_a(t) \tag{6.2}$$

$h_a(t)$ is the amplitude shaping pulse, and $\{x_k\}$ is a complex symbol sequence. In general, ASK and PSK waveforms are included, but most FSK waveforms are not.

Suppose that the signal in (6.2) is transmitted over a channel having a time-invariant complex low-pass impulse response $c(t)$. The received complex envelope is

$$w(t) = \sum_{k=0}^{\infty} x_k h(t - kT) + z(t) \tag{6.3}$$

where

$$h(t) = \int_{-\infty}^{\infty} h_a(\tau) c(t - \tau) d\tau \tag{6.4}$$

is the convolution of the transmitted pulse $h_a(t)$ and the channel impulse response $c(t)$, and $z(t)$ is a zero-mean complex additive white Gaussian noise (AWGN) with a power spectral density of N_o watts/Hz. Since the physical channel is causal, the lower limit of integration in (6.4) can be replaced by zero, resulting in

$$h(t) = \int_{0}^{\infty} h_a(\tau) c(t - \tau) d\tau \qquad t \geq 0 \ . \tag{6.5}$$

Finally, the overall pulse $h(t)$ is assumed to have a finite duration so that $h(t) = 0$ for $t \leq 0$ and $t \geq LT$, where L is some positive integer. We will show in Section 6.3 that the maximum likelihood receiver consists of an analog filter $h^*(-t)$ that is matched to the received pulse $h(t)$, followed by a symbol- or T-spaced sampler. Assuming that a matched filter has been implemented, the complex low-pass signal at the output of the matched filter is

$$y(t) = \sum_{k=-\infty}^{\infty} x_k f(t - kT) + \nu(t) \tag{6.6}$$

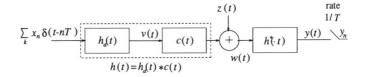

Figure 6.1 Digital signaling on an ISI channel. The receiver implements a filter that is matched to the received pulse followed by a symbol rate sampler.

where

$$f(t) = \int_{-\infty}^{\infty} h^*(\tau)h(\tau + t)d\tau \qquad (6.7)$$

and

$$\nu(t) = \int_{-\infty}^{\infty} h^*(\tau)z(t - \tau)d\tau \qquad (6.8)$$

is the filtered noise. Note that the overall pulse response $f(t)$ accounts for the transmit filter, channel, and receive filter. The overall system as described above is shown in Fig. 6.1.

Sampling the matched filter output every T seconds yields the sample sequence

$$
\begin{aligned}
y_n = y(nT) \; &\overset{\Delta}{=} \; \sum_{k=-\infty}^{\infty} x_k f(nT - kT) + \nu(nT) \\
&\equiv \; \sum_{k=-\infty}^{\infty} x_k f_{n-k} + \nu_n \\
&= \; x_n f_0 + \sum_{\substack{k=-\infty \\ k \neq n}}^{\infty} x_k f_{n-k} + \nu_n \qquad (6.9)
\end{aligned}
$$

where the definitions of f_n and ν_n are implicit. The first term in (6.9) is the desired term, the second term is the ISI, and the last term is the noise at the output of the matched filter. It follows that the overall discrete-time system in Fig. 6.1 can be represented by a discrete-time transversal filter with coefficients

$$\mathbf{f} = (f_{-L},\ f_{-L+1},\ \dots,\ f_{-1},\ f_0,\ f_1,\ \dots,\ f_{L-1},\ f_L) \ . \qquad (6.10)$$

This representation is depicted in Fig. 6.2.

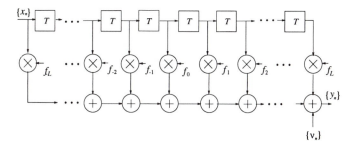

Figure 6.2   Discrete-time model for digital signaling on an ISI channel.

Conditions for ISI-free Transmission

From (6.9), the condition for ISI-free transmission is

$$f_k = \delta_{k0} f_0 \tag{6.11}$$

in which case

$$y_n = x_n f_0 + \nu_n \ . \tag{6.12}$$

An equivalent requirement in the frequency domain for ISI-free transmission can be stated in the following theorem.

Theorem 6.1: ──────────────────────────────────────

The pulse $f(t)$ satisfies $f_k = \delta_{k0} f_0$ if and only if

$$F_\Sigma(f) \triangleq \frac{1}{T} \sum_{n=-\infty}^{\infty} F\left(f + \frac{n}{T}\right) = f_0 \ . \tag{6.13}$$

That is, the **folded spectrum** $F_\Sigma(f)$ must be flat.

Proof:

First, note that

$$
\begin{aligned}
f_k &= \int_{-\infty}^{\infty} F(f) e^{j 2\pi f kT} df \\
&= \sum_{n=-\infty}^{\infty} \int_{(2n-1)/2T}^{(2n+1)/2T} F(f) e^{j 2\pi k f T} df
\end{aligned}
$$

$$= \sum_{n=-\infty}^{\infty} \int_{-1/2T}^{1/2T} F\left(f' + \frac{n}{T}\right) e^{j2\pi k\left(f' + \frac{n}{T}\right)T} df'$$

$$= \int_{-1/2T}^{1/2T} e^{j2\pi f kT} \left[\sum_{n=-\infty}^{\infty} F\left(f + \frac{n}{T}\right)\right] df \ .$$

To prove sufficiency, the term in parenthesis inside the integral is $f_0 T$. Therefore,

$$\int_{-1/2T}^{1/2T} e^{j2\pi f kT} f_0 T df \ = \ \frac{\sin \pi k}{\pi k} f_0$$

$$= \ \delta_{k0} f_0 \ .$$

To prove necessity, from (6.13)

$$f_k = T \int_{-1/2T}^{1/2T} F_\Sigma(f) e^{j2\pi k fT} \ .$$

Hence, f_k and $F_\Sigma(f)$ are a Fourier transform pair, i.e.,

$$F_\Sigma(f) = \sum_{k=-\infty}^{\infty} f_k e^{-j2\pi k fT} \ .$$

Therefore, if $f_k = f_0 \delta_{k0}$, then

$$F_\Sigma(f) = \frac{1}{T} \sum_{n=-\infty}^{\infty} F\left(f + \frac{n}{T}\right) = f_0 \ .$$

For ISI-free transmission, the pulse $f(t)$ can be any function that has equally spaced zero crossings. For example, $f(t) = (\sin 2\pi f_o t)/2\pi f_o t$ has equally spaced zero crossing at $t = \pm n/(2f_o)$ and can be used for ISI-free transmission. Since $f(t) = h_a(t) * c(t) * h^*(-t)$, we note that the channel impulse response $c(t)$ must be known to achieve ISI-free transmission.

Vector Representation of Received Signals

As discussed in Section 5.1, a Gram-Schmidt orthonormalization procedure can be used to express the received signal as

$$w(t) = \sum_{n=1}^{N} w_n \varphi_n(t) \tag{6.14}$$

where the $\{\varphi_n(t)\}$ form a complete set of complex orthonormal basis functions. It can be readily shown that

$$w_n = \sum_{k=-\infty}^{\infty} x_k h_{nk} + z_n \tag{6.15}$$

where

$$
\begin{aligned}
h_{nk} &= \int_0^T h(t - kT)\varphi_n^*(t)dt \\
z_n &= \int_0^T z(t)\varphi_n^*(t)dt \ .
\end{aligned}
\tag{6.16}
$$

Since the z_n are complex Gaussian random variables with zero-mean and covariance $\frac{1}{2}E[z_k^* z_m] = N_o \delta_{km}$, the vector $\mathbf{w} = (w_1, w_2, \cdots, w_N)$ has the multivariate Gaussian distribution

$$p(\mathbf{w}|\mathbf{x}, \mathbf{H}) = \prod_{n=1}^{N} \frac{1}{\pi N_o} \exp\left\{ -\frac{1}{N_o}\left| w_n - \sum_{k=-\infty}^{\infty} x_k h_{nk} \right|^2 \right\} \tag{6.17}$$

where

$$\mathbf{H} = (\mathbf{h}_1, \ \mathbf{h}_2, \ \ldots, \ \mathbf{h}_N)^T \tag{6.18}$$

and

$$\mathbf{h}_n = (\cdots, h_{n,-3}, h_{n,-2}, h_{n,-1}, h_{n,0}, h_{n,1}, h_{n,2}, h_{n,3} \cdots) \ . \tag{6.19}$$

6.3 OPTIMUM RECEIVER FOR ISI CHANNELS WITH AWGN

The maximum likelihood receiver decides in favor of the symbol sequence $\mathbf{x}$ that maximizes the likelihood function $p(\mathbf{w}|\mathbf{x}, \mathbf{H})$ or the log-likelihood function $\log p(\mathbf{w}|\mathbf{x}, \mathbf{H})$, i.e.,

$$\text{choose} \ \ \mathbf{x} \ \ \text{if} \quad \log p(\mathbf{w}|\mathbf{x}, \mathbf{H}) > \log p(\mathbf{w}|\hat{\mathbf{x}}, \mathbf{H}) \quad \forall \ \hat{\mathbf{x}} \neq \mathbf{x} \ . \tag{6.20}$$

For an AWGN channel, $p(\mathbf{w}|\mathbf{x}, \mathbf{H})$ has the form in (6.17) and the decision rule in (6.20) is equivalent choosing that $\mathbf{x}$ which maximizes the quantity

$$\mu(\mathbf{x}) = -\sum_{n=1}^{N} \left| w_n - \sum_{k=-\infty}^{\infty} x_k h_{nk} \right|^2$$

$$= -\sum_{n=1}^{N} |w_n|^2 + \sum_{n=1}^{N} \left(w_n^* \sum_k x_k h_{nk} + w_n \sum_k x_k^* h_{nk}^* \right)$$

$$-\sum_{n=1}^{N} \left(\sum_k x_k h_{nk} \right) \left(\sum_m x_m^* h_{nm}^* \right) . \tag{6.21}$$

Since the term $\sum_{n=1}^{N} |w_n|^2$ is independent of $\mathbf{x}$, it may be discarded so that the maximum likelihood receiver chooses $\mathbf{x}$ to maximize

$$\mu(\mathbf{x}) = 2\text{Re} \left\{ \sum_k x_k^* \sum_{n=1}^{N} w_n h_{nk}^* \right\} - \sum_k \sum_m x_k x_m^* \sum_{n=1}^{N} h_{nk} h_{nm}^* \tag{6.22}$$

where $\text{Re}\{z\}$ denotes the real part of z. To proceed further, note that

$$\sum_{n=1}^{N} w_n h_{nk}^* = \int_{-\infty}^{\infty} w(\tau) h^*(\tau - nT) d\tau = y_n \tag{6.23}$$

$$\sum_{n=1}^{N} h_{nk} h_{nm}^* = \int_{-\infty}^{\infty} h(\tau - kT) h^*(\tau - mT) d\tau = f_{k-m} \tag{6.24}$$

where y_n and f_k were introduced earlier. The variables $\{y_n\}$ are obtained by passing the received low-pass signal $w(t)$ through a filter $h^*(-t)$ that is matched to the received pulse $h(t)$, and sampling the output. Note that the T-spaced samples at the output of the matched filter must be obtained with the correct timing phase, implying perfect symbol synchronization. The $\{f_n\}$ are called the **ISI coefficients** and have the property that $f_n = f_{-n}^*$. By using (6.23) and (6.24) in (6.22) we have the final form

$$\mu(\mathbf{x}) = 2\text{Re} \left\{ \sum_k x_k^* y_k \right\} - \sum_k \sum_m x_k x_m^* f_{k-m} . \tag{6.25}$$

The noise samples at the matched filter output are, from (6.8),

$$\nu_n = \int_0^{\infty} h^*(-\tau) z(nT - \tau) d\tau$$

$$= \int_{-\infty}^{0} h^*(\tau) z(nT + \tau) d\tau \tag{6.26}$$

and their discrete autocorrelation function is

$$\phi_{\nu\nu}(n) = \frac{1}{2}\mathrm{E}[\nu_{k+n}\nu_k^*] = N_o f_n \ . \tag{6.27}$$

6.3.1 Discrete-Time White Noise Channel Model

The correlation between the noise samples poses some complications when implementing the various equalization schemes. To overcome this difficulty, a **noise whitening filter** can be employed to process the sampled sequence $\{y_n\}$ as described below, resulting in an equivalent **discrete-time white noise channel model**. The z-transform of the vector **f** is

$$F(z) = \sum_{n=-L}^{L} f_n z^{-n} \ . \tag{6.28}$$

Using the property $f_n^* = f_{-n}$ we can write

$$F^*(1/z^*) = F(z) \ . \tag{6.29}$$

It follows that $F(z)$ has $2L$ roots with the factorization

$$F(z) = G(z)G^*(1/z^*) \tag{6.30}$$

where $G(z)$ and $G^*(1/z^*)$ are polynomials of degree L having conjugate reciprocal roots. There are 2^L possible choices for the roots of $G^*(1/z^*)$ and any one will suffice for a noise whitening filter. However, some equalization techniques such as RSSE and DDFSE require that the polynomial of the overall response $G(z)$ have **minimum-phase** . In this case, we can choose the unique $G(z)$ that has minimum phase, i.e., all the roots of $G(z)$ are inside the unit circle. With this choice of $G(z)$, the noise whitening filter $1/G^*(1/z^*)$ is a stable but noncausal filter. In practice, such an noncausal noise whitening filter can be implemented by using an appropriate delay. If the overall response $G(z)$ need not have minimum phase, then we can choose $G^*(1/z^*)$ to have minimum phase, i.e., all the roots of $G^*(1/z^*)$ are inside the unit circle. This choice ensures that the noise whitening filter $1/G^*(1/z^*)$ is both causal and stable.

Example 6.1———————————————————————————

Consider a simple T-spaced two-ray channel where the received pulse is

$$h(t) = h_a(t) + ah_a(t - T)$$

and the transmitted pulse $h_a(t)$ is normalized to have unit energy. The ISI coefficients are

$$
\begin{aligned}
f_n &= \int_{-\infty}^{\infty} h^*(t)h(t + nT)dt \\
&= \begin{cases} 1 + |a|^2 & n = 0 \\ a & n = 1 \\ a^* & n = -1 \end{cases}
\end{aligned}
$$

and, hence,

$$
\begin{aligned}
F(z) &= az + (1 + |a|^2) + az^{-1} \\
&= (az^{-1} + 1)(a^*z + 1) \ .
\end{aligned}
$$

There are two possible choices for the noise whitening filter. Under the assumption that $|a| > 1$, suppose that the zero of $G^*(1/z^*)$ is chosen to be inside the unit circle. That is,

$$
\begin{aligned}
G(z) &= 1 + az^{-1} \\
G^*(1/z^*) &= 1 + a^*z \ .
\end{aligned}
$$

In this case, the noise whitening filter is stable and causal, and the overall system is characterized by the non-minimum phase polynomial

$$G(z) = 1 + az^{-1} \ .$$

Again, under the assumption that $|a| > 1$, suppose that the zero of $G^*(1/z^*)$ is chosen to be outside the unit circle. That is,

$$
\begin{aligned}
G(z) &= 1 + a^*z \\
G^*(1/z^*) &= 1 + az^{-1} \ .
\end{aligned}
$$

In this case, the noise whitening filter is stable and noncausal, and the overall system is characterized by the minimum phase polynomial

$$G(z) = 1 + a^*z \ .$$

For any choice of noise whitening filter, the filter output is

$$V(z) = (X(z)F(z) + \nu(z))\frac{1}{G^*(1/z^*)}$$
$$= X(z)G(z) + \nu(z)\frac{1}{G^*(1/z^*)} \; . \tag{6.31}$$

From (6.27), the power spectral density of the noise at the input to the noise whitening filter is

$$S_{\nu\nu}(f) = N_o F(e^{j2\pi fT}) \; , \qquad |f| \leq \frac{1}{2T} \; . \tag{6.32}$$

Therefore, the power spectral density of the noise at the output of noise whitening filter is

$$S_{\eta\eta}(f) = N_o \frac{F(e^{j2\pi fT})}{|G^*(e^{j2\pi fT})|^2}$$
$$= N_o \frac{G(e^{j2\pi fT})G^*(e^{j2\pi fT})}{G(e^{j2\pi fT})G^*(e^{j2\pi fT})}$$
$$= N_o \; , \qquad |f| \leq \frac{1}{2T} \tag{6.33}$$

which is clearly white. The above development leads to the system shown in Fig. 6.3, with the equivalent discrete-time white noise channel model shown in Fig. 6.4. The discrete-time samples at the output of the noise whitening filter are

$$v_k = \sum_{n=0}^{L} g_n x_{k-n} + \eta_k \; . \tag{6.34}$$

It follows that the effective overall channel impulse response can be described by the **channel vector**

$$\mathbf{g} = (g_0, \, g_1, \ldots, \, g_L)^T \; . \tag{6.35}$$

The symbol energy-to-noise ratio is

$$\gamma_s = \frac{E}{N_o} = \frac{E[|x_k|^2]\sum_{i=0}^{L}|g_i|^2}{2N_o} \tag{6.36}$$

and the bit energy-to-noise ratio is $\gamma_b = \gamma_s / \log_2 M$ where M is the modulation alphabet size.

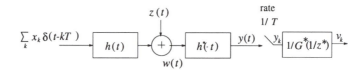

Figure 6.3   Block diagram of system that implements a filter matched to $h(t)$
followed by a discrete-time noise whitening filter.

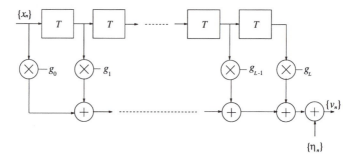

Figure 6.4   Discrete-time white noise channel model.

Time Varying Channels with Diversity Reception

For time-varying channels with D-branch diversity, the corresponding discrete-
time white noise channel model is shown in Fig. 6.5. At epoch k, the tap gains
associated with diversity branch d are described by the vector

$$\mathbf{g}_d(k) = (g_{0,d}(k),\ g_{1,d}(k),\ \ldots\ g_{L,d}(k))^T \tag{6.37}$$

The $\{g_{i,d}(k)\}$ are complex Gaussian random processes having covariance matrix

$$\Phi_{\mathbf{g}_d}(m) = \frac{1}{2}\mathrm{E}[\mathbf{g}_d(k+m)\mathbf{g}_d^H(k)] \tag{6.38}$$

where H denotes Hermitian transposition. If spatial diversity is used with sufficient antenna separation to generate uncorrelated taps and the approximation
is made that the tap coefficients associated with each diversity branch are uncorrelated, then $\Phi_{\mathbf{g}_d}(m)$ is a diagonal matrix. If we further assume 2-D isotropic
scattering, then from (2.20)

$$\Phi_{\mathbf{g}_d}(m) = J_0(2\pi f_m\ mT)\Sigma_d^2 \tag{6.39}$$

where $J_0(\cdot)$ is a zero-order Bessel function of the first kind and f_m is the
maximum Doppler frequency, and

$$\Sigma_d \overset{\Delta}{=} \mathrm{diag}[\sigma_{0,d},\ \sigma_{1,d},\ldots,\ \sigma_{L,d}\,] \tag{6.40}$$

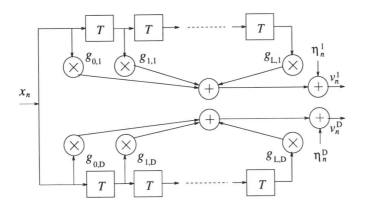

Figure 6.5 Discrete-time white noise channel model with D-branch diversity, from [281].

where $\sigma_{i,d} = \frac{1}{2}\mathrm{E}[|g_{i,d}|^2]$. The received sample on branch d at epoch k is

$$v_{k,d} = \sum_{i=0}^{L} g_{i,d}(k)x_{k-i} + \eta_{k,d} \ . \tag{6.41}$$

The $\eta_{k,d}$ are independent complex zero-mean white Gaussian noise samples with variance $\frac{1}{2}\mathrm{E}[|\eta_{k,d}|^2] = N_o$. The average received symbol energy-to-noise ratio for branch d is

$$\bar{\gamma}_s^d = \frac{\bar{E}}{N_o} = \frac{\mathrm{E}[|x_k|^2]\sum_{i=0}^{L}|g_{i,d}|^2}{2N_o} \ . \tag{6.42}$$

Usually the branches are balanced so that $\bar{\gamma}_s^d = \bar{\gamma}_s$, $d = 1,\ldots,D$. The averaged received branch bit energy-to-noise ratio is $\bar{\gamma}_c = \bar{\gamma}_s/\log_2 M$. Finally, in this chapter we assume that the average power in the signal constellation in normalized so that $\mathrm{E}[|x_k|^2] = 1$ in (6.36) and (6.42).

Fractionally-Spaced Receiver

In practice the the matched filter outputs are often **oversampled** for the purpose of extracting timing information and to mitigate the effects of timing errors. One important example that will be considered at various points in this chapter is when the output of the receiver filter, $y(t)$, is sampled with rate $2/T$. In this case the overall channel impulse response and sampler can be represented by a $T/2$-spaced discrete-time transversal filter with coefficients

$$\mathbf{f}^{(2)} = (\ f_{-2L}^{(2)},\ f_{-2L+1}^{(2)},\ \ldots,\ f_{-1}^{(2)},\ f_0^{(2)},\ f_1^{(2)},\ \ldots,\ f_{2L-1}^{(2)},\ f_{2L}^{(2)}\ ) \ . \tag{6.43}$$

where $(\,\cdot\,)^{(2)}$ indicates rate $2/T$ sampling. If it so happens that the samples in (6.43) are obtained with the correct timing phase, i.e., $f_n^{(2)} = f(nT/2)$ and $f_n^{(2)} = \left(f_{-n}^{(2)}\right)^*$, then

$$
\begin{aligned}
\mathbf{f} &= (\,f_{-L},\ f_{-L+1},\ \ldots,\ f_{-1},\ f_0,\ f_1,\ \ldots,\ f_{L-1},\ f_L\,) &\qquad (6.44) \\
&= (\,f_{-2L}^{(2)},\ f_{-2L+2}^{(2)},\ \ldots,\ f_{-2}^{(2)},\ f_0^{(2)},\ f_2^{(2)},\ \ldots,\ f_{2L-2}^{(2)},\ f_{2L}^{(2)}\,)
\end{aligned}
$$

where $f_n = f_{2n}^{(2)}$. More details on timing phase sensitivity will be provided in Section 6.7.4.

The $T/2$-spaced noise samples at the matched filter output have the autocorrelation

$$
\phi_{\nu\nu}(n) = N_o f_n^{(2)} \ . \tag{6.45}
$$

The z-transform of $\mathbf{f}^{(2)}$, denoted as $F^{(2)}(z)$, has $4L$ roots with the factorization

$$
F^{(2)}(z) = G^{(2)}(z)(G^{(2)}(1/z^*))^* \tag{6.46}
$$

where $G^{(2)}(z)$ and $(G^{(2)}(1/z^*))^*$ are polynomials of degree $2L$ having conjugate reciprocal roots. The $T/2$-spaced correlated noise samples can be whitened by using a filter with transfer function $1/(G^{(2)}(1/z^*))^*$. Once again, $(G^{(2)}(1/z^*))^*$ can be chosen such that all its roots are inside the unit circle, yielding a stable and causal noise whitening filter. On the other hand, we could choose the overall response $G^{(2)}(z)$ to have minimum phase, if necessary. The output of the noise whitening filter is

$$
v_n^{(2)} = \sum_{k=0}^{2L} g_k^{(2)} x_{n-k}^{(2)} + \eta_n^{(2)} \tag{6.47}
$$

where $\{\eta_n^{(2)}\}$ is a $T/2$-spaced white Gaussian noise sequence with variance $\frac{1}{2}\mathrm{E}[|\eta_n^{(2)}|^2] = N_o$ and the $\{g_n^{(2)}\}$ are the coefficients of a $T/2$-spaced discrete-time transversal filter having a transfer function $G^{(2)}(z)$. The sequence $\{x_n^{(2)}\}$ is the corresponding $T/2$-spaced input symbol sequence and is given by

$$
x_n^{(2)} = \begin{cases} x_{n/2}\,, & n = 0, 2, 4, \ldots \\ 0\,, & n = 1, 3, 5, \ldots \end{cases} \tag{6.48}
$$

The overall system and equivalent discrete-time models are shown in Figs. 6.6 and 6.7, respectively.

Comparing (6.30) and (6.46), we have

$$
\sum_{k=0}^{2L} |g_k^{(2)}|^2 = \sum_{k=0}^{L} |g_k|^2 = f_0^{(2)} = f_0 \ . \tag{6.49}
$$

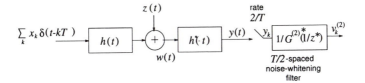

Figure 6.6 Block diagram of system that implements a filter matched to $h(t)$ followed by a $T/2$-spaced sampler and a $T/2$-spaced noise whitening filter.

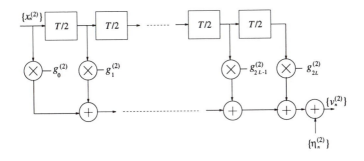

Figure 6.7 Discrete-time white noise channel model with rate $2/T$ sampling.

Notice that the samples $v_{2n}^{(2)}$ and $v_{2n+1}^{(2)}$ correspond to the nth received baud, where

$$v_{2n}^{(2)} = \sum_{k=0}^{L} g_{2k}^{(2)} x_{n-k} + \eta_{2n}^{(2)} \tag{6.50}$$

$$v_{2n+1}^{(2)} = \sum_{i=0}^{L-1} g_{2k+1}^{(2)} x_{n-k} + \eta_{2n+1}^{(2)} \ . \tag{6.51}$$

Finally, by comparing (6.34) and (6.51), we note that $v_{2n}^{(2)}$ is not necessarily equal to v_n because a different noise whitening filter is used to whiten the $T/2$-spaced samples.

6.4 SYMBOL-BY-SYMBOL EQUALIZERS

There are two broad categories of symbol-by-symbol equalizers, linear forward equalizers and nonlinear decision feedback equalizers. As shown in Fig. 6.8, a linear forward equalizer consists of a transversal filter with adjustable tap coefficients. The tap co-efficients of the equalizer are denoted by the column vector

$$\mathbf{c} = (c_0, \ c_1, \ \cdots, \ c_{N-1})^T \tag{6.52}$$

where N is the number of equalizer taps. Assuming that the equalizer is preceded by a whitened matched filter, the output of the equalizer is

$$\tilde{x}_k = \sum_{j=0}^{N-1} c_j v_{k-j} \tag{6.53}$$

where the v_n are given by (6.34). The equalizer output $\tilde{x}_k$ is quantized to the nearest (in Euclidean distance) information symbol to form the decision $\hat{x}_k$.

Observe that the overall channel and equalizer can be represented by a single filter having the sampled impulse response

$$\mathbf{q} = (q_0, q_1, \ \ldots, \ q_{N+L-1})^T \tag{6.54}$$

where

$$
\begin{aligned}
q_n &= \sum_{j=0}^{N-1} c_j g_{n-j} \\
&= \mathbf{c}^T \mathbf{g}_n
\end{aligned}
\tag{6.55}
$$

with

$$\mathbf{g}_n = (g_n, g_{n-1}, g_{n-2}, \ldots, g_{n-N+1})^T \tag{6.56}$$

and $g_i = 0, i < 0, i > L$. That is, $\mathbf{q}$ is the discrete convolution of $\mathbf{g}$ and $\mathbf{c}$.

Let the component of $\mathbf{g}$ of greatest magnitude be denoted by g_{d_1}. Note that any choice of noise whitening filter that does not result in an overall response $G(z)$ with minimum phase may have $d_1 \neq 0$. Also, let the number of equalizer taps be equal to $N = 2d_2 + 1$ where d_2 is an integer. Perfect equalization means that

$$\mathbf{q} = \mathbf{e}_d = (0, \ 0, \ \ldots, \ 0, \ 1, \ 0, \ \ldots, \ 0, \ 0)^T \tag{6.57}$$

where $d - 1$ zeroes precede the "1" and d is an integer representing the overall delay. Unfortunately, perfect equalization is difficult to achieve and does not always yield the best performance.

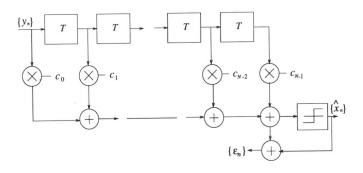

Figure 6.8 Linear transversal equalizer with adjustable T-spaced taps.

6.4.1   Linear Equalization

Zero-Forcing Equalizer

With a zero-forcing (ZF) equalizer, the tap coefficients **c** are chosen to minimize the peak distortion of the equalized channel, defined as

$$D_{\mathrm{p}} = \frac{1}{|q_d|} \sum_{\substack{n=0 \\ n \neq d}}^{N+L-1} |q_n - \hat{q}_n| \tag{6.58}$$

where $\hat{q} = (\hat{q}_0, \ldots, \hat{q}_{N+L-1})^T$ is the *desired equalized channel* and the delay d is a positive integer optimized to have the value $d = d_1 + d_2$ [47]. Lucky showed that if the initial distortion without equalization is less than unity, i.e.,

$$D = \frac{1}{|g_{d_1}|} \sum_{\substack{n=0 \\ n \neq d_1}}^{L} |g_n| < 1 , \tag{6.59}$$

then D_p is minimized by those N tap values which simultaneously cause $q_j = \hat{q}_j$ for $d - d_2 \leq j \leq d + d_2$. However, if the initial distortion before equalization is greater than unity, the ZF criterion is not guaranteed to minimize the peak distortion. For the case when $\hat{q} = e_d$ the equalized channel is given by

$$q = (q_0, \ldots, q_{d_1-1}, 0, \ldots, 0, 1, 0, \ldots, 0, q_{d_1+N}, \ldots, q_{N+L-1})^T . \tag{6.60}$$

In this case the equalizer forces zeroes into the equalized channel and, hence, the name "zero-forcing equalizer."

Equalizer Tap Solution

For a known channel impulse response, the tap gains of the ZF equalizer can be found by the direct solution of a simple set of linear equations [47]. To do so, we form the matrix

$$\mathbf{G} = [\mathbf{g}_{d_1}, \ldots, \mathbf{g}_d, \ldots, \mathbf{g}_{N+d_1-1}] \qquad (6.61)$$

and the vector

$$\tilde{\boldsymbol{q}} = \left(\hat{q}_{d_1}, \ldots, \hat{q}_d, \ldots, \hat{q}_{N+d_1-1}\right)^T . \qquad (6.62)$$

Then the vector of optimal tap gains, $\mathbf{c}_{\mathrm{op}}$, satisfies

$$\mathbf{c}_{\mathrm{op}}^T \mathbf{G} = \tilde{\boldsymbol{q}}^T \longrightarrow \mathbf{c}_{\mathrm{op}}^T = \tilde{\boldsymbol{q}}^T \mathbf{G}^{-1} . \qquad (6.63)$$

Example 6.2

Suppose that a system has the channel vector $\mathbf{g} = (0.90, -0.15, 0.20, 0.10, -0.05)^T$, where $g_i = 0, i < 0, i > 4$. The initial distortion before equalization is

$$D = \frac{1}{|g_0|} \sum_{n=0}^{4} |g_n| = 0.5555$$

and, therefore, the minimum distortion is achieved with the ZF solution. Suppose that we wish to design a 3-tap ZF equalizer. Since g_0 is the component of $\mathbf{g}$ having the largest magnitude, $d_1 = 0$ and the optimal equalizer delay is $d = 1$. Suppose that the desired response is $\hat{q}_0 = \mathbf{e}_0$ so that $\tilde{\mathbf{q}} = (1, 0, 0)$. We then construct the matrix

$$
\begin{aligned}
\mathbf{G} &= [\mathbf{g}_0, \mathbf{g}_1, \mathbf{g}_2] \\
&= \begin{bmatrix} 0.90 & -0.15 & 0.20 \\ 0.00 & 0.90 & -0.15 \\ 0.00 & 0.00 & 0.90 \end{bmatrix}
\end{aligned}
$$

and obtain the optimal tap solution

$$\mathbf{c}_{\mathrm{op}} = (\mathbf{G}^{-1})^T \tilde{\mathbf{q}} = (1.1111, 0.1852, -0.2160)^T .$$

The overall response of the channel and equalizer is

$$\boldsymbol{q} = (1.0, 0.0, 0.0, 0.1806, -0.080, -0.031, -0.011, 0, \ldots)^T .$$

Hence, the minimum distortion with this equalizer is

$$D_{\min} = \frac{1}{|q_0|} \sum_{n=1}^{6} |q_n - \hat{q}_n| = 0.30247 .$$

Adaptive Solution

In practice, the channel impulse response is unknown to the receiver so that a known finite length sequence $\mathbf{x}$ must be used to train the equalizer. During the training mode, the equalizer taps can be obtained by using the following steepest-descent recursive algorithm:

$$c_j^{k+1} = c_j^k + \alpha \epsilon_k x_{k-j-d_1}^* , \qquad j = 0, \ldots, N-1 \qquad (6.64)$$

where

$$\begin{aligned}
\epsilon_k &= x_{k-d} - \tilde{x}_k \\
&= x_{k-d} - \sum_{i=0}^{N-1} c_i v_{k-i} \qquad (6.65)
\end{aligned}$$

is the error sequence, $\{c_j^k\}$ is the set of equalizer tap gains at epoch k, and α is an adaptation step-size that can be optimized to provide rapid convergence. Notice that adaptation rule in (6.64) attempts to force the crosscorrelations $\epsilon_k x_{k-j-d_1}^* , j = 0, \ldots, N-1$, to zero. To see that (6.64) leads to the desired solution we use (6.34) and (6.65), along with the fact that the symbol sequence $\{x_n\}$ is uncorrelated with the noise sequence $\{\eta_n\}$ to obtain

$$\begin{aligned}
\mathrm{E}[\epsilon_k x_{k-j-d_1}^*] &= \mathrm{E}[x_{k-d} x_{k-j-d_1}^*] - \sum_{i=0}^{N-1} \sum_{\ell=0}^{L} g_\ell \mathrm{E}[x_{k-i-\ell} x_{k-j-d_1}^*] \\
&= \delta_{d_2-j} - \sum_{i=0}^{N-1} c_i g_{j+d_1-i} \\
&= \delta_{d_2-j} - q_{j+d_1} , \qquad j = 0, 1, \ldots, N-1 . \quad (6.66)
\end{aligned}$$

Therefore, the conditions $\epsilon_k x_{k-j-d_1}^* = 0, j = 0, \ldots, N-1$ are satisfied by $q_d = 1$ and $q_i = 0$ for $d_1 \leq i < d$ and $d < i \leq d + d_2$.

After training the equalizer, a decision-feedback mechanism is typically employed where the sequence of symbol decisions $\hat{\mathbf{x}}$ is used to update the tap coefficients. This mode is called the data mode and allows the equalizer to track variations in the channel vector $\mathbf{g}$. In the data mode, the error ϵ_k in (6.65) becomes

$$\epsilon_k = \hat{x}_{k-d} - \sum_{i=0}^{N-1} c_i v_{k-i} \qquad (6.67)$$

where, again, $\hat{x}_k$ is the decision on the equalizer output $\tilde{x}_k$.

Performance of the ZF Equalizer

If the ZF equalizer has an infinite number of taps it is possible to select the tap weights so that $D_p = 0$, i.e., $q = \hat{q}$. Assuming that $\hat{q}_n = \delta_{n0}$ this condition means that

$$Q(z) = 1 = C(z)G(z) \ . \tag{6.68}$$

Therefore,

$$C(z) = \frac{1}{G(z)} \tag{6.69}$$

and the ideal ZF equalizer has a discrete transfer function that is simply the inverse of overall channel $G(z)$. The cascade of the noise whitening filter with transfer function $1/G^*(1/z^*)$ and the ZF equalizer with transfer function $1/G(z)$ results in an equivalent equalizer with transfer function

$$C'(z) = \frac{1}{G^*(1/z^*)G(z)} = \frac{1}{F(z)} \ . \tag{6.70}$$

Recall from (6.32) that the noise at the input to the equivalent equalizer $C'(z)$ has autocorrelation $\phi_{\nu\nu}(k) = N_o f_k$ and power spectral density (psd)

$$S_{\nu\nu}(f) = N_o F(e^{j2\pi fT}) \ , \qquad |f| \le \frac{1}{2T} \ . \tag{6.71}$$

Therefore, the psd of the noise sequence $\{\zeta_n\}$ at the output of the equalizer is

$$S_{\zeta\zeta}(f) = \frac{N_o}{F(e^{j2\pi fT})} \ , \qquad |f| \le \frac{1}{2T} \ . \tag{6.72}$$

and the noise samples have variance

$$\begin{aligned}
\sigma_\zeta^2 &= T \int_{-1/2T}^{1/2T} S_{\zeta\zeta}(f) df \\
&= T \int_{-1/2T}^{1/2T} \frac{N_o}{F(e^{j2\pi fT})} df \ .
\end{aligned} \tag{6.73}$$

If $E[|x_k|^2] = 1$ and $\hat{q}_n = \delta_{n0}$, then the signal-to-noise ratio at the output of the infinite-tap equalizer is

$$\gamma_\infty = \frac{E[|x_k|^2]}{\sigma_\zeta^2} = \frac{1}{\sigma_\zeta^2} \ . \tag{6.74}$$

Finally, we can show that (see Problem 6.2)

$$F(e^{j2\pi fT}) = F_\Sigma(f) \ , \qquad |f| \le \frac{1}{2T} \tag{6.75}$$

where $F_\Sigma(f)$ is the folded spectrum of $F(f)$ defined in (6.13), and $F(f) = |H(f)|^2$ is the Fourier transform of the pulse $f(t) = h(t) * h^*(-t)$. It is clear from (6.73) that ZF equalizers are unsuitable for channels that have severe ISI, where the folded spectrum $F_\Sigma(f)$ has spectral nulls or very small values. Under these conditions, the equalizer tries to compensate for nulls in the folded spectrum by introducing infinite gain at these frequencies. Unfortunately, this results in severe noise enhancement at the output of the equalizer. Mobile radio channels often exhibit spectral nulls and, therefore, ZF equalizers are typically not used for mobile radio applications.

Mean-Square-Error Equalizer

The mean-square-error (MSE) equalizer is more robust and superior to the ZF equalizer in its performance and convergence properties [257, 256, 259]. By defining the vector

$$\mathbf{v}_k = (v_k, \ v_{k-1}, \ \ldots, \ v_{k-N+1}) \tag{6.76}$$

the output of the equalizer in (6.53) can be expressed in the form

$$\tilde{x}_k = \mathbf{c}^T \mathbf{v}_k = \mathbf{v}_k^T \mathbf{c} \ . \tag{6.77}$$

A MSE equalizer adjusts the tap coefficients to minimize the MSE

$$\begin{aligned} J &\overset{\Delta}{=} \mathrm{E}[|x_{k-d} - \tilde{x}_k|^2] \\ &= \mathrm{E}\left[\mathbf{c}^T \mathbf{v}_k \mathbf{v}_k^H \mathbf{c}^* - 2\mathrm{Re}\{\mathbf{v}_k^H \mathbf{c}^* x_{k-d}\} + |x_{k-d}|^2\right] \ . \end{aligned} \tag{6.78}$$

Equalizer Tap Solution

If the channel impulse response is known, the optimum equalizer taps can be obtained by direct solution. Define

$$\begin{aligned} \mathbf{M}_v &\overset{\Delta}{=} \mathrm{E}[\mathbf{v}_k \mathbf{v}_k^H] \\ \mathbf{v}_x^H &\overset{\Delta}{=} \mathrm{E}[\mathbf{v}_k^H x_{k-d}] \end{aligned} \tag{6.79}$$

where $\mathbf{M}_v$ is an $N \times N$ Hermitian matrix and $\mathbf{v}_x$ is a length N column vector. Using these definitions and assuming that $\mathrm{E}[|x_{k-d}|^2] = 1$, the mean-square-error is

$$J = \mathbf{c}^T \mathbf{M}_v \mathbf{c}^* - 2\mathrm{Re}\{\mathbf{v}_x^H \mathbf{c}^*\} + 1 \ . \tag{6.80}$$

The tap vector $\mathbf{c}$ that minimizes the mean square error can obtained by equating the gradient $\nabla_\mathbf{c} J$ to zero. It can be shown that (see Problem 6.15)

$$\nabla_\mathbf{c} J = \left(\frac{\partial J}{\partial c_0}, \ \cdots, \ \frac{\partial J}{\partial c_{N-1}}\right) = 2\mathbf{c}^T \mathbf{M}_v - 2\mathbf{v}_x^H \ . \tag{6.81}$$

Setting $\nabla_c J = 0$ gives

$$\mathbf{c}_{op} = (\mathbf{M}_v^T)^{-1}\mathbf{v}_x^* \quad . \tag{6.82}$$

By using the identity $(\mathbf{A}^{-1})^T = (\mathbf{A}^T)^{-1}$ and the fact that $\mathbf{M}_v$ is Hermitian, the minimum mean-square-error (MMSE) is

$$
\begin{aligned}
J_{\min} &= \mathbf{c}_{op}^T \mathbf{M}_v \mathbf{c}_{op}^* - 2\mathrm{Re}\{\mathbf{v}_x^H \mathbf{c}_{op}^*\} + 1 \\
&= 1 - \mathbf{v}_x^H \mathbf{M}_v^{-1} \mathbf{v}_x \quad .
\end{aligned} \tag{6.83}
$$

Since the overall channel and equalizer can be represented as a single filter with impulse response $\mathbf{q}$ in (6.54) it follows that the MMSE can also be expressed in the form

$$J_{\min} = \|\mathbf{q} - \hat{\mathbf{q}}\|^2 + N_o \|\mathbf{c}\|^2 \quad . \tag{6.84}$$

Example 6.3_____

Consider a system having the same channel vector as in Example 6.2. Suppose that we wish to design a 3-tap MSE equalizer. In this case

$$\mathbf{v}_x^H = (-0.15,\ 0.90,\ 0.00)$$

and

$$
\mathbf{M}_v = \begin{bmatrix}
\beta & -0.1500 & 0.1550 \\
-0.1500 & \beta & -0.1500 \\
0.1550 & -0.1500 & \beta
\end{bmatrix}
$$

where $\beta = 0.8850 + N_o$. The inverse of $\mathbf{M}_v$ is

$$\mathbf{M}_v^{-1} = \frac{\mathrm{adj}(\mathbf{M}_v)}{\det(\mathbf{M}_v)}$$

where $\det(\mathbf{M}_v) = \beta(\beta^2 - 0.069025) + 0.006975$ and

$$
\mathrm{adj}(\mathbf{M}_v) = \begin{bmatrix}
\beta^2 - 0.0225 & 0.15\beta - 0.02325 & 0.0225 - 0.155\beta \\
0.15\beta - 0.02325 & \beta^2 - 0.024025 & 0.15\beta - 0.02325 \\
0.0225 - 0.155\beta & 0.15\beta - 0.02325 & \beta^2 - 0.0225
\end{bmatrix}.
$$

Hence,

$$
\mathbf{c}_{op} = \frac{1}{\det(\mathbf{M}_v)} \begin{pmatrix}
-0.15\beta^2 + 0.135\beta - 0.1755 \\
0.90\beta^2 - 0.0225\beta - 0.018135 \\
0.15825\beta - 0.0243
\end{pmatrix} \quad .
$$

With this tap solution,

$$J_{\min} = 1 - \frac{1}{\det(\mathbf{M}_v)}\left(0.8325\beta^2 - 0.013689\right)$$

and as $N_o \to 0$, $J_{\min} = 0.001089424$.

Adaptive Solution

In practice, the channel impulse response is unknown. However, the equalizer taps can be obtained by using the stochastic gradient algorithm

$$c_j^{k+1} = c_j^k + \alpha \epsilon_k v_{k-j}^* \qquad j = 0, \ldots, N-1 \qquad (6.85)$$

where ϵ_k is given in (6.65). To show that (6.85) leads to the desired solution, note from (6.82) that

$$
\begin{aligned}
\nabla_{\mathbf{c}} J &= 2\mathrm{E}[\mathbf{c}^T \mathbf{v}_k \mathbf{v}_k^H - x_{k-d} \mathbf{v}_k^H] \\
&= 2\mathrm{E}[(\mathbf{c}^T \mathbf{v}_k - x_{k-d}) \mathbf{v}_k^H] \\
&= 2\mathrm{E}[\epsilon_k \mathbf{v}_k^H] = 0 \ .
\end{aligned}
\qquad (6.86)
$$

It follows that

$$\mathrm{E}[\epsilon_k v_{k-j}^*] = 0, \qquad j = 0, \ldots, N-1 \ . \qquad (6.87)$$

Performance of the MSE Equalizer

The performance of an MSE equalizer having an infinite number of taps provides some useful insight. In this case

$$
\begin{aligned}
\mathbf{c} &= (c_{-\infty}, \ldots, c_0, \ldots, c_\infty) \\
\mathbf{v}_k &= (v_{k+\infty}, \ldots, v_k, \ldots, c_{k-\infty}) \ .
\end{aligned}
$$

Since the delay d with an infinite-tap equalizer is irrelevant we can choose $d = 0$ so that

$$\mathrm{E}[x_k v_{k-j}^*] = \begin{cases} g_{-j}^* &, \quad -L \le j \le 0 \\ 0 &, \quad \text{otherwise} \end{cases} \qquad (6.88)$$

The equation for the optimal tap gain vector $\mathbf{c}^T \mathbf{M}_v = \mathbf{v}_x^H$ can be written in the form

$$\sum_{i=-\infty}^{\infty} c_i \left(f_{j-i} + N_o \delta_{ij} \right) = g_{-j}^* \qquad -\infty < j < \infty \ . \qquad (6.89)$$

Taking the z-transform of both sides of (6.89) gives

$$C(z) \left(G(z) G(1/z^*) + N_o \right) = G(1/z^*) \qquad (6.90)$$

and, therefore,

$$C(z) = \frac{G(1/z^*)}{G(z)G(1/z^*) + N_o} . \tag{6.91}$$

The equivalent MSE equalizer that includes the noise whitening filter $1/G(1/z^*)$ is

$$C'(z) = \frac{1}{G(z)G(1/z^*) + N_o} = \frac{1}{F(z) + N_o} . \tag{6.92}$$

Notice that $C'(z)$ has the same form as the ZF equalizer in (6.70), except for the noise term N_o in the denominator. Clearly, the ZF and MSE criterion lead to the same solution in the absence of noise.

The most meaningful measure of performance is the bit error probability. However, for many equalization techniques, the bit error probability is a highly nonlinear function of the equalizer co-efficients. One possibility with the MSE equalizer is to evaluate the MMSE [256]

$$\begin{aligned} J_{\min} &= 1 - \frac{1}{T} \int_{-1/2T}^{1/2T} \frac{F_\Sigma(f)}{F_\Sigma(f) + N_o} df \\ &= T \int_{-1/2T}^{1/2T} \frac{N_o}{F_\Sigma(f) + N_o} df . \end{aligned} \tag{6.93}$$

Note that $0 \le J_{\min} \le 1$, and that $J_{\min} = 0$ when there is no ISI or noise and $J_{\min} = 1$ when the folded spectrum $F_\Sigma(f)$ exhibits a spectral null.

Another measure of the effectiveness of linear equalization is the *signal-to-noise-plus-interference ratio* (SNIR) defined as

$$\text{SNIR} = \frac{|q_d|^2}{\sum_{\substack{j=0 \\ j \ne d}}^{N+L-1} |q_j|^2 + N_o \sum_{j=0}^{N-1} |c_j|^2} . \tag{6.94}$$

Although the MSE equalizer accounts for the effects of noise, satisfactory performance still cannot be achieved for channels with severe ISI or spectral nulls, because of the noise enhancement at the output of the equalizer [256], [89]. Another problem with a linear equalizer is the adaptation of the equalizer during data mode. This problem is especially acute for systems that use trellis-coded modulation, because the equalizer-based decisions are unreliable and inferior to those in uncoded systems due to the reduced separation between the points in the signal constellation. This problem can be partially alleviated by using periodic training, where the equalizer taps are allowed to converge in the periodic training modes. When the equalizer has converged, the updating algorithm

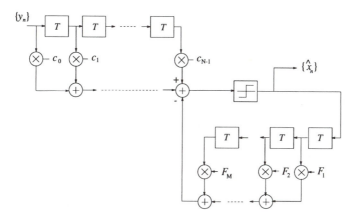

Figure 6.9 Decision feedback equalizer.

is disabled [74]. However, this approach is only suitable for fixed channels or channels with very slow variations such as voiceband data channels.

6.4.2 Decision Feedback Equalization

The deleterious effects of noise enhancement that degrade the performance of linear equalizers can be mitigated by using a nonlinear decision feedback equalizer (DFE). The DFE consists of two sections; a feedforward section and a feedback section. A typical DFE is illustrated in Fig. 6.9. The DFE is nonlinear because the feedback path includes a decision device. The feedforward section is identical to the linear forward equalizer discussed earlier, and its purpose is to reduce the precursor ISI. Decisions made on the equalizer outputs are propagated through the feedback filter, which is used to estimate the ISI contributed by these symbols. The coefficients of the feedback filter are the sampled impulse response of the tail of the system impulse response including the forward part of the DFE.

The output of the DFE is

$$\tilde{x}_n = \sum_{i=0}^{N-1} c_{N-1-i} v_{n+i} - \sum_{i=1}^{M} F_i \hat{x}_{n-i} \qquad (6.95)$$

where $\{c_i\}$ and $\{F_i\}$ are the tap coefficients of the feedforward and feedback filters, respectively, and $\{\hat{x}_i\}$ is the sequence of previously detected symbols.

Recall that the overall channel and feedforward portion of the equalizer can be represented by the sampled impulse response in (6.54). By using (6.34), the DFE output can be written as

$$
\tilde{x}_n = \sum_{i=-L}^{N-1} q_{N-1-i} x_{n+i} - \sum_{i=1}^{M} F_i \tilde{x}_{n-i} + n_n
$$

$$
= x_n q_{N-1} + \sum_{i=1}^{M} (x_{n-i} - \tilde{x}_{n-i}) q_{N-1+i} + \sum_{i=1}^{M} (q_{N-1+i} - F_i) \tilde{x}_{n-i}
$$

$$
+ \sum_{i=1}^{N-1} q_{N-1+i} x_{n+i} + \sum_{i=M+1}^{L} q_{N-1+i} x_{n-i} + n_n \qquad (6.96)
$$

If we choose

$$
F_i = q_{N-1+i} = \mathbf{c}^T \mathbf{g}_{N-1+i}, \quad i = 1, 2, \ldots, M \qquad (6.97)
$$

so that the second summation is zero and if correct decisions are made so that the first summation is zero, then

$$
\tilde{x}_n = x_n q_{N-1} + \sum_{i=1}^{N-1} q_{N-1+i} x_{n+i} + \sum_{i=M+1}^{L} q_{N-1+i} x_{n-i} + n_n . \qquad (6.98)
$$

The first and second summations in (6.98) are the residual ISI associated with the feedforward and feedback filters, respectively. Note that feedback coefficients in (6.97) result in the complete removal of ISI from the previously detected symbols if $L \leq M$.

Equalizer Tap Solution

The co-efficients $\{c_i\}$ and $\{F_i\}$ can be adjusted simultaneously to minimize the mean square error. Define

$$
\mathbf{c} = (c_0, c_1, \ldots, c_{N-1})^T \qquad (6.99)
$$

$$
\mathbf{v}_k = (v_{k+N-1}, v_{k+N-2}, \ldots, v_k)^T \qquad (6.100)
$$

$$
\hat{\mathbf{x}}_k = (\hat{x}_{k-1}, \hat{x}_{k-2}, \ldots, \hat{x}_{k-M})^T \qquad (6.101)
$$

$$
\mathbf{F} = (F_1, F_2, \ldots, F_M)^T \qquad (6.102)
$$

and

$$
\tilde{\mathbf{c}} = (\mathbf{c}^T, \mathbf{F}^T)^T \qquad (6.103)
$$

$$
\tilde{\mathbf{v}}_k = (\mathbf{v}_k^T, \hat{\mathbf{x}}_k^T)^T \qquad (6.104)
$$

Then the MSE can be expressed as

$$\begin{aligned} J &= \mathrm{E}[|x_k - \tilde{x}_k|^2] \\ &= \mathrm{E}\left[\tilde{\mathbf{c}}^T \tilde{\mathbf{v}}_k \tilde{\mathbf{v}}_k^H \tilde{\mathbf{c}}^* - 2\mathrm{Re}\{\tilde{\mathbf{v}}_k^H \tilde{\mathbf{c}}^* x_k\} + |x_k|^2\right] \; . \end{aligned} \tag{6.105}$$

Since (6.105) and (6.78) have the same form it follows that the optimal tap solution can be obtained by defining

$$\tilde{\mathbf{M}}_v \triangleq \mathrm{E}[\tilde{\mathbf{v}}_k \tilde{\mathbf{v}}_k^H] \tag{6.106}$$

$$\tilde{\mathbf{v}}_x^H \triangleq \mathrm{E}[\tilde{\mathbf{v}}_k^H x_k] \tag{6.107}$$

$$\tilde{\mathbf{c}}_{\mathrm{op}} = (\tilde{\mathbf{M}}_v^T)^{-1} \tilde{\mathbf{v}}_x^* \; . \tag{6.108}$$

Adaptive Solution

The feedforward taps of the DFE can be adjusted by using

$$c_j^{k+1} = c_j^k + \alpha \epsilon_k v_{k+j}^* \qquad j = 0, \ldots, N-1 \tag{6.109}$$

while the feedback coefficients can be adjusted according to

$$F_j^{k+1} = F_j^k + \alpha \epsilon_k \hat{x}_{k-j}^* \qquad j = 1, \ldots, M \; . \tag{6.110}$$

To see that this leads to the desired solution, observe that $\nabla_{\mathbf{c}} J = 2\mathrm{E}[\epsilon_k \tilde{\mathbf{v}}_k^H] = 0$ implies that

$$\mathrm{E}[\epsilon_k v_{k+j}^*] = 0, \qquad j = 0, \ldots, N-1 \tag{6.111}$$

$$\mathrm{E}[\epsilon_k x_{k-j}^*] = 0, \qquad j = 1, \ldots, M \; . \tag{6.112}$$

Performance of the DFE

Since the feedback section of the DFE eliminates the postcursor residual ISI at the output of the forward filter, it is apparent that the optimum setting for the forward filter for an infinite length DFE is identical to a stable, non-causal, noise whitening filter that results in a overall channel with a minimum phase response [259]. The MMSE for the infinite length DFE is [272]

$$J_{\min} = \exp\left\{ T \int_{-1/2T}^{1/2T} \ln\left[\frac{N_o}{F(e^{j2\pi fT}) + N_o} \right] df \right\} \tag{6.113}$$

where $0 \leq J_{\min} \leq 1$.

6.4.3   Numerical Comparison of Symbol-by-Symbol Equalizers

The typical steady-state performance for the various symbol-by-symbol equalizers is now illustrated for 4-PSK modulation on the static ISI channels shown in Fig. 6.10. Channel A is an 11-tap typical data-quality telephone channel with [337]

$$
\begin{aligned}
\mathbf{g}_A \;=\; & (0.0000 + j0.0000,\; 0.0485 + j0.0194, \hspace{2cm} (6.114)\\
& 0.0573 + j0.0253,\; 0.0786 + j0.0282,\; 0.0874 + j0.0447, \\
& 0.9222 + j0.3031,\; 0.1427 + j0.0349,\; 0.0835 + j0.0157, \\
& 0.0621 + j0.0078,\; 0.0359 + j0.0049,\; 0.0214 + j0.0019)\;.
\end{aligned}
$$

Channels B and C have [256]

$$
\begin{aligned}
\mathbf{g}_B \;&=\; (0.407,\; 0.815,\; 0.407) & (6.115)\\
\mathbf{g}_C \;&=\; (0.227,\; 0.460,\; 0.688,\; 0.460,\; 0.227)\;. & (6.116)
\end{aligned}
$$

Channels B and C have severe ISI, with Channel C having the worst spectral characteristics because of the in-band spectral null.

Fig. 6.11 shows the performance of the ZF and MSE linear equalizers for Channel A. The equalizers have 21 taps and the tap gains are obtained using the previously discussed iterative techniques. The ZF and MSE linear equalizers have about the same performance for Channel A.

Fig. 6.12 shows the performance for Channel B. With linear equalization, the optimum tap weights are obtained from a direct solution that assumes a known channel response. Obviously, the ZF equalizer is not suitable for Channel B and the MSE does not perform much better. The performance of a MSE DFE with 11-tap forward section and 10-tap feedback section is also shown. The DFE offers much better performance than the ZF or MSE equalizers for the same complexity. Likewise, Fig. 6.13 shows the performance on Channel C. Again, both the ZF and MSE equalizers perform quite poorly, while the DFE offers much better performance.

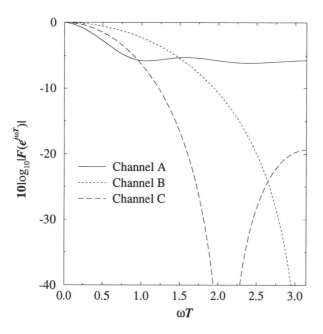

Figure 6.10   Amplitude spectrum for several static ISI channels.

6.5   SEQUENCE ESTIMATION

6.5.1   MLSE and the Viterbi Algorithm

Recall that the overall channel can be modeled by collection of D transversal filters that are T-spaced and have $(L + 1)$-taps, as shown in Fig. 6.5. From Fig. 6.5, it can be seen that the channel has a finite number of states. If the size of the signal constellation is 2^n, there are total of $N_S = 2^{nL}$ **states**. The state at epoch k is

$$s_k = (x_{k-1}, x_{k-2}, \cdots, x_{k-L}) \ . \tag{6.117}$$

Assume that k symbols have been transmitted over the channel. Let $\mathbf{V}_n = (v_{n,1}, \ v_{n,2}, \ \cdots, \ v_{n,D})$ denote the vector of signals received on all diversity branches at epoch n. After receiving the sequence $\{\mathbf{V}_n\}_{n=1}^k$, the ML receiver decides in favor of the sequence $\{x_n\}_{n=1}^k$ that maximizes the **likelihood function**

$$p(\mathbf{V}_k, \ \cdots, \ \mathbf{V}_1 | \ x_k, \ \cdots, \ x_1) \tag{6.118}$$

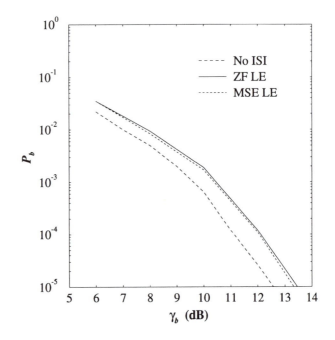

Figure 6.11   Performance of 21-tap ZF and MSE linear equalizers for 4-PSK
signaling Channel A in Fig. 6.10.

or, equivalently, the **log-likelihood function**

$$\log p(\mathbf{V}_k, \ \cdots, \ \mathbf{V}_1 |\ x_k, \ \cdots, \ x_1) \ . \tag{6.119}$$

Since the noise samples $\{\eta_{n,d}\}$ in (6.34) are independent, and $\mathbf{V}_n$ depends only
on the L most recent transmitted symbols, the log-likelihood function (6.119)
can be rewritten as

$$\log p(\mathbf{V}_k, \cdots, \mathbf{V}_1 | x_k, \cdots, x_1) =$$

$$\log p(\mathbf{V}_k | x_k, \cdots, x_{k-L}) + \log p(\mathbf{V}_{k-1}, \cdots, \mathbf{V}_1 | x_{k-1}, \cdots, x_1) \tag{6.120}$$

where $x_{k-L} = 0$ for $k - L \leq 0$. If the second term on the right side of (6.120)
has been calculated previously at epoch $k - 1$ then only the first term, called
the **branch metric** , has to be computed for each incoming signal vector $\mathbf{V}_k$
at epoch k.

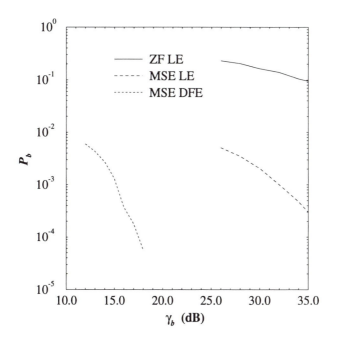

Figure 6.12 Performance of the 21-tap ZF and MSE linear equalizers and a 21-tap DFE equalizers for 4-PSK signaling on Channel B in Fig. 6.10.

The model in Fig. 6.5 gives the conditional pdf

$$p(\mathbf{V}_k | x_k, \cdots, x_{k-L}) = \frac{1}{(\pi N_o)^D} \exp \left\{ -\frac{1}{N_o} \sum_{d=1}^{D} \left| v_{k,d} - \sum_{i=0}^{L} g_{i,d} x_{k-i} \right|^2 \right\}$$

(6.121)

so that $\log p(\mathbf{V}_k | x_k, \cdots, x_{k-L})$ yields the branch metric

$$\mu_k = -\sum_{d=1}^{D} \left| v_{k,d} - \sum_{i=0}^{L} g_{i,d} x_{k-i} \right|^2 .$$

(6.122)

Note that the receiver requires knowledge of the channel vectors $\{\mathbf{g}_d\}$ to compute the branch metrics.

Based on the recursion in (6.120) and the branch metric in (6.122), the well-known **Viterbi algorithm** [322] can be used to implement the ML receiver by searching through the N_S-state trellis for the most likely transmitted sequence $\mathbf{x}$. This search process is called maximum likelihood sequence estimation (MLSE). Here, we give a very brief outline of the Viterbi algorithm

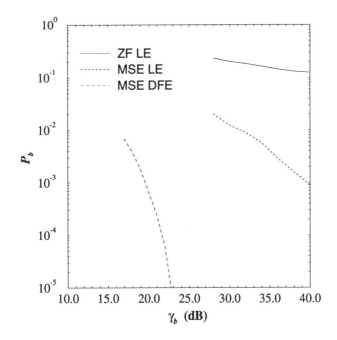

Figure 6.13   Performance of the 21-tap ZF and MSE linear equalizers and a
21-tap DFE equalizers for 4-PSK signaling on Channel C in Fig. 6.10.

followed by an example. At epoch k, assume that the algorithm has stored
N_S **surviving sequences** $\check{x}(s_k^{(i)})$ (paths through the trellis) along with their
associated **path metrics** $\Gamma(s_k^{(i)})$ (distances from the received sequence) that
terminate at state $s_k^{(i)}$, $i = 0, \cdots, N_S - 1$. The path metric is defined as

$$\Gamma(s_k^{(i)}) = \sum_{\{k\}} \mu_k \qquad (6.123)$$

where $\{\mu_k\}$ is the set of branch metrics along the surviving path $\check{x}(s_k^{(i)})$. After
the vector $\mathbf{V}_k$ has been received, the Viterbi algorithm executes the following
steps for each state $s_{k+1}^{(j)}$, for $j = 0, \cdots, N_S - 1$:

1. Compute the set of path metrics $\Gamma(s_k^{(i)} \to s_{k+1}^{(j)}) = \Gamma(s_k^{(i)}) + \mu(s_k^{(i)} \to s_{k+1}^{(j)})$
   for all possible paths through the trellis that terminate in state $s_{k+1}^{(j)}$.

2. Find $\Gamma(s_{k+1}^{(j)}) = \max_i \Gamma(s_k^{(i)} \rightarrow s_{k+1}^{(j)})$ where, again, the maximization is over all possible paths through the trellis that terminate in state $s_{k+1}^{(j)}$.

3. Store $\Gamma(s_{k+1}^{(j)})$ and its associated surviving sequence $\check{\mathbf{x}}(s_{k+1}^{(j)})$. Drop all other paths.

In Step 1 above, $\mu(s_k^{(i)} \rightarrow s_{k+1}^{(j)})$ is the branch metric associated with the transition $s_k^{(i)} \rightarrow s_{k+1}^{(j)}$ and is computed according to the following variation of (6.122)

$$\mu(s_k^{(i)} \rightarrow s_{k+1}^{(j)}) = -\sum_{d=1}^{D} \left| v_{k,d} - g_{0,d} x_k(s_k^{(i)} \rightarrow s_{k+1}^{(j)}) - \sum_{m=1}^{L} g_{m,d} x_{k-m}(s_k^{(i)}) \right|^2$$

(6.124)

where $x_k(s_k^{(i)} \rightarrow s_{k+1}^{(j)})$ is a symbol that is uniquely determined by the transition $s_k^{(i)} \rightarrow s_{k+1}^{(j)}$, and the L most recent symbols $\{x_{k-m}(s_k^{(i)})\}_{m=1}^{L}$ are uniquely specified by the previous state $s_k^{(i)}$.

After all states have been processed, the time index k is incremented and the whole algorithm repeats. As implied in (6.120), the ML receiver waits until the entire sequence $\{\mathbf{V}_n\}_{n=1}^{\infty}$ has been received before making a decision. In practice, such a long delay (maybe infinite) is intolerable and, therefore, a decision about x_{k-Q} is usually made when $\mathbf{V}_k$ is received and processed. It is well known that if $Q > 5L$, the performance degradation caused by the resulting path metric truncation is negligible [322]. MLSE and the Viterbi algorithm is best explained by example and one follows.

Example 6.4_____

Suppose that the binary sequence $\mathbf{x}$, $x_n \in \{-1, +1\}$, is transmitted over a three-tap static ISI channel with channel vector $\mathbf{g} = (1, 1, 1)$. In this case there are four states, and the system can be described the **state diagram** shown in Fig. 6.14. Note that there are two branches entering and leaving each state. In general there are 2^n such branches.

The system state diagram can be used to construct the **trellis diagram** shown in Fig. 6.15, where the initial zero state is assumed to be $s_0^{(0)} = (-1, -1)$. State transitions with a solid line correspond to an input $+1$, while those with a dashed line correspond to an input -1.

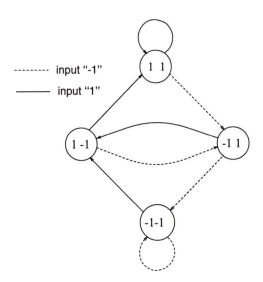

Figure 6.14   State diagram for binary signaling on a three-tap ISI channel.

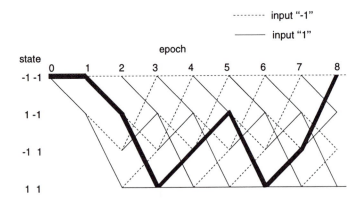

Figure 6.15   Trellis diagram for binary signaling on a three-tap ISI channel.

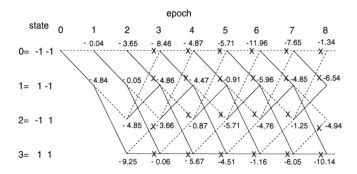

Figure 6.16 Cumulative path metrics and surviving sequences with the Viterbi algorithm.

Suppose that the data sequence $\mathbf{x} = (-1, 1, 1, -1, 1, 1, -1, -1, \ldots)$ is transmitted. Then the state sequence follows the shaded path in Fig. 6.15. The noiseless received sequence is $\mathbf{v} = (v_0, v_1, v_2, v_3, v_4, \ldots)$ where

$$
\begin{aligned}
v_n &= g_0 x_n + g_1 x_{n-1} + g_2 x_{n-2} \\
&= x_n + x_{n-1} + x_{n-2}
\end{aligned}
$$

Hence, for the data sequence $\mathbf{x} = (-1, 1, 1, -1, 1, 1, -1, -1, \ldots)$ the noiseless received sequence is $\mathbf{v} = (-3, -1, 1, 1, 1, 1, 1, -1, \ldots)$.

Suppose that the noisy received sequence is

$$
\begin{aligned}
\mathbf{v} &= (v_0, v_1, v_2, v_3, v_4, \ldots) \\
&= (-3.2, -1.1, 0.9, 0.1, 1.2, 1.5, 0.7, -1.3, \ldots)
\end{aligned}
$$

The Viterbi algorithm is initialized with $\Gamma(s_0^{(i)}) = 0$ for $i = 0, \cdots, N_S - 1$. The initial state is assumed to be $s_0^{(0)} = (-1, -1)$. Executing the Viterbi algorithm yields the result shown in Fig. 6.16, where the X's on the branches in the trellis denote dropped paths and the numbers in the trellis are the path metrics corresponding to the surviving sequences. The path metrics are equal to the square Euclidean distance between the surviving sequence $\check{\mathbf{x}}(s_k^{(i)})$ and the corresponding received sequence $\mathbf{v}$.

Adaptive MLSE Receiver

The Viterbi algorithm requires knowledge of the channel vectors $\mathbf{g}_d$ to compute the branch metrics in (6.122) so that an adaptive channel estimator is needed. Various channel estimators have been proposed in the literature [50], [201], [81]. Usually, a transversal digital filter with the LMS algorithm is used for this purpose, because of its good performance, numerical stability, and simplicity in implementation [141], [201]. Another possible adaptation algorithm is the Recursive Least Squares (RLS) or the Kalman algorithm [141]. The RLS algorithm has a very fast convergence rate as compared to the LMS algorithm. However, it is very complicated to implement and it is sensitive to roundoff noise that accumulates due to recursive computations which may cause numerical instabilities in the algorithm [256]. It has also been reported that the tracking properties of the LMS algorithm for the purpose of channel estimation in a fast varying environment are quite similar to those of the RLS algorithm [81], [187], [285]. For these reasons the LMS algorithm is commonly used during the tracking mode in adaptive MLSE receivers. During the training mode, it is possible that the RLS algorithm could offer better performance than the LMS algorithm.

A straightforward method for adaptive channel estimation with an MLSE receiver is to use the final decisions at the output of the Viterbi algorithm to update the channel estimator during the tracking mode. With the LMS algorithm , the tap coefficients are updated according to

$$\hat{g}_{i,d}(k+1) = \hat{g}_{i,d}(k) + \alpha \epsilon_{k-Q,d}\hat{x}^*_{k-i-Q} \, , \quad \begin{aligned} i &= 0, \ldots, L \\ \ell &= 1, \ldots, D \end{aligned} \quad (6.125)$$

where α is the adaptation step size, and

$$\epsilon_{k-Q,d} = v_{k-Q,d} - \sum_{i=0}^{L} \hat{g}_{i,d}(k)\hat{x}_{k-i-Q} \qquad (6.126)$$

is the error associated with branch d at epoch k. A major problem with this channel estimator is that it lags behind the true channel vector by the decision delay Q that is used in the Viterbi algorithm. To see this, we can write

$$v_{k-Q,d} = \sum_{i=0}^{L} g_{i,d}(k-Q)x_{k-i-Q} + \eta_{k-Q,d} \qquad (6.127)$$

so that

$$\epsilon_{k-Q,d} = \sum_{i=0}^{L} \left(g_{i,d}(k-Q) - \hat{g}_{i,d}(k) \right) x_{k-i-Q} + \eta_{k-Q,d} \, . \qquad (6.128)$$

Hence, channel time variations over the decision delay Q will cause the terms $\{g_{i,d}(k - Q) - \hat{g}_{i,d}(k)\}$ to be non-zero, and this will degrade the tracking performance. The decision delay Q could be reduced but this will also reduce the reliability of the decisions $\hat{x}_{k-i-Q}$ that are used to update the channel estimates in (6.125). Since decision errors will also degrade the performance of the estimator, the overall performance improvement obtained by reducing Q is often minimal.

One solution to this problem is to use **per-survivor processing** [279], [280], [261], [188], where *each state* has its own channel estimator that tracks the channel. In this case, the tap coefficients are updated according to

$$\hat{g}_{i,d}(k + 1) = \hat{g}_{i,d}(k) + \alpha\epsilon_{k,d}\check{x}^*_{k-i} , \quad \begin{aligned} i &= 0, \ldots, L \\ \ell &= 1, \ldots, D \end{aligned} \quad (6.129)$$

where $\check{x}$ is the surviving sequence associated with *each state*. Notice that the individual channel estimators for each state use zero-delay symbols in their adaptation algorithm and, therefore, good channel tracking performance is expected. These zero-delay symbols are uniquely defined by the state transitions in the trellis diagram.

$T/2$-spaced MLSE Receiver

Suppose that the matched filter output is sampled at rate $2/T$ and the $T/2$-spaced samples are processed with a $T/2$-spaced noise whitening filter as shown in Fig. 6.6. Once again, the channel can be modeled as a finite-state machine with the states defined in (6.117). The Viterbi decoder searches for the most likely path in the trellis based on the $T/2$-spaced received sequence. For each state transition $s_k^{(j)} \to s_{k+1}^{(i)}$ at epoch k, the samples $v_{2k}^{(2)}$ and $v_{2k+1}^{(2)}$ are used by the Viterbi algorithm to evaluate the branch metric[1]

$$\gamma_k(s_k^{(i)} \to s_{k+1}^{(j)}) = \left| v_{2k}^{(2)} - g_0^{(2)} x_k(s_k^{(i)} \to s_{k+1}^{(j)}) - \sum_{m=1}^{L} g_{2m}^{(2)} x_{k-m}(s_k^{(i)}) \right|^2$$
$$+ \left| v_{2k+1}^{(2)} - g_1^{(2)} x_k(s_k^{(i)} \to s_{k+1}^{(j)}) - \sum_{m=1}^{L-1} g_{2m+1}^{(2)} x_{k-m}(s_k^{(i)}) \right|^2$$
$$(6.130)$$

---

[1]For notational simplicity we assume $D = 1$.

6.5.2   Delayed Decision-Feedback Sequence Estimation (DDFSE)

Unfortunately, the complexity of the MLSE receiver grows exponentially with the channel memory length. When the channel memory length becomes large, the MLSE receiver becomes impractical. One solution is to reduce the receiver complexity by truncating the effective channel memory to μ terms, where μ is an integer that can be varied from 0 to L. Thus, a suboptimum decoder is obtained with complexity controlled by parameter μ. This is the basic principle of DDFSE.

Let the z transform of the channel vector, $G(z)$, be represented as a rational function $\beta(z)/\Psi(z)$, where $\beta(z)$ and $\Psi(z)$ are polynomials. It is assumed that $\Psi(z)$ has degree n_2 and $\Psi_0 = 1$. If L is finite then $\Psi(z) = 1$. The polynomial $G(z)$ can be written as

$$G(z) = G_\mu(z) + z^{-(\mu+1)}G^+(z) \tag{6.131}$$

where

$$G_\mu(z) \;=\; \sum_{i=0}^{\mu} g_i z^{-i} \tag{6.132}$$

$$G^+(z) \;=\; \sum_{i=0}^{L-\mu-1} g_{i+\mu+1} z^{-i} \;. \tag{6.133}$$

From (6.131), $G^+(z)$ is a rational function that can be written as $\beta^+(z)/\Psi(z)$ where $\beta^+(z)$ is a polynomial of degree n_1 satisfying the equality

$$\beta^+(z) = [\beta(z) - G_\mu(z)\Psi(z)]z^{\mu+1} \;. \tag{6.134}$$

Let $W(z) = G_+(z)X(z)$, where $X(z)$ is the z-transform of the input sequence. Then

$$w_k = \left\{ \begin{array}{ll} \sum_{i=0}^{n_1} \beta_i^+ x_{k-i} - \sum_{i=1}^{n_2} \Psi_i w_{k-i} & , \; L = \infty \\ \sum_{i=0}^{L-\mu-1} g_{i+\mu+1} x_{k-i} & , \; L < \infty \; \text{or} \; n_2 = 0 \end{array} \right. \tag{6.135}$$

and

$$v_k = \sum_{i=0}^{\mu} g_i x_{k-i} + w_{k-\mu-1} + \eta_k \;. \tag{6.136}$$

From (6.135) and (6.136), the system state at epoch k can be decomposed into the state

$$s_k^\mu = (x_{k-1}, \; \ldots, \; x_{k-\mu}) \tag{6.137}$$

and a partial state

$$\kappa_k = \begin{cases} x_{k-\mu-1}, \ldots, x_{k-\mu-n_1-1}, w_{k-\mu-2}, \ldots, w_{k-\mu-n_2-2} & ; L = \infty \\ x_{k-\mu-1}, \ldots, x_{k-L} & , L < \infty \end{cases} \tag{6.138}$$

There are $N_\mu = 2^{n\mu}$ states in (6.137).

The DDFSE receiver can be viewed as a combination of the Viterbi algorithm and a decision feedback detector. For each state transition $s_k^{\mu(i)} \rightarrow s_{k+1}^{\mu(j)}$, the DDFSE receiver stores N_μ estimates of the partial states κ_k associated with $s_k^{\mu(i)}$. The DDFSE receiver uses the following branch metric

$$\mu_k(s_k^{\mu(i)} \rightarrow s_{k+1}^{\mu(j)}) = -\left| y_k - g_0 x_k(s_k^{\mu(i)} \rightarrow s_{k+1}^{\mu(j)}) \right.$$
$$\left. - \sum_{l=1}^{\mu} g_l x_{k-l}(s_k^{\mu(i)}) - \hat{w}_{k-\mu-1} \right|^2 . \tag{6.139}$$

The estimate $\hat{w}_{k-\mu-1}$ of $w_{k-\mu-1}$ is obtained from the estimate of the partial state using (6.135). For finite length channels, the DDFSE branch metric can be written as

$$\mu_k(s_k^{\mu(i)} \rightarrow s_{k+1}^{\mu(j)}) = -\left| y_k - g_0 x_k(s_k^{\mu(i)} \rightarrow s_{k+1}^{\mu(j)}) \right.$$
$$\left. - \sum_{l=1}^{\mu} g_l x_{k-l}(s_k^{\mu(i)}) - \sum_{l=\mu+1}^{L} g_l \breve{x}_l(s_k^{\mu(i)}) \right|^2 \tag{6.140}$$

where $\breve{x}_l(s_k^{\mu(i)})$ is the lth component of the surviving sequence $\breve{\mathbf{x}}(s_k^{\mu(i)})$. Since each path uses decision-feedback based on its own history, the DDFSE receiver avoids using a single unreliable decision for feedback. Hence, error propagation with a DDFSE receiver is not a severe as with a DFE receiver. When $\mu = 0$ the DDFSE receiver is equivalent to Driscoll's decoder [74] and when $\mu = L$ the DDFSE receiver is equivalent to the MLSE receiver.

Finally, since only the μ most recent symbols are represented by the state in (6.137), it is important to have most of the signal energy contained in these terms. Hence, it is very important that the noise whitening filter be selected so that the overall channel $G(z)$ has minimum phase. This requirement can present some practical problems. For example, if one of the zeros is close to the unit circle, then the noncausal noise whitening filter has a very long impulse response and will be hard to approximate. Also, when the channel is time-varying or unknown, the receiver cannot ensure that $G(z)$ will have minimum

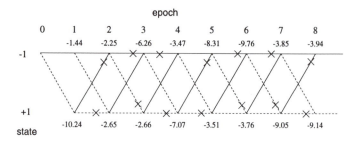

Figure 6.17   Cumulative path metrics and surviving sequences with DDFSE.

phase. Without $G(z)$ having minimum phase, these techniques do not work well. This point will be repeated again in Chapter 7.

Example 6.5_____

Consider again the system in Example 6.4, where the received sequence is

$$\mathbf{v} = (v_0,\ v_1,\ v_2,\ v_3,\ v_4, \ldots)$$
$$= (-3.2,\ -1.1,\ 0.9,\ 0.1,\ 1.2,\ 1.5,\ 0.7,\ -1.3, \ldots)$$

Recall that $s_k^{(i)} = (x_{k-1},\ x_{k-2})$ so there are 4 system states. However, we wish to apply DDFSE with the state $s_k^{\mu(i)} = x_{k-1}$, $i = 1, 2$. The initial state is assumed to $\Gamma(s_0^{(0)}) = -1$. Since the channel has finite length, (6.140) gives the branch metric

$$\mu_k\big(s_k^{\mu(i)} \to s_{k+1}^{\mu(j)}\big) = \left| y_k - x_k\big(s_k^{\mu(i)} \to s_{k+1}^{\mu(j)}\big) - x_{k-l}\big(s_k^{\mu(i)}\big) - \check{x}_2\big(s_k^{\mu(i)}\big) \right|^2 .$$

Similar to MLSE, the Viterbi algorithm is first initialized with $\Gamma(s_0^{(i)}) = 0$ for $i = 0, 1$. Applying DDFSE with the Viterbi algorithm gives the result shown in Fig. 6.17. Once again, the $X's$ on the branches in the trellis denote dropped paths and the numbers in the trellis denote the path metrics.

6.5.3   Reduced-State Sequence Estimation

For large signal constellations the number of states with DDFSE, $2^{n\mu}$, is substantial even for small μ. One possible remedy is to reduce the number of states by using Ungerboeck-like set partitioning principles. As described in [91], for each element x_{k-n} in s_k^μ, a set partitioning $\Omega(n), 1 \leq n \leq \mu \leq L$ is defined where the signal set is partitioned into J_i subsets in a way of increasing intrasubset minimum Euclidean distance[2].

The subset in the partitioning $\Omega(i)$ to which x_{k-i} belongs is denoted by $c_i(x_{k-i})$. The subset partitioning is constrained such that $\Omega(i)$ is a finer partition of $\Omega(i+1)$, $1 \leq i \leq \mu - 1$ and $J_1 \geq J_2 \ldots \geq J_\mu$. In this case this following **subset-state** can be defined

$$t_k^\mu = (c_1(x_{k-1}), \; c_2(x_{k-2}), \; \ldots, \; c_\mu(x_{k-\mu})) \; . \tag{6.141}$$

Note that the RSSE subset-state does not completely specify the μ most recent symbols $\{x_{k-i}\}_{i=1}^\mu$. Rather, the subset-state only specifies the subsets to which these symbols belong. The constraints on the subset partitioning insures a properly defined **subset-trellis** . Given the current subset-state t_k^μ and the subset $c_1(x_k)$ to which the current symbol x_k belongs, the next subset-state t_{k+1}^μ is uniquely determined. Since $c_i(x_{k-i})$ can only assume J_i possible values, there are $\prod_{i=1}^\mu J_i$ subset-states which could be much less than 2^{nL}. Note that if $J_1 < 2^n$, there are parallel transitions associated with each **subset-transition**. The number of the parallel transitions is equal to the number of symbols in the corresponding subset.

The Viterbi algorithm used to search the subset-trellis is the same one used for MLSE except for a different branch metric and the possibility of parallel transitions associated with the subset-transitions.[3] When there are parallel transitions, the Viterbi algorithm chooses the parallel transition with the maximum branch metric first[4] and then execute steps for the Viterbi algorithm as defined in Section 6.5.1.

With RSSE, the branch metric in (6.122) is not uniquely determined by the associated pair of subset-states. This is solved by introducing a decision feedback mechanism for the branch metric calculation [91, 76]. The RSSE branch metric for a particular parallel transition associated with the subset-transition $(t_k^{\mu(i)} \to t_{k+1}^{\mu(j)})$ is

$$\gamma_k(t_k^{\mu(i)} \to t_{k+1}^{\mu(j)}) = -\left| y_k - g_0 x_k(t_k^{\mu(i)} \to t_{k+1}^{\mu(j)}) - \sum_{l=1}^{L} g_l \check{x}_l(t_k^{\mu,(i)}) \right|^2 \tag{6.142}$$

---

[2] If $J_1 = J_2 = \cdots = J_\mu = M$ and $\mu < L$, then RSSE becomes DDFSE.

[3] With DDFSE there are no parallel transitions.

[4] If the signal constellation has some symmetries, this step can be easily done by using a slicing operation [91].

where $x_k(t_k^{\mu(i)}) \rightarrow x_k(t_{k+1}^{\mu(j)})$ is the source symbol corresponding to the particular parallel transition, and $\check{x}_l(t_k^{\mu,(i)})$ is the lth component of the source symbol sequence $\check{x}(t_k^{\mu,(i)})$ that corresponds to the surviving path leading to the subset-state $t_k^{(i)}$. Similar to DDFSE, each path uses decision-feedback based on its own history.

6.6 ERROR PROBABILITY FOR MLSE ON ISI CHANNELS

Let $\mathbf{x}$ and $\hat{\mathbf{x}}$ be the transmitted and estimated symbol sequences, respectively. For every pair $\mathbf{x}$ and $\hat{\mathbf{x}}$, the error sequence $\boldsymbol{\epsilon} = \{\epsilon_i\}$ can be formed by defining $\epsilon_i = x_i - \hat{x}_i$. We arbitrarily assume that the bit error probability at epoch k_1 is of interest, so that $\epsilon_{k_1} \neq 0$ for all error sequences that are considered. For each error sequence $\boldsymbol{\epsilon}$, define the following useful error events.

$\mathcal{E}'(\varepsilon)$:   The sequence $\mathbf{x} - \boldsymbol{\epsilon}$ is the maximum likelihood sequence.

$\mathcal{E}(\boldsymbol{\epsilon})$:   The sequence $\mathbf{x} - \boldsymbol{\epsilon}$ has a larger path metric than sequence $\mathbf{x}$.

It is also convenient to define the events

$$\mathcal{E}'_G = \bigcup_{\boldsymbol{\epsilon} \in G} \mathcal{E}'(\varepsilon) \tag{6.143}$$

and

$$\mathcal{E}_F = \bigcup_{\boldsymbol{\epsilon} \in F} \mathcal{E}(\boldsymbol{\epsilon}) \tag{6.144}$$

where G is the set of all possible error sequences having $\epsilon_{k_1} \neq 0$ and $F \subset G$ is the set of error sequences containing no more than $L-1$ consecutive zeroes amid nonzero elements.

Let $\mathbf{s} = \{s_k\}$ and $\hat{\mathbf{s}} = \{\hat{s}_k\}$ be the system state sequences corresponding to the symbol sequences $\mathbf{x}$ and $\hat{\mathbf{x}}$, respectively. An **error event** occurs between j_1 and j_2, of length $j_2 - j_1$, if

$$s_{j_1} = \hat{s}_{j_1}, \; s_{j_2} = \hat{s}_{j_2}, \text{ and } s_k \neq \hat{s}_k \text{ for } j_1 < k < j_2 \tag{6.145}$$

where $j_1 \leq k_1 \leq j_2$. The symbol error probability at epoch k_1 is

$$P_s(k_1) \triangleq \Pr(x_{k_1} \neq \hat{x}_{k_1})$$

$$= \Pr(\mathcal{E}'_G)$$

$$= \sum_{\epsilon \in G} \sum_{\mathbf{x} \in \mathcal{X}(\epsilon)} \Pr(\mathcal{E}'(\epsilon)|\mathbf{x})\Pr(\mathbf{x}) \qquad (6.146)$$

where $\mathcal{X}(\epsilon)$ is the set of symbol sequences that can have ϵ as the error sequence. For different ϵ, the set $\mathcal{X}(\epsilon)$ might be different. The third equation in (6.146) is obtained by using the property that the events $\mathcal{E}'(\epsilon)$ are disjoint for $\epsilon \in G$. Unfortunately, (6.146) does not admit an explicit expression and, hence, upper bounding techniques are needed for the performance evaluation. A union bound on the error probability will be employed in our analysis.

To obtain a tighter union bound, we now prove the following

Theorem 6.2 _____

The symbol error probability at epoch k_1 is

$$P_s(k_1) = \Pr(\mathcal{E}_F) \ . \qquad (6.147)$$

Proof

Consider the typical trellis diagram as shown in Fig. 6.18, where $\mathbf{x}$ denotes the transmitted symbol sequence, and $\tilde{\mathbf{x}}^{(1)}$ and $\tilde{\mathbf{x}}^{(2)}$ denote two different symbol sequences. It can be seen that the error sequence $\epsilon^{(1)}$ associated with $\tilde{\mathbf{x}}^{(1)}$ and the error sequence $\epsilon^{(2)}$ associated with $\tilde{\mathbf{x}}^{(2)}$ belong to sets F and $G \setminus F$, respectively. For every $\epsilon^{(2)} \in G \setminus F$ there always exists an $\epsilon^{(1)} \in F$. If the sequence $\mathbf{x} - \epsilon^{(2)}$ is the ML sequence, i.e., the event $\mathcal{E}'_G$ has occurred, then the sequence $\mathbf{x} - \epsilon^{(1)}$ has a larger path metric than the sequence $\mathbf{x}$, i.e., the event $\mathcal{E}_F$ has occurred. This means that $\mathcal{E}'_G$ implies $\mathcal{E}_F$. On the other hand, if $\epsilon^{(1)} \in F$ and the sequence $\mathbf{x} - \epsilon^{(1)}$ has a larger path metric than sequence $\mathbf{x}$, then there exists a sequence $\epsilon \in G$ such that the sequence $\mathbf{x} - \epsilon$ is the ML sequence. Therefore, $\mathcal{E}_F$ implies $\mathcal{E}'_G$, and (6.147) is proven.

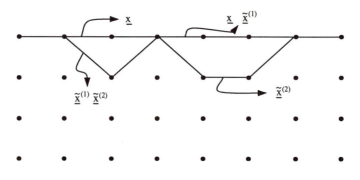

Figure 6.18   A typical Error state trellis diagram

The union bound on (6.147) yields

$$P_s(k_1) \leq \sum_{\epsilon \in F} \mathrm{P_r}(\mathcal{E}(\epsilon))$$

$$= \sum_{\epsilon \in F} \sum_{\mathbf{x} \in \mathcal{X}(\epsilon)} \mathrm{P_r}(\mathcal{E}(\epsilon)|\mathbf{x})\mathrm{P_r}(\mathbf{x}) \qquad (6.148)$$

or, equivalently,

$$P_s(k_1) \leq \sum_{\epsilon \in E} w_s(\epsilon) \sum_{\mathbf{x} \in \mathcal{X}(\epsilon)} \mathrm{P_r}(\mathcal{E}(\epsilon)|\mathbf{x})\mathrm{P_r}(\mathbf{x}) \qquad (6.149)$$

where $E \in F$ is the set of error sequences that have the first non-zero element starting at time k_1, and $w_s(\epsilon)$ is the number of symbol errors associated with the error sequence ϵ. To obtain (6.149), we have used the following observations; i) there are $w_s(\epsilon)$ places for the error sequence ϵ to start such that $\epsilon_{k_1} \neq 0$, and ii) the error probability $\mathrm{P_r}\{\mathcal{E}(\epsilon)|\mathbf{x}\}$ is independent of the place where the error sequence ϵ starts. If the transmitted symbol sequence is long enough, then the symbol error probability $P_s(k_1)$ is independent of the time index k_1 and, therefore, the time index will be omitted hereafter. Finally, for a given transmitted symbol sequence $\mathbf{x}$, the events $\{\mathcal{E}(\epsilon)\}$ for $\epsilon \in F$ in (6.148) might overlap. The reason is that there may be multiple symbol sequences that simultaneously have a larger path metric than the path metric of the transmitted symbol sequence. When the system is operating at a low SNR, there are more overlapping events $\mathcal{E}(\epsilon)$ and, hence, the union bound (6.148) becomes looser.

From the definition of event $\mathcal{E}(\epsilon)$, the union bound (6.149) becomes

$$P_s \leq \sum_{\epsilon \in E} w_s(\epsilon) \sum_{\mathbf{x} \in \mathcal{X}(\epsilon)} \mathrm{P_r}(\Gamma(\mathbf{x}-\epsilon) \geq \Gamma(\mathbf{x})|\mathbf{x})\mathrm{P_r}(\mathbf{x}) \ . \tag{6.150}$$

To obtain the bit error probability, (6.150) can be easily modified as

$$P_b \leq \frac{1}{n} \sum_{\epsilon \in E} w_b(\epsilon) \sum_{\mathbf{x} \in \mathcal{X}(\epsilon)} \mathrm{P_r}(\Gamma(\mathbf{x}-\epsilon) \geq \Gamma(\mathbf{x})|\mathbf{x})\mathrm{P_r}(\mathbf{x}) \tag{6.151}$$

where n is the number bits transmitted per unit time, and $w_b(\epsilon)$ is the number of bit errors associated with the error sequence ϵ. The probability

$$\mathrm{P_r}(\Gamma(\mathbf{x}-\epsilon) \geq \Gamma(\mathbf{x})|\mathbf{x}) \tag{6.152}$$

is called **pairwise error probability** .

We will see in the following two sections that the pairwise error probability is independent of the transmitted symbol sequence $\mathbf{x}$. Therefore, the union bounds (6.150) and (6.151) simplify to

$$\mathrm{P}_s \leq \sum_{\epsilon \in E} w_s(\epsilon)\mathrm{P_r}\left(\Gamma(\mathbf{x}-\epsilon) \geq \Gamma(\mathbf{x})|\mathcal{X}(\epsilon)\right)\mathrm{P_r}\left(\mathcal{X}(\epsilon)\right) \tag{6.153}$$

and

$$\mathrm{P}_b \leq \frac{1}{n} \sum_{\epsilon \in E} w_b(\epsilon)\mathrm{P_r}\left(\Gamma(\mathbf{x}-\epsilon) \geq \Gamma(\mathbf{x})|\mathcal{X}(\epsilon)\right)\mathrm{P_r}\left(\mathcal{X}(\epsilon)\right) \tag{6.154}$$

respectively. The expressions in (6.153) and (6.154) are easier to calculate than those in (6.150) and (6.151), because not all of the symbol sequences have to be considered in the calculation.

6.6.1   Static ISI Channels

The pairwise error probability associated with the error event of length ℓ in (6.145) is (see Problem 6.16)

$$\mathrm{P_r}(\Gamma(\mathbf{x}-\epsilon) \geq \Gamma(\mathbf{x})|\mathbf{x}) = Q\left(\sqrt{\frac{\Delta^2}{4N_o}}\right) \tag{6.155}$$

where

$$\Delta^2 = \sum_{k=k_1}^{k_1+\ell-1} \left|\sum_{i=0}^{L} g_i \epsilon_{k-i}\right|^2 \tag{6.156}$$

and Δ^2 is the squared Euclidean path distance. At high signal-to-noise ratios the error event probability is approximately

$$P_e \approx N_{\min} Q \left(\sqrt{\frac{d_{\min}^2}{4N_o}} \right) \qquad (6.157)$$

where $d_{\min}^2$ is the minimum value of Δ^2 and $N_{\min}$ denotes the average number of error events at distance $d_{\min}$.

The squared Euclidean path distance in (6.156) can be rewritten as

$$\Delta^2 = \sum_{k=k_1}^{k_1+\ell-1} \Delta_k^2 \qquad (6.158)$$

where

$$\Delta_k^2 = \mathbf{g}^H \mathbf{E}_k \mathbf{g} \qquad (6.159)$$

is the squared branch distance and

$$\mathbf{E}_k = [(e_{mn})_k] \qquad (6.160)$$

is the $(L+1) \times (L+1)$ **branch distance matrix** having elements $(e_{mn})_k = \epsilon_{k-m+1}^* \epsilon_{k-n+1}$. Define the error vector $\boldsymbol{\varepsilon}_k = (\epsilon_k^*, \epsilon_{k-1}^*, \ldots, \epsilon_{k-L}^*)^T$. It follows that $\mathbf{E}_k = \boldsymbol{\varepsilon}_k \boldsymbol{\varepsilon}_k^H$ and, hence, $\mathbf{E}_k$ has rank one. Note that $\mathbf{E}_k \boldsymbol{\varepsilon}_k = (\boldsymbol{\varepsilon}_k^H \boldsymbol{\varepsilon}_k) \boldsymbol{\varepsilon}_k$ and, therefore, $\boldsymbol{\varepsilon}_k$ is an eigenvector of $\mathbf{E}_k$ and the only eigenvalue of $\mathbf{E}_k$ is $\lambda(k) = \sum_{i=0}^{L} |\epsilon_{k-i}|^2$. The path distance matrix of the length ℓ error event in (6.145) is defined as

$$\mathbf{E} \stackrel{\Delta}{=} \sum_{k=k_1}^{k_1+\ell-1} \mathbf{E}_k \; . \qquad (6.161)$$

Using (6.117) and (6.145), the elements of $\mathbf{E}$ are

$$e_{mn} = r_\ell(n-m) \qquad (6.162)$$

where

$$r_\ell(i) = \begin{cases} \sum_{k=k_1}^{k_1+\ell-(L+1)-i} \epsilon_k \epsilon_{k+i}^* & i \geq 0 \\ r_\ell^*(-i) & i < 0 \end{cases} . \qquad (6.163)$$

It follows that (6.158) has the Hermitian form $\Delta^2 = \mathbf{g}^H \mathbf{E} \mathbf{g}$. Since $\Delta^2 > 0$, $\mathbf{E}$ is a positive definite matrix with all eigenvalues being real and positive. The matrix $\mathbf{E}$ depends on the signal constellation and the length of the channel $L+1$.

By using (6.36) and the normalization $E[|x_k|^2] = 1$, the squared Euclidean path distance can be expressed in the form

$$\Delta^2 = 2E_s \frac{\mathbf{g}^H \mathbf{E} \, \mathbf{g}}{\mathbf{g}^H \mathbf{g}} \, . \tag{6.164}$$

The ratio of the Hermitian form $\mathbf{g}^H \mathbf{E} \, \mathbf{g}$ to the inner product $\mathbf{g}^H \mathbf{g}$ is called the **Rayleigh quotient** of the vector $\mathbf{g}$ and is denoted $R(\mathbf{g})$ [141]. The eigenvalues of $\mathbf{E}$ are equal to the Rayleigh quotient of the corresponding eigenvectors. The Rayleigh quotient of $\mathbf{E}$ satisfies $\lambda_{\min} \leq R(\mathbf{g}) \leq \lambda_{\max}$. The minimum value of $R(\mathbf{g})$ occurs when $\mathbf{g} = \mathbf{v}_{\min}$ and the maximum value occurs when $\mathbf{g} = \mathbf{v}_{\max}$. The eigenvalues of $\mathbf{E}$ are bounded by [141]

$$\lambda_{\max} \leq \sum_{i=0}^{L} |r_\ell(i)| \quad \text{and} \quad \lambda_{\min} \geq r_\ell(0) - \sum_{i=1}^{L} |r_\ell(i)| \, . \tag{6.165}$$

The condition number of $\mathbf{E}$ is defined as $c(\mathbf{E}) \triangleq \lambda_{\max} / \lambda_{\min}$.

6.6.2 Fading ISI Channels

For fading ISI channels with D-branch diversity reception and maximal ratio combining, the pairwise error probability is still given by (6.155) but the squared Euclidean path distance associated with an error event of length ℓ is [281]

$$\Delta^2 = \sum_{d=1}^{D} \Delta_d^2 \tag{6.166}$$

where

$$\Delta_d^2 = \sum_{k=k_1}^{k_1+\ell-1} \left| \sum_{i=0}^{L} g_{i,d}(k) \epsilon_{k-i} \right|^2 \, . \tag{6.167}$$

The above expression can be written in the form

$$\Delta_d^2 = \sum_{k=k_1}^{k_1+\ell-1} \mathbf{g}_d^H(k) \mathbf{E}_k \mathbf{g}_d(k) \, . \tag{6.168}$$

In general, the covariance matrix $\Phi_{\mathbf{g}_d}(0)$ defined in (6.38) is not diagonal when the actual channel rays are not T-spaced, as discussed in Section 2.3.3. A non-diagonal $\Phi_{\mathbf{g}_d}(0)$ matrix leads to considerable analytical difficulty and loss of insight. However, if $\Phi_{\mathbf{g}_d}(0)$ is diagonal, then a normalized channel vector $\mathbf{f}_d(k)$ can be defined such that $\Phi_{\mathbf{f}_d}(0) = \mathbf{I}_{L+1}$. As a result, (6.168) can be rewritten as

$$\Delta_d^2 = \sum_{k=k_1}^{k_1+\ell-1} \mathbf{f}_d^H(k) \mathbf{A}_{k,d} \mathbf{f}_d(k) \qquad (6.169)$$

where

$$\mathbf{A}_{k,d} = \mathbf{\Sigma}_d \mathbf{E}_k \mathbf{\Sigma}_d \qquad (6.170)$$

and $\mathbf{\Sigma}_d$ is defined in (6.40). It follows that $\mathbf{A}_{k,d} = \mathbf{u}_{k,d} \mathbf{u}_{k,d}^H$ where $\mathbf{u}_{k,d} = \mathbf{\Sigma}_d \boldsymbol{\varepsilon}_k$ and, hence, $\mathbf{A}_{k,d}$ is a rank one matrix and $\mathbf{u}_{k,d}$ is an eigenvector of $\mathbf{A}_{k,d}$. The only nonzero eigenvalue of $\mathbf{A}_{k,d}$ is $\lambda_d = \sum_{i=0}^{L} \sigma_{i,d}^2 |\epsilon_{k-i}|^2$, where $\sigma_{i,d} = \mathrm{E}[|g_{i,d}|^2]$.

For slowly time-variant channels it is reasonable to assume that $\mathbf{g}_d(k)$ remains constant over the length of the dominant error events, i.e., $\mathbf{g}_d(k) \equiv \mathbf{g}_d$. This assumption holds even for relatively large Doppler frequencies and error event lengths. For example, if the channel exhibits 2-D isotropic scattering and $f_m T = 0.0025$, then error events up to length 20 have $J_0(2\pi f_m |k|T) \geq J_0(2\pi f_m 10T) = 0.9984 \approx 1$. By using the above assumption, (6.169) can be written as

$$\Delta_d^2 = \mathbf{f}_d^H \mathbf{A}_d \mathbf{f}_d \qquad (6.171)$$

where

$$\begin{aligned} \mathbf{A}_d &= \sum_{k=k_1}^{k_1+\ell-1} \mathbf{A}_{k,d} \\ &= \mathbf{\Sigma}_d \mathbf{E} \mathbf{\Sigma}_d \ . \end{aligned} \qquad (6.172)$$

The matrix $\mathbf{A}_d$ is also positive definite with all its eigenvalues real and positive. The elements of $\mathbf{A}_d$ are given by $[(a_{mn})]_d = \sigma_{m-1,d} \ \sigma_{n-1,d} \ r_\ell(n - m)$ where $r_\ell(i)$ is given by (6.163). The trace of the matrix $\mathbf{A}_d$ is

$$\mathrm{tr}(\mathbf{A}_d) = \sum_{i=1}^{L+1} \lambda_{i,d} = r_\ell(0) \sum_{i=0}^{L} \sigma_{i,d}^2 = \bar{E} r_\ell(0) \qquad (6.173)$$

where the $\lambda_{i,d}$, $i = 1, \ldots, L+1$ are the eigenvalues of $\mathbf{A}_d$. The last equality in (6.173) is obtained by using (6.42) along with the normalization $\mathrm{E}[|x_k|^2] = 1$. Since $\mathbf{A}_d$ is Hermitian, there exists a diagonalization $\mathbf{A}_d = \mathbf{U}_d \mathbf{\Lambda}_d \mathbf{U}_d^H$ such that $\mathbf{U}_d$ is a unitary matrix and $\mathbf{\Lambda}_d$ is a diagonal matrix consisting of the eigenvalues of $\mathbf{A}_d$. Let $\boldsymbol{\omega}_d = \mathbf{U}_d^H \mathbf{f}_d$ be the corresponding diagonal transformation. It follows that

$$\Delta_d^2 = \boldsymbol{\omega}_d^H \mathbf{\Lambda}_d \boldsymbol{\omega}_d = \sum_{i=1}^{L+1} \lambda_{i,d} |\omega_{i,d}|^2 \qquad (6.174)$$

where $\frac{1}{2}E[\boldsymbol{\omega}_d \boldsymbol{\omega}_d^H] = \mathbf{I}_{L+1}$ so that the $\{\omega_{i,d}\}$ are independent zero-mean unit-variance Gaussian random variables. Using (6.166) and (6.174) gives

$$\Delta^2 = \sum_{d=1}^{D} \sum_{i=1}^{L+1} \alpha_{i,d} \qquad (6.175)$$

where $\alpha_{i,d} = \lambda_{i,d} |\omega_{i,d}|^2$. Since the $\alpha_{i,d}$ are chi-square distributed with 2 degrees of freedom and, therefore, the characteristic function of Δ^2 is

$$\boldsymbol{\Psi}_{\Delta^2}(z) = \prod_{i=1}^{D(L+1)} \frac{1}{1 - \overline{\alpha}_{i,d} z} \qquad (6.176)$$

where $\overline{\alpha}_{i,d} = 2\lambda_{i,d}$. Finally, the pairwise error probability is

$$P_r(\Gamma(\mathbf{x} - \boldsymbol{\epsilon}) \geq \Gamma(\mathbf{x})|\mathbf{x}) = \int_0^\infty Q\left(\sqrt{2x}\right) f_{\Delta^2}(x)\, dx \qquad (6.177)$$

where $f_{\Delta^2}(x)$ is the probability density function of Δ^2. Note that if some of the eigenvalues $\lambda_{i,d}$ are the same, then there will be repeated poles in the characteristic function in (6.176). This can be expected to be the case for balanced diversity branches, and will also be the case if the channel has equal strength taps. Consider the case where D-branch antenna diversity is used and the channel taps are not of equal strength. In this case, $\lambda_{i,d} \equiv \lambda_i$, $d = 1, \ldots, D$ and the characteristic function in (6.176) has the form

$$
\begin{aligned}
\boldsymbol{\Psi}_{\Delta^2}(z) &= \prod_{i=1}^{L+1} \frac{1}{(1 - z\overline{\alpha}_i)^D} \\
&= \sum_{i=1}^{L+1} \sum_{d=1}^{D} \frac{A_{id}}{(1 - z\overline{\alpha}_i)^d} \qquad (6.178)
\end{aligned}
$$

where

$$A_{id} = \frac{1}{(D-d)!(-\overline{\alpha}_i)^{D-d}} \left\{ \frac{d^{D-d}}{dz^{D-d}} (1 - z\overline{\alpha}_i)^D\, \boldsymbol{\Psi}_{\chi}(z) \right\}_{z\,=\,1/\overline{\alpha}_i} \qquad (6.179)$$

and $\overline{\alpha}_i = 2\lambda_i$. The pdf of Δ^2 is

$$f_{\Delta^2}(x) = \sum_{i=1}^{L+1} \sum_{d=1}^{D} A_{id}\, \frac{1}{(d-1)!(\overline{\alpha}_i)^d}\, x^{d-1}\, e^{-x/\overline{\alpha}_i} \ . \qquad (6.180)$$

From (6.177) and (6.180), the exact pairwise error probability is

$$
P_r(\Gamma(\mathbf{x} - \boldsymbol{\epsilon}) \geq \Gamma(\mathbf{x})|\mathbf{x}) = \sum_{i=1}^{L+1} \sum_{d=1}^{D} A_{id} \left(\frac{1 - \mu_i}{2}\right)^d
$$

$$
\times \sum_{m=0}^{d-1} \binom{d-1+m}{m} \left(\frac{1 + \mu_i}{2}\right)^m \quad (6.181)
$$

where

$$
\mu_i = \sqrt{\frac{\bar{\alpha}_i}{1 + \bar{\alpha}_i}} . \quad (6.182)
$$

From (6.173), the $\bar{\alpha}_{i,d}$ have the sum value constraint

$$
\sum_{i=1}^{L+1} \bar{\alpha}_{i,d} = 2 \sum_{i=1}^{L+1} \lambda_{i,d} = 2\bar{E}r_\ell(0) . \quad (6.183)
$$

Define $S \subseteq R^{L+1}$ as the set of all $(L + 1)$-component vectors $\{\boldsymbol{\gamma} : \sum_{i=1}^{L+1} \gamma_i = 2Er_\ell(0)\}$. The set S is convex, since for any pair of vectors $\boldsymbol{\gamma}_i$ and $\boldsymbol{\gamma}_j$ the convex combination $\theta\boldsymbol{\gamma}_i + (1 - \theta)\boldsymbol{\gamma}_j$ is contained in S for any $0 \leq \theta \leq 1$. If the pairwise error probability is treated as a mapping from S to R, then it is a convex function of $\boldsymbol{\gamma}$ and, hence, has a unique minimum. For example, Fig. 6.19 shows the pairwise error probability for a three-tap channel ($L = 2$, $D = 1$) with equal strength taps ($\gamma_1 = \gamma_2 = \gamma_3$). Note that the value of γ_3 is determined uniquely by the values of γ_1 and γ_2, and that is why a three dimensional graph is used. By using variational calculus, it is shown in Appendix 6A that the pairwise error probability is minimized when the $\bar{\alpha}_{i,d}$ are all equal, i.e., $\lambda_{i,d} = \lambda = r_\ell(0)E/(L+1)$, resulting in the minimum pairwise error probability

$$
P_{\min} = \left(\frac{1 - \mu}{2}\right)^{D(L+1)} \sum_{m=0}^{D(L+1)-1} \binom{D(L+1) - 1 + m}{m} \left(\frac{1 + \mu}{2}\right)^m \quad (6.184)
$$

where

$$
\mu = \sqrt{\frac{\lambda/4N_o}{1 + \lambda/4N_o}} . \quad (6.185)
$$

For a given error event, the pairwise error probability is minimized when $\mathbf{A}_d$ is perfectly conditioned, i.e., $c(\mathbf{A}_d) = 1$. Recall that $c(\mathbf{A}_d) = c(\boldsymbol{\Sigma}_d \mathbf{E} \boldsymbol{\Sigma}_d) \leq (c(\boldsymbol{\Sigma}_d))^2 c(\mathbf{E})$, where $(c(\boldsymbol{\Sigma}_d))^2$ represents the ratio of the maximum and minimum channel tap variances $(\sigma_d^2)_{\max}/(\sigma_d^2)_{\min}$. We have seen that $\mathbf{E}$ depends

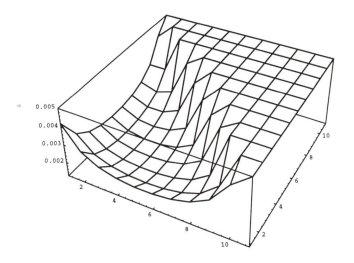

Figure 6.19 Pairwise error probability for a three-tap fading ISI channel, from [136].

only the signal constellation being used and the channel vector length $L + 1$. However, $\mathbf{A}_d$ has information about the signal constellation and power distribution of the fading ISI channel. It follows that $c(\mathbf{A}_d) \leq c(\mathbf{E})$ with equality if and only if the channel has equal strength taps. This means that any system has the best performance when the fading ISI channel has equal strength taps.

6.6.3 Computing the Union Bound

Many algorithms have been suggested for evaluating the union bound on the error probability. One technique is to obtain a union-Chernoff bound by finding the transfer function of the error-state diagram and imposing a Chernoff upper bound on the complementary error function appearing in (6.177). This approach has three draw backs i) the Chernoff bound is very loose when the channel exhibits a deep fade, ii) the transfer function is difficult to obtain for large-state systems, and iii) if the exact pairwise error probability is available, then the transfer function approach cannot be used. To overcome these difficulties, a method based on the error-state transition matrix can be used for calculating the upper bound [4], but it demands a very large amount of computer memory. Here we discuss another alternative that uses an error-state diagram with a one-directional stack algorithm. Note that other types of algorithms could also be used for this calculation [270].

Error-State Diagram

To evaluate the upper bound, the error-state diagram must be defined. Assume a system with N_V error-states, $V_0, V_1, \cdots, V_\nu$. By splitting the zero state, an $(N_V + 1)$-node error-state diagram can be constructed such that the initial and final nodes, V_0 and V_{N_V}, respectively, are zero-error states and the intermediate nodes are non-zero error states. Let t_{ij} denote the branch-weight associated with the v_i to v_j transition, and define t_{ij} as follows:

$$t_{ij} = P_{ij} Z_1^{u_{ij}} Z_2^{\mathbf{A}_{ij}} \qquad (6.186)$$

where

- Z_1 and Z_2 are intermediate (dummy) variables.

- P_{ij} is the fraction of correct symbols x_k such that the transition from v_i to v_j is possible.

- u_{ij} is the number of bit errors associated with the transition from v_i to v_j.

- $\mathbf{A}_{ij}$ is given by $(6.170)^5$, but we emphasize that it is a function of the v_i to v_j transition.

From the definition of t_{ij} in (6.186), the weight of a particular path in the error-state diagram is

$$\prod_{\{(i,j)\}} P_{ij} Z_1^{\sum_{\{(i,j)\}} u_{ij}} Z_2^{\sum_{\{(i,j)\}} \mathbf{A}_{ij}} \qquad (6.187)$$

where $\{(i,j)\}$ denotes the set of state transitions associated with the path under consideration. Note that each path beginning at the initial node and ending at the final node in the error-state diagram represents an error sequence $\epsilon \in E$, where the set E is defined in (6.149). From (6.187)

---

^{5}Here we assume equal diversity branches, i.e., $\mathbf{\Sigma}_d \equiv \mathbf{\Sigma}$, $\forall\, d$, so that $\mathbf{A}_{k,d} \equiv \mathbf{A}_k$, $\forall\, d$.

$$P_r(\mathcal{X}(\varepsilon)) = \prod_{\{(i,j)\}} P_{ij} \tag{6.188}$$

$$w(\epsilon) = \sum_{\{(i,j)\}} u_{ij} \tag{6.189}$$

and

$$\mathbf{A} = \sum_{\{(i,j)\}} \mathbf{A}_{ij} \ . \tag{6.190}$$

These values are required in the calculation of (6.153) or (6.154).

The Stack Algorithm

The union bounds in (6.153) and (6.154) require the calculation of an infinite series. In practice, the mathematical rigor must be sacrificed by truncating the series at an appropriate point. The basic idea of the stack algorithm is to include the R error sequences $\epsilon \in E$ that correspond to the R largest terms in (6.153) or (6.154). The value of R is chosen so that the rest of the terms in the union bound are insignificant. Alternatively, the union bound can be truncated by excluding all paths that have a pairwise error probability P_I less than a threshold P_T.

The stack algorithm maintains a stack with each path (entry) containing the following information; terminal node, $\prod_{\{(i,j)\}} p_{ij}$, $\sum_{\{(i,j)\}} u_{ij}$, $\sum_{\{(i,j)\}} \mathbf{A}_{ij}$, and the intermediate bit error probability P_I. Here, P_I is calculated by

$$P_I = \prod_{\{(i,j)\}} P_{ij} \cdot \sum_{\{(i,j)\}} u_{ij} \cdot \check{P}_r\{\Gamma(\mathbf{x} - \epsilon) \geq \Gamma(\mathbf{x})|\mathbf{x}\} \tag{6.191}$$

where $\check{P}_r\{\Gamma(\mathbf{x} - \epsilon) \geq \Gamma(\mathbf{x})|]\mathbf{x}\}$ is calculated by using (6.181) along with the eigenvalues associated with the matrix $\sum_{\{(i,j)\}} \mathbf{A}_{ij}$.

The stack is ordered (from top to bottom) in order of decreasing intermediate bit error probability P_I. The algorithm first checks if the top path has terminated at the final node. If it has, then the algorithm outputs P_I which is one of the R terms that will be included in the calculation of (6.153) or (6.154); otherwise, the top path is extended and the stack is reordered. Since the top path has the largest P_I, it is likely that the extensions of this path will correspond to one or more of the R dominant terms that are of interest. All paths with the same P_I can be grouped together for easier sorting of the stack. The complete algorithm is given in Fig. 6.20 and is described as follows.

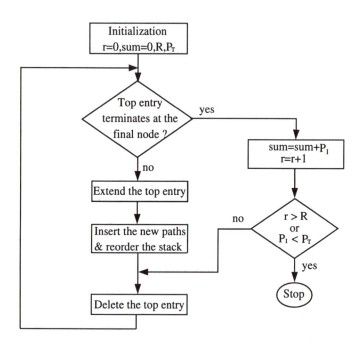

Figure 6.20 The stack algorithm for computing the error probability union bound with MLSE, from [281].

1. Load the stack with the initial node, set all the parameters equal to zero, and input the threshold value P_T (described below) or R.

2. Determine if the top path terminates at the final node. If it does, then go to Step 3; otherwise go to Step 4.

3. Output P_I, and determine if the algorithm should be terminated. If yes, then terminate the algorithm; otherwise delete the top path and go to Step 2.

4. Extend the top path and calculate $\prod_{\{(i,j)\}} P_{ij}$, $\sum_{\{(i,j)\}} u_{ij}$, $\sum_{\{(i,j)\}} \mathbf{A}_{ij}$, and P_I for all of the extension paths.

5. Delete the top path.

6. Insert the new extension paths and rearrange the stack in the order of increasing intermediate bit error probability P_I.

7. Go to Step 2.

6.6.4 Examples

Union bounds will now be evaluated and compared with computer simulations for two example systems by using the above procedure. In the simulations, $\sigma_{i,d}^2 = \sigma^2$, $\forall\ i,\ d$. The tap coefficients are generated by passing independent complex white Gaussian noise through a digital Butterworth filter with a normalized 3-dB cut off frequency equal to $fT = 8.333 \times 10^{-5}$ Hz, typical of an HF channel [191]. All analytical results are obtained by setting the threshold $P_T = 10^{-3} \cdot P_{\max}$, where $P_{\max}$ is the maximum term in the upper bound in (6.154).

Example 6.6 BPSK Modulated System————————————————

A three-tap channel with BPSK modulation is analyzed in this example, where $x_k \in \{-1, +1\}$. There are three different error symbols in this case, i.e., $\epsilon_k \in \{0,\ \pm 2\}$. The error-state diagram is shown in Fig. 6.21. Observe that the error-state diagram is symmetrical in that there are always two paths having the same set of parameters $\Pr\{\mathcal{X}(\varepsilon)\}$, $w(\varepsilon)$, and $\mathbf{A}$. Combining all such pairs of paths together, results in the simplified error-state diagram shown in Fig. 6.22. For equal strength taps, $\mathbf{A}_{ij} = \sigma^2 \mathbf{E}_{ij}$, where $\mathbf{E}_{ij}$ is given by (6.160). The branch weights for the error-state diagram are defined in Table 6.22. Since $\mathbf{E}_{ij}$ is Hermitian, only the lower triangular elements of the matrix $\mathbf{E}_{ij}$ are given.

Fig. 6.23 compares the union bound with simulation results. The received branch bit energy-to-noise ratio $\bar{\gamma}_c$, can be obtained from (6.42). For $D = 1$, the union bound is loose by about 2 dB for bit error probabilities less than 10^{-3}. However, for $D = 2$, the union bound is tight to within 1 dB. This is reasonable because the channel is unlikely to experience a deep fade on both diversity branches where the union bound becomes loose. In general, the bound is tighter for larger $\bar{\gamma}_c$ and D.

Example 6.7 QPSK Modulated System

This example considers QPSK on a two-tap channel model. The x_k are complex taking on the values $\exp\{j(\pi/4 \pm k\pi/2)\}$, $k = 0,\ 1,\ 2,\ 3$. There are nine different error symbols in this case, i.e., $\epsilon_k \in \{0,\ \pm\sqrt{2},\ \pm j\sqrt{2},\ \pm\sqrt{2} \pm j\sqrt{2}\}$. It is left as an exercise to the reader that Fig. 6.24 represents a simplified error state diagram. The branches labeled with "2" represent two error-state transitions. For example, the branch b_{12} represents the error-state transitions

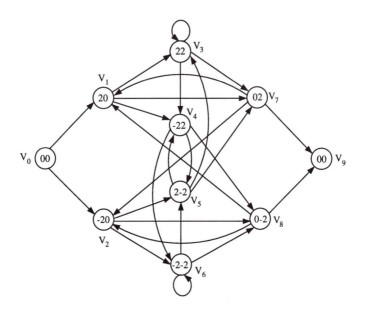

Figure 6.21 Error-state diagram for the BPSK system, from [281].

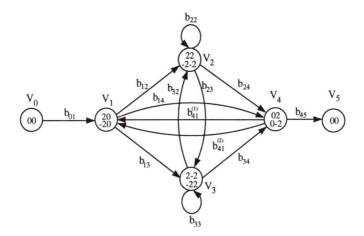

Figure 6.22 Simplified error-state diagram for the BPSK system, from [281].

branch	P_{ij}	u_{ij}	e_{11}	e_{21}	e_{22}	e_{31}	e_{32}	e_{33}
b_{01}	1.0	1.0	4.0	0.0	0.0	0.0	0.0	0.0
b_{12}	0.5	1.0	4.0	4.0	4.0	0.0	0.0	0.0
b_{13}	0.5	1.0	4.0	-4.0	4.0	0.0	0.0	0.0
b_{14}	0.5	0.0	0.0	0.0	4.0	0.0	0.0	0.0
b_{22}	0.5	1.0	4.0	4.0	4.0	4.0	4.0	4.0
b_{23}	0.5	1.0	4.0	-4.0	4.0	-4.0	4.0	4.0
b_{24}	0.5	0.0	0.0	0.0	4.0	0.0	4.0	4.0
b_{32}	0.5	1.0	4.0	4.0	4.0	-4.0	-4.0	4.0
b_{33}	0.5	1.0	4.0	-4.0	4.0	4.0	-4.0	4.0
b_{34}	0.5	0.0	0.0	0.0	4.0	0.0	-4.0	4.0
$b_{41}^{(1)}$	0.5	1.0	4.0	0.0	0.0	4.0	0.0	4.0
$b_{41}^{(2)}$	0.5	1.0	4.0	0.0	0.0	-4.0	0.0	4.0
b_{45}	0.5	0.0	0.0	0.0	0.0	0.0	0.0	4.0

Table 6.1 Branch weights of BPSK modulated system, from [281]

$\epsilon_{12} = \pm(\sqrt{2}, j\sqrt{2})$ and $\epsilon_{12} = \pm(\sqrt{2}, -j\sqrt{2})$. The transition-gains are shown in Table 6.24, where only the lower triangular elements of $\mathbf{E}_{ij}$ are given. Fig. 6.25 compares the union bound with simulation results. For $D = 1$, the difference is about 4 dB. However, for $D = 2$ the difference is only 1.5 dB.

6.7 ERROR PROBABILITY FOR FRACTIONALLY-SPACED MLSE RECEIVERS

Referring to Fig. 6.6, let $X(z)$, $V(z)$, and $V^{(2)}(z)$ be the z-transforms of the input sequence $\mathbf{x}$, the T-spaced received sequence $\mathbf{v}$ and the $T/2$-spaced received sequence $\mathbf{v}^{(2)}$, respectively. The mappings from $X(z)$ to $V(z)$ and from $X(z)$ to $V^{(2)}(z)$ are one-to-one and both the T-spaced and $T/2$-spaced MLSE receivers operate on noisy sequences that are corrupted by noise samples with variance N_o. Therefore, we only need to compare the Euclidean distances between

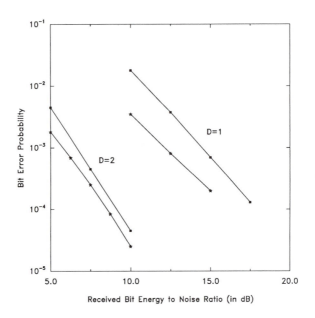

Figure 6.23 Comparison of union bounds with simulation results for BPSK on a three-equal-ray T-spaced fading ISI channel, [281].

allowed sequences of channel outputs to determine the relative performance of the T- and $T/2$-spaced receivers.

6.7.1 T-spaced MLSE Receiver

From the definition of the error event in (6.145), the z-transform of the error sequence is

$$\mathcal{E}(z) = \epsilon_{k_1} + \epsilon_{k_1+1}z^{-1} + \ldots + \epsilon_{k_2-L-1}z^{L-\ell+1} \tag{6.192}$$

where $\epsilon_k = x_k - \hat{x}_k$. The z-transform of the signal error sequence associated with the error event is

$$\mathcal{E}_v(z) = (v_{k_1} - \hat{v}_{k_1}) + (v_{k_1+1} - \hat{v}_{k_1+1})z^{-1} + \ldots + (v_{k_2-L-1} - \hat{v}_{k_1+\ell-1})z^{-\ell+1} \tag{6.193}$$

and we have

$$\mathcal{E}_v(z) = \mathcal{E}_x(z)G(z) \ . \tag{6.194}$$

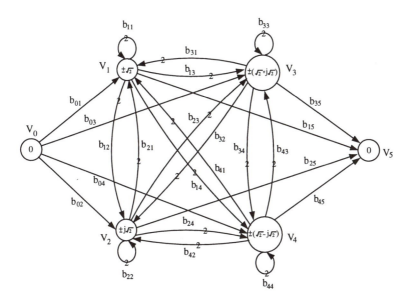

Figure 6.24   Simplified error-state diagram for QPSK system, from [281].

From (6.156), the squared Euclidean distance Δ^2 of the error event in (6.145) is [100]

$$
\begin{aligned}
\Delta^2 &= \sum_{k=k_1}^{k_1+\ell-1} \left| \sum_{i=0}^{L} g_i \epsilon_{k-i} \right|^2 \\
&= [\, \mathcal{E}_v(z)\mathcal{E}_v^*(1/z^*)\,]_0 \\
&= [\, \mathcal{E}(z)F(z)\mathcal{E}^*(1/z^*)\,]_0
\end{aligned}
\qquad (6.195)
$$

where $[\,\cdot\,]_0$ is the coefficient of z^0.

6.7.2 $T/2$-spaced MLSE Receiver

For the same error event described in (6.145), the corresponding z-transform of the $T/2$-spaced error sequence is

$$
\mathcal{E}_{X^{(2)}}(z) = \epsilon_{k_1}^{(2)} + \epsilon_{k_1+1}^{(2)} z^{-2} + \ldots + \epsilon_{k_2-L-1}^{(2)} z^{2(L-\ell+1)} \ . \qquad (6.196)
$$

branch	P_{lm}	u_{lm}	e_{11}	e_{21}	e_{22}
b_{01}	1.0	1.0	$2.0 + j0.0$	$0.0 + j0.0$	$0.0 + j0.0$
b_{02}	1.0	1.0	$2.0 + j0.0$	$0.0 + j0.0$	$4.0 + j0.0$
$b_{11}^{(1)}$	0.5	1.0	$2.0 + j0.0$	$2.0 + j0.0$	$2.0 + j0.0$
$b_{11}^{(2)}$	0.5	1.0	$2.0 + j0.0$	$-2.0 + j0.0$	$2.0 + j0.0$
$b_{23}^{(1)}$	0.25	2.0	$4.0 + j0.0$	$2.0 - j2.0$	$2.0 + j0.0$
$b_{23}^{(2)}$	0.25	2.0	$4.0 + j0.0$	$-2.0 + j2.0$	$2.0 + j0.0$
$b_{34}^{(1)}$	0.25	2.0	$4.0 + j0.0$	$0.0 - j4.0$	$4.0 + j0.0$
$b_{34}^{(2)}$	0.25	2.0	$4.0 + j0.0$	$-4.0 + j0.0$	$4.0 + j0.0$
b_{35}	1.0	0.0	$0.0 + j0.0$	$0.0 + j0.0$	$4.0 + j0.0$
b_{45}	1.0	0.0	$0.0 + j0.0$	$0.0 + j0.0$	$4.0 + j0.0$

Table 6.2   Transition-gain examples of QPSK system, from [281]

Notice that $\epsilon_k^{(2)} = x_k^{(2)} - \hat{x}_k^{(2)}$ is zero for even k. Therefore, $\mathcal{E}^{(2)}(z) = \mathcal{E}(z^2)$. The corresponding z-transform of the $T/2$-spaced signal error sequence associated with the error event in (6.145) is

$$\mathcal{E}_v^{(2)}(z) = \mathcal{E}^{(2)}(z)G^{(2)}(z) \ . \tag{6.197}$$

From (6.156), the squared Euclidean distance of the error event in (6.145) is

$$
\begin{aligned}
\left(\Delta^{(2)}\right)^2 &= \sum_{k=2k_1}^{2(k_1+\ell-1)} \left| \sum_{i=0}^{2L} b_i^{(2)} \epsilon_{k-i}^{(2)} \right|^2 \\
&= [\ \mathcal{E}_v^{(2)}(z)\mathcal{E}_v^{(2)*}(1/z^*)\ ]_0 \\
&= [\ \mathcal{E}^{(2)}(z)F^{(2)}(z)\mathcal{E}^{(2)*}(1/z^*)\ ]_0 \\
&= [\ \mathcal{E}(z^2)F^{(2)}(z)\mathcal{E}^*(1/z^{*2})\ ]_0 \ . \tag{6.198}
\end{aligned}
$$

Note that polynomial $\mathcal{E}(z^2)\ \mathcal{E}^*(1/z^{*2})$ has the property that the odd powers of z have zero coefficients. Therefore, the contributions to the coefficient $[\ \mathcal{E}(z^2)F^{(2)}(z)\mathcal{E}^*(1/z^{*2})\ ]_0$ arise only from the coefficients of $F^{(2)}(z)$ associated with even powers of z. Note also from (6.43) and (6.44) that the coefficients $f_{2k}^{(2)}$ of $F^{(2)}(z)$ associated with even powers of z are equal to the coefficients f_k of $F(z)$, i.e., $f_{2k}^{(2)} = f_k$. Therefore,

$$\left(\Delta^{(2)}\right)^2 = [\ \mathcal{E}(z^2)F^{(2)}(z)\mathcal{E}^*(1/z^{*2})\ ]_0 = [\ \mathcal{E}(z)F(z)\mathcal{E}^*(1/z^*)\ ]_0 = \Delta^2 \ . \tag{6.199}$$

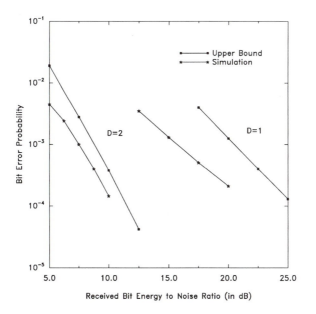

Figure 6.25 Comparison of union bound and simulation results for a QPSK on a two-equal-ray T-spaced channel, from [281].

Consequently, the error probability performance of the T- and $T/2$-spaced MLSE receivers are identical.

Example 6.8

Let

$$
\begin{aligned}
\mathcal{E}(z) &= \epsilon_0 + \epsilon_1 z^{-1} + \epsilon_2 z^{-2} \\
F^{(2)}(z) &= f_{-2}^{(2)} z^2 + f_{-1}^{(2)} z + f_0^{(2)} + f_1^{(2)} z^{-1} + f_2^{(2)} z^{-2} \ .
\end{aligned}
$$

Then

$$
\begin{aligned}
F(z) &= f_{-1} z + f_0 + f_1 z^{-1} = f_{-2}^{(2)} z + f_0^{(2)} + f_2^{(2)} z^{-1} \\
\mathcal{E}^{(2)}(z) &= \epsilon_0 + \epsilon_1 z^{-2} + \epsilon_2 z^{-4} \ .
\end{aligned}
$$

Therefore,

$$
\begin{aligned}
[\, \mathcal{E}(z) F(z) \mathcal{E}^*(1/z^*)\,]_0 &= \left(|\epsilon_0|^2 + |\epsilon_1|^2 + |\epsilon_2|^2 \right)^2 f_0 \\
&\quad + \epsilon_0 \epsilon_1^* f_1 + \epsilon_1 \epsilon_0^* f_{-1}
\end{aligned}
$$

and

$$[\,\mathcal{E}^{(2)}(z)F^{(2)}(z)\mathcal{E}^{(2)^*}(1/z^*)\,]_0 \;=\; \left(|\epsilon_0|^2 + |\epsilon_1|^2 + |\epsilon_2|^2\,\right)^2 f_0^{(2)}$$
$$+\epsilon_0\epsilon_1^* f_2^{(2)} + \epsilon_1\epsilon_0^* f_{-2}^{(2)}\;.$$

Hence, $\Delta^2 = (\Delta^{(2)})^2$.

6.7.3   Practical $T/2$-spaced MLSE Receiver

The receivers in Figs. 6.3 and 6.6 use a filter that is matched to the received pulse $h^*(-t)$. Since this filter requires knowledge of the unknown channel impulse response, such receivers are impractical. One solution is to implement an 'ideal' low-pass filter with a cutoff frequency of $1/T$ and sample the output at rate $2/T$. The noise samples at the output of this filter will be uncorrelated and, therefore, the $T/2$-spaced MLSE receiver can be implemented. Vachula and Hill [315] showed that this receiver is optimum; however, it has some drawbacks. First, it is not suitable for bandwidth efficient systems that are affected by adjacent channel interference such as the North American IS-54 and Japanese PDC systems, because the cutoff frequency of the low-pass filter will extend significantly into the adjacent band. Second, the ideal low-pass filter is nonrealizable and difficult to approximate. One solution is to use a receiver filter that is matched to the transmitted pulse $h_a(t)$ Hamied and Stüber [137]. If the transmitted signals have a most 100% excess bandwidth, then rate $2/T$ sampling satisfies the sampling theorem and the $T/2$-spaced samples provide sufficient statistics.

Let $H_a^{(2)}(z)$, $C^{(2)}(z)$, and $H^{(2)}(z)$ be the z-transforms of the $T/2$-spaced discrete-time signals corresponding to $h_a(t)$, $c(t)$, and $h(t)$, respectively. The z-transform of the autocorrelation function of the noise samples at the output of the receive filter $h_a^*(-t)$ is $N_o F_h^{(2)}(z)$ where $F_h^{(2)}(z) = H_a^{(2)}(z)\left(H_a^{(2)}(1/z^*)\right)^*$. Using the factorization

$$F_h^{(2)}(z) = G_h^{(2)}(z)\left(G_h^{(2)}(1/z^*)\right)^* \tag{6.200}$$

the $T/2$-spaced noise sequence can be whitened by using a filter with transfer function $1/\left(G_h^{(2)}(1/z^*)\right)^*$. The resulting system is shown in Fig. 6.26. We now show that the receivers in Figs. 6.6 and 6.26 yield identical performance.

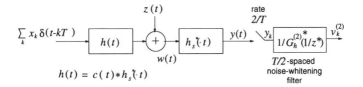

Figure 6.26 Block diagram of system that implements a filter matched to $h_a(t)$ followed by a $T/2$-spaced noise whitening filter. The structure of the noise whitening filter depends only on the pulse $h_a(t)$.

The z-transform of the overall $T/2$-spaced discrete-time channel that includes the noise-whitening filter is

$$
\begin{aligned}
G_{eq}^{(2)}(z) &= H_a^{(2)}(z)C^{(2)}(z)\left(H_a^{(2)}(1/z^*)\right)^* / \left(G_h^{(2)}(1/z^*)\right)^* \\
&= C^{(2)}(z)G_h^{(2)}(z) \ .
\end{aligned}
\tag{6.201}
$$

On the other hand, referring to the conventional system shown in Fig. 6.6, we have

$$
H^{(2)}(z) = H_a^{(2)}(z)\,C^{(2)}(z)
\tag{6.202}
$$

and

$$
F^{(2)}(z) = H_a^{(2)}(z)\left(H_a^{(2)}(1/z^*)\right)^* C^{(2)}(z)\left(C^{(2)}(1/z^*)\right)^* \ .
\tag{6.203}
$$

Let

$$
C^{(2)}(z)\left(C^{(2)}(1/z^*)\right)^* = G_c^{(2)}(z)\left(G_c^{(2)}(1/z^*)\right)^*
\tag{6.204}
$$

be a factorization of $C^{(2)}(z)\big(C^{(2)}(1/z^*)\big)^*$ such that $\left(G_c^{(2)}(1/z^*)\right)^*$ has minimum phase. Using (6.200), (6.203) and (6.204) yields

$$
F^{(2)}(z) = G_h^{(2)}(z)\left(G_h^{(2)}(1/z^*)\right)^* G_c^{(2)}(z)\left(G_c^{(2)}(1/z^*)\right)^* \ .
\tag{6.205}
$$

The transfer function of the noise-whitening filter must be chosen as

$$
1/\left(\left(G_h^{(2)}(1/z^*)\right)^* \left(G_c^{(2)}(1/z^*)\right)^* \right) \ .
\tag{6.206}
$$

Therefore, the overall transfer function at the output of the noise whitening filter is

$$
G^{(2)}(z) = G_h^{(2)}(z)G_c^{(2)}(z) \ .
\tag{6.207}
$$

The equivalent response $G_{\mathrm{eq}}^{(2)}(e^{j\omega})$ in (6.201) has the same amplitude as $G^{(2)}(e^{j\omega})$ but different phase. Also

$$G_{\mathrm{eq}}^{(2)}(z)\left(G_{\mathrm{eq}}^{(2)}(1/z^*)\right)^* = F^{(2)}(z) \ . \tag{6.208}$$

Therefore, the Euclidean distance between sequences of channel outputs for the system in Fig. 6.26 is the same as those for the T-spaced MLSE receiver and, therefore, the system shown in Fig. 6.26 has maximum likelihood performance. The main advantage of the system in Fig. 6.26 is that the noise-whitening filter does not depend on the unknown channel and has a fixed structure. The channel estimation can be performed after the noise-whitening filter and the Viterbi algorithm can be implemented using the metric in (6.130). Although the number of computations needed in the $T/2$-spaced MLSE receiver is twice that of a T-spaced receiver, the latter can not be implemented for unknown channels. Moreover, a T-spaced MLSE receiver has poor performance when it is implemented with a matched filter that is derived from an inaccurate channel estimate [239].

6.7.4   Timing Phase Sensitivity

The conventional MLSE receiver based on T-spaced sampling at the output of the matched filter suffers from sensitivity to the sampler timing phase [259]. We now show that a $T/2$-spaced MLSE receiver is not sensitive to the sampler timing phase.

For a given a timing offset t_0, the sampled impulse response at the output of the matched filter is represented by the vector $\mathbf{f}_{t_0}^{(2)}$, where $f_{t_0,k}^{(2)} = f(kT' \pm t_0)$ and $T' = T/2$. Note that $f_{t_0,n}^{(2)} \neq \left(f_{t_0,-n}^{(2)}\right)^*$ in this case. The discrete-time of Fourier transform of $\mathbf{f}_{t_0}^{(2)}$ is

$$F_{t_0}^{(2)}(e^{j\omega}) = F^{(2)}(e^{j\omega})e^{\pm j\omega\tau_0} \tag{6.209}$$

where $\tau_0 = t_0/T'$. If the sampler phase is known, then a discrete-time filter with response $e^{\pm\omega\tau_0}$ after the sampler will give the symmetric signal $\mathbf{f}^{(2)}$ at its output. However, as we now show, there is no need to correct the phase.

The power spectrum of the noise at the output of the matched filter is independent of the timing offset t_0 and is given by

$$S_{\nu\nu}(f) = N_o F^{(2)}(e^{j\omega}) \ . \tag{6.210}$$

Since the discrete-time Fourier transform of the noise-whitening filter is

$$1/(G^{(2)}(1/z^*))^*|_{z=e^{j\omega}} = 1/(G^{(2)}(e^{j\omega}))^* \tag{6.211}$$

and we have

$$F^{(2)}(e^{j\omega}) = G^{(2)}(e^{j\omega})(V^{(2)}(e^{j\omega}))^* = |G^{(2)}(e^{j\omega})|^2 \tag{6.212}$$

it follows from (6.210) that the noise is white at the output of the noise-whitening filter. The discrete-time Fourier transform of the message signal at the output of the noise-whitening filter is

$$G_{t_0}^{(2)}(e^{j\omega}) = G^{(2)}(e^{j\omega})e^{\pm j\omega\tau_0} \tag{6.213}$$

and we have

$$\sum_i |g_i^{(2)}|^2 = \sum_i |g_{t_0,i}^{(2)}|^2 = \frac{1}{2\pi}\int_{-\pi}^{\pi} |G^{(2)}(e^{j\omega})|^2 d\omega . \tag{6.214}$$

This means that

$$G_{t_0}^{(2)}(z)(G_{t_0}^{(2)})^*(1/z^*) = F^{(2)}(z) . \tag{6.215}$$

Therefore, the distances between allowed sequences of channel outputs with the $T/2$-spaced MLSE receiver is not sensitive to the sampler phase $e^{\pm\omega\tau_0}$. Since the noise remains white the performance is also insensitive to the sampler phase.

Appendix 6A
DERIVATION OF EQUATION (6.184)

Assume that (6.176) has M different poles γ_1, γ_2, $\ldots,\gamma_M$. Then the pairwise error probability is equal to

$$P(\boldsymbol{\gamma}) = \sum_{i=1}^{M} \left\{ \left(\frac{1}{2} - \frac{1}{2}\sqrt{\frac{\gamma_i}{1+\gamma_i}}\right)\prod_{j\neq i}\left(1 - \frac{\gamma_j}{\gamma_i}\right)^{-1} \right\} . \tag{6A.1}$$

Define the function $\phi(\boldsymbol{\gamma}) = \sum_{i=1}^{M} \gamma_i - C = 0$, where C is a constant. The method of Lagrange multipliers suggests that

$$\frac{\partial P}{\partial \gamma_i} + \lambda\frac{\partial \phi}{\partial \gamma_i} = 0 \quad i = 1, \ldots, M \tag{6A.2}$$

for any real number λ. It can be shown by induction that

$$
\begin{aligned}
\frac{\partial P}{\partial \gamma_k} = {} & -\left(\frac{1}{2} - \frac{1}{2}\sqrt{\frac{\gamma_k}{1+\gamma_k}}\right) \sum_{i\neq k}\left\{\frac{\gamma_i}{\gamma_k^2}\left(1-\frac{\gamma_i}{\gamma_k}\right)^{-2}\prod_{j\neq i,k}\left(1-\frac{\gamma_j}{\gamma_k}\right)^{-1}\right\} \\
& +\sum_{i\neq k}\left\{\frac{1}{\gamma_i}\left(1-\frac{\gamma_k}{\gamma_i}\right)^{-2}\left(\frac{1}{2}-\frac{1}{2}\sqrt{\frac{\gamma_i}{1+\gamma_i}}\right)\prod_{j\neq i,k}\left(1-\frac{\gamma_j}{\gamma_i}\right)^{-1}\right\} \\
& -\left(\frac{1}{4\gamma_k^{1/2}}\frac{1}{(1+\gamma_k)^{3/2}}\right)\prod_{j\neq k}\left(1-\frac{\gamma_j}{\gamma_k}\right)^{-1}.
\end{aligned}
\tag{6A.3}
$$

By solving (6A.2) and observing the symmetry of $P(\boldsymbol{\gamma})$ and the derivative (6A.3) with respect to the permutations of $\boldsymbol{\gamma}$, it is apparent that the minimum of $P(\boldsymbol{\gamma})$ is achieved when $\gamma_1 = \gamma_2 = \ldots = \gamma_M$.

Problems

6.1. Assume that a received signal is given by

$$
y(t) = \sum_{i=-\infty}^{\infty} x_i f(t - iT)
$$

where $x_k = \pm 1$, and $f(t)$ is a the minimum bandwidth pulse satisfying Nyquist's criterion for zero ISI, i.e.,

$$
F(f) = \begin{cases} T & |f| \leq 1/2T \\ 0 & |f| > 1/2T \end{cases}
$$

and

$$
f(t) = \frac{\sin(\pi t/T)}{\pi t/T} .
$$

There are two problems associated with this pulse shape. One is the problem of realizing a pulse having the rectangular spectral characteristic $F(f)$ given above. The other problem arises from the fact that the tails in $f(t)$ decay as $1/t$. Consequently, a sampling timing error results in an infinite series of ISI components. Such a series is not absolutely summable and, hence, the sum of the resulting interference does not converge.

Assume that $f(t) = 0$ for $|t| > NT$, where N is a positive integer. In spite of the restriction that the channel is band-limited, this assumption holds in all practical communication systems.

a) Due to a slight timing error, the received signal is sampled at $t = kT + t_0$, where $t_0 < T$. Calculate the response for $t = kT + t_0$. Separate the response into two components, the desired term and the ISI term.

b) Assume that the polarities of x_i are such that every term in the ISI is positive, i.e., worst case ISI. Under this assumption show that the ISI term is

$$\text{ISI} = \frac{2}{\pi} \sin(\pi t_0 / T) \sum_{n=1}^{N} \frac{n}{n^2 - t_0^2 / T^2} \; .$$

and, therefore, ISI $\to \infty$ as $N \to \infty$.

6.2. Starting with

$$f_k = \int_{-\infty}^{\infty} h^*(\tau) h(\tau + kT) d\tau$$

show that

$$F(e^{j 2\pi f T}) = F_\Sigma(f) \; .$$

6.3. Suppose that the impulse response of an overall channel consisting of the transmit filter, channel, and receive filter, is

$$F(f) = \begin{cases} 1 & , \; |f| \le f_\ell \\ \frac{f_u - |f|}{f_u - f_\ell} & , \; f_\ell \le |f| \le f_u \end{cases} \; .$$

a) Find the overall impulse response $f(t)$.

b) Is it possible to transmit data without ISI?

c) How do the magnitudes of the tails of the overall impulse response decay with large values of t?

d) Suppose that binary signaling is used with this pulse shape so that the noiseless signal at the output of the receive filter is

$$y(t) = \sum_{n} x_n f(t - nT)$$

where $x_n \in \{-1, +1\}$. What is the maximum possible magnitude that $y(t)$ can achieve?

6.4. Show that the ISI coefficients $\{f_n\}$ may be expressed in terms of the channel vector coefficients $\{g_n\}$ as

$$f_n = \sum_{k=0}^{L-n} g_k^* g_{k+n} \qquad n = 0, 1, 2, \ldots, L \; .$$

6.5. Suppose that BPSK is used on a static ISI channel. The complex envelope has the form

$$v(t) = A \sum_{k=-\infty}^{\infty} x_k h_a(t - kT)$$

where $x_k \in \{-1, +1\}$ and $h_a(t)$ is the amplitude shaping pulse. The non-return-to-zero pulse $h_a(t) = u_T(t)$ is used and the impulse response of the channel is

$$c(t) = c_0 \delta(t) - c_1 \delta(t - \tau)$$

where c_0 and c_1 are complex numbers and $0 < \tau < T$.

a) Find the received pulse $h(t)$.

a) What is the filter matched to $h(t)$.

c) What are the ISI coefficients $\{f_i\}$

6.6. Suppose that BPSK signaling is used on a static ISI channel having impulse response

$$c(t) = \delta(t) + 0.1\delta(t - T)$$

The receiver employs a filter that is matched to the transmitted pulse $h_a(t)$, and the sampled outputs of the matched filter are

$$y_n = x_n q_0 + \sum_{k \neq n} x_k q_{n-k} + \eta_n$$

where $x_n \in \{-1, +1\}$. Decisions are made on the $\{y_n\}$ without any equalization.

a) What is the variance of noise term η_n?

b) What are the values of the $\{q_n\}$?

c) What is the probability of error in terms of the average received bit-energy-to-noise ratio?

6.7. A typical receiver for digital signaling on an ISI channel consists of a matched filter followed by an equalizer. The matched filter is designed to minimize the effect of random noise, while the equalizer is designed to minimize the effect of intersymbol interference. By using mathematical arguments, show that i) the matched filter tends to accentuate the effect of ISI, and ii) the equalizer tends to accentuate the effect of random noise.

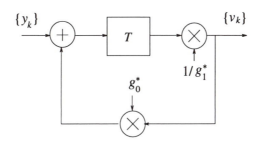

Figure 6.27   Digital filter for Problem 6.8.

6.8. Consider an ISI channel, where $f_n = 0$ for $|n| > 1$. Suppose that the receiver uses a filter matched to the received pulse $h(t) = h_a(t) * c(t)$, and the T-spaced samples at the output of the matched filter, $\{y_k\}$, are filtered as shown in Fig. 6.27. The values of g_0 and g_1 are chosen to satisfy

$$|g_0|^2 + |g_1|^2 = f_0$$
$$g_0 g_1^* = f_1$$

Find an expression for the filter output v_k in terms of g_0, g_1, x_k, x_{k-1}, and the noise component at the output of the digital filter, η_k.

6.9. The z-transform of the channel vector $\mathbf{g}$ of a communication system is equal to

$$G(z) = 0.1 + 1.0z^{-1} - 0.1z^{-2} \ .$$

A binary sequence $\mathbf{x}$ is transmitted, where $x_k = \in \{-1, +1\}$. The received samples at the output of the noise whitening filter are

$$v_n = \sum_{k=0}^{2} g_k x_{n-k} + \eta_n$$

where $\{\eta_n\}$ is a white Gaussian noise sequence with variance $\sigma_\eta^2 = N_o$.

a) Evaluate the probability of error if the demodulator ignores ISI.

b) Design a 3-tap zero-forcing equalizer for this system.

c) What is the response $\{v_k\}$ for the input sequence $\{x_k\} = (-1)^k$, $k = 0, 1, 2, 3$? What is the response at the output of the equalizer?

d) Evaluate the probability of error for the equalized channel.

6.10. Suppose that a system is characterized by the received pulse

$$h(t) = \sqrt{2a}e^{-at}, \qquad 0 \le t \le \infty .$$

A receiver implements a filter matched to $h(t)$ and generates T-spaced samples at the output of the filter. Note that the matched filter is actually noncausal.

a) Find the ISI co-efficients f_i.

b) What is the transfer function of the noise whitening filter that yields a system having an overall minimum phase response.

c) Find the transfer function of the equivalent zero-forcing equalizer $C'(z)$.

d) Find the noise power at the output of the zero-forcing equalizer, and find the condition when the noise power becomes infinite.

6.11. Consider M-PAM on a static ISI channel, where the receiver employs a filter that is matched to the received pulse. The sampled outputs of the matched filter are

$$y_n = x_n f_0 + \sum_{k \ne n} x_k f_{n-k} + \nu_n$$

where the source symbols are from the set $\{\pm 1, \pm 3, \ldots, \pm(M-1)\}$. Decisions are made on the $\{y_n\}$ without any equalization by using a threshold detector. The ℓth ISI pattern can be written as

$$D(\ell) = \sum_{k \ne n} x_{\ell,k} f_{n-k}$$

and $D(\ell)$ is maximum when $\text{sgn}(x_{\ell,k}) = \text{sgn}(f_{n-k})$ and each of the $x_{\ell,k}$ takes on the maximum signaling level, i.e., $x_{\ell,k} = (M-1)d$ for M even. The *maximum distortion* is defined as

$$D_{\max} = \frac{1}{f_0} \sum_{n \ne 0} |f_n| .$$

a) Discuss and compare error performance M-ary signaling $(M > 2)$ with binary signaling $(M = 2)$, using $D_{\max}$ as a parameter.

b) Suppose that the channel has ISI coefficients

$$
\begin{aligned}
f_i &= 0.0 , \quad |i| \ge 3 \\
f_2 &= f_{-2} = 0.1 \\
f_1 &= f_{-1} = -0.2 \\
f_0 &= 1.0 .
\end{aligned}
$$

Plot the probability of error against the signal-to-noise ratio and compare with the ideal channel case, i.e., $f_0 = \delta_{n0}$. Show your results for $M = 2$ and 4.

6.12. Consider a linear MSE equalizer and suppose that the tap gain vector **c** satisfies

$$\mathbf{c} = \mathbf{c}_{\mathrm{op}} + \mathbf{c}_e$$

where $\mathbf{c}_e$ is the tap gain error vector. Show that the mean square error that is achieved with the tap gain vector **c** is

$$J = J_{\min} + \mathbf{c}_e^T \mathbf{M}_v \mathbf{c}_e^* \ .$$

6.13. The matrix $\mathbf{M}_v$ has an eigenvalue λ_k and eigenvector $\mathbf{x}_k$ if

$$\mathbf{x}_k \mathbf{M}_v = \lambda_k \mathbf{x}_k \qquad\qquad k = 1, \ldots, N \ .$$

Prove that the eigenvectors are orthogonal, i.e.,

$$\mathbf{x}_i \mathbf{x}_j^T = \delta_{ij} \ .$$

6.14. Show that the relationship between the output SNR and $J_{\min}$ for an infinite-tap mean-square error linear equalizer is

$$\gamma_\infty = \frac{1 - J_{\min}}{J_{\min}}$$

where the subscript ∞ on γ indicates that the equalizer has an infinite number of taps. Note that this relationship between γ_∞ and $J_{\min}$ holds when there is residual intersymbol interference in addition to the noise.

6.15. In this question, we will show in steps that

$$\nabla_\mathbf{c} J = 2\mathbf{c}^T \mathbf{M}_v - 2\mathbf{v}_x^H \ .$$

Define

$$\begin{aligned}
\mathbf{M}_v &= \mathbf{M}_{v_R} + j\mathbf{M}_{v_I} \\
\mathbf{c} &= \mathbf{c}_R + j\mathbf{c}_I \\
\mathbf{v}_x &= \mathbf{v}_{x_R} + j\mathbf{v}_{x_I}
\end{aligned}$$

a) By using the Hermitian property $\mathbf{M}_v = \mathbf{M}_v^H$ show that

$$\mathbf{M}_{v_R} = \mathbf{M}_{v_R}^T \text{ and } \mathbf{M}_{v_I} = -\mathbf{M}_{v_I}^T \ .$$

b) Show that

$$
\begin{aligned}
\nabla_{\mathbf{c}_R} \mathrm{Re}\{\mathbf{v}_x^H \mathbf{c}^*\} &= \mathbf{v}_{x_R}^T \\
\nabla_{\mathbf{c}_I} \mathrm{Re}\{\mathbf{v}_x^H \mathbf{c}^*\} &= -\mathbf{v}_{x_I}^T \\
\nabla_{\mathbf{c}_R} \mathbf{c}^T \mathbf{M}_v \mathbf{c}^* &= 2\mathbf{c}_R^T \mathbf{M}_{v_R} - 2\mathbf{c}_I^T \mathbf{M}_{v_I} \\
\nabla_{\mathbf{c}_I} \mathbf{c}^T \mathbf{M}_v \mathbf{c}^* &= 2\mathbf{c}_I^T \mathbf{M}_{v_R} + 2\mathbf{c}_R^T \mathbf{M}_{v_I}
\end{aligned}
$$

where $\nabla_{\mathbf{x}}$ is the gradient with respect to vector $\mathbf{x}$.

c) If we define the gradient of a real-valued function with respect to a complex vector $\mathbf{c}$ as

$$\nabla_{\mathbf{c}} = \nabla_{\mathbf{c}_R} + j\nabla_{\mathbf{c}_I}$$

show that

$$
\begin{aligned}
\nabla_{\mathbf{c}} \mathrm{Re}\{\mathbf{v}_x^H \mathbf{c}^*\} &= \mathbf{v}_x^H \\
\nabla_{\mathbf{c}} \mathbf{c}^H \mathbf{M}_v \mathbf{c}^* &= 2\mathbf{c}^T \mathbf{M}_v .
\end{aligned}
$$

6.16. Show that the pairwise error probability for digital signaling on an ISI channel is given by (6.155).

6.17. Consider the transmission of the binary sequence $\mathbf{x}$, $x_n \in \{-1, +1\}$ over the equivalent discrete-time white noise channel model shown in Fig. 6.28. The received sequence is

$$
\begin{aligned}
v_0 &= .70x_0 + \eta_1 \\
v_1 &= .70x_1 - .60x_0 + \eta_2 \\
v_2 &= .70x_2 - .60x_1 + \eta_3 \\
&\vdots \\
v_k &= .70x_k - .60x_{k-1} + \eta_k
\end{aligned}
$$

a) Draw the state diagram for this system.

b) Draw the trellis diagram.

c) Suppose that the received sequence is

$$\{v_i\}_{i=0}^6 = \{1.0, -1.5, 0.0, 1.5, 0.0, -1.5, 1.0\}$$

Show the surviving paths and their associated path metrics after v_6 has been received.

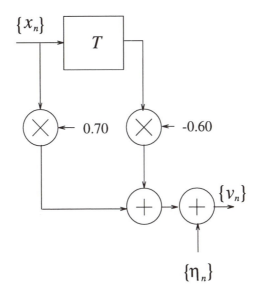

Figure 6.28 Discrete-time white noise channel model for Problem 6.17.

6.18. Suppose that BPSK signaling is used on a frequency selective fading channel. The discrete-time system consisting of the transmit filter, channel, receiver filter, and baud-rate sampler can be described by the polynomial

$$F(z) = \frac{5}{16} - \frac{1}{8}z^{-1} - \frac{1}{8}z .$$

The samples at the output of the receiver filter are processed by a noise whitening filter such that the overall discrete-time white noise channel model $G(z)$ has minimum phase.

a) Find $G(z)$

b) Draw the state diagram and the trellis diagram for the discrete-time white noise channel model.

a) A block of 10 symbols $\mathbf{x} = \{x_i\}_{i=0}^{9}$ is transmitted over the channel and it is known that $x_9 = -1$. Assume that $x_i = 0, i < 0$ and the suppose that the sampled sequence at the output of the matched filter is

$$\mathbf{y} = (y_0, y_1, y_2, y_3, \cdots y_9)$$
$$= (1/2, 1/4, -3/4, 3/4, -3/4, -1/4, 3/4, -3/4, -1/4, -1/4)$$

What sequence $\mathbf{x}$ was most likely transmitted?

7

BANDWIDTH EFFICIENT CODING

Channel coding and interleaving techniques have long been recognized as an effective technique for combating the deleterious effects of noise, interference, jamming, fading, and other channel impairments. The basic idea of channel coding is to introduce controlled redundancy into the transmitted signals that is exploited at the receiver to correct channel induced errors by means of forward error correction. Channel coding can also be used for error detection in schemes that employ automatic repeat request (ARQ) strategies. ARQ strategies must have a feedback channel to relay the retransmission requests from the receiver back to the transmitter when errors are detected. ARQ schemes require buffering at the transmitter and/or receiver and, therefore, are suitable for data applications but are not suitable for delay sensitive voice applications. Hybrid ARQ schemes use both error correction and error detection; the code is used to correct the most likely error patterns, and to detect the more infrequently occurring error patterns. Upon detection of errors a retransmission is requested. Although ARQ schemes are essential for data transmission over wireless channels they nevertheless employ error correction codes. Therefore, the emphasis in this chapter is on error correction coding.

There are many different types of error correcting codes, but historically they have been classified into block codes and convolutional codes. To generate a code word of an (n, k) block code, a block of k data bits is appended by $n - k$ redundant parity bits that are algebraically related to the k data bits, thereby producing a code word consisting of n code bits. The ratio $R_c = k/n$ is called the code rate, where $0 < R_c \leq 1$. Convolutional codes, on the other hand, are generated by the discrete-time convolution of the input data sequence with the impulse response of the encoder. The memory of the encoder is measured by the duration of the impulse response. While block encoder operates on k-bit

341

blocks of data bits, a convolutional encoder accepts a continuous sequence of input data bits.

While both block codes and convolutional codes find potential applications mobile radio systems, our emphasis will be on the treatment of convolutional codes and trellis-coded modulation. Most of the second generation digital cellular standards (e.g., GSM, PDC, IS-54) use convolutional codes, the reason being that there exist very simple decoding algorithms (e.g., the Viterbi algorithm) that can obtain a soft decision decoding gain and achieve optimal theoretical performance. Therefore, a significant coding gain can be obtained with a relatively simple convolutional codes. Nevertheless, block codes are sure to find application in future mobile data systems, especially when computationally efficient soft decision block decoders become available.

In the early application of coding to digital communications, the modulator and coder were treated a separate entities. Hence, a block code or a convolutional code was employed to obtain a coding gain at the cost of bandwidth expansion or data rate. Although this may be a feasible approach for power limited channels where bandwidth resources are plentiful, it is undesirable and sometimes not even possible for bandwidth limited applications such as cellular radio. If no sacrifices of data rate or bandwidth can be made, then schemes that separate the operations of coding and modulation require a very powerful code just to break even with an uncoded system. In 1974, Massey[206] suggested that the performance of a coded digital communication system could be improved by treating coding and modulation as a single entity. Ungerboeck, later developed the basic principles of trellis-coded modulation (TCM) [312] and identified classes of trellis codes that provide substantial coding gains on bandwidth limited additive white Gaussian noise (AWGN) channels.

TCM schemes combine the operations of coding and modulation and can be viewed as a generalization of convolutional codes. While convolutional codes attempt to maximize the minimum Hamming distance between allowed code symbol sequences, trellis-codes attempt to maximize the Euclidean distance between allowed code symbol sequences. By jointly designing the encoder and modulator Ungerboeck showed that, for an AWGN channel, coding gains of 3-6 dB could be obtained relative to an uncoded system by using trellis codes with 4-128 encoder states, without sacrificing bandwidth or data rate. This property makes TCM very attractive for cellular radio applications where high spectral efficiency is needed due to limited bandwidth resources and good power efficiency is needed to extend battery life in portable radios. TCM experienced an almost immediate and widespread application into high-speed power-efficient and bandwidth-efficient digital modems. In 1984, a variant of the Ungerboeck

8-state 2-D trellis code was adopted by CCITT for both 14.4 kb/s leased-line modems and the 9.6 kb/s switched-network modems [28]. In 1985, a TCM-based modem operating at 19.2 kb/s was introduced by Codex [313].

Ungerboeck's work [312] captured the attention of the coding community and laid the foundation for intensified research. Calderbank and Mazo introduced an analytic description of trellis codes [31]. They showed how to realize the two operations (coding and mapping) in Ungerboeck's codes by using a single-step procedure. Calderbank and Sloan [32], and Wei [328], proposed multi-dimensional trellis codes. Spaces with larger dimensionality are attractive, because the signals are spaced at larger Euclidean distance [28]. Calderbank and Sloan [32], and Forney [101], made the observation that the signal constellation should be regarded as a finite set of points taken from an infinite lattice, and the partitioning of the constellation into subsets corresponds to the partitioning of the lattice into a sub-lattice and its cosets. They then developed a new class of codes, called coset codes, based on this principle.

Many studies have examined the performance of TCM on interleaved flat fading channels [69], [70], [77], [33]. Divsalar and Simon [70], [71] constructed trellis codes that are effective for interleaved flat Ricean and Rayleigh fading channels. Interleaving randomizes the channel with respect to the transmitted symbol sequence and has the effect of reducing the channel memory. Consequently, interleaving improves the performance of codes that have been designed for memoryless channels. Moreover, trellis codes that are designed for flat fading channels exhibit time diversity when combined with interleaving of sufficient depth. It was reported in [33] that interleaving with reasonably long interleaving depths is almost as good as ideal infinite interleaving. The design of trellis codes for interleaved flat fading channels is not guided by the minimum Euclidean distance used for AWGN channels, but rather by the minimum product squared Euclidean distance and the minimum built-in time diversity between any two allowed code symbol sequences. Wei [329] introduced an additional design parameter called the minimum decoding depth, and proposed a set of efficient codes for interleaved flat Rayleigh fading channels.

Many studies have also considered the effect of intersymbol interference (ISI) on the performance of trellis codes that have been designed for AWGN channels [301], [337], [74], [89]. The coded performance on *static* ISI channels may be significantly degraded compared to that on ISI-free channels. Receivers for trellis-coded modulation on static ISI channels typically use a linear forward equalizer followed by a soft decision Viterbi decoder. For channels with severe ISI, a more appropriate approach is to use a decision feedback equalizer (DFE) in front of the TCM decoder to avoid the problems of noise enhancement. However, the feedback section of the DFE requires that decisions be available with zero delay. Since the zero-delay decisions are unreliable, the performance improvement by using the DFE is marginal [38]. It is possible that the performance can be improved if equalization and decoding is performed in a joint manner by using maximum likelihood sequence estimation (MLSE) or some other form of sequence estimator. However, the complexity of an MLSE receiver grows exponentially with the number of encoder states and the length of the channel vector.

Several studies have investigated the performance of TCM on *fading* ISI channels, by using trellis codes that have been designed for interleaved flat fading channels. Like the case of static ISI channels there are generally two approaches to the receiver design when using TCM on fading ISI channels. The first approach is to separate the equalizer and decoder so that interleaving can be used in a straight forward fashion to enhance the performance of the trellis codes, keeping in mind that the trellis codes have been designed for interleaved flat fading channels. Again, effective equalization of the channel is very difficult to achieve prior to decoding and, therefore, the coded performance suffers on fading ISI channels due to the failure of the equalizer.

As with static ISI channels, the second approach to the receiver design for TCM on fading ISI channels is to use joint equalization and decoding, such as MLSE. However, it is generally difficult to use interleaving in a straight forward fashion with joint equalization and decoding, because the deinterleaver destroys the channel memory that is needed to update the adaptive channel estimator; knowledge of the channel vector is needed to compute the branch metrics. Furthermore, if interleaving is not used on fading ISI channels, then almost all known trellis codes will fail, meaning that the equivalent uncoded system will outperform the trellis-coded system even if an MLSE receiver is employed. The problem in this case is not with the receiver structure, but rather with the code design itself.

The above problems associated with the use of TCM on fading ISI channels are not insurmountable, and the trick is to employ cleaver interleaving techniques that still allow joint equalization and decoding. If such receivers are used, then trellis codes that are designed for interleaved flat fading channels may be useful. In any case, it is clear that more effective receiver structures and/or trellis codes are needed so that applications such as cellular radio can benefit from bandwidth efficient TCM. Later in the chapter we will take steps in this direction.

The remainder of the chapter is organized as follows. Sections 7.1 and 7.2 introduce convolutional codes and trellis codes. This is followed by a consideration of the design and performance analysis of trellis codes for various types of channels that are found in mobile radio applications. These include the AWGN channels in Section 7.3, interleaved flat fading channels in Section 7.4, and non-interleaved fading ISI channels in Section 7.5. The evaluation of error probability upper bounds is important for performance prediction and Section 7.6 presents a technique for union bounding the error probability of TCM on a fading ISI channel; flat fading channels and static ISI channels can be treated as special cases. Finally, Section 7.7 compares a variety of receiver structures for TCM on fading ISI channels.

7.1 CONVOLUTIONAL CODES

7.1.1 Encoder Description

The encoder for a rate-$1/n$ binary convolutional code can be viewed as a **finite-state machine** (FSM) that consists of an ν-stage binary shift register with connections to n modulo-2 adders, and a multiplexer that converts the adder outputs to serial code words. The **constraint length** of a convolutional code is defined as the number of shifts through the FSM over which a single input data bit can affect the encoder output. For an encoder having a ν-stage shift register, the constraint length is equal to $K = \nu + 1$. A very simple rate-1/2, constraint length-3, binary convolutional encoder is shown in Fig. 7.1.

The above concept can be generalized to rate-k/n binary convolutional code by using k shift registers, n modulo-2 adders, along with input and output multiplexers. For a rate-k/n code, the k-bit information vector $\mathbf{a}_\ell = (a_\ell^{(1)}, \cdots, a_\ell^{(k)})$ is input to the encoder at epoch ℓ to generate the n-bit code vector $\mathbf{b}_\ell = (b_\ell^{(1)}, \cdots, b_\ell^{(n)})$. If K_i denotes the constraint length of the ith shift register, then the overall constraint length is defined as $K = \max_i K_i$. Fig. 7.2 shows a simple rate-2/3, constraint length-1 convolutional encoder.

A convolutional encoder can be described by the set of impulse responses, $\{\mathbf{g}_i^{(j)}\}$, where $\mathbf{g}_i^{(j)}$ is the jth output sequence $\mathbf{b}^{(j)}$ that results from the ith input sequence $\mathbf{a}^{(i)} = (1, 0, 0, 0, \ldots)$. The impulse responses can have a duration of at most K and have the form $\mathbf{g}_i^{(j)} = (g_{i,0}^{(j)}, g_{i,1}^{(j)}, \ldots, g_{i,K-1}^{(j)})$. Sometimes the $\{\mathbf{g}_i^{(j)}\}$

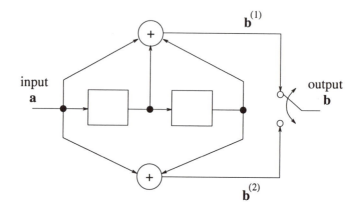

Figure 7.1 Binary convolutional encoder; $R_c = 1/2$, $K = 3$.

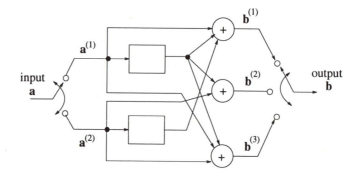

Figure 7.2 Binary convolutional encoder; $R_c = 2/3$, $K = 1$.

are called **generator sequences** . For the encoder in Fig. 7.1

$$
\begin{aligned}
\mathbf{g}^{(1)} &= (1,\ 1,\ 1) \\
\mathbf{g}^{(2)} &= (1,\ 0,\ 1)
\end{aligned}
\tag{7.1}
$$

and for the encoder in Fig. 7.2

$$
\begin{aligned}
\mathbf{g}_1^{(1)} &= (1,\ 1), \quad \mathbf{g}_1^{(2)} = (0,\ 1), \quad \mathbf{g}_1^{(3)} = (1,\ 1) \\
\mathbf{g}_2^{(1)} &= (0,\ 1), \quad \mathbf{g}_2^{(2)} = (1,\ 0), \quad \mathbf{g}_2^{(3)} = (1,\ 0)\ .
\end{aligned}
\tag{7.2}
$$

It follows that the jth output, $\mathbf{b}_i^{(j)}$, corresponding to the ith input sequence $\mathbf{a}^{(i)}$ is the discrete convolution $\mathbf{b}_i^{(j)} = \mathbf{a}^{(i)} \circledast \mathbf{g}_i^{(j)}$, where $\circledast$ denotes modulo-2 convolution. The time domain convolutions can be conveniently replaced by polynomial multiplications in a D-transform domain according to

$$\mathbf{b}_i^{(j)}(D) = \mathbf{a}^{(i)}(D)\mathbf{g}_i^{(j)}(D) \tag{7.3}$$

where

$$\mathbf{a}^{(i)}(D) = \sum k = 0^\infty a_{i,k}D^k \tag{7.4}$$

is the ith input data polynomial,

$$\mathbf{b}_i^{(j)}(D) = \sum_{k=0}^{\infty} b_{i,k}^{(j)} D^k \tag{7.5}$$

is the jth output polynomial corresponding to the ith input, and

$$\mathbf{g}_i^{(j)}(D) = \sum_{k=0}^{K-1} g_{i,k}^{(j)} D^k \tag{7.6}$$

is the associated **generator polynomial** . It follows that the jth output sequence is

$$\mathbf{b}^{(j)}(D) = \sum_{i=1}^{k} \mathbf{b}_i^{(j)}(D) = \sum_{i=1}^{k} \mathbf{a}^{(i)}(D)\mathbf{g}_i^{(j)}(D) \ . \tag{7.7}$$

After multiplexing the outputs, the final code word has the polynomial representation

$$\mathbf{b}(D) = \sum_{j=1}^{n} D^{j-1}\mathbf{b}^{(j)}(D^n) \ . \tag{7.8}$$

7.1.2   State and Trellis Diagrams, and Weight Enumeration

Since the convolutional encoder is a FSM, its operation can be described by a state-diagram and trellis diagram in a manner very similar to the treatment of ISI channels in Chapter 6. The state of the encoder is defined by the shift register contents. For a rate-k/n code, the ith shift register contains ν_i previous information bits. The state of the encoder at epoch ℓ is defined as

$$\boldsymbol{\sigma}_\ell = \left(a_{\ell-1}^{(1)}, \cdots, a_{\ell-\nu_1}^{(1)} ; \cdots ; a_{\ell-1}^{(k)}, \cdots, a_{\ell-\nu_m}^{(k)} \right) \ . \tag{7.9}$$

There are a total of $N_S = 2^{\nu_T}$ encoder states, where $\nu_T \triangleq \sum_{i=1}^{k} \nu_i$ is defined as the **total encoder memory** . For a rate-$1/n$ code, the encoder state at epoch ℓ is simply $\boldsymbol{\sigma}_\ell = (a_{\ell-1}, \cdots, a_{\ell-\nu})$.

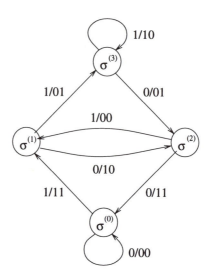

Figure 7.3   State diagram for the binary convolutional encoder in Fig. 7.1.

The state diagrams for codes shown in Figs. 7.1 and 7.2 are shown in 7.3 and 7.4, respectively. The states are labeled using the convention $\sigma^{(i)}, i = 0, \ldots, \nu_T - 1$, where $\sigma^{(i)}$ represents the encoder state $(c_0, \ldots, c_{\nu_T - 1})$ corresponding to the integer $i = \sum_{j=0}^{\nu_T - 1} c_j 2^j$. In general, for a rate-k/n code there are 2^k branches entering and leaving each state. The branches in the state diagram are labeled with the convention $\mathbf{a}/\mathbf{b} = (a^{(1)}, a^{(2)}, \ldots, a^{(k)})/(b^{(1)}, b^{(2)}, \ldots, b^{(n)})$. For example, the state transition $\sigma^{(1)} \to \sigma^{(3)}$ in Fig. 7.3 has the label 1/01. This means that the input $\mathbf{a} = 1$ to the encoder in Fig. 7.1 with state $\sigma^{(1)} = (01)$ gives the output $\mathbf{b} = (01)$ and leaves the encoder in state $\sigma^{(3)} = (11)$.

Convolutional codes are linear codes, meaning that the sum of any two code words is another code word and the all-zeroes sequence a code word. In order to achieve good error probability performance, convolutional codes are designed to maximize the minimum **Hamming distance** between code words, defined as the number of positions in which they differ. Because of the linearity property, the set of Hamming distances between the all-zeroes code word and the set of all code words is the same as the set of Hamming distances between any other code word and the set of all code words. This set of Hamming distances is called the **weight enumeration** of the code.

The weight enumeration and other distance properties of a convolutional code can be obtained from the state diagram. Consider, for example, the encoder in

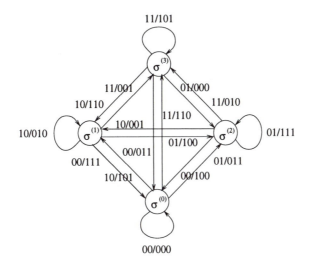

Figure 7.4   State diagram for the binary convolutional encoder in Fig. 7.2.

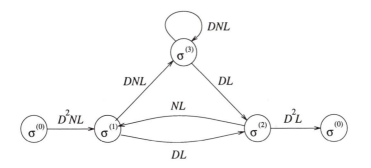

Figure 7.5   Modified state diagram for the binary convolutional encoder in Fig. 7.1.

Fig. 7.1 along with its state diagram in Fig. 7.3. Since the self-loop at the zero state $\sigma^{(0)}$ corresponds to the all-zeroes code word, we can split the zero state $\sigma^{(0)}$ into two nodes, representing the input and output of the state diagram. This leads to the **modified state diagram** shown in Fig. 7.5. The branches in the modified state diagram have labels of the form $D^i N^j L$, where i is the number of 1s in the encoder output sequence corresponding to a particular state transition, and j is the number of input 1s into the encoder for that transition. Every branch is labeled with the letter L, and the exponent of L is unity because each branch has length one.

The distance properties of a convolutional code can be obtained by computing the transfer function $T(D, N, L)$ of the modified state diagram. Any appropriate technique can be used to obtain the transfer function, such as Mason's formula [205]. For the example shown in Fig. 7.5, the transfer function is

$$
\begin{aligned}
T(D, N, L) &= \frac{D^5 N L^3}{1 - DNL(L + 1)} \\
&= D^5 L^3 N + D^6 N^2 L^4 (L + 1) + D^7 N^3 L^5 (L + 1)^2 \\
&\quad + \cdots + D^{k+5} N^{k+1} L^{k+3} (L + 1)^k + \cdots \quad (7.10)
\end{aligned}
$$

The term $D^{k+5} N^{k+1} L^{k+3} (L + 1)^k$ appearing in the transfer function means there are 2^k paths at Hamming distance $k + 5$ from the all-zeroes path, caused by $k + 1$ input ones. Of these 2^k paths, $\binom{k}{n}$ have length $k + n + 3$.

Sometimes the transfer function can be simplified if we are only interested in extracting certain distance properties of the convolutional code. For example, the weight enumeration of the code can be obtained by setting $N = 1$ and $L = 1$ in the transfer function. For the particular transfer function in (7.10) this leads to

$$
\begin{aligned}
T(D) &= \frac{D^5}{1 - 2D} \\
&= D^5 + 2D^6 + 4D^7 + \cdots + 2^k D^{5+k} + \cdots \quad (7.11)
\end{aligned}
$$

meaning that there are 2^k code words at Hamming distance $5 + k$ from the all-zeroes code word. Notice that there does not exist any non-zero code word with a Hamming distance less than 5 from the all-zeroes code word. This fact can also be seen by inspecting the trellis diagram in Fig. 7.6. A solid line in the trellis diagram corresponds to an input 1 and a dashed line corresponds to an input 0. The branches in the trellis diagram are labeled with the encoder output bits that correspond to the various state transitions.

The minimum Hamming distance between any two code words is called the **minimum free distance** of the code, denoted by d_{free}. For the code in Fig. 7.1, we see that $d_{\text{free}} = 5$ from (7.11). Most convolutional codes are designed to have the largest possible d_{free} for a given code rate and total encoder memory. Tabulation of convolutional codes that are optimal in this sense can be found in many references, e.g., Proakis [256], Lin and Costello [189], and Clark and Cain [51].

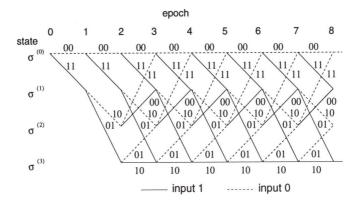

Figure 7.6 Trellis diagram for the binary convolutional encoder in Fig. 7.1.

7.2 TRELLIS CODED MODULATION

7.2.1 Encoder Description

Ungerboeck showed that a coding gain can be achieved without sacrificing data rate or bandwidth by using a rate-$m/(m+r)$ convolutional encoder, and mapping the coded bits onto signal points $\{x_k\}$ through a technique called **mapping by set partitioning** [312]. This combination of coding and modulation, called trellis coded modulation (TCM), has three basic features;

1. An expanded signal constellation is used that is larger than the one necessary for uncoded modulation at the same data rate. The additional signal points allow redundancy to be inserted without sacrificing data rate or bandwidth.

2. The expanded signal constellation is partitioned such that the intra-subset minimum squared Euclidean distance is maximized for each step in the partition chain.

3. Convolutional encoding and signal mapping is used so that only certain sequences of signal points are allowed.

Fig. 7.7 shows the basic encoder structure for Ungerboeck's trellis codes. The n-bit information vector $\mathbf{a}_k = (a_k^{(1)}, \cdots, a_k^{(n)})$ is transmitted at epoch k. At each epoch, $m \leq n$ information bits are encoded into $m+r$ code bits by using a rate-$m/(m+r)$ linear convolutional encoder. The $m+r$ code bits select one of 2^{m+r}

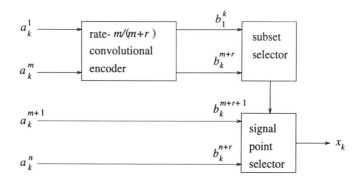

Figure 7.7   Ungerboeck trellis encoder.

subsets of a 2^{n+r}-point signal constellation. The uncoded $n-m$ information bits select one of the 2^{n-m} signal points within the selected subset. This principle is best explained by example, and Fig. 7.8 shows a 4-state 8-PSK Ungerboeck trellis code. The equivalent uncoded system is 4-PSK which has a bit rate of 2 bits/symbol. The 4-state 8-PSK code uses a rate-1/2 convolutional code along with one uncoded bit to select signal points in an expanded 8-PSK signal constellation. Note that the overall rate is still 2 bits/symbol. Fig. 7.9 shows an 8-state 8-PSK Ungerboeck trellis code. The equivalent uncoded system is again 4-PSK with 2 bits/symbol. The 8-state 8-PSK code uses a rate-2/3 convolutional code to select one of the points in an expanded 8-PSK signal constellation so that the overall rate is again 2 bits/symbol.

7.2.2   Mapping by Set Partitioning

The critical step in the design of Ungerboeck's codes is the method of mapping the outputs of the convolutional encoder to points in the expanded signal constellation. Fig. 7.10 shows how the 8-PSK signal constellation is partitioned into subsets such that the intra-subset minimum squared Euclidean distance is maximized for each step in the partition chain. In the 8-PSK signal constellation there are 8 signal points equally spaced around a circle of unit radius. Notice that the minimum Euclidean distance between signal points in the 8-PSK signal constellation is $\Delta_0 = 0.765$, while the minimum Euclidean distances between signal points in the first and second level partitions are $\Delta_1 = \sqrt{2}$ and $\Delta_2 = 2$, respectively. The minimum Euclidean distance increases at each level of partitioning.

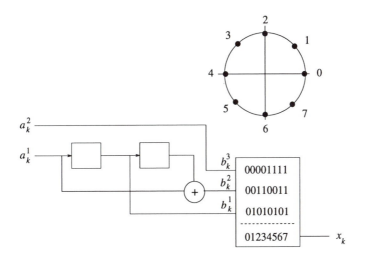

Figure 7.8 Encoder and signal mapping for the 4-state 8-PSK Ungerboeck trellis code.

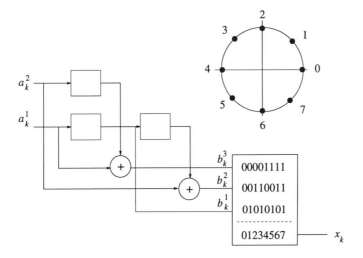

Figure 7.9 Encoder and signal mapping for the 8-state 8-PSK Ungerboeck trellis code.

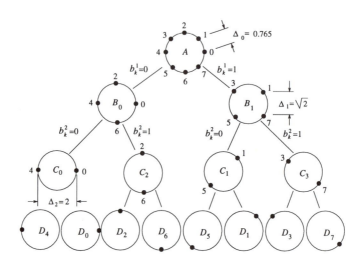

Figure 7.10 Set partitioning for an 8-PSK signal constellation.

The advantages of using TCM can most easily be seen by considering the trellis diagram. For both the 4-state and 8-state 8-PSK trellis codes the equivalent uncoded system is 4-PSK. The trellis diagram for uncoded 4-PSK is shown in Fig. 7.11. The trellis only has one state and there are 4 **parallel transitions** between the states. The subsets D_0, D_2, D_4, and D_6 are used as the signal points. The label D_0, D_2, D_4, D_6 means that the branches in the trellis diagram are labeled from top to bottom with signal points taken from the sets D_0, D_2, D_4, D_6. The minimum Euclidean distance between any two paths through the trellis is $d_{\min} = \sqrt{2}$.

The trellis diagram for the 4-state 8-PSK code is shown in Fig. 7.12. Each branch in the 4-state trellis is labeled with one of the four subsets C_0, C_1, C_2, and C_3. Again, the label $C_i C_j$ associated with a state means the that the branches in the trellis diagram originating from that state are labeled from top to bottom with the subsets C_i and C_j. As shown in Fig. 7.10, each subset C_i contains two signal points. Thus, each branch in the trellis diagram actually contains two parallel transitions. For example branches with the label C_0 have two parallel transitions that are labeled with the signal points 0 and 4. For the 4-state 8-PSK code, it is possible that two coded sequences could differ by just a single parallel transition and, hence, their Euclidean distance is $d = 2$. Also, any two signal paths that diverge from a state and remerge with the same state after more that one transition have a minimum Euclidean distance of $d = \sqrt{\Delta_1^2 + \Delta_0^2 + \Delta_1^2} = 2.141$. For example, the closest non-parallel code

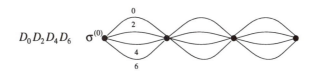

Figure 7.11 Trellis diagram for uncoded 4-PSK.

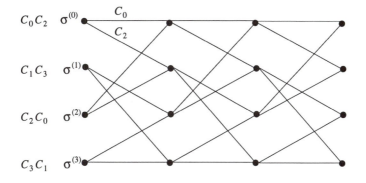

Figure 7.12 Trellis diagram for 4-state 8-PSK Ungerboeck trellis code.

sequence to the all-zeroes sequence $\mathbf{x} = (0, 0, 0)$ is the sequence $\mathbf{x} = (2, 1, 2)$ at distance $d = 2.141$. Hence, the minimum Euclidean distance of the code over all parallel and non-parallel pairs of sequences for the 4-state 8-PSK code is $d_{\min} = 2$.

At high signal-to-noise ratio (SNR), the bit error rate performance on an AWGN channel is dominated by the minimum Euclidean distance error events. The pairwise error probability between two coded sequences $\mathbf{x}$ and $\hat{\mathbf{x}}$ at distance $d_{\min}$ is

$$\mathrm{P_r}(\mathbf{x} \rightarrow \hat{\mathbf{x}}) = Q\left(\sqrt{\frac{d_{\min}^2}{4N_o}}\right) \tag{7.12}$$

The **asymptotic coding gain** is defined by [28]

$$G_a = 10\log_{10} \frac{(d_{\min,\text{coded}}^2/E_{\text{av,coded}})}{(d_{\min,\text{uncoded}}^2/E_{\text{av,uncoded}})} \quad \text{dB} \tag{7.13}$$

where E_{av} is the average energy per symbol in the signal constellation. For the 4-state 8-PSK code, the asymptotic coding gain is $G_a = 3$ dB.

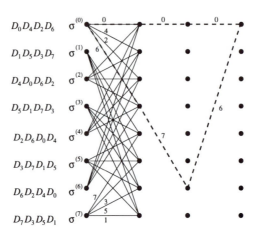

Figure 7.13   Trellis diagram for 8-state 8-PSK Ungerboeck trellis code. The dashed lines show two minimum distance paths.

The concept of mapping by set partitioning was developed by Ungerboeck as a method for maximizing the minimum Euclidean distance of a code so as to optimize the performance on an AWGN channel. Ungerboeck's construction of the optimum 4-state 8-PSK code was based on the following heuristic rules [313];

1. Parallel transitions (when they occur) are assigned signal points having the maximum Euclidean distance between them.

2. The transition starting or ending in any state is assigned the subsets (C_0, C_2) or (C_1, C_3) which have a maximum distance between them.

3. All signal points are used in the trellis diagram with equal frequency.

It is clear that the performance of the 4-state 8-PSK code is limited by the parallel transitions. Larger asymptotic coding gains can be obtained by introducing more code states so that the parallel transitions can be eliminated. For example, the above design rules can be applied to the 8-state 8-PSK code to obtain the code trellis shown in Fig. 7.13. In this case, the minimum Euclidean distance is $d_{\min} = \sqrt{\Delta_1^2 + \Delta_0^2 + \Delta_1^2} = 2.141$. This yields an asymptotic coding gain of 3.6 dB over uncoded 4-PSK.

7.3 CODED PERFORMANCE ON AWGN CHANNELS

Viterbi originally exploited the trellis structure of convolutional codes and developed the Viterbi algorithm for ML decoding of convolutional codes [321]. Given the similarity between the trellis structures of ISI channels, convolutional codes, and trellis codes (e.g., compare Figs. 6.15, 7.6 and 7.12), it is not surprising that the union bounding techniques that were developed to evaluate the error probability of digital signaling on ISI channels with an MLSE receiver in Section 6.6 can also be applied, with some modification, for evaluating the error probability of convolutional and trellis codes an MLSE receiver.

To develop the union bound, let $\mathbf{a} = \{\mathbf{a}_k\}$ denote the transmitted information sequence. For any other sequence $\hat{\mathbf{a}} \neq \mathbf{a}$, define the corresponding error sequence as $\mathbf{e} = \{e_k\} = \mathbf{a} \oplus \hat{\mathbf{a}}$, where $\oplus$ denotes modulo-2 addition. Since the bit error probability at epoch k_1 is of interest, $e_{k_1} \neq 0$ for all error sequences. An error event occurs between j_1 and j_2 of length $j_2 - j_1$, if $\sigma_{j_1} = \hat{\sigma}_{j_1}$ and $\sigma_{j_2} = \hat{\sigma}_{j_2}$, but $\sigma_k \neq \hat{\sigma}_k$ for $j_1 < k < j_2$, where $j_1 \leq k_1 < j_2$, and $\sigma = \{\sigma_k\}$ and $\hat{\sigma} = \{\hat{\sigma}_k\}$ are the system state sequences associated with $\mathbf{a}$ and $\hat{\mathbf{a}}$, respectively. Let $\mathbf{E}$ be the set of error sequences corresponding to all possible error events at epoch k_1. Then, the average bit error probability is bounded by

$$P_b \leq \frac{1}{n} \sum_{\mathbf{e} \in \mathbf{E}} w_b(\mathbf{e}) \sum_{\mathbf{a}} \mathrm{P_r}\left(\mathbf{a}\right) \mathrm{P_r}\left(\Gamma(\mathbf{a} \oplus \mathbf{e}) \geq \Gamma(\mathbf{a}) \middle| \mathbf{a}\right) \qquad (7.14)$$

where $\Gamma(\mathbf{a})$ is the path metric of $\mathbf{a}$, and $w_b(\mathbf{e})$ is the number of bit errors associated with $\mathbf{e}$. The factor $1/n$ appears in front of the first summation, because n information bits are transmitted per epoch (or per branch in the trellis diagram). The second summation is over all possible information sequences, because each sequence $\mathbf{a}$ can have $\mathbf{e}$ as the error sequence. This is necessary for trellis codes because the signal mapping and, hence, the codes are nonlinear.

Another way of writing the bound on the bit error probability in (7.14) is

$$P_b \leq \sum_{\mathbf{x} \in \mathbf{C}} \sum_{\hat{\mathbf{x}} \in \mathbf{C}} w_b(\mathbf{x}, \hat{\mathbf{x}}) \mathrm{P_r}(\mathbf{x}) \mathrm{P_r}(\mathbf{x} \rightarrow \hat{\mathbf{x}}) \qquad (7.15)$$

where $\mathbf{C}$ is the set of all coded symbol sequences, $w_b(\mathbf{x}, \hat{\mathbf{x}})$ is the number of bit errors that occur when the sequence $\mathbf{x}$ is transmitted and the sequence $\hat{\mathbf{x}} \neq \mathbf{x}$ is chosen by the decoder, $\mathrm{P_r}(\mathbf{x})$ is the *a priori* probability of transmitting $\mathbf{x}$, and $\mathrm{P_r}(\mathbf{x} \rightarrow \hat{\mathbf{x}})$ is the pairwise error probability.

For convolutional codes the upper bound in (7.14) simplifies because the codes
are linear, meaning that the sum of any two code words is another code word
and that all-zeroes sequence is a code word [189]. Because of this property, we
can assume that $\mathbf{a} = \mathbf{0}$, so that the union bound becomes

$$P_b \leq \frac{1}{k} \sum_{\mathbf{e} \in \mathbf{E}} w_b(\mathbf{e}) \, \mathrm{P_r} \left(\Gamma(\mathbf{e}) \geq \Gamma(\mathbf{0}) \right) \tag{7.16}$$

Note that we divide by k rather than n in front of the summation, because a
convolutional code transmits k bits per epoch whereas a trellis code transmits
n bits per epoch.

7.3.1 Evaluating the Upper Bound

Evaluation of the error probability upper bound for trellis codes is complicated
by the fact that trellis codes are nonlinear and, therefore, all possible correct
sequences must be considered when computing the upper bound. We will defer
treatment of the coded error probability upper bound for trellis codes until
Section 7.6, where we will consider the more general case of TCM on a fading
ISI channel. In this section we will show how the error probability upper bound
can be computed for convolutional codes with Viterbi decoding.

For convolutional codes the set $\mathbf{E}$ in (7.16) consists of all sequences that begin
and end at the zero-state in the state diagram. The enumeration of these sequences (or code words) along with their associated Hamming distances, information weights, and lengths, was obtained earlier by by computing the transfer
function, $T(D, N, L)$, of the augmented state diagram. When a particular incorrect path through the trellis is selected over the all-zeroes path at a given
node in the trellis, the corresponding number of bits errors, $w_b(\mathbf{e})$, is given by
the exponent of N in the transfer function. Multiplying $w_b(\mathbf{e})$ by the pairwise
error probability $\mathrm{P_r} \left(\Gamma(\mathbf{e}) \geq \Gamma(\mathbf{0}) \right)$ for that path and dividing by the number
of input bits per branch, n, gives the bit error rate for that path. Summing
over the set of all possible incorrect sequences E yields a union bound on the
bit error probability.

In general, the transfer function $T(D, N)$ for a convolutional code has the form

$$T(D, N) = \sum_{d=d_{\text{free}}}^{\infty} a_d D^d N^{f(d)} \tag{7.17}$$

where $f(d)$ is the exponent of N as a function of d. For the example in (7.10),
$a_d = 2^{d-5}$ and $f(d) = d - 4$. Differentiating $T(D, N)$ with respect to N and

setting $N = 1$ gives

$$\frac{dT(D, N)}{dN}\Big|_{N=1} = \sum_{d=d_{\text{free}}}^{\infty} a_d f(d) D^d \tag{7.18}$$

Once again, for the example in (7.10) this leads to

$$\frac{dT(D, N)}{dN}\Big|_{N=1} = \sum_{d=d_{\text{free}}}^{\infty} 2^{d-5}(d-4)D^d \ . \tag{7.19}$$

The pairwise error probability in (7.16) depends on the type of modulation, detection, and decoding that is employed. The code bits are mapped onto symbols taken from a signal constellation, and transmitted over the channel. The sampled output of the receiver matched filter at epoch k is

$$y_k = x_k + \eta_k \tag{7.20}$$

where x_k is one of the M *low-pass* points in the signal constellation and η_k is a zero-mean complex-valued Gaussian random variable with variance N_o. For convolutional codes, two types of decoding can be used, **hard decision decoding** and **soft decision decoding** . Soft decision decoders do not make symbol by symbol decisions on the received symbols, rather, the decoder operates directly on the sequence of matched filter outputs **y**. For an AWGN channel, the MLSE receiver searches for the symbol sequence $\hat{\mathbf{x}}$ that is closest in Euclidean distance to the received sequence **y**. Following the same argument in Section 5.2, the MLSE receiver decides in favor of the sequence $\hat{\mathbf{x}}$ that maximizes the metric

$$\mu(\mathbf{x}) = -\|\mathbf{y} - \mathbf{x}\|^2 \ . \tag{7.21}$$

The sequence $\hat{\mathbf{x}}$ corresponds to a unique sequence $\hat{\mathbf{a}}$ that is the final estimate of the transmitted information sequence **a**.

In general, the pairwise error probability for an AWGN channel that is associated with an error event of length ℓ beginning at epoch k_1 is

$$P_2(\ell) = Q\left(\sqrt{\frac{\Delta^2}{4N_o}}\right) \tag{7.22}$$

where

$$\Delta^2 = \sum_{k=k_1}^{k_1+\ell+1} \delta_k^2 \tag{7.23}$$

$$\delta_k^2 = |x_k - \hat{x}_k|^2 \tag{7.24}$$

and $\mathbf{x} = \{x_k\}$ and $\hat{\mathbf{x}} = \{\hat{x}_k\}$ are the symbol sequences corresponding to the information sequences $\mathbf{a}$ and $\hat{\mathbf{a}}$, respectively. The parameter δ_k^2 is the **squared branch Euclidean distance** associated with branch k, and Δ^2 is the **squared path Euclidean distance** associated with the error event. Clearly, the pairwise error probability depends on the particular mapping between the encoder output bits and the points in the signal constellation. Suppose for example that code bits are mapped onto a BPSK signal constellation. Then the pairwise error probability between the two code words $\mathbf{b}$ and $\hat{\mathbf{b}}$ that differ in d positions is

$$P_2(d) = Q(\sqrt{2R_c d\gamma_b}) \tag{7.25}$$

where γ_b is the received bit energy-to-noise ratio[1]. Therefore, the union bound on bit error probability becomes

$$P_b \leq \frac{1}{k} \sum_{d=d_{\text{free}}}^{\infty} a_d f(d) P_2(d) \ . \tag{7.26}$$

Note that we have explicitly shown the pairwise error probability to be a function of the Hamming distance between the code words in (7.25). However, it is very important to realize that this property does not apply to all convolutionally encoded systems. For example, suppose that the outputs of the rate-2/3 convolutional encoder in Fig. 7.2 are mapped onto symbols from an 8-PSK signal constellation. In this case, the pairwise error probability depends not only on the Hamming distance between code words, but also upon the particular mapping between the 8-PSK symbols and the encoder outputs.

Hard decision decoders make symbol by symbol decisions on the received sequence of matched filter outputs $\mathbf{y} = \{y_k\}$ to yield the received symbol sequence $\tilde{\mathbf{x}}$. A **minimum distance decoder** decides in favor of the symbol sequence $\hat{\mathbf{x}}$ that is closest in Hamming distance to the received symbol sequence $\tilde{\mathbf{x}}$. Again, the pairwise error probability depends on the particular mapping between the encoder outputs and the points in the signal constellation. If BPSK signaling is used for example, then the pairwise error probability between two code words $\mathbf{b}$ and $\hat{\mathbf{b}}$ at Hamming distance d is

$$P_2(d) = \begin{cases} \sum_{k=(d+1)/2}^{d} \binom{d}{k} p^k (1-p)^{d-k} \ , & d \text{ odd} \\ \sum_{k=d/2+1}^{d} \binom{d}{k} p^k (1-p)^{d-k} + \frac{1}{2}\binom{d}{d/2} p^{d/2}(1-p)^{d/2} \ , & d \text{ even} \end{cases} \tag{7.27}$$

where

$$p = Q(\sqrt{2R_c\gamma_b}) \tag{7.28}$$

---

[1] The received symbol energy-to-noise ratio is $\gamma_s = R_c\gamma_b$

is the probability of symbol error. Once again, the pairwise error probability for BPSK is a function of the Hamming distance between the code words.

The union bound in (7.26) can be simplified by imposing a Chernoff bound (see Appendix A) on the pairwise error probability. First consider the case of soft decision decoding. Suppose that sequence $\mathbf{x}$ is transmitted and $\mathbf{y}$ is the received sequence. Then the pairwise error probability between sequences $\mathbf{x}$ and $\hat{\mathbf{x}}$ with an ML receiver can be Chernoff bounded by

$$
\begin{aligned}
\mathrm{P_r}(\mathbf{x} \to \hat{\mathbf{x}}) &= \mathrm{P_r}\left(\|\mathbf{y} - \hat{\mathbf{x}}\|^2 < \|\mathbf{y} - \mathbf{x}\|^2\right) \\
&\leq \mathrm{E}\left[\exp\left\{\lambda\left(\|\mathbf{y} - \mathbf{x}\|^2 - \|\mathbf{y} - \hat{\mathbf{x}}\|^2\right)\right\}\big| \mathbf{x}\right] \quad . \quad (7.29)
\end{aligned}
$$

Substituting $\mathbf{y} = \mathbf{x} + \boldsymbol{\eta}$, taking the expectation over the Gaussian random vector $\boldsymbol{\eta}$, and simplifying gives

$$
\mathrm{P_r}(\mathbf{x} \to \hat{\mathbf{x}}) \leq \exp\left\{-\lambda\|\mathbf{x} - \hat{\mathbf{x}}\|^2(1 - \lambda 2N_o)\right\} \quad . \quad (7.30)
$$

This bound is optimized with $\lambda^* = 1/(4N_o)$ yielding

$$
\mathrm{P_r}(\mathbf{x} \to \hat{\mathbf{x}}) \leq \exp\left\{\frac{-\|\mathbf{x} - \hat{\mathbf{x}}\|^2}{8N_o}\right\} \quad . \quad (7.31)
$$

Finally, if the signal constellation is *normalized* so that $\mathrm{E}[|x_i|^2] = 1$, then the Chernoff bound can be written in the form

$$
\mathrm{P_r}(\mathbf{x} \to \hat{\mathbf{x}}) \leq \exp\left\{-\frac{\gamma_s}{4}\|\mathbf{x} - \hat{\mathbf{x}}\|^2\right\} \quad (7.32)
$$

where γ_s is the received symbol energy-to-noise ratio.

For the case of BPSK signaling on an AWGN channel, the Chernoff bound on the pairwise error probability becomes

$$
P_2(d) \leq e^{-R_c d \gamma_b} \quad . \quad (7.33)
$$

Likewise, if BPSK signaling is used with hard decision decoding, then the pairwise error probability has the Chernoff bound

$$
P_2(d) \leq [4p(1 - p)]^{d/2} \quad . \quad (7.34)
$$

Notice that the Hamming distance d appears in the exponent of the pairwise error probability. The resulting upper bound on bit error probability is called a **union-Chernoff bound** and has the simple form

$$
P_b \leq \frac{1}{k} \frac{dT(D, N)}{dN}\bigg|_{N=1, D=Z} \quad (7.35)
$$

where

$$Z = \begin{cases} \sqrt{4p(1-p)}\ , & \text{hard decision decoding} \\ e^{-R_c\gamma_b}\ , & \text{soft decision decoding} \end{cases} \qquad . \qquad (7.36)$$

At high SNR, the performance is dominated by the error events with minimum Hamming distance. Since the minimum distance error events are not necessarily mutually exclusive, the bit error probability at high SNR is approximately

$$\begin{aligned} P_b &\approx \frac{1}{k} a_{d_{\text{free}}} f(d_{\text{free}}) P_2(d_{\text{free}}) \\ &\leq \frac{1}{k} a_{d_{\text{free}}} f(d_{\text{free}}) Z^{d_{\text{free}}}\ . \end{aligned} \qquad (7.37)$$

The above procedure for upper bounding the error probability is called the **transfer function approach**, because it relies upon the transfer function of the state diagram. The transfer function approach, however, has its limitations. First, if the number of encoder states is large, then obtaining the transfer function $T(D, N)$ quickly becomes intractable. Second, if the pairwise error probability is not just a function of the Hamming distance between allowable code sequences, then the branch labeling in the augmented state diagram must be done differently and the Chernoff bound cannot be employed. These problems can be overcome by a using a different approach to compute the upper bound, such as the stack algorithm presented in Section 7.6.

7.4 CODED PERFORMANCE ON INTERLEAVED FLAT FADING CHANNELS

Fig. 7.14 is a block diagram of a coded communication system operating on an interleaved flat fading channel. The information sequence **a** is encoded and mapped onto a signal set to generate the symbol sequence **x** by using either convolutional coding or trellis coded modulation. The symbol sequence is then interleaved (or scrambled), and the resulting sequence $\check{\mathbf{x}}$ is filtered for spectral shaping and transmitted over the channel. The receiver employs a filter that is matched to the transmitted pulse and symbol- or T-spaced samples are taken at the output of the matched filter. With hard decision decoding, these outputs are applied to a decision device and deinterleaved to yield the received code sequence $\tilde{\mathbf{b}}$. With soft decision decoding, the received samples $\check{\mathbf{y}}$ are

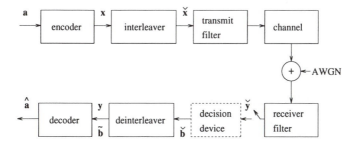

Figure 7.14 Digital communication on an interleaved flat fading channel.

deinterleaved to generate the sequence **y** which is then input to the Viterbi decoder.

The channel is characterized by flat Ricean fading, so that the sampled output of the matched filter is

$$\check{y}_k = \check{\alpha}_k \check{x}_k + \check{\eta}_k \tag{7.38}$$

where the pdf of $\check{\alpha}_k$ is

$$p_{\check{\alpha}}(x) = \frac{2x(1+K)}{\Omega_p} \exp\left\{-K - \frac{(K+1)x^2}{\Omega_p}\right\} I_0\left(2x\sqrt{\frac{K(K+1)}{\Omega_p}}\right), \quad x \geq 0 \tag{7.39}$$

where $\Omega_p = \mathrm{E}[\check{\alpha}_k^2]$. The $\check{\eta}_k$ are independent zero-mean complex Gaussian random variables with variance N_o.

The interleaver serves to reduce the correlation between the fades experienced by successive source symbols that are transmitted over the channel. There are a variety of interleaver structures [51], and the interleaver that we consider here is a **block interleaver** . A block interleaver can be regarded as a buffer with J rows and M columns, where J represents the interleaving depth and M represents the interleaving span. The length of the interleaver is JM symbols. Source symbols are fed into the buffer in successive rows and transmitted out of the buffer in columns. The deinterleaver performs the reverse operation. In practise, the interleaver depth J should be chosen so that successive source symbols, which are actually transmitted J symbol durations apart, are independently faded. In a 2-D isotropic scattering environment, the fades experienced at two different locations separated by a half wavelength are approximately uncorrelated. If the signals are received by a mobile station (MS) traveling with a speed of v km/h relative to the base station (BS), then the spatial distance associated with one symbol duration T is equal to vT. Therefore, we should have $JvT > 0.5\lambda$, where λ is the carrier wavelength. For example, a carrier frequency of 900 MHz yields

$$J > 0.6\frac{R}{v} \text{ (symbols)} \qquad (7.40)$$

where R is the signaling rate (symbols/s), v is the vehicle speed (km/h) and J is in units of symbols. Observe that the required interleaving depth is inversely proportional to the speed and, therefore, slow moving MSs require large interleaving depths. For $R = 24$ ks/s and $v = 30$ km/h, $J = 478$ symbols.

The basic objective of any interleaver is to at least separate any $L_D + 1$ successive source symbols as far as possible, where L_D is the decoding depth. Hence, M should be at least equal to $L_D + 1$ yielding an interleaving delay of

$$t_d = 0.6 \, \frac{(L_D + 1)}{v} \text{ (seconds)} \; . \qquad (7.41)$$

For example, with $L_D = 13$ symbols and a MS speed of 30 km/h, the interleaving delay should be at least 280 ms. This delay is quite large, especially for voice applications, and the problem is exasperated by lower MS speeds. One possible solution is to design codes that minimize the decoding depth, L_D. The other solution is to use better interleaving techniques so that the effective interleaving length is longer than the actual interleaving length JM.

For analytical purposes, an infinite interleaving depth is often assumed so that the deinterleaved sequence $\{\alpha_k\}$ is a sequence of independent random variables. In this case the conditional density of $\mathbf{y}$ has the product from

$$p(\mathbf{y}|\boldsymbol{\alpha} \cdot \mathbf{x}) = \prod_k p(y_k|\alpha_k x_k) \; . \qquad (7.42)$$

Suppose that sequence $\mathbf{x}$ is transmitted and the vector $\mathbf{y} = \boldsymbol{\alpha} \cdot \mathbf{x} + \boldsymbol{\eta}$ is received. An ML receiver having perfect knowledge of $\boldsymbol{\alpha}$ chooses the sequence $\hat{\mathbf{x}}$ that maximizes the metric

$$\mu(\hat{\mathbf{x}}) = -\|\mathbf{y} - \boldsymbol{\alpha} \cdot \hat{\mathbf{x}}\|^2 \; . \qquad (7.43)$$

The pairwise error probability between the sequences $\mathbf{x}$ and $\hat{\mathbf{x}}$ has the Chernoff bound

$$\mathrm{P_r}(\mathbf{x} \rightarrow \hat{\mathbf{x}}) \leq \exp\left\{ -\frac{\|\boldsymbol{\alpha} \cdot (\mathbf{x} - \hat{\mathbf{x}})\|^2}{4N_o} \right\} \; . \qquad (7.44)$$

Once again, if we assume the normalization $\mathrm{E}[|x_k|^2] = 1$ then the Chernoff bound becomes

$$\mathrm{P_r}(\mathbf{x} \rightarrow \hat{\mathbf{x}}) \leq \exp\left\{ -\frac{E}{4N_o} \|\boldsymbol{\alpha} \cdot (\mathbf{x} - \hat{\mathbf{x}})\|^2 \right\} \; . \qquad (7.45)$$

Averaging (7.45) over the probability density function in (7.39) gives [69]

$$P(\mathbf{x} \to \hat{\mathbf{x}}) \leq \prod_{i \in A} \frac{1 + K}{1 + K + \frac{\bar{\gamma}_s}{4}|x_i - \hat{x}_i|^2} \exp \left\{ -\frac{K\frac{\bar{\gamma}_s}{4}|x_i - \hat{x}_i|^2}{1 + K + \frac{\bar{\gamma}_s}{4}|x_i - \hat{x}_i|^2} \right\} \quad (7.46)$$

where $\bar{\gamma}_s = \mathrm{E}[\alpha^2]E/N_o$ is the average received symbol energy-to-noise ratio, and $A = \{i | x_i \neq \hat{x}_i\}$. At sufficiently high $\bar{\gamma}_s$, (7.46) simplifies to

$$\mathrm{P_r}(\mathbf{x} \to \hat{\mathbf{x}}) \leq \prod_{i \in A} \frac{4(1 + K)}{\bar{\gamma}_s |x_i - \hat{x}_i|^2} e^{-K} . \quad (7.47)$$

It follows that the bound in (7.14) will be dominated by the error event path having the smallest number of elements in set A. Divsalar and Simon [69] [70] called this path the **shortest error event path** and defined its length as $L_{\min}$. Based on previous arguments, the bit error probability can be approximated as

$$P_b \simeq C \left(\frac{(1 + K)e^{-K}}{\bar{\gamma}_s} \right)^{L_{\min}} \quad \bar{\gamma}_s \gg K \quad (7.48)$$

where C is a constant that depends on the distance structure of the code. Observe that P_b varies inversely with $(\bar{\gamma}_s)^{L_{\min}}$, yielding a diversity effect of order $L_{\min}$. Wei [329] called $L_{\min}$ the **minimum built-in time diversity** (MTD). The MTD dominates the performance of TCM on an interleaved flat fading channel, and the maximization of the MTD is the major design criterion for TCM on interleaved flat fading channels.

The pairwise error probability in (7.46) can be written in the form

$$\mathrm{P_r}(\mathbf{x} \to \hat{\mathbf{x}}) \leq e^{-\frac{\bar{\gamma}_s}{4}d^2} \quad (7.49)$$

where

$$\begin{aligned} d^2 &= \sum_{i \in A} \frac{|x_i - \hat{x}_i|^2 K}{1 + K + \frac{\bar{\gamma}_s}{4}|x_i - \hat{x}_i|^2} + \left(\frac{\bar{\gamma}_s}{4} \right)^{-1} \ln \left(\frac{1 + K + \frac{\bar{\gamma}_s}{4}|x_i - \hat{x}_i|^2}{1 + K} \right) \\ &= \sum_{i \in A} d_{1i}^2 + d_{2i}^2 \end{aligned} \quad (7.50)$$

Two special cases are associated with (7.50), $K = \infty$ and $K = 0$. For $K = \infty$ (no fading),

$$d_{1i}^2 = |x_i - \hat{x}_i|^2, \qquad d_{2i}^2 = 0 \quad (7.51)$$

and, therefore, d^2 becomes the sum of the squared Euclidean distances over the error event path. Maximizing d^2 under this condition is the TCM design criterion for AWGN channels.

For $K = 0$ (Rayleigh fading),

$$d_{1i}^2 = 0, \qquad d_{2i}^2 = \left(\frac{\bar{\gamma}_s}{4}\right)^{-1} \ln\left(1 + \frac{\bar{\gamma}_s}{4}|x_i - \hat{x}_i|^2\right) \ . \tag{7.52}$$

For reasonably large SNR, d^2 is the sum of the logarithms of the squared Euclidean distances, each weighted by $\bar{\gamma}_s$. In this case, the pairwise error probability is given by

$$\mathrm{P_r}(\mathbf{x} \rightarrow \hat{\mathbf{x}}) \leq \left(\prod_{i \in A} \frac{\bar{\gamma}_s}{4}|x_i - \hat{x}_i|^2\right)^{-1} \tag{7.53}$$

which is inversely proportional to the product of the squared Euclidean distances along the error event path. The **minimum product squared Euclidean distance** (MPSD) between any two valid sequences,

$$\min_{\mathbf{x},\hat{\mathbf{x}}} \prod_{i \in A} |x_i - \hat{x}_i|^2 \tag{7.54}$$

is another design parameter for Rayleigh fading channels. For values of K between 0 and ∞, the equivalent squared Euclidean distance of (7.50) becomes a mixture of the two limiting cases given above.

If interleaving is not used, then the assumption that the fading is independent from symbol to symbol is no longer valid. If the fading is slow enough to be considered constant over the duration of the minimum distance error event path, then for coherent detection with a Gaussian metric the bit error probability at high SNR is, approximately,

$$P_b \simeq C_1 \mathrm{E}\left[\exp\left\{-\frac{\gamma_s}{4}d_{\min}^2\right\}\right] \tag{7.55}$$

where C_1 is a constant, $\gamma_s = \alpha^2 E/N_o$ is the received symbol energy-to-noise ratio, $d_{\min}^2$ is the minimum Euclidean distance of the code, and the averaging is over the density in (7.39). Performing this average gives

$$P_b \simeq C_1 \frac{1 + K}{1 + K + d_{\min}^2 \frac{\bar{\gamma}_s}{4}|x_i - \hat{x}_i|^2} \exp\left\{-\frac{K d_{\min}^2 \frac{\bar{\gamma}_s}{4}|x_i - \hat{x}_i|^2}{1 + K + d_{\min}^2 \frac{\bar{\gamma}_s}{4}|x_i - \hat{x}_i|^2}\right\} \tag{7.56}$$

which can be approximated at large $\bar{\gamma}_s$ by

$$P_b \simeq 4C_1 \frac{1 + K}{d_{\min}^2 \bar{\gamma}_s} e^{-K} \ . \tag{7.57}$$

Observe that without interleaving, P_b is asymptotically inverse linear with $\bar{\gamma}_s$, independent of the trellis code. If follows that interleaving is required to achieve diversity with TCM on a flat fading channel.

7.4.1 Design Rules for TCM on Flat Fading Channels

According to the previous section, when TCM is used on Ricean fading channel with interleaving/deinterleaving, the design of the code for optimum performance is guided by the minimum built-in time diversity (MTD) of the code. For Rayleigh fading channels, the design of the code is also guided by the minimum squared product distance (MPSD) of the code. The minimum Euclidean distance, which is the principal design criterion for trellis coded modulation AWGN channels, plays a less significant role on Ricean fading channels as the K factor decreases, and no role for Rayleigh fading channels ($K = 0$). A third design criterion is to minimize the decoding depth of the code.

The design of trellis codes for interleaved flat fading channels is based on Unger-boeck's principle of mapping by set partitioning, but now the partitioning is done to maximize the MTD and MPSD of the code. This can be accomplished by maximizing the intra-subset MTD and MPSD, but it should be pointed out that large MTD and MPSD can be sometimes achieved even if the partitioning is done to maximize the minimum Euclidean distance as in Ungerboeck's codes for AWGN channels.

In general, the following guidelines are followed when designing trellis codes for interleaved flat fading channels;

1. All signals occur with equal frequency and with fair amount of regularity and symmetry.

2. Transitions originating from the even numbered states are assigned signals from the first subset of the first partitioning level.

3. Transitions originating from odd numbered states are assigned signals from the second subset of the first partitioning level.

4. Whenever possible, the transitions joining in the same state receive signals either from the first subset or the second subset of the first partitioning level.

5. Parallel transitions receive signals from the same subset of the finest partitioning level.

6. The state transitions originating from each current state and going to even-numbered next states are assigned signals from subsets whose inter-subset

MTD and MPSD are maximized. The same applies for the transition originating from each current state and going to odd-numbered next states.

The first five rules are similar to those suggested by Ungerboeck [312], but now the subsets used may be different. The sixth rule is used to reduce the **decoding depth** of the code. Wei [329] developed several codes based on minimizing the decoding depth of a code. He defined two minimum decoding depths (MDD1, MDD2) to characterize a code. MDD1+1 is defined as the length (in symbols) of the longest valid sequence of signal points, say x_1, which originates from the same state as another valid sequence x and merges into the same last state as x and whose Hamming distance from x is the same as the MTD of the code. Note that the performance of a code is mainly governed by the pairs of sequences which determine the MTD of the code. Each such pair of sequences differ in at most MDD1+1 successive symbols. The farther these symbols are separated, the better the performance of the code. Hence, to benefit from the MTD of the code, the interleaver should separate the symbols in each sequence of MDD1+1 input symbols as far as possible.

MDD2 is defined as the length of the longest unmerged valid sequence of signal points, say x_2, which originates from the same state as another valid sequence, say x, and whose Hamming distance from x is not greater than the MTD of the code. In the case the Hamming distance between the two sequences is equal to the MTD of the code, the squared product distance between the two sequences must be less than the MPSD of the code. Since MDD2 is greater than MDD1, the decoding depth should be at least equal to MDD2 to realize the MTD and MPSD of a code. It suffices if the decoding depth is few symbols longer than MDD2. Finally, to benefit from both the MTD and MPSD of a code, the interleaver should separate the symbol in each sequence of MDD2+1 input symbols as far as possible.

Multidimensional TCM

Recall that the length of the shortest error event with conventional trellis codes (one symbol per trellis branch) is equal to the number of branches along that error event path. If the trellis code has parallel transitions, then MTD = 1. Unfortunately, parallel transitions are inevitable when the size of the signal constellation exceeds the number of states. In this case, the bit error probability for Rayleigh fading channels has an inverse linear dependency on the bit energy-to-noise ratio. To solve this problem, multidimensional TCM techniques can be used.

Multidimensional TCM uses signal spaces having a larger dimensionality so as to increase the minimum Euclidean distance between signal points. Another feature of multidimensional trellis codes is noticed when comparing the coding gain of these codes to 1- or 2-D codes. When the size of the signal constellation is doubled with respect to uncoded modulation, the average signal energy may also increase. For example, doubling the size of a 2-D M-QAM constellation implies a 3 dB increase in average signal energy. However, if this increase in average signal energy could be avoided, then the TCM coding gain would be greater. This 3 dB cost falls to 1.5 dB when four dimensions are used, and to 0.75 dB when eight dimensions are used [28]. Multidimensional TCM is also attractive for fading channels. 4-D TCM schemes are special because they can be implemented in radio communications without any increase in bandwidth, by transmitting on the same carrier frequency with two spatially orthogonal electric field polarizations [28].

A $2N$-D constellation is formed by first selecting a constituent 2-D constellation and then concatenating N such constellations together in the time domain. If the size of the $2N$-D constellation is larger than needed, then some of the *less desirable* points are deleted [329]. The resulting constellation is then partitioned into a chain of increasingly large numbers of subsets. The partitioning is performed first to maximize the intra-subset MTD, and then to maximize the MPSD between any pair of $2N$-D signal points within the same subset having that MTD.

When $N > 1$, an MTD of at least two is easily achieved for each subset in the finest partition. Fig. 7.15 shows a 32-point 4-D 8-PSK constellation that is partitioned into 8-subsets. The 4-D constellation is formed by concatenating a pair of 2-D 8-PSK constellations in the time domain and deleting those points having the form (even,odd),(odd,even). The intra-subset MTD within each of the finest partitions is 2 with an intra-subset MPSD of 4. Fig. 7.16 shows a rate-2/3, 4-D, 4-state, 8-PSK trellis code with 2 bits/symbol. The bits b_1, b_2, b_3 are used to select one of the 8 subsets in Fig. 7.15 and bits b_4, b_5 are used to select one of the four 4-D elements within each subset. The MTD and MPSD of the code are the same as the intra-subset MTD and MPSD and, hence, are maximized for the partitioning in Fig. 7.15.

Fig. 7.16 also shows the trellis diagram of the code, along with examples of the longest sequences which determine the values of MDD1, and MDD2. Note that MDD1 and MDD2 are measured in units of 2-D symbols. Since there are parallel transitions of length 2 symbols, MDD1 = 1. To find MDD2, suppose that the all zeroes sequence is the reference sequence. Note that the 2-D sequence $\{1, 5, 0, 0, 0\}$ associated with the 4-D sequence $\{(1,5),(0,0),(0,2)\}$

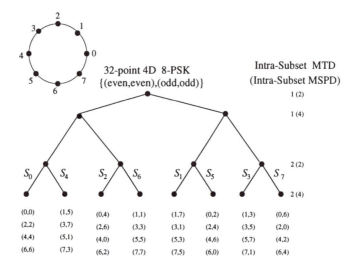

Figure 7.15   Partition of the 32-point 4-D 8-PSK constellation into 8 subsets.

has a time diversity of 2 (which is equal to the MTD) and an MPSD of $d^2(0,1) \times d^2(0,5) = 0.5857 \times 3.414 < 4$ and, hence, MDD2 = 5. Wei [329] investigated different multi-dimensional codes. He found that multi-dimensional TCM requires longer decoding depths than 2-D TCM. This longer decoding depth has proven to be very detrimental and, therefore, Wei considered only 4-D codes.

Multiple TCM (MTCM)

Multiple TCM is implemented by using a rate-b/s encoder, where the encoder outputs are mapped onto k M-ary symbols in each transmission interval, as shown in Fig. 7.17. The s encoder output symbols are divided into k groups of $m = \log_2 M$ symbols each, in this case $s = k \log_2 M$. Another method is to divide the s binary symbols into k groups of m_i symbols where each group now corresponds to a signal constellation of different size M_i. If $m_i = \log_2 M_i$ for the ith group, then $s = \sum_{i=1}^{k} m_i$. Notice that $k = 1$ corresponds to conventional Ungerboeck trellis codes. MTCM codes can be designed with parallel transitions, while still achieving an asymptotic bit error probability on fading channels that decays faster than an inverse linear rate with the average received symbol energy-to-noise ratio $\bar{\gamma}_s$.

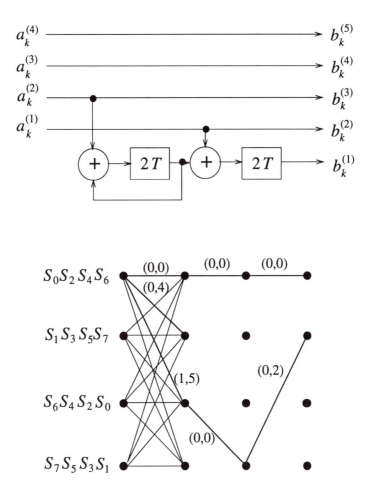

Figure 7.16   4-D 4-state 8-PSK rate-2/3 trellis code with 2 bits/symbol.

Fig. 7.18 shows the code trellis for the rate-4/6, 4-state, 8-PSK MTCM code reported in [71]. The signal point sets are obtained using the above method and they are the same as those used in the 4-D code and shown in Fig. 7.15. There are 16 paths emanating from each node and, hence, there are four parallel paths between nodes. This code has the same MTD and MPSD as the previous 4-D code. However, MDD1 = 5 and MDD2 = 5 and, therefore, the previous 4-D code remains a better choice since MDD1 is smaller.

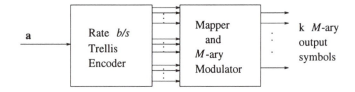

Figure 7.17   MTCM encoder.

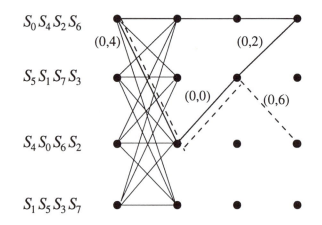

Figure 7.18   Code trellis for the rate-4/5 4-state 8-PSK MTCM trellis code.

2-D Trellis Codes

It is surprising that Ungerboeck's 2-D 8-state and 16-state 8-PSK trellis codes that were originally designed for AWGN channels, remain good for interleaved flat fading channels. In fact, Ungerboeck's 2-D, rate-2/3, 8-state 8-PSK trellis code shown in Fig. 7.19, has an MTD of two (the thick lines) and the corresponding product squared Euclidean distance is 8 which is better than the previous codes. Note that the shortest error event is not necessarily the minimum distance error event. In Fig. 7.19, the minimum squared Euclidean distance is 4.585, corresponding to an error event of length 3 (the dashed path). Note also that MDD1 = 3 and MDD2 = 3. It is obvious that the set partitioning for this code was intended to maximize the minimum squared Euclidean distance. Finally, we note that good 2-D TCM codes for interleaved flat fading channels will not have parallel transitions and the connectivity between the states will be as low as possible.

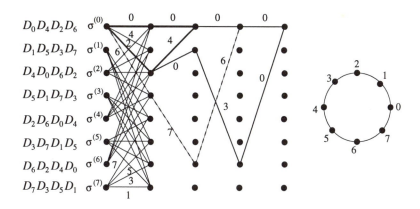

Figure 7.19 Trellis diagram for Ungerboeck's rate-2/3, 8-state, 8-PSK trellis code.

7.5 CODED PERFORMANCE ON FADING ISI CHANNELS

Fig. 7.20 shows a model for digital signaling on a *non-interleaved* fading ISI channel. Section 6.3.1 showed that the overall channel with D-branch diversity can be replaced with the model in Fig. 6.5, consisting of D $(L + 1)$-tap transversal filters, the outputs of which are corrupted by AWGN samples $\{\eta_k^d\}$ with $\sigma_{\eta_k^d}^2 = N_o \ \forall \ k, \ d$. For TCM, this leads to the equivalent discrete-time model shown in Fig. 7.21. As discussed in Section 6.3.1, the tap gains are modeled as uncorrelated complex Gaussian processes, and with 2-D isotropic scattering the tap gain vector $g_d(k) = (g_{0,d}(k), g_{1,d}(k), \ldots, g_{L,d}(k))^T$ has covariance matrix $\Phi_{g_d}(m) = J_0(2\pi f_m mT)\Sigma_d^2$ where $J_0(\ \cdot\ )$ is the zero-order Bessel function of the first kind and f_m is the maximum Doppler frequency, and $\Sigma_d = \text{diag}[\sigma_{0,d}, \sigma_{1,d}, \ldots, \sigma_{L,d}]$. Here, we assume the $g_{i,d}(k)$ have zero-mean so that the magnitudes $|g_{i,d}(k)|$ are Rayleigh distributed. Assuming that the branches are balanced, the average received branch bit energy-to-noise ratio is

$$\bar{\gamma}_c = \frac{E[|x_k|^2] \sum_{i=0}^{L} E[|g_{i,d}|^2]}{2nN_o} \tag{7.58}$$

where n is the number of bits per symbol.

As discussed earlier, the rate-$m/(m + r)$ linear convolutional encoder contains m shift registers and is characterized by a set of generator polynomials $g_i^{(j)}$, $1 \leq i \leq m$, $1 \leq j \leq m + r$. The length of the ith shift register is ν_i and the total number of memory elements in the encoder is $\nu_T = \sum_{i=1}^{m} \nu_i$. Since both the

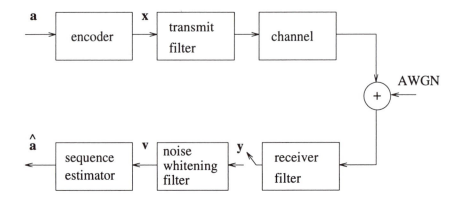

Figure 7.20   Digital communication on a non-interleaved fading ISI channel.

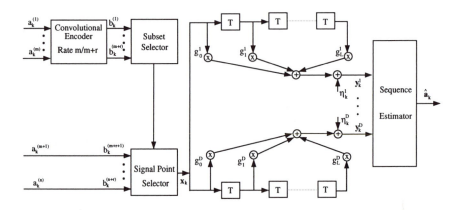

Figure 7.21   TCM-coded system with an equivalent discrete-time white noise channel model, from [284].

trellis encoder and ISI channel are finite state machines, it follows that the combined trellis encoder and ISI channel is also a finite state machine having **super-states**

$$s_k^{(i)} = (\boldsymbol{\sigma}_k; x_{k-1}, \ldots, x_{k-L}) \ . \tag{7.59}$$

There are $N_S = 2^{\nu_T} 2^{nL} = 2^{\nu_S}$ super-states and $\boldsymbol{\sigma}_k$ is the encoder state defined in (7.9). An equivalent definition of the super-state is [39]

$$\mathbf{s}_k^{(i)} = \left(\boldsymbol{\sigma}_{k-L}; \mathbf{a}_{k-1}, \ldots, \mathbf{a}_{k-L} \right)$$

$$= \left(a_{k-1}^{(1)}, \cdots, a_{k-\nu_1-L}^{(1)} ; \cdots ; a_{k-1}^{(m)}, \cdots, a_{k-\nu_m-L}^{(m)} ; \right.$$

$$\left. a_{k-1}^{(m+1)}, \cdots, a_{k-L}^{(m+1)} ; \cdots ; a_{k-1}^{(n)}, \cdots, a_{k-L}^{(n)} \right) \ . \qquad (7.60)$$

The overall system also has state diagram and trellis diagram descriptions. With MLSE, the Viterbi algorithm searches for the most likely path in the **super-trellis** based on the sequence of received samples $\mathbf{v}$ at the output of the matched filter. An error event of length $\ell = k_2 - k_1$ occurs between epochs k_1 and k_2 in the super-trellis if the actual super-state $\mathbf{s}_k$ and the estimated super-state $\hat{\mathbf{s}}_k$ satisfy

$$s_{k_1} = \hat{s}_{k_1}, \quad s_{k_2} = \hat{s}_{k_2} \quad \text{and} \quad s_k \neq \hat{s}_k, \quad k_1 < k < k_2 \ . \qquad (7.61)$$

Associated with every error event in the super-trellis is a pairwise error probability $P_r(\mathbf{x} \rightarrow \hat{\mathbf{x}})$, representing the probability that the receiver chooses sequence $\hat{\mathbf{x}}$ when $\mathbf{x}$ is transmitted.

7.5.1   TCM on Static ISI Channels

As shown in Section 6.6.1, the pairwise error probability is a function of the path distance matrix $\mathbf{E}$ defined in (6.161). The matrix $\mathbf{E}$ depends only on the trellis code and the length of the channel $L + 1$. Equations (6.164) and (6.165) provide a guideline for designing trellis codes for static ISI channels. The design criterion should be to maximize $\lambda_{\min}$ for the dominant error event. If the dominant error event has length ℓ, then this criterion implies that $\min\{r_\ell(0)\}$ is maximized, where $r_\ell(0)$ is the squared Euclidean distance between two allowable sequences of length ℓ in the super-trellis. The other design criterion should be to minimize the condition number $c(\mathbf{E})$. The matrix $\mathbf{E}$ is perfectly conditioned, or $c(\mathbf{E}) = 1$, only when $r_\ell(i) = 0, i \neq 0$. In this case $\mathbf{E}$ has the form $r_\ell(0)\mathbf{I}_{L+1}$, where $\mathbf{I}$ is the identity matrix. AWGN channels of the form $\mathbf{g} = \mathbf{e}_i, \ i \in \{0, \ldots, L\}$ represent the eigenvectors of this matrix.

If the dominant error event of a trellis code has a perfectly conditioned path distance matrix $\mathbf{E}$, then the asymptotic performance of the code is the same for any channel vector $\mathbf{g}$, including the AWGN channel. An interesting phenomenon occurs when the dominant error event associated with a trellis code does not have a perfectly conditioned path distance matrix. In this case the asymptotic performance of the trellis code over the channel described by the vector $\mathbf{g} = \mathbf{v}_{\max}$ is better than its asymptotic performance on an AWGN channel! The code has the worst performance on the channel described by the vector $\mathbf{g} = \mathbf{v}_{\min}$. Therefore, as an alternative to finding trellis codes that are suitable for a particular set of channel vectors, we might as well find the set of channel vectors that are suitable for a particular trellis code.

7.5.2 TCM on Fading ISI Channels

As mentioned before, $c(\mathbf{E}) = 1$ if and only if $r_\ell(i) = 0$, $i \neq 0$. It is impossible to obtain a code where $c(\mathbf{E}) = 1$ for all error events. The cross terms $r_\ell(i), i \neq 0$ in the path distance matrix $\mathbf{E}$ of any error event will degrade the performance. The next section shows an example where the cross terms cause an error event with a smaller $r_\ell(0)$ to have a lower pairwise error probability than another error event having a larger $r_\ell(0)$. It is very difficult to control the cross terms of the dominant error events and, therefore, a less stringent criterion would be useful for predicting the performance of a trellis code. We now show that if the squared Euclidean distance $\min\{r_\ell(0)\}$ does not increase linearly with ℓ, then a trellis code will not have good performance on a non-interleaved fading ISI channel. We require the following definition and two theorems to develop this criterion.

Define $\Omega(k_1, k_1 + i)$ as the set of all distinct pairs of sequences in the code-trellis that originate from the same state at epoch k_1 and merge into the same state at epoch $k_1 + i$. Each pair of sequences in set $\Omega(k_1, k_1 + i)$ may also merge into the same state between epochs k_1 and $k_1 + i$ and possibly stay merged for at most $L - 1$ branches in the code-trellis and then diverge, thus forming one or more error events in the code-trellis.

Theorem 7.1: ⎯⎯⎯⎯⎯⎯⎯⎯⎯⎯⎯⎯⎯⎯⎯⎯⎯⎯⎯⎯⎯⎯⎯⎯

If an error event of length i having a squared Euclidean distance d^2 occurs in the code-trellis, then an error event of length $\ell = L + i$ having a squared Euclidean distance $r_{L+i}(0) = d^2$ occurs in the super-trellis. Conversely, if an error event occurs in the super-trellis between epochs k_1 and k_2, having length $k_2 - k_1 = L + i$ and a corresponding squared path distance $r_{L+i}(0)$, then there exists a pair of sequences $(\mathbf{x}_1, \mathbf{x}_2) \in \Omega(k_1, k_1 + i)$ in the code-trellis having a squared Euclidean distance equal to $r_{L+i}(0)$.

Proof:

Using the definition of the error event in the super-trellis in (7.61), and using the two equivalent forms of the super-state (7.59) and (7.60), it follows directly that an error event in the code-trellis $\sigma_{k_1} = \hat{\sigma}_{k_1}$, $\sigma_{k_1+i} = \hat{\sigma}_{k_1+i}$, and $\sigma_k \neq \hat{\sigma}_k$ for $k_1 < k < k_1 + i$ implies that $s_{k_1} = \hat{s}_{k_1}$, $s_{k_1+i+L} = \hat{s}_{k_1+i+L}$, and $s_k \neq \hat{s}_k$ for $k_1 < k < k_1 + i + L$. It also follows that $\epsilon_k = 0$, $k \geq k_1 + i$ and, hence, $r_{L+i}(0) = d^2$.

Conversely, suppose that an error event occurs in the super-trellis between epochs k_1 and $k_1 + L + i$. It follows directly from (7.61), and the fact that $x_k = f_1(\sigma_k, a_k)$ and $\sigma_{k+1} = f_2(\sigma_k, a_k)$, that there exists a pair of sequences in the code-trellis $(\mathbf{x}_1, \mathbf{x}_2)$ that originate from the same state at epoch k_1, i.e., $\epsilon_{k_1} \neq 0$, and merge into the same state at epoch $k_1 + i$, i.e., $\epsilon_{k_1+i-1} \neq 0$. The two sequences may merge into the same state between epochs k_1 and $k_1 + i$, say at epoch $k_1 + m$ ($m < i$) and possibly stay merged for at most $L - 1$ branches; otherwise, $s_{k_1+L+m} = \hat{s}_{k_1+L+m}$ and the length of error event is shorter than $L + i$. Hence, $(\mathbf{x}_1, \mathbf{x}_2) \in \Omega(k_1, k_1 + i)$. It also follows from (6.163) that the squared Euclidean distance between $\mathbf{x}_1$ and $\mathbf{x}_2$ is equal to $r_{L+i}(0)$.

Theorem 7.2: ──

Let $d^2_{\min}(i)$ denote the squared minimum distance of all error events in the code-trellis of length i. Then $\min\{r_{L+i}(0)\} \leq d^2_{\min}(i)$.

Proof:

An error event in the super-trellis between epochs k_1 and $k_1 + L + i$ and the corresponding pair of sequences $(\mathbf{x}_1, \mathbf{x}_2)$ may result from either a single error event in the code-trellis between epochs k_1 and $k_1 + i$, or multiple error events of shorter lengths, e.g., i error events of length one (parallel transitions) or, in general, m error events of lengths $j_1, \ldots, j_m$. Note that m can take any value between one and i. Also, for any m, the lengths $j_1, \ldots, j_m$ can assume different values with the constraint $\sum j_m \leq i$ and $(\mathbf{x}_1, \mathbf{x}_2) \in \Omega(k_1, k_1 + i)$. It follows that

$$\min\{r_{L+i}(0)\} = \min\left\{ \sum d^2_{\min}(j_m) \right\} \leq d^2_{\min}(i) \qquad (7.62)$$

where the minimization goes over all $m \leq i$ and all $j_1, \ldots, j_m$ with the above constraint. If there exists a single error event of length i in the code-trellis, then the theorem is immediate. Also, if there are no error events of length i

in the code-trellis then, by definition, $d_{\min}(i) = \infty$ and the above inequality in (7.62) is satisfied.

If we treat the uncoded system as a trellis-coded system with a single state and parallel transitions, then, for two-tap channels $\min\{r_{i+L}(0)\} = i d^2_{\min}(0)$, $i = 1, 2, \ldots$, where $d^2_{\min}(0)$ is the minimum squared Euclidean distance in the signal constellation of the uncoded system. The important point is that $r_{L+i}(0)$ for the uncoded system grows linearly with the length of the error event. Theorems 7.1 and 7.2 suggest that if a trellis code has a dense distance spectrum [270], then the set of $\min\{r_\ell(0)\}$, $\ell \geq L$ will also have a dense spectrum. Therefore, if the minimum squared Euclidean distance of a trellis code does not grow linearly with the length of the error events, then the uncoded system is expected to outperform the trellis-coded system for a non-interleaved fading ISI channel.

7.5.3   Examples

This section illustrates the above concepts by focusing on the 4-state 8-PSK and the 8-state 8-PSK Ungerboeck codes [312]. The corresponding uncoded system is 4-PSK. We have seen earlier that the 8-state code is suitable for interleaved flat fading channels, having MTD = 2 and MPSD = 8. Two ISI channels are considered i) a two-tap static ISI channel and, ii) a two-tap Rayleigh fading ISI channel.

Static ISI Channels

For a two-tap static ISI channel, the path distance matrix is

$$\mathbf{E} = \begin{pmatrix} r_\ell(0) & r_\ell(1) \\ r_\ell^*(1) & r_\ell(0) \end{pmatrix} . \tag{7.63}$$

Tables 7.1, 7.2, and 7.3, tabulate the values associated with matrix $\mathbf{E}$ for all error events of up to length 8 for the uncoded system and the two coded systems. Notice that the minimum squared Euclidean distance is $r_2(0) = 2.00$ for the uncoded system, $r_2(0) = 4.00$ for the 4-state trellis code, and $r_4(0) = r_5(0) = 4.59$ for the 8-state trellis code. Also, the matrix $\mathbf{E}$ associated with the minimum distance error event is perfectly conditioned for both the uncoded system and the 4-state trellis code, but not for the 8-state trellis code.

ℓ	$\min\{r_\ell(0)\}$	$|r_\ell(1)|$	λ_1	λ_2	$w(\mathbf{x}, \hat{\mathbf{x}})$	$P(\mathbf{x} \to \hat{\mathbf{x}})$
2	2.00	0.00	2.00	2.00	1	0.1705E-01
3	4.00	2.00	6.00	2.00	2	0.6786E-02
4	6.00	4.00	10.00	2.00	3	0.4238E-02
5	8.00	6.00	14.00	2.00	4	0.3081E-02
6	10.00	8.00	18.00	2.00	5	0.2420E-02
7	12.00	10.00	22.00	2.00	6	0.1993E-02
8	14.00	12.00	26.00	2.00	7	0.1694E-02

Table 7.1 Error events in the super-trellis for uncoded 4-PSK system over a two-tap channel

ℓ	$\min\{r_\ell(0)\}$	$|r_\ell(1)|$	λ_1	λ_2	$w(\mathbf{x}, \hat{\mathbf{x}})$	$P(\mathbf{x} \to \hat{\mathbf{x}})$
2	4.00	0.00	4.00	4.00	1	0.5528E-02
3	8.00	4.00	12.00	4.00	2	0.2033E-02
4	4.59	2.16	6.75	2.42	2	0.5261E-02
5	5.17	2.72	7.89	2.46	4	0.3831E-02
6	5.17	2.16	7.34	3.01	3	0.3790E-02
7	5.76	2.72	8.47	3.04	5	0.3173E-02
8	5.76	2.00	7.76	3.76	3	0.3227E-02

Table 7.2 Error events in the super-trellis for 4-state 8-PSK code over a two-tap channel

We now consider the coded performance for seven different channels with impulse responses listed in Table 7.4. Channels G, A1, and A2 were chosen arbitrarily and have the best spectral characteristics. Channels B1 and B2 are equal to the eigenvectors associated with the minimum and maximum eigenvalues, respectively, for one of the length 3 and 4 error events in the uncoded system, and one of the error events of length 4 in the 8-state trellis code. Channels C1 and C2 are equal to the eigenvectors associated with the maximum and minimum eigenvalues, respectively, for some of the minimum distance error events that are associated with the 8-state trellis code.

Simulation results for the uncoded 4-PSK system, the 4-state trellis code, and the 8-state trellis code are shown in Figs. 7.22, 7.23, and 7.24, respectively. Although channel C2 has an in-band spectral null, making it perhaps the most

ℓ	$\min\{r_\ell(0)\}$	$|r_\ell(1)|$	λ_1	λ_2	$w(\mathbf{x}, \hat{\mathbf{x}})$	$P(\mathbf{x} \to \hat{\mathbf{x}})$
3	6.00	2.83	8.83	3.17	1	0.3301E-02
4	4.59	2.16	6.75	2.42	3	0.5106E-02
5	4.59	1.08	5.67	3.50	3	0.4556E-02
6	5.17	2.16	7.34	3.01	6	0.4094E-02
7	5.17	1.08	6.25	4.09	6	0.3654E-02
8	5.17	1.08	6.25	4.09	4	0.3654E-02

Table 7.3 Error events in the super-trellis for 8-state 8-PSK code over a two-tap channel

channel	g_0	g_1
G	0.9747	0.223
A1	0.866	0.500
A2	0.500	0.866
B1	-0.7071	0.7071
B2	0.7071	0.7071
C1	0.7071	-0.65333+j0.2705
C2	0.7071	0.65333-j0.2705

Table 7.4 Static ISI channels used to evaluate the performance of trellis codes.

difficult channel to equalize, the performance of uncoded 4-PSK on this channel is better than that on channels B1 and B2. The 4-state trellis code performs better on channels B1 and B2 than on channels C1 and C2, although channel B1 has a band-edge null while channel B1 does not. The 8-state trellis code performs better on channel C1 than C2. This makes sense because channel C1 is the eigenvector associated with $\lambda_{\max}$ for one of the dominant error events of the 8-state code. In general, Tables 7.1, 7.2, and 7.3 show that the coded systems have a larger $\lambda_{\min}$ than the uncoded systems. Therefore, it is reasonable that the coded systems have better performance than the uncoded system, although $c(\mathbf{E})$ for the dominant error event is greater for the 8-state code.

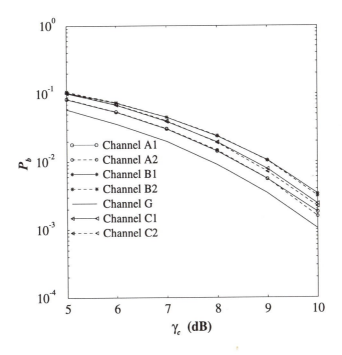

Figure 7.22 Performance of uncoded 4-PSK over static ISI channels.

Multipath Fading ISI Channels

Consider a 2-tap fading ISI channel with matrix

$$\mathbf{A} = \Sigma_d \mathbf{E} \Sigma_d = \begin{pmatrix} \sigma_0^2 r_\ell(0) & \sigma_0 \sigma_1 r_\ell(1) \\ \sigma_0 \sigma_1 r_\ell^*(1) & \sigma_1^2 r_\ell(0) \end{pmatrix} \tag{7.64}$$

and corresponding eigenvalues

$$\lambda_{1,2} = r_\ell(0) \frac{\bar{E}}{2} \pm \frac{1}{2} \sqrt{r_\ell(0)^2 \left(\sigma_0^2 - \sigma_1^2\right)^2 + 4\sigma_0^2 \sigma_1^2 |r_\ell(1)|^2} \ . \tag{7.65}$$

It is obvious that $\lambda_1 = \lambda_2$ implies that $\sigma_0 = \sigma_1$ and $r_\ell(1) = 0$. For the case when $\sigma_0 = \sigma_1$, the eigenvalues are given by $\frac{\bar{E}}{2}[r_\ell(0) \pm |r_\ell(1)|]$.

Fig 7.25 plots the pairwise error probability of the 8-state code against the normalized energy in the first tap for different values of $r_\ell(0)$ and $|r_\ell(1)|$ as described in Table 7.3. Notice that the pairwise error probability is minimized for equal energy taps. Fig. 7.25 also shows how the pairwise error probability

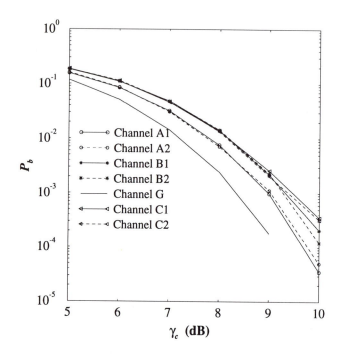

Figure 7.23 Performance of 4-state 8-PSK TCM over static ISI channels.

decreases with an increase in the squared Euclidean distance, $r_\ell(0)$. Finally, Fig. 7.25 shows the effect of the eigenvalue spread. For example, the pairwise error probability associated with the error event having $r_6(0) = 5.76$ and $|r_6(1)| = 1.08$ $(c = 1.46)$ is better than that with $r_3(0) = 6.00$ and $|r_3(1)| = 2.83$ $(c = 2.79)$, although the squared Euclidean distance is larger for the latter.

Fig. 7.26 plots the bit error probability of the uncoded 4-PSK system, and the 4- and 8-state 8-PSK trellis codes on a two-equal-strength-tap fading channel. The performance is completely reversed from that on an AWGN channel. The uncoded system outperforms either trellis-coded system and, moreover, the 4-state trellis code outperforms the 8-state trellis code. This behavior is consistent with the parameters listed in Tables 7.1, 7.2 and 7.3. Although the uncoded system has a smaller squared Euclidean distance, $r_\ell(0)$ for the uncoded systems grows faster with the length of the error events than either the 4- or 8-state trellis codes. By comparing the parameters of the 4- and 8-state trellis codes in Tables 7.2 and 7.3, respectively, it is not surprising that the 4-state trellis code outperforms the 8-state trellis code.

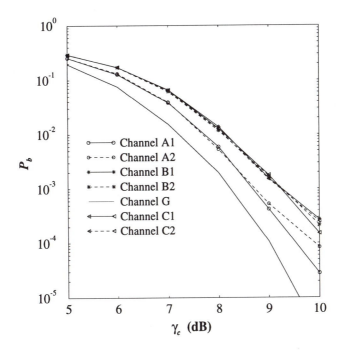

Figure 7.24 Performance of 8-state 8-PSK TCM over static ISI channels.

As a final example, consider the simple rate-1/2, 2-state, 4-PSK trellis code shown in Fig 7.27. The equivalent uncoded system is BPSK having a minimum squared Euclidean distance growth given by the values $\{d^2_{\min}(i)/E\} = \{4.0, 8.0, 12.0, 16.0, 20.0, 24.0, 28.0, \ldots\}$. Table 7.5 lists the parameters of the code. Note that the code has a minimum squared Euclidean distance that grows linearly with the length of the error event but at a slower rate than that of the uncoded system. Fig. 7.28 shows the performance of the code. Unlike the Ungerboeck codes, the code at least offers slightly better performance than the equivalent uncoded system despite its simplicity.

7.6 EVALUATION OF UNION BOUNDS FOR TCM

The pair-state approach is one method for evaluating the error probability upper bound for TCM on intersymbol interference (ISI) channels [72], [27].

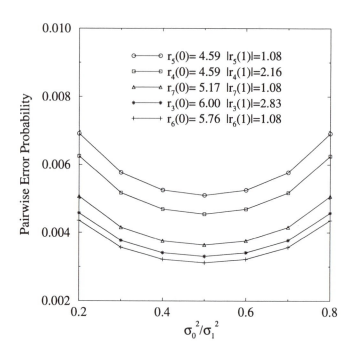

Figure 7.25   Pairwise error probability for several error events for the 8-state
8-PSK trellis code at $\bar{\gamma}_c = 10$ dB. See Table 7.3 for corresponding eigenvalues.

Unfortunately, there are $(N_S)^2$ pair-states, where N_S is the number of super-states. A simpler method that uses the transfer function of an N_S-state error diagram has been proposed for linear filter channels [193]. Both of these techniques require a Chernoff bound on the pairwise error probability which can be loose, especially for fading channels.[2] Here we describe a method for evaluating the union bound that uses an error-state diagram and a one-directional stack algorithm. The proposed method does not require the transfer function and, therefore, i) an exact expression for the pairwise error probability can be used yielding a tighter upper bound, and ii) the method is useful for large-state systems.

The bit error probability for TCM on an ISI channel has the bound in (7.14), where $\mathbf{E}$ is the set of error sequences that correspond to all error events in the super-trellis at epoch k_1. For a static ISI channel, the pairwise error probability

---

[2] The union bound may also be loose for fading channels.

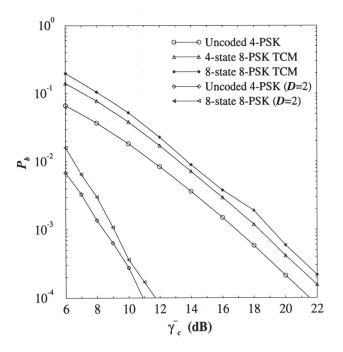

Figure 7.26 Performance over a two-equal-ray T-spaced fading ISI channel.

is given by (6.155) with the squared Euclidean path distance in (6.158). By following the same procedure as in Section 6.6.3, the parameters needed to evaluate the upper bound for a static ISI channel are the probability $P\{\mathbf{a}\}$, the number of bit errors $w_b(\mathbf{e})$, and the squared Euclidean path distance Δ^2. Likewise, for a fading ISI channel with equal diversity branches, the exact pairwise error probability is given by (6.181), and the parameters needed to evaluate the union bound are $P_r\{\mathbf{a}\}$, $w_b(\mathbf{e})$, and the matrix $\mathbf{A} \equiv \mathbf{A}_d$, where $\mathbf{A}_d$ is defined in (6.172).

The overall system has super states $\mathbf{s}_k^{(i)}$ for $i = 0, \cdots, N_S - 1$, where $\mathbf{s}_k^{(i)}$ is defined in (7.60). Define the **error state** as $\mathbf{v}_l = \mathbf{s}_k^{(i)} \oplus \mathbf{s}_k^{(j)}$ for some i and j. An error-state diagram can be constructed such that the initial and final nodes in the error-state diagram are zero-error states and each intermediate node represents a distinct non-zero error-state. A directed line from $\mathbf{v}_l$ to

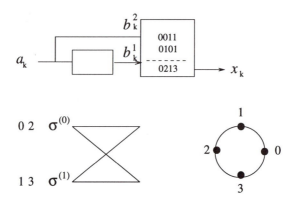

Figure 7.27   Simple rate-1/2, 2-state, 4-PSK trellis code.

i	$d_{\min}^2(i)/E$	$w(\mathbf{x}, \hat{\mathbf{x}})$	N_{eq}
1	∞	0	0
2	6.0	1	1
3	8.0	2	1
4	10.0	3	1
5	12.0	4	1
6	14.0	5	1
7	16.0	6	1
8	18.0	7	1

Table 7.5   Error events in the code-trellis for the rate-1/2, 2-state, 4-PSK code in Fig. 7.27; N_{eq} is the number of error events of length i having a squared Euclidean distance of $d_{\min}^2(i)$.

$\mathbf{v}_m$ indicates an allowable error-state transition $(\mathbf{v}_l, \mathbf{v}_m)$. There is one-to-one correspondence between the set $\mathbf{E}$ in (7.14) and the set of paths from the initial to final node in the error-state diagram.

To evaluate the union bound (7.14), we define an appropriate transition-gain for each transition in the error-state diagram as follows. Given an error-state transition $(\mathbf{v}_l, \mathbf{v}_m)$, each branch from $\mathbf{s}_k^{(i)}$ to $\mathbf{s}_{k+1}^{(j)}$ in the overall trellis diagram is assigned the appropriate branch distance (or branch distance matrix), $\mathbf{C}_{lm}(i, j)$ and number of bit errors $u_{lm}(i, j)$. This assignment can be conveniently de-

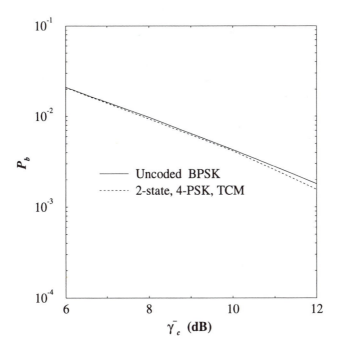

Figure 7.28 Performance of the rate-1/2, 2-state, 4-PSK trellis code on a two-equal-ray T-spaced fading ISI channel.

scribed by an $N_S \times N_S$ transition matrix $\mathbf{B}_{lm} = [b_{lm}(i,j)]$ with elements

$$b_{lm}(i,j) \triangleq \begin{cases} Z_1^{u_{lm}} Z_2^{\mathbf{C}_{lm}(i,j)} & , \ s_k^{(i)} \text{ to } s_{k+1}^{(j)} \text{ transition possible} \\ 0 & , \ \text{otherwise} \end{cases} \qquad (7.66)$$

Note that u_{lm} in (7.66) does not depend on the branch from $s_k^{(i)}$ to $s_{k+1}^{(j)}$, but only on the error-state transition $(\mathbf{v}_l, \mathbf{v}_m)$.

Consider the following simple example, consisting of a two-state 4-PAM trellis code with a two-tap channel ($L = 1$). The encoder has generators $\mathbf{g}^{(1)} = (1,0)$ and $\mathbf{g}^{(2)} = (0,1)$, and the signal mapping is $x_k = 4a_k + 2a_{k-1} - 3$. Fig. 7.29 shows the error-state diagram. As an example of how to obtain $\mathbf{B}_{lm}$, consider the error sequence $\{e_k = 1, \ e_{k+1} = 1, \ e_{k+2} = 1, \ e_{k+3} = 0, \ e_{k+4} = 0\}$ in Fig. 7.30. The error sequence corresponds to the path $\{\mathbf{v}_0, \mathbf{v}_2, \mathbf{v}_3, \mathbf{v}_3, \mathbf{v}_1, \mathbf{v}_4\}$ in the error-state diagram. Fig. 7.30 shows the super-trellis, along with the symbol error ε_k for $k_1 \leq k \leq k_1 + 4$. Note that all branches merging at the same node in Fig. 7.30 have the same symbol error ε_k. Given the pair $(\varepsilon_{k-1}, \ \varepsilon_k)$, the squared branch distance Δ_k^2 can be calculated by using (6.158)

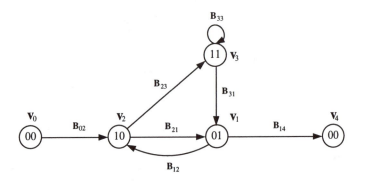

Figure 7.29 Error-state diagram for 2-state, 4-PAM, TCM on a two-tap ISI channel.

and (6.159), and the matrix $\mathbf{A}_k$ in (6.170) can be calculated from the branch distance matrix $\mathbf{E}_k$ in (6.160)[3]. For example, consider a static ISI channel with $g_0 = 1$ and $g_1 = 1$. The transition-gains $\mathbf{B}_{02}$ and $\mathbf{B}_{33}$ are, respectively,

$$\mathbf{B}_{02} = Z_1 \cdot \begin{bmatrix} Z_2^{16} & 0 & Z_2^{16} & 0 \\ Z_2^{16} & 0 & Z_2^{16} & 0 \\ 0 & Z_2^{16} & 0 & Z_2^{16} \\ 0 & Z_2^{16} & 0 & Z_2^{16} \end{bmatrix}, \tag{7.67}$$

and

$$\mathbf{B}_{33} = Z_1 \cdot \begin{bmatrix} Z_2^{144} & 0 & Z_2^{16} & 0 \\ Z_2^{64} & 0 & Z_2^{0} & 0 \\ 0 & Z_2^{0} & 0 & Z_2^{64} \\ 0 & Z_2^{16} & 0 & Z_2^{144} \end{bmatrix}. \tag{7.68}$$

Likewise, for a two-tap fading ISI channel, the transition gain $\mathbf{B}_{02}$ is

$$\mathbf{B}_{02} = Z_1 \cdot \begin{bmatrix} Z_2^{\mathbf{A}_{02}} & 0 & Z_2^{\mathbf{A}_{02}} & 0 \\ Z_2^{\mathbf{A}_{02}} & 0 & Z_2^{\mathbf{A}_{02}} & 0 \\ 0 & Z_2^{\mathbf{A}_{02}} & 0 & Z_2^{\mathbf{A}_{02}} \\ 0 & Z_2^{\mathbf{A}_{02}} & 0 & Z_2^{\mathbf{A}_{02}} \end{bmatrix} \tag{7.69}$$

where

$$\mathbf{A}_{02} = \mathbf{\Sigma} \begin{bmatrix} 16 & 0 \\ 0 & 0 \end{bmatrix} \mathbf{\Sigma} . \tag{7.70}$$

---

[3] In general, the squared branch distance and branch distance matrix are calculated using $(\varepsilon_{k-L}, \varepsilon_{k-L+1}, \cdots, \varepsilon_k)$.

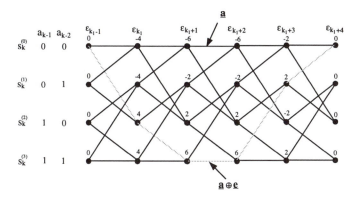

Figure 7.30 Super-trellis showing symbol errors for the error sequence $\{e_{k_1} = 1,\ e_{k_1+1} = 1,\ e_{k_1+2} = 1,\ e_{k_1+3} = 0,\ e_{k_1+4} = 0\}$.

For the error-state transition $(\mathbf{v}_l, \mathbf{v}_m)$, the **distance polynomial** [270] associated with the node $s_k^{(i)}$ is the sum of the ith-row elements of the matrix $1/2^n \cdot \mathbf{B}_{lm}$, where we have assumed that the information vector $\mathbf{a}_k$ is transmitted with equal probability $1/2^n$. If the sum of row elements is the same for every row, then the matrix is **row-uniform** [193]. If the matrix $\mathbf{B}_{lm}$ is row-uniform for all error-state transitions, then the trellis code has the **uniform property**. For AWGN channels, many trellis codes including the Ungerboeck codes have the uniform property, meaning that the error probability can be evaluated by just considering the set of information sequences that originate from a particular state, say $s_{k_1}^{(0)}$. However, for ISI channels the uniform property is lost and all possible information sequences must be considered. For example, the matrix $\mathbf{B}_{33}$ does not have the row-uniform property in our example system, although the trellis-code is uniform for AWGN channels. Nevertheless, some symmetry properties of the trellis code and trellis structure can still be exploited to simplify the performance analysis. Considering again the error sequence $\{e_k = 1,\ e_{k+1} = 1,\ e_{k+2} = 1,\ e_{k+3} = 0,\ e_{k+4} = 0\}$ in Fig. 7.30, the following observations can be made:

Observation 7.1: Only half of the sequences originating from each state $s_{k_1}^{(i)}$ must be considered, because there is always a pair of correct sequences having the same probability $P_r\{\mathbf{a}\}$, number of bit errors $w_b(\mathbf{e})$, and path distance (or path distance matrix), e.g., the information sequences $\mathbf{a}$ and $\mathbf{a} \oplus \mathbf{e}$. This symmetry property exists for every TCM system $\square$

Observation 7.2: Only the information sequences originating from states $s_{k_1}^{(0)}$ and $s_{k_1}^{(2)}$ must be considered, because for each information sequence originating from the state $s_{k_1}^{(1)}$ (or $s_{k_1}^{(3)}$) there always exists an information sequence originating from the state $s_{k_1}^{(0)}$ (or $s_{k_1}^{(2)}$) having the same set of parameters. This type of symmetry usually exists but depends on the particular trellis code. The algorithm discussed below exploits this type of symmetry by combining together all paths attached to the same node that have the same set of parameters □

The stack algorithm maintains an ordered stack where each entry represents one or more paths in the error-state diagram and contains the following information: terminal node, terminal state s_k, the number of branches H, $\sum_{(l,m)} u_{lm}$, $\sum_{(l,m)} \Delta_{lm}^2$ (or $\sum_{(l,m)} \mathbf{A}_{lm}$), and the *intermediate* bit error probability P_I. The set $\{(l,m)\}$ is the set of error-state transitions associated with the path under consideration. P_I is calculated according to

$$P_I = \frac{1}{n} \cdot \frac{1}{N_S \cdot 2^{nH}} \sum_{\{(l,m)\}} u_{lm} \cdot \hat{P}\left\{ \Gamma(\mathbf{a} \oplus \mathbf{e}) \geq \Gamma(\mathbf{a}_k) \big| \mathbf{a} \right\} \tag{7.71}$$

where $\hat{P}\left\{ \Gamma(\mathbf{a} \oplus \mathbf{e}) \geq \Gamma(\mathbf{a}) \big| \mathbf{a} \right\}$ is computed by using the squared path distance $\sum_{\{(l,m)\}} \Delta_{lm}^2$ for a static ISI channel and the matrix $\sum_{\{(l,m)\}} \mathbf{A}_{lm}$ for a fading ISI channel. The stack is ordered according to decreasing P_I. For a path from the initial to final node, we have $P\{\mathbf{a}\} = 1/(N_S \cdot 2^{nH})$, $w_b(\mathbf{e}) = \sum_{\{(l,m)\}} u_{lm}$, and $P\left\{ \Gamma(\mathbf{a} \oplus \mathbf{e}) \geq \Gamma(\mathbf{a}) \big| \mathbf{a} \right\} = \hat{P}\left\{ \Gamma(\mathbf{a} \oplus \mathbf{e}) \geq \Gamma(\mathbf{a}) \big| \mathbf{a} \right\}$. A key feature of the stack algorithm is that paths having the same terminal node, terminal state s_k, number of branches H, $\sum_{\{(l,m)\}} u_{lm}$, and $\sum_{\{(l,m)\}} \Delta_{lm}^2$ (or $\sum_{\{(l,m)\}} \mathbf{A}_{lm}$) are combined together, because from that point on they can be treated as a single path. These combinations reduce the computation required to evaluate the upper bound as discussed in Observation 7.2. The number of paths represented by a stack entry is called the path multiplicity, M. The detailed stack algorithm is shown in Fig. 7.31 an operates much the same as the stack algorithm described in Section 6.6.3.

Example 7.1───────────────────────────────────

Consider a system that uses the 4-state 8-PSK Ungerboeck trellis code in Fig. 7.32 on a two-tap multipath-fading channel. In the simulations, the tap coefficients $\{g_{i,d}\}$ are generated by passing independent complex white Gaussian noise through a digital Butterworth filter having a 3-dB cut-off frequency equal to 0.4 Hz. The transmission rate is assumed to be 2400 symbols/sec and $\sigma^2 = \frac{1}{2}E[|g_{i,d}|^2] = 1$ for all i and d. Once again, the analytical results are obtained by setting the threshold $P_T = 10^{-3} \cdot P_{max}$, where P_{max} is the largest

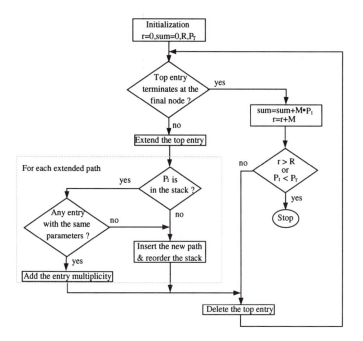

Figure 7.31 The stack algorithm.

term in the upper bound. For TCM on multipath-fading channels, the average received bit energy-to-noise ratio per diversity branch is given by (7.58). Fig. 7.33 compares analytical and simulation results for this system. Without diversity $(D = 1)$, the difference is about 2 dB for $P_b < 10^{-3}$. However, for two-branch diversity $(D = 2)$, the difference is within 1 dB.

7.7 TCM ON EQUALIZED ISI CHANNELS

7.7.1 Symbol-by-Symbol Equalization with Interleaving

Receivers that use symbol-by-symbol equalization, such as the linear forward equalizers and the nonlinear decision-feedback and lattice equalizers, are the

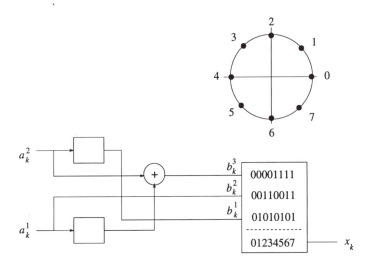

Figure 7.32   Encoder and signal mapping for the 4-state 8-PSK Ungerboeck code.

simplest. The equalizer in these receivers is followed by a separate soft-decision Viterbi decoder, or other sequence estimator, that works on the code trellis under the assumption that perfect equalization is achieved.

Linear equalizers exhibit noise enhancement on channels with spectral nulls. Since spectral nulls are typical of fading ISI channels, linear equalizers perform poorly. To mitigate the effect of noise enhancement, nonlinear equalizers such as the decision feedback equalizer (DFE) can be used. The feedback section of the DFE subtracts the ISI caused by previously detected symbols, provided that correct decisions are available with zero delay. Usually, the zero-delay decisions are made by a threshold detector that makes *hard decisions* on the equalized samples.

It is apparent that interleaving/deinterleaving is straight forward for receivers that use symbol by symbol equalization and, hence, a major advantage follows. TCM codes that are designed for flat fading channels, with their powerful built-in time diversity, can be applied to fading ISI channels. The input sequence is interleaved and transmitted, and the received samples are equalized, deinterleaved, and fed into a sequence estimator for decoding. Unfortunately, this class of receivers has a major drawback. The equalizer is *unaware* of the sequence constraints imposed by the trellis code, and thus it ignores the distance between sequences. The performance of the threshold detector in the equalizer

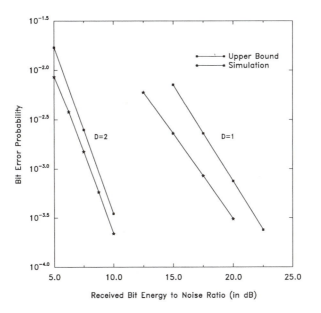

Figure 7.33 Comparison of analytical and simulation results for 4-state, 8-PSK, TCM on a two-equal-ray T-spaced fading ISI channel, from [284].

is determined by the minimum separation between signal points in the signal constellation which is decreased when trellis coding is used. In other words, the threshold detector operates at a lower effective signal-to-noise ratio than the corresponding uncoded system and, hence, it is more likely to make wrong decisions. This ruins the performance of the equalizer adaptation algorithm. Moreover, with a DFE, the error propagation in the feedback path of the DFE resulting from incorrect equalizer-based decisions ruins the performance of the combination of the DFE and the sequence estimator. Still another drawback is that perfect equalization cannot be achieved and, hence, the sequence estimator can be affected by the residual ISI from the equalizer. For these reasons, it is expected that uncoded systems will outperform trellis-coded systems if symbol by symbol equalization is used.

7.7.2   Equalization and Decoding with Separate Sequence Estimators and Interleaving

When joint equalization and decoding is used without interleaving, the decoder searches through the super-trellis with the super-state defined in (7.59). Interleaving, however, allows trellis codes that are effective for flat fading channels to be used on fading ISI channels. The sequence of symbols **x** is interleaved and the interleaved sequence x̌ is transmitted over the fading ISI channel. The receiver first performs equalization by using a sequence estimator that works on the sequence of interleaved symbols that are corrupted by ISI, without knowledge of the sequence constraints imposed by the trellis code. This sequence estimator gives reliable hard decisions on the transmitted symbols in the expanded signal set. These hard decisions are used to cancel the ISI and generate the sequence of soft outputs **z**, where $z_k = y_k - \sum_{i=1}^{L} g_i x'_{k-i}$, which is then deinterleaved and fed into a second sequence estimator that works on the code trellis only. The state observed by the first sequence estimator is

$$u_k = (\check{x}_{k-1}, \ldots, \check{x}_{k-L}) \; . \tag{7.72}$$

Recall that the $\{\check{x}_k\}$ are independent complex symbols taken from the expanded signal set used by the trellis encoder and, hence, the number of states observed by the first sequence estimator is $(2^{m+r})^L$.

The above receiver is similar to the receiver described in Section 7.7.1 except that the unreliable zero-delay decisions generated by the DFE are replaced with reliable hard decisions taken from the inner Viterbi algorithm. The soft input to the trellis decoder ignores the energy in the ISI terms and, therefore, this receiver is a suboptimal version of the MAP equalizer described in [145] but has a lower complexity. A reduced complexity equalizer may be implemented by using either RSSE or DDFSE. The complexity of the overall receiver, assuming no state reduction in the equalizer, is proportional to $S + (2^{m+r})^L$. For channels with a short impulse response, the complexity of the overall receiver is much less than a receiver employing joint MLSE equalization and decoding on the super-trellis. Furthermore, it can perform better as we shall see in the next section. For example, the number of super-states when an 8-state 32-QAM trellis code is used on a two- and three-tap channel is 128 and 2048, respectively. On the other hand, when a separate MLSE equalizer and MLSE decoder is used the corresponding number of states is 40 and 1032, respectively.

7.7.3 DDFSE with Interleaving

Many interleaving algorithms exist for flat fading channels [89], [144], [329]. For fading ISI channels, these conventional interleaving techniques can be applied only if the deinterleaver is preceded by an equalizer. We now describe an interesting interleaver/deinterleaver pair that can be used for ISI channels without the need for separate equalization and decoding.

Consider the interleaver/deinterleaver used in [89]. The deinterleaver introduces the periodic sequence of delays

$$d_k = [k]_P D, \quad k = 0, 1, \dots \tag{7.73}$$

where $[k]_P$ is the residue of k modulo P, D is a positive integer, and P is the period. If P and $D+1$ are relatively prime, then $k_1 + d_{k_1} = k_2 + d_{k_2}$ implies that $k_1 = k_2$ and two samples are never delivered to the output of the deinterleaver at the same time. The interleaver introduces the following sequence of delays

$$d'_k = [(k+1)R]_P D, \quad k = 0, 1, \dots \tag{7.74}$$

where R is the unique integer satisfying

$$[(D+1)R]_P = P - 1 \quad \text{and} \quad 1 \le R < P . \tag{7.75}$$

The total delay of the interleaver/deinterleaver pair is $d = (P - 1)D$ symbol durations.

The symbol x_k is fed to the interleaver at epoch k and transmitted over the channel at epoch $k + d'_k$. Therefore, $\breve{x}_{k+d'_k} = x_k$. Consider an overall discrete-time channel with two taps. The received sample at epoch k is

$$y_k = g_0(k)\breve{x}_k + g_1(k)\breve{x}_{k-1} + \eta_k , \quad k = 0, 1, \dots \tag{7.76}$$

These received samples are fed into the deinterleaver. It can be shown that the output of the deinterleaver is

$$\breve{y}_k = \breve{g}_0(k)x_{k-d} + \breve{g}_1(k)x_{k-d-(D+1)} + \breve{\eta}_k \quad \text{for } [k]_P \ne 0 \tag{7.77}$$

and

$$\breve{y}_k = \breve{g}_0(k)x_{k-d} + \breve{g}_1(k)x_{k-1} + \breve{\eta}_k \quad \text{for } [k]_P = 0 \tag{7.78}$$

where $\breve{g}_0(k)$, $\breve{g}_1(k)$, and $\breve{\eta}_k$ are the deinterleaved versions of the channel vector and noise sequence.

The receiver works as follows. The transmitter periodically inserts a symbol at epoch $iP - 1$, $i = 1, 2, \ldots$ that is known *a priori* to the receiver.[4] The receiver uses this symbol along with (7.78) to decode the symbol $x_{\ell P - d}$ at epoch $k = \ell P$, $(\ell + 1)P > d$. This happens periodically, once every P received samples. At all other epochs, the received sample is described by (7.77). If $D \leq L_D$, where L_D is the decision delay used in the Viterbi algorithm, then DDFSE can be used as the decoding algorithm where the Viterbi algorithm searches through a trellis that is equivalent to the code trellis. When evaluating the branch metric, each survivor path uses decision-feedback based on its own history to cancel the ISI term $x_{k-d-(D+1)}$. If $D > L_D$, then DDFSE can not be implemented because $x_{k-d-(D+1)}$ is no longer available in the survivor sequences when the branch metric is evaluated. However, the decoded symbols at the output of the Viterbi algorithm can be buffered, thereby allowing the ISI term $x_{k-d-(D+1)}$ to be canceled by using the corresponding decoded symbol taken from this buffer. In this case, the same ISI term is used when evaluating the branch metric for all states.

7.7.4   Comparison of Receiver Structures for Fading ISI Channels

Computer simulations were carried out to evaluate the performance of the above receiver structures. Unless otherwise specified, the channel used in the simulations consists of two equal strength Rayleigh faded rays having a differential delay of τ s. A signaling rate of 24 ks/s is assumed, the same rate as specified in IS-54. The tap coefficients of the overall discrete-time channel are assumed to be known to the receiver in order to eliminate the effects of errors in the channel estimates so as to isolate the effects of the equalization, interleaving, and decoding algorithms.

Fig. 7.34 shows the performance of three trellis codes and the equivalent uncoded 4-PSK system with joint MLSE equalization and decoding. The first code is the 4-state, 8-PSK, Ungerboeck code [312] and the second code is the 8-state, 8-PSK, Ungerboeck code [312]. The third code is a 4-D, 4-state, 8-PSK, Wei code with 2 bits/symbol taken from [329]. This code is designed for flat fading channels and has MTD = 2 and MPSD = 4. Fig. 7.34 shows that these codes exhibit a complete reversal in behavior when compared to their performance on AWGN channels. In particular, the uncoded system performs

---

[4] Usually P is quite large, e.g., $P = 80$, so that the loss incurred by transmitting known symbols is negligible.

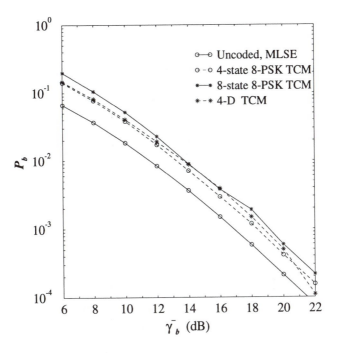

Figure 7.34 Performance of joint MLSE equalization and decoding for the 4-state and 8-state 8-PSK trellis codes and equivalent uncoded 4-PSK on a two-equal-ray Rayleigh fading channel ($\tau = T, f_m = 31.48$ Hz).

the best, and the 4-state, 8-PSK, Ungerboeck code outperforms the 8-state, 8-PSK, Ungerboeck code.

Fig. 7.35 shows the performance of the 8-state 8-PSK Ungerboeck code and the equivalent uncoded 4-PSK system when joint MLSE and DDFSE equalization and decoding is used without interleaving for $\tau = T$ and $f_m = 31.48$ Hz. The effect of using (or not using) a minimum phase noise whitening filter on the performance of DDFSE is also shown, as is the effect of using two-branch diversity with uncorrelated branches. Once again, the uncoded system outperforms the trellis-coded system for both MLSE and DDFSE, with and without diversity.

Fig. 7.36 shows the performance of the nonlinear 8-state, 32-QAM, trellis code specified in CCITT Recommendation V.32 and the equivalent uncoded 16-QAM system, for $\tau = T$ and $f_m = 31.48$ Hz. This code has an asymptotic gain of 4.8 dB when compared to uncoded 16-QAM on an AWGN channel

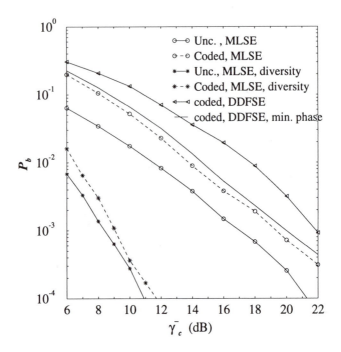

Figure 7.35 Performance of joint MLSE and DDFSE equalization and decoding for the 8-state 8-PSK trellis code and equivalent uncoded 4-PSK on a two-equal-ray Rayleigh fading channel ($\tau = T, f_m = 31.48$ Hz).

[314]. The uncoded system also outperforms the trellis-coded system when joint MLSE equalization and decoding is used without interleaving.

Fig. 7.37 shows the performance of the 8-state, 8-PSK, Ungerboeck code and the uncoded 4-PSK system for the typical urban channel described in Tab. 2.1. The overall T-spaced discrete-time channel response is truncated to one tap. Once again, the uncoded system outperforms the trellis-coded system. Fig. 7.37 also shows the performance on a flat Rayleigh fading channel. The delay spread of the channel described by Tab. 2.1 is much less than the symbol duration (41.16 μ s) and, hence, the performance on that channel is similar to that on a flat fading channel.

Fig. 7.38 shows the performance of different equalization techniques with and without interleaving when the 8-state, 8-PSK Ungerboeck code is used, for $\tau = T$ and $f_m = 31.48$ Hz. The interleaver used in the simulations is a block interleaver with 120 rows and 10 columns. The resulting interleaver/deinterleaver

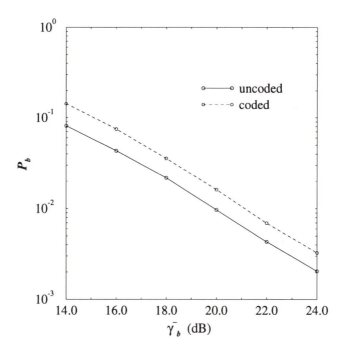

Figure 7.36 Performance of joint MLSE equalization and decoding without interleaving for trellis coded 8-state 32-QAM and equivalent uncoded 16-QAM on a two-equal-ray Rayleigh fading channel ($\tau = T, f_m = 31.48$ Hz).

delay is 98.76 ms. When simulating the DFE, we use a feedback section only with tap coefficients equal to the tap gains of the overall discrete-time channel. The reason for choosing this structure is to isolate the effect of error propagation and the zero-delay decisions made by the threshold detector on the performance of the trellis-coded system. The trellis-coded system that employs interleaving along with a DFE in the receiver gives the worst performance. The system that employs joint MLSE equalization and decoding without spatial diversity performs better than same trellis-coded system with interleaving and a receiver using a DFE with two-branch spatial diversity. An ideal DFE (correct symbols are fed back) with interleaving and without diversity outperforms joint MLSE equalization and decoding without interleaving for both trellis-coded and uncoded systems. Although this situation is very ideal, it further motivates us to examine the performance of powerful trellis codes designed for interleaved flat fading channels when they are applied to fading ISI channels.

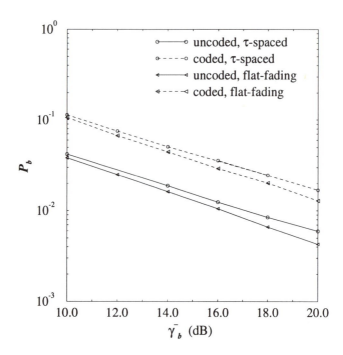

Figure 7.37 Performance of joint MLSE equalization and decoding for trellis coded 8-state, 8-PSK, and equivalent uncoded 4-PSK for the channel described in Tab. 2.1, ($f_m = 31.48$ Hz).

Figs. 7.39 through 7.42 show the performance of different receivers for $\tau = T/4$, $T/2$, $3T/4$, and T, and $f_m = 78.7$ Hz. In all cases, we assume pulse shaping with a truncated raised-cosine pulse having a duration of $6T$. With this model there are eight taps in the overall T-spaced discrete-time channel impulse response. The various receivers under consideration only operate on the two dominant taps, while the remaining 6 taps contribute to uncorrected residual ISI. The receiver described in Section 7.7.2 is labeled as Receiver C. Notice that the uncoded system outperforms the trellis-coded system when joint MLSE equalization and decoding is used for all the values of τ considered. For this receiver, we also observe that the relative performance of the trellis-coded and equivalent uncoded system is insensitive to the differential delay τ. Receiver C outperforms the joint MLSE equalizer and decoder, for $\tau = T$ and $\tau = T/4$. The joint MLSE equalizer and decoder works on the super-trellis with 32 states, while Receiver C has two Viterbi algorithms each working on an 8-state trellis. The performance of Receiver C can be improved further by increasing the number of states in the equalizer, but this implies more complexity. The receiver de-

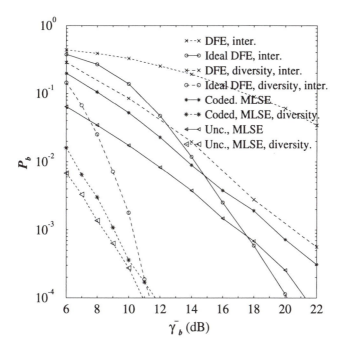

Figure 7.38 Performance of different equalization techniques for trellis coded 8-state 8-PSK and equivalent uncoded 4-PSK on a two-equal-ray Rayleigh fading channel with and without interleaving ($\tau = T, f_m = 31.48$ Hz).

scribed in Section 7.7.3 is labeled as Receiver D. Notice that Receiver D has the best performance for all values of τ under consideration, if the noise whitening filter is selected to give an overall channel impulse response having minimum phase. The minimum phase condition is relatively easy to achieve for a 2-tap channel, because the receiver just needs to identify $g_0(t)$ as the tap having the largest magnitude while the other tap is $g_1(t)$. In the simulations, the values of $P = 80$ and $D = 10$ were chosen, resulting in an interleaving/deinterleaving delay of 32.5 ms. It is interesting that the relative performance of Receiver D on a minimum phase channel is insensitive to the differential delay τ. The trellis-coded system that employs interleaving with a DFE in the receiver gives the worst performance for $\tau = T/2$, $\tau = 3T/4$, and $\tau = T$. An ideal DFE with interleaving always outperforms joint MLSE equalization and decoding without interleaving for all values of τ. This again demonstrates the power of trellis codes designed for flat fading channels with interleaving when applied to fading ISI channels.

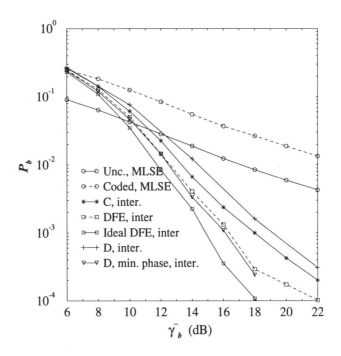

Figure 7.39   Performance of different receiver structures for trellis coded 8-state 8-PSK and equivalent uncoded 4-PSK on a two-equal-ray Rayleigh fading channel ($\tau = T/4, f_m = 78.7$ Hz).

7.8   SUMMARY

This chapter has shown that trellis coded modulation is very effective for interleaved flat fading channels. Good trellis codes for interleaved flat have been identified based on the MTD and MPSD of the code. Interleaving is essential for achieving diversity on a flat fading channel. Trellis coding for equalized ISI channels is much more complicated. For non-interleaved fading ISI channels with joint MLSE equalization and decoding, most of the well known trellis codes actually fail, meaning that the equivalent uncoded system outperforms the trellis coded system. In this case, the problem lies with the code design. Several criteria have been developed in Section 7.5.2 that can be used to identify codes that will have good performance. It was shown that the minimum squared Euclidean distance of the trellis code must grow linearly with the length of the error events if the trellis code is the achieve good performance. Unfortunately, most trellis codes have a dense distance spectrum [270], meaning that

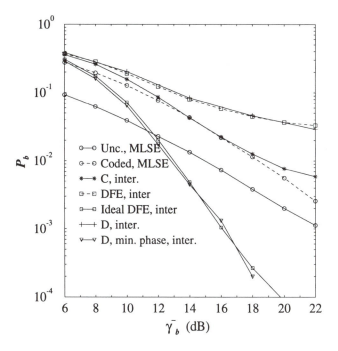

Figure 7.40 Performance of different receiver structures for trellis coded 8-state 8-PSK and equivalent uncoded 4-PSK on a two-equal-ray Rayleigh fading channel ($\tau = T/2, f_m = 78.7$ Hz).

this condition is difficult to satisfy. The identification of effective trellis codes for non-interleaved fading ISI channels is still an open problem.

Interleaving techniques remain attractive for fading ISI channels. If the equalizer and decoder are separated, then interleaving can be applied in a straight forward manner. However, symbol-by-symbol equalization techniques tend to perform poorly with this approach, due to the increased size of the signal constellation when TCM is used. One potentially viable solution is to use an MLSE equalizer, followed by a deinterleaver, and an MLSE decoder. One difficulty with this approach will be the degradation caused by errors in the channel estimates that are used in the MLSE equalizer. The most effective strategy appears to use cleaver interleaving techniques that still allow joint equalization and decoding. In this case, trellis codes that are effective for interleaved flat fading channels can be applied to fading ISI channels. The combination of periodic interleaving and the DDFSE receiver in Section 7.7.3 is one such example.

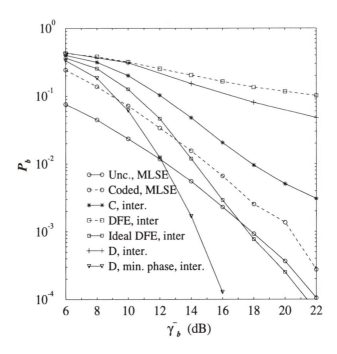

Figure 7.41   Performance of different receiver structures for trellis coded 8-state 8-PSK and equivalent uncoded 4-PSK on a two-equal-ray Rayleigh fading channel ($\tau = 3T/4, f_m = 78.7$ Hz).

Block interleaving strategies have also been identified in [160] that allow joint equalization and decoding to be used with good channel tracking.

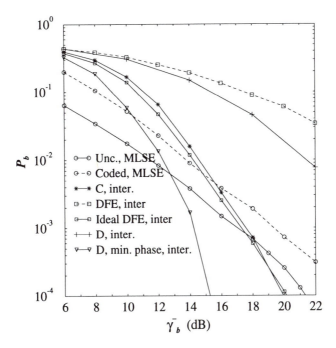

Figure 7.42 Performance of different receiver structures for trellis coded 8-state 8-PSK and equivalent uncoded 4-PSK on a two-equal-ray Rayleigh fading channel ($\tau = T, f_m = 78.7$ Hz).

Problems

7.1. Consider a rate-1/3 convolutional code with generators $\mathbf{g}^{(1)} = (111)$, $\mathbf{g}^{(2)} = (111)$, and $\mathbf{g}^{(3)} = (101)$.

a) Draw a block diagram of the encoder structure.

b) Draw the state diagram and trellis diagram.

c) Determine the output sequence corresponding to the input sequence 1110101.

7.2. The output of a rate-1/3 convolutional encoder with constraint length 3 to the input $\mathbf{a} = (1, 1, 0, \ldots)$ is $\mathbf{b} = (111, 110, 010, \ldots)$

a) Determine the transfer function $T(D, N, L)$.

b) Determine the number of paths through the state diagram or trellis that diverge from the all-zeroes state and remerge with the all-zeroes state 7 branches later.

c) Determine the number of paths of Hamming distance 20 from the all zeroes sequence.

7.3. Consider the rate-1/3 code in Problem 7.1.

a) Determine the transfer function $T(D, N, L)$ of the code. What is the free Hamming distance d_{free}?

b) Assuming the use of BPSK signaling and an AWGN channel, derive a union-Chernoff bound on the decoded bit error probability with i) hard decision decoding and ii) soft decision decoding.

c) Repeat part b) assuming an interleaved flat Rayleigh fading channel, where the receiver has perfect knowledge of the channel.

7.4. Consider the 8-PAM and 32-CROSS signal constellations in Fig. 7.43.

a) Construct the partition chain as in Fig. 7.10 and compute the minimum Euclidean distance between signal points at each step in the partition chain.

b) What is the average symbol energy for each of the signal constellations.

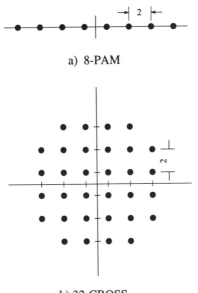

a) 8-PAM

b) 32-CROSS

Figure 7.43   Signal constellations for Problem 7.4.

7.5. Consider the 2-state, rate-1/2, trellis encoder shown in Fig. 7.44. By using this encoder with a 4-PAM and 8-PAM signal constellation we can construct a TCM systems having bandwidth efficiencies of 1 bit/s/Hz and 2 bits/s/Hz, respectively.

a) Determine the appropriate partitions for the signal constellation for the 2-state, 4-PAM and 8-PAM trellis codes.

b) Construct and label the trellis diagrams for the 2-state 4-PAM and 8-PAM trellis codes.

c) Determine the minimum Euclidean distance for each trellis code, and the asymptotic coding gain on an AWGN channel relative to the equivalent uncoded systems.

7.6. Construct and label the trellis diagram for a two-state MTCM system using 8-PSK. What is the asymptotic coding gain for this system on an AWGN channel relative to the equivalent uncoded system.

7.7. For the MTCM code shown in Fig. 7.18, show how the values of MDD1 and MDD2 are determined. Repeat for the 2-D code shown in Fig. 7.19.

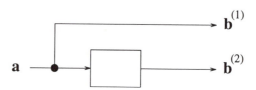

Figure 7.44   Trellis encoder for Problem 7.5.

7.8. To simplify the calculation of performance bounds a Chernoff bound is often imposed on the pairwise error probability.

a) Derive the Chernoff bound on the pairwise error probability for an AWGN channel with soft decision decoding, given by (7.32).

b) Derive the Chernoff bound on the pairwise error probability for an AWGN channel with hard decision decoding, given by (7.34).

c) Derive the Chernoff bound on the pairwise error probability for an interleaved flat fading channel with soft decision decoding, given by (7.46).

7.9. Suppose that the 2-state, 4-PAM trellis code in Problem 7.5 is used on a 2-tap ISI channel characterized by the channel vector $\mathbf{g} = (g_0, g_1)$.

a) Construct the super-trellis diagram.

b) What is the minimum distance error event in the super-trellis?

c) Determine the condition number of the path distance matrix $\mathbf{E}$ for the minimum distance error event found in part b).

d) Determine the channel vectors that minimize and maximize the pairwise error probability.

8

CODE DIVISION MULTIPLE ACCESS

Spread spectrum systems were originally developed for military applications, to provide antijam and low probability of intercept communications by spreading a signal over a large frequency band and transmitting it with a low power per unit bandwidth [73], [247], [286]. Recently, code division multiple access (CDMA) based on spread spectrum technology has been recognized as a viable alternative to both frequency division multiple access (FDMA) and time division multiple access (TDMA) for cellular systems. During the late 1980s and early 1990s, Qualcomm, Inc.'s efforts, along with those of many other organizations such as Motorola and AT&T, have lead to the North American IS-95 cellular standard [79]. A detailed description of the IS-95 CDMA cellular approach can be found in a number of papers, including those by Lee [181] and Gilhousen *et al.* [118]. The book by Viterbi [324] provides a good coverage of the spread spectrum concepts that form the foundation of the IS-95 CDMA cellular system.

Spread spectrum signals have the distinguishing characteristic that the bandwidth used is much greater than the message bandwidth. This band spread is achieved by using a spreading code or pseudo-noise (PN) sequence that is *independent* of the message and is known to the receiver. The receiver uses a synchronized replica of the PN sequence to despread the received signal allowing recovery of the message. Since the PN sequence is independent of the message, the bandwidth expansion does not combat additive white Gaussian noise (AWGN), unlike some other modulation techniques such as wide band analog FM. Nevertheless, the wide band character of spread spectrum signals can be utilized to mitigate the effects of interference and multipath fading.

While there are many different types of spread spectrum systems, the two predominant types are direct sequence (DS) spread spectrum and frequency hopped (FH) spread spectrum. DS spread spectrum achieves the band spread by using the PN sequence to introduce rapid phase transitions into the carrier containing the data, while FH spread spectrum achieves the band spread by using the PN sequence to pseudo-randomly hop the carrier frequency throughout a large band. An excellent tutorial treatment of spread spectrum can be found in the books by Simon *et al.* [286] and Ziemer and Peterson [354]. Some of the early proposals that applied CDMA to cellular radio, such as the system proposed by Cooper and Nettleton [53], were based on FH spread spectrum while most of the recent standards, such as IS-95, favor DS spread spectrum. As a result, the focus of this chapter is on DS CDMA.

While it appears that any cellular system can be suitably optimized to yield a competitive spectral efficiency regardless of the multiple access technique being used, CDMA offers a number of advantages along with some disadvantages. The advantages of CDMA for cellular applications include i) universal one-cell frequency reuse, ii) narrow band interference rejection, iii) inherent multipath diversity in DS CDMA, iv) ability to exploit silent periods in speech voice activity, v) soft hand-off capability, vi) soft capacity limit, and vii) inherent message privacy. The disadvantages of CDMA include i) stringent power control requirements with DS CDMA , ii) hand-offs in dual mode systems, and iii) difficulties in determining the base station (BS) power levels for deployments that have cells of differing sizes, and iv) pilot timing.

This chapter begins with an introduction to DS and FH spread spectrum in Section 8.1, along with a comparison between these two types of spread spectrum systems. Such a comparison is important if we are to determine the best CDMA approach for a given environment. PN sequences are fundamental to all spread spectrum systems and are the subject of Section 8.2. The remainder of the chapter concentrates on DS spread spectrum. Section 8.3 discusses the performance of point to point DS spread spectrum on frequency selective fading channels and shows how a RAKE receiver can be used to gain multipath diversity. Error probability upper and lower bounds and approximations are essential for predicting the performance of CDMA systems. Section 8.4 considers an accurate analysis of the error probability of DS CDMA on AWGN channels. Several Gaussian approximations to the error probability are derived. The chapter concludes with a performance evaluation of cellular DS CDMA. Unfortunately, DS CDMA cellular systems are very complex systems with intricate interactions between system functions. Therefore, the analytical evaluation of system capacity typically requires simplifying assumptions, while focusing on a particular parameter or effect. Usually we can obtain relative performance comparisons, while the true capacity of a suitably optimized CDMA system in a realistic deployment scenario remains elusive.

8.1 INTRODUCTION TO DS AND FH SPREAD SPECTRUM

8.1.1 DS Spread Spectrum

A simplified quadrature DS/QPSK spread spectrum system is shown in Fig. 8.1. The PN sequence generator produces the **spreading waveform**

$$a(t) = \sum_k a_k h_a(t - kT_c) \ , \tag{8.1}$$

where $\mathbf{a} = \{a_k : a_k \in \{\pm 1 \pm j\}\}$ is the complex spreading sequence, T_c is the PN symbol or **chip** duration, and $h_a(t)$ is a real chip amplitude shaping function. The energy per chip is

$$E_c = \frac{1}{2} \int_0^{T_c} h_a^2(t) dt \ . \tag{8.2}$$

The data sequence can be represented by the waveform

$$x(t) = A \sum_n x_n u_T(t - nT) \ , \tag{8.3}$$

where $\mathbf{x} = \{x_n : x_n \in \{\pm 1 \pm j\}\}$ is the complex source symbol sequence, A is the amplitude, and T is the symbol duration. It is necessary that T be an integer multiple of T_c, and the ratio $G = T/T_c$ is called the **processing gain**, defined as the ratio of spread to unspread bandwidth. The complex envelope is obtained by multiplying $a(t)$ and $d(t)$, viz

$$v(t) = A \sum_n \sum_{k=1}^{G} x_n a_{nG+k} h_a(t - (nG + k)T_c) \ . \tag{8.4}$$

This waveform is applied to a quadrature modulator to produce the bandpass waveform

$$s(t) = \sum_n \sum_{k=1}^{G} \left\{ x_n^R a_{nG+k}^R h_a(t - (nG + kT_c) \cos(2\pi f_c t) \right.$$
$$\left. - x_k^I a_{nG+k}^I h_a(t - (nG + kT_c) \sin(2\pi f_c t) \right\} \tag{8.5}$$

where

$$a_k = a_k^R + j a_k^I \tag{8.6}$$
$$x_n = x_n^R + j x_n^I \ . \tag{8.7}$$

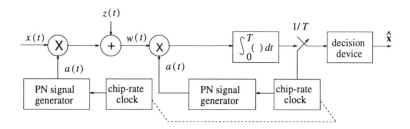

Figure 8.1 Simplified quadrature DS system operating on an AWGN channel.

The complex envelope $v(t)$ appears like that for ordinary QPSK, except that the signaling rate is G times faster. If the sequences $\mathbf{a}$ and $\mathbf{x}$ are completely random, then the power spectral density (psd) of $v(t)$ can be obtained directly from (4.226) as

$$S_{vv}(f) = \frac{A^2}{T_c} |H_a(f)|^2 \ . \tag{8.8}$$

Fig. 8.1 also shows a simplified DS/QPSK receiver. A variety of receivers structures can be constructed that are functionally equivalent to the one shown in Fig. 8.1. In general, the DS spread spectrum receiver must perform three functions; synchronize with the incoming spreading sequence, despread the signal, and detect the data. Multiplying the received complex envelope $w(t) = v(t) + z(t)$ by $a(t)$, integrating over the nth data symbol interval, and sampling, yields the decision variable

$$
\begin{aligned}
\mu &= x_n \int_0^T \sum_{k=1}^{G} h_a^2(t-(nG+k)T_c)dt + \int_0^T z(t) \sum_{k=1}^{G} h_a(t)(t-(nG+k)T_c)dt \\
&= 2GE_c x_n + z_n \\
&= 2E x_n + z_n
\end{aligned}
\tag{8.9}
$$

where $E = GE_c$ and z_n is a zero-mean Gaussian random variable with variance $\frac{1}{2}\text{E}[|z_n|^2] = 2N_o E$. Since $x_n \in \{\pm 1 \pm j\}$, it follows that the probability of decision error is exactly the same as QPSK on an AWGN channel, i.e.,

$$P_b = Q(\sqrt{2\gamma_b}) \tag{8.10}$$

where $\gamma_b = E_b/N_o$ is the received bit energy-to-noise ratio. Note that the use of spread spectrum signaling does not improve the bit error rate performance on an AWGN channel. However, spread spectrum signaling will be shown to offer significant performance gains against interference, multipath-fading, and other types of channel impairments.

8.1.2   FH Spread Spectrum

A conceptual FH spread spectrum transmitter is shown in Fig. 8.2. Frequency hopping is accomplished by using a digital frequency synthesizer that is driven by a PN sequence generator. Each information symbol is transmitted on one or more hops. The most commonly used modulation with frequency hopping is M-ary frequency shift keying (MFSK). . With MFSK, the complex envelope is

$$v(t) = A \sum_n e^{x_n 2\pi \Delta_f t} u_T(t - nT) \qquad (8.11)$$

where $x_n \in \{\pm1, \ \pm3, \ \ldots, \ \pm M - 1\}$ and ϕ is an arbitrary random phase. Usually, the frequency separation $\Delta_f = 1/2T$ is chosen so that the waveforms

$$v_i(t) = A e^{x_n 2\pi \Delta_f t} \ , \ \ 0 \le t \le T \qquad (8.12)$$

are orthogonal. A FH/MFSK signal is generated by using a PN sequence to select a set of carrier frequency shifts. There are two basic types of FH spread spectrum, fast frequency hop (FFH) and slow frequency hop (SFH). With SFH one or more (in general L) source symbols are transmitted per hop. The complex envelope in this case can be written as

$$v(t) = A \sum_n \sum_i e^{x_{n,i} 2\pi \Delta_f t + 2\pi f_n t} u_T(t - nT) \qquad (8.13)$$

where f_n is the nth hop frequency and $x_{n,i}$ is the ith source symbol that is transmitted on the nth hop. FFH systems, on the other hand transmit the same source symbol on multiple hops. In this case, the complex envelope is

$$v(t) = A \sum_n \sum_i e^{x_n 2\pi \Delta_f t + 2\pi f_{n,i} t} u_T(t - nT) \qquad (8.14)$$

where $f_{n,i}$ is the ith hop frequency for the nth source symbol.

Detection of FH/MFSK is usually performed noncoherently using a square-law detector. With SFH, the error probability on an AWGN channel is given by

$$P_b = \frac{1}{2} e^{-\gamma_b/2} \ . \qquad (8.15)$$

8.1.3   Comparison of DS and FH Spread Spectrum

Both DS and FH spread spectrum have been proposed for cellular radio applications, and one has a number of advantages and disadvantages with respect

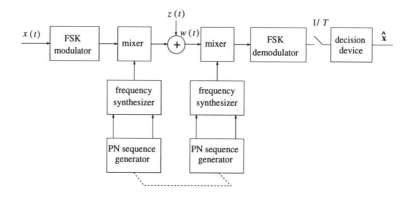

Figure 8.2   Simplified FH system operating on an AWGN channel.

to the other. We now compare the advantages and disadvantages of DS and FH spread spectrum as they are applied to cellular radio.

DS Spread Spectrum

Advantages

1. *Radiolocation:* Subscriber location information can be provided through radiolocation. Radiolocation methods typically use time difference of arrival information of a signal from a mobile station (MS) that is received at multiple BSs that are synchronized to a common clock. Coarse acquisition timing information can provide a time difference of arrival estimate that is accurate to within a half chip duration; for a 10 MHz chip rate this corresponds to about 15 m. Further accuracy in the location estimates can be obtained by using the PN code tracking loop.

2. *Detection:* Coherent or differentially coherent detection can be used. These detection schemes can offer a 1 to 3 dB performance advantage over non-coherent detection that may be necessary in FH spread spectrum.

3. *Processing Gain:* The bandwidth is limited by the clock rate of the PN sequence. This is not a problem for cellular and PCS applications, since the bandwidth used is not that large anyway. Today, the maximum achievable chip rate is about 100 Mchips/s, representing a bandwidth of about 100 MHz.

4. *Electromagnetic Compatibility:* The signal energy is spread throughout the entire bandwidth. This minimizes the interference that a spread spectrum system can cause to narrow-band systems occupying the same band.

Disadvantages

1. *Power Control:* Power control is essential for combating the near-far effect on the reverse (mobile-to-base) channel. The near-far problem occurs when a transmitter that is close-in to the BS swamps out the weaker signals from more distant MSs. Hence, the power levels from all MSs linked to a BS must be adjusted so that the BS receives the same power from each of them.

2. *Multiple-access Interference:* Multiple-access interference is accentuated by multipath. This applies to both the synchronous forward channel and the asynchronous reverse channel. This reduces system performance, but can be offset by exploiting multipath diversity as discussed below.

3. *Coding gain:* Interleaving must be used to reduce fading induced channel memory. However, the achievable interleaving depth is limited by the delay constraints imposed by voice communication.

4. *Flexibility:* DS spread spectrum cannot be easily included as an add-on feature to other typed of digital cellular radio systems, e.g., those using TDMA.

Other

1. *Diversity:* For frequency selective fading channels, DS spread-spectrum can obtain diversity by exploiting the correlation properties of the spreading sequences to resolve and combine the signal replicas that are received over multiple independently faded paths. Sometimes this is called multipath diversity or spread spectrum diversity. In practice, multipath diversity is obtained by using a RAKE receiver.

2. *Rejection of Narrow-band Interference:* During the despreading operation, unwanted narrow-band interference is spread throughout the spread spectrum bandwidth which will reduce its effect on the desired signal.

FH Spread Spectrum

Advantages

1. *Power Control:* Power control is unnecessary in FH spread spectrum systems. Because the interfering signals overlap for only a short period of time, the affect of a close-in MS on a more distant MS is limited.

2. *Multiple-Access Interference:* Multiple-access interference is not accentuated by multipath because the instantaneous bandwidth is usually smaller than the coherence bandwidth of the channel.

3. *Coding Gain:* Large coding gains are possible provided that independent fades are experienced from hop-to-hop. Successive code symbols can be transmitted on separate hops, thereby reducing the channel memory that is introduced by fading and multiple-access interference. Since most error correction codes are designed for memoryless channels, this will make the codes more effective.

4. *Flexibility:* Frequency hopping can be a simple add-on feature to TDMA cellular systems. The GSM system can use SFH as an option to mitigate the effects of co-channel interference, where the carrier is hopped during the guard times between data bursts.

Disadvantages

1. *Radiolocation:* Radiolocation techniques cannot be applied.

2. *Detection:* FFH spread spectrum is usually restricted to differentially coherent or noncoherent detection. It is more difficult to achieve coherent detection with FH spread spectrum, because the receiver must acquire the carrier phase after each hop.

3. *Processing Gain:* The bandwidth is not limited by the clock speed. Processing gain is only limited by the available bandwidth, and the bandwidth does not even have to be contiguous. The hop rate, however, is limited by the clock speed. Modern numerically controlled oscillators can provide about 1 million hops/s over a 20 MHz bandwidth.

4. *Electromagnetic Compatibility:* The instantaneous bandwidth is small. Therefore, when the signal hops into a bandwidth that is occupied by another signal, it may cause significant interference to the other signal.

Other

1. *Diversity:* For frequency-selective fading channels, FFH can obtain frequency diversity provided that the channel coherence bandwidth is much greater that the instantaneous bandwidth of the FH signal. Under this condition, FFH transmits the same data bit on multiple independently faded hops. FFH can also reduce the effect of multiple-access interference, because multiple hops have to be hit to destroy a data bit.

2. *Rejection of Narrow-band Interference:* The action of hopping from one carrier frequency to the next places a limit on the amount of interference that a narrow-band signal can inflict on the spread spectrum signal. That is, frequency hopping rejects narrow-band interference by avoidance.

8.2 SPREADING SEQUENCES AND THEIR CORRELATION FUNCTIONS

CDMA systems achieve their multiple-access capability by using large sets spreading sequences that are chosen to have three desirable attributes; i) the sequences are balanced so that each element of the sequence alphabet occurs with equal frequency, ii) the autocorrelations have small off-peak values, to allow for rapid sequence acquisition at the receiver and to minimize self interference due to multipath, iii) the crosscorrelations are small at all delays, to minimize multiple access interference.

Spreading sequences are often described in terms of their correlation properties. Let the spreading sequence for the kth user be denoted by $\mathbf{a}^{(k)}$. The **full period autocorrelation** of the sequence $\mathbf{a}^{(k)}$ is defined as

$$\phi_{k,k}(n) = \frac{1}{2N} \sum_{i=1}^{N} a_i^{(k)} a_{i+n}^{(k)*} \tag{8.16}$$

and the **full period crosscorrelation** between the sequences $\mathbf{a}^{(k)}$ and $\mathbf{a}^{(m)}$ is defined as

$$\phi_{k,m}(n) = \frac{1}{2N} \sum_{i=1}^{N} a_i^{(k)} a_{i+n}^{(m)*} \tag{8.17}$$

where N is the length or period of the spreading sequences. Sometimes the **aperiodic crosscorrelation** between $\mathbf{a}^{(k)}$ and $\mathbf{a}^{(m)}$ is of interest, defined by

$$
\phi_{k,m}^a(n) = \begin{cases} \frac{1}{2N} \sum_{i=1}^{N-n} a_i^{(k)} a_{i+n}^{(m)^*} & , \quad 0 \leq n \leq N-1 \\ \frac{1}{2N} \sum_{i=1}^{N+n} a_{i-n}^{(k)} a_i^{(m)^*} & , \quad -N+1 \leq n \leq 0 \\ 0 & , \quad |n| \geq N \end{cases} \tag{8.18}
$$

The **aperiodic autocorrelation function** $\phi_{k,k}^a(n)$ has a similar definition.

In many cases, the period of the spreading sequence is much larger that the processing gain, i.e., duration of the sequence is much larger than the duration of a data bit. In this case, **partial period correlations** are of interest, because the correlations in the receiver take place over a time interval that is equal to the duration of a data bit, i.e., $G < N$. The partial period auto- and cross-correlations are equal to

$$
\phi_{k,k}^p(n) \quad = \quad \frac{1}{2G} \sum_{i=1}^{G} a_i^{(k)} a_{i+n}^{(m)^*} \tag{8.19}
$$

$$
\phi_{k,m}^p(n) \quad = \quad \frac{1}{2G} \sum_{i=1}^{G} a_i^{(k)} b_{i+n}^{(m)^*} \quad . \tag{8.20}
$$

The partial period correlations are not only a function of the delay n, but also depend upon the point in the sequence(s) where the summation actually starts. The exact values of the partial period correlations are difficult to derive, except for certain types of sequences. Therefore, we often resort to a statistical treatment under the assumption that the sequences are randomly chosen sequences of independent, identically distributed random variables, e.g., for binary sequences $\mathrm{P_r}\{a_n^{(k)} = +1\} = \mathrm{P_r}\{a_n^{(k)} = -1\} = 1/2$. For random sequences

$$
\frac{1}{2}\mathrm{E}[a_n^{(k)}] = 0 \qquad \frac{1}{2}\mathrm{E}[|a_n^{(k)}|^2] = 1 \qquad \frac{1}{2}\mathrm{E}[a_n^{(k)} a_n^{(m)^*}] = 0 \quad . \tag{8.21}
$$

Hence, the mean value of the partial period autocorrelation is

$$
\mu_{\phi_{k,k}^p}(n) = \mathrm{E}[\phi_{k,k}^p(n)] \quad = \quad \frac{1}{2G} \sum_{i=1}^{G} \mathrm{E}[a_i^{(k)} a_{i+n}^{(k)^*}]
$$

$$
= \quad \begin{cases} 1 & , \quad n = \ell N \\ 0 & , \quad n \neq \ell N \end{cases} \tag{8.22}
$$

where ℓ is an integer. The variance of the partial period autocorrelation is

$$\sigma^2_{\phi^p_{k,k}(n)} = E[(\phi^p_{k,k}(n))^2] - \mu^2_{\phi^p_{k,k}(n)} = \frac{1}{(2G)^2} \sum_{i=1}^{G} \sum_{j=1}^{G} E[a_i^{(k)} a_{i+n}^{(k)^*} a_j^{(k)} a_{j+n}^{(k)^*}]$$

$$= \begin{cases} 0 & , \quad n = \ell N \\ 1/G & , \quad n \neq \ell N \end{cases} . \qquad (8.23)$$

Likewise, the mean and variance of the partial period crosscorrelation are

$$\mu_{\phi^p_{k,m}(n)} = E[\phi^p_{k,m}(n)] = 0 \qquad (8.24)$$

$$\sigma^2_{\phi^p_{k,m}(n)} = E[(\phi^p_{k,m}(n))^2] - \mu^2_{\phi^p_{k,m}(n)} = 1/G . \qquad (8.25)$$

8.2.1 Spreading Waveforms

The full period crosscorrelation between two spreading waveforms $a^{(k)}(t)$ and $a^{(m)}(t)$ is

$$R_{k,m}(\tau) = \frac{1}{T} \int_0^T a^{(k)}(t) \, a^{(m)^*}(t + \tau)dt$$

$$= \frac{1}{T} \sum_{i=-\infty}^{\infty} \sum_{j=-\infty}^{\infty} a_i^{(k)} a_j^{(m)^*} \int_0^T h_a(t - iT_c)h_a(t + \tau - jT_c)dt .$$

$$(8.26)$$

The integral in (8.26) is nonzero only where the chip pulses $h_a(t - iT_c)$ and $h_a(t + \tau - jT_c)$ overlap. Since the delay τ can assume any value let $\tau = \ell T_c + \delta$, where $\ell = \lfloor \tau/T_c \rfloor$ is an integer and $0 \leq \delta < T_c$. If the chip pulses are chosen to have duration T_c and $\tau = \ell T_c + \delta$, then the chip pulses overlap only for $i = \ell + j$ and $i = \ell + j + 1$, so that

$$R_{k,m}(\tau) = \frac{1}{N} \sum_{i=0}^{N-1} a_i^{(k)} a_{\ell+i}^{(m)^*} \frac{1}{T_c} \int_0^{T_c-\delta} h_a(t')h_a(t' + \delta)dt'$$

$$+ \frac{1}{N} \sum_{i=0}^{N-1} a_i^{(k)} a_{\ell+i+1}^{(m)^*} \frac{1}{T_c} \int_{T_c-\delta}^{T_c} h_a(t')h_a(t' - T_c + \delta)dt' . (8.27)$$

The **continuous-time partial autocorrelation functions** of the chip waveform $h_a(t)$ are defined as [258]

$$R_h(\delta) \;=\; \frac{1}{T_c} \int_0^{T_c - \delta} h_a(t') h_a(t' + \delta) dt' \qquad (8.28)$$

$$\hat{R}_h(\delta) \;=\; \frac{1}{T_c} \int_{T_c - \delta}^{T_c} h_a(t') h_a(t' - T_c + \delta) dt' \qquad (8.29)$$

allowing us to write

$$R_{k,m}(\tau) = \phi_{k,m}(\ell) R_h(\delta) + \phi_{k,m}(\ell + 1) \hat{R}_h(\delta) \qquad (8.30)$$

where $\phi_{k,m}(\ell)$ is the full period crosscorrelation defined in (8.17). If $h_a(t) = u_{T_c}(t)$, then

$$R_{k,m}(\tau) = \phi_{k,m}(\ell) \left(1 - \frac{\delta}{T_c} \right) + \phi_{k,m}(\ell + 1) \frac{\delta}{T_c} \; . \qquad (8.31)$$

Likewise, the full period autocorrelation of $a^{(k)}(t)$ with $h_a(t) = u_{T_c}(t)$ is

$$R_{k,k}(\tau) = \phi_{k,k}(\ell) \left(1 - \frac{\delta}{T_c} \right) + \phi_{k,k}(\ell + 1) \frac{\delta}{T_c} \qquad (8.32)$$

where $\phi_{k,k}(\ell)$ is defined in (8.19).

When $G < N$, the partial correlations in (8.19) and (8.20) must be used. In this case the crosscorrelation in (8.30) becomes a random variable that (for random spreading sequences) has mean and variance

$$\mu_{R_{k,m}(\tau)} \;=\; \mu_{\phi_{k,m}(\ell)} R_h(\delta) + \mu_{\phi_{k,m}(\ell+1)} \hat{R}_h(\delta) = 0 \qquad (8.33)$$

$$\sigma^2_{R_{k,m}(\tau)} \;=\; \sigma^2_{\phi_{k,m}(\ell)} R_h^2(\delta) + \sigma^2_{\phi_{k,m}(\ell+1)} \hat{R}_h^2(\delta)$$

$$\;=\; \frac{1}{G} \left(R_h^2(\delta) + \hat{R}_h^2(\delta) \right) \; . \qquad (8.34)$$

Likewise, the autocorrelation is also a random variable that (for random spreading sequences) has mean and variance

$$\mu_{R_{k,k}(\tau)} \;=\; \mu_{\phi_{k,k}(\ell)} R_h(\delta) + \mu_{\phi_{k,k}(\ell+1)} \hat{R}_h(\delta)$$

$$\;=\; \begin{cases} R_h(\delta) \,, & \ell = iG \\ \hat{R}_h(\delta) \,, & \ell + 1 = iG \\ 0 \,, & \text{elsewhere} \end{cases} \qquad (8.35)$$

$$\sigma^2_{R_{k,k}(\tau)} \;=\; \sigma^2_{\phi_{k,k}(\ell)} R_h^2(\delta) + \sigma^2_{\phi_{k,k}(\ell+1)} \hat{R}_h^2(\delta)$$

$$\;=\; \begin{cases} R_h^2(\delta) \,, & \ell = iG \\ \hat{R}_h^2(\delta) \,, & \ell + 1 = iG \\ 1/G \,, & \text{elsewhere} \end{cases} \qquad (8.36)$$

where i is an integer.

8.2.2 m-sequences

One very well known class of spreading sequences are the maximal length sequences or m-sequences. An m-sequence of length $N = 2^m - 1$ is generated by using a linear feedback shift register (LFSR) of length m having a characteristic polynomial that is a primitive polynomial of degree m over GF(2). This polynomial has the form

$$p(x) = 1 + p_1 x + p_2 x^2 + p_3 x^3 + \cdots + p_{m-1} x^{m-1} + x^m . \tag{8.37}$$

Tables of primitive polynomials are tabulated in many texts, e.g., [189]. A typical m-sequence generator is shown in Fig. 8.3. The m-sequences have many remarkable properties. It is well known that all $2^m - 1$ cyclic shifts of an m-sequence along with the zero vector form a $(2^m - 1, m)$ linear cyclic block code, called a simplex code. This code has $2^m - 1$ codewords of weight 2^{m-1} and one codeword of weight zero. The autocorrelation of an m-sequence is

$$\phi_{aa}(n) = \begin{cases} 1 & , \quad n = \ell N \\ -1/N & , \quad n \neq \ell N \end{cases} . \tag{8.38}$$

For large values of N, $\phi_{aa}(n) \approx \delta(n)$ so that m-sequences are almost ideal when viewed in terms of their autocorrelation. The mean and variance of the partial period autocorrelation of an m-sequence can be obtained in a straight forward fashion by replacing the expectations in (8.22) and (8.23) with averages over all possible starting positions. This gives

$$\mu_{\phi_{aa}(n)} = \begin{cases} 1 & , \quad n = \ell N \\ -1/G & , \quad n \neq \ell N \end{cases} \tag{8.39}$$

$$\sigma^2_{\phi_{ab}(n)} = \begin{cases} 0 & , \quad n = \ell N \\ \frac{1}{G}\left(1 + \frac{1}{N}\right)\left(1 - \frac{G}{N}\right) & , \quad n \neq \ell N \end{cases} . \tag{8.40}$$

Unfortunately, m-sequences have a number of undesirable properties for CDMA applications. First, the number of m-sequences that can be generated by a LFSR of length m is equal to the number of primitive polynomials of degree m over GF(2), and is given by $\Phi(2^m - 1)/m$, where $\Phi(\,\cdot\,)$ is the Euler Totient function

$$\Phi(n) = n \prod_{p|n} \left(1 - \frac{1}{p}\right) \tag{8.41}$$

where the product is over all primes p that divide n. Hence, there are relatively few m-sequences for a given m. Second, only for certain values of m, do there exist a few pairs of m-sequences with good crosscorrelations. In general, m-sequences do not have good crosscorrelation properties, making them unsuitable

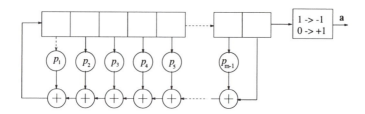

Figure 8.3 General m-sequence generator.

for CDMA applications. However, two well known classes of PN sequences that have excellent correlation properties, making them suitable for CDMA applications, are the **Gold sequences** [122] and **Kasami sequences** [158], [159].

8.2.3   Gold Sequences

A set of Gold sequences consists of $2^m + 1$ sequences each with a period of $N = 2^m - 1$ that are generated by using a **preferred pair** of m-sequences obtained as follows. Let $GF(2^m)$ be an extension field of $GF(2)$. Let α be a primitive Nth root of unity in the extension field $GF(2^m)$, where $N = 2^m - 1$. Let $p_1(x)$ and $p_2(x)$ be a pair of primitive polynomials over $GF(2)$ each having degree m such that $p_1(\alpha) = 0$ and $p_2(\alpha^d) = 0$ for some integer d. Consider the case when $m \neq 0 \bmod 4$. If $d = 2^h + 1$ or $d = 2^{2h} - 2^h + 1$ and if $e = GCD(m, h)$ is such that m/e is odd, then $p_1(x)$ and $p_2(x)$ constitute a preferred pair of polynomials. The two m-sequences **a** and **b** that are generated by using $p_1(x)$ and $p_2(x)$ are known as a preferred pair of m-sequences. Their crosscorrelation function is three-valued with the values $\{-1, -t(m), t(m) - 2\}$ where

$$t(m) = \left\{ \begin{array}{ll} 2^{(m+1)/2} + 1 & , \ m \text{ odd} \\ 2^{(m+2)/2} + 1 & , \ m \text{ even} \end{array} \right. \tag{8.42}$$

By using the preferred pair of sequences **a** and **b**, we can construct a set of Gold sequences by taking the sum of **a** with all cyclically shifted versions of **b** or vice versa. This procedure yields N new sequences each with period $N = 2^m - 1$. These sequences along with the original two sequences gives a set of $2^m + 1$ sequences. A typical Gold sequence generator is shown in Fig. 8.4, where the preferred pair of polynomials are $p_1(x) = 1 + p^2 + p^5$ and $p_2(x) = 1 + p + p^2 + p^4 + p^5$.

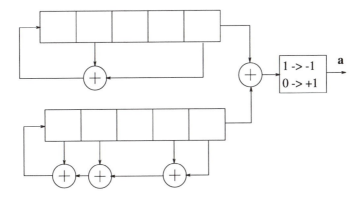

Figure 8.4 A Gold sequence generator with $p_1(x) = 1 + p^2 + p^5$ and $p_2(x) = 1 + p + p^2 + p^4 + p^5$. This sequence generator can produce 32 Gold sequences of length 31.

With the exception of the preferred pair of sequences **a** and **b**, Gold sequences are not m-sequences and, therefore, their autocorrelations are not two-valued. However, Gold showed that Gold sequences have three-valued off-peak auto-correlations and crosscorrelations, with possible values $\{-1, -t(m), t(m) - 2\}$ where $t(m)$ is defined in (8.42).

8.2.4   Kasami Sequences

Let m be even. Let $p_1(x)$ be a primitive polynomial over the binary field GF(2) with degree m and α as a root, and let $p_2(x)$ be the irreducible minimal polynomial of α^d where $d = 2^{m/2} + 1$. Let **a** and **b** represent the two m-sequences of periods $2^m - 1$ and $2^{m/2} - 1$ that are generated by $p_1(x)$ and $p_2(x)$, respectively. The set of Kasami sequences is generated by using the two m-sequences in a fashion similar to the generation of Gold sequences, i.e., the set of Kasami sequences consists of the long sequence **a** and the sum of **a** with all $2^{m/2} - 1$ cyclic shifts of the short sequence **b**. The number of Kasami sequences in the set is $2^{m/2}$, each having period $N = 2^m - 1$. A typical Kasami sequence generator is shown in Fig. 8.5 with generator polynomials $p_1(x) = 1 + x + x^6$ and $p_2(x) = 1 + x + x^3$.

Like Gold sequences, the off-peak autocorrelation and crosscorrelation functions of Kasami sequences are also three-valued, however, the possible values are

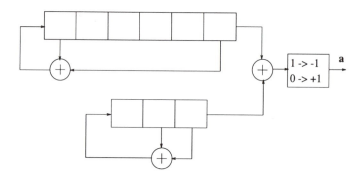

Figure 8.5   A Kasami sequence generator with $p_1(x) = 1 + x + x^6$ and $p_2(x) = 1 + x + x^3$. This sequence generator can produce 8 Kasami sequences of length 63.

$\{-1, -s(m), s(m) - 2\}$ where

$$s(m) = 2^{m/2} + 1 \ . \tag{8.43}$$

8.2.5 Walsh-Hadamard Sequences

Walsh-Hadamard sequences are obtained by selecting as sequences the rows of a **Hadamard matrix** $\mathbf{H}_M$. For $M = 2$ the Hadamard matrix is

$$\mathbf{H}_2 = \begin{bmatrix} +1 & +1 \\ +1 & -1 \end{bmatrix} \ . \tag{8.44}$$

Larger Hadamard matrices are obtained by using the recursion

$$\mathbf{H}_{2M} = \begin{bmatrix} \mathbf{H}_M & \mathbf{H}_M \\ \mathbf{H}_M & -\mathbf{H}_M \end{bmatrix} \ . \tag{8.45}$$

For example,

$$\mathbf{H}_8 = \begin{bmatrix} +1 & +1 & +1 & +1 & +1 & +1 & +1 & +1 \\ +1 & -1 & +1 & -1 & +1 & -1 & +1 & -1 \\ +1 & +1 & -1 & -1 & +1 & +1 & -1 & -1 \\ +1 & -1 & -1 & +1 & +1 & -1 & -1 & +1 \\ +1 & +1 & +1 & +1 & -1 & -1 & -1 & -1 \\ +1 & -1 & +1 & -1 & -1 & +1 & -1 & +1 \\ +1 & +1 & -1 & -1 & -1 & -1 & +1 & +1 \\ +1 & -1 & -1 & +1 & -1 & +1 & +1 & -1 \end{bmatrix} \ . \tag{8.46}$$

Observe that all rows in the Hadamard matrix are orthogonal to each other. This orthogonality property has important implications for the forward channel of CDMA cellular systems. Since all signals are available at the BS in trunked form, the forward channel transmissions can be perfectly synchronized and, hence, are orthogonal. Channel time dispersion due to multipath, however, will destroy the orthogonality of the received waveforms because the Walsh-Hadamard sequences have very poor autocorrelation characteristics. This will introduce a floor into the error rate performance.

8.3 DS SPREAD SPECTRUM ON FREQUENCY-SELECTIVE FADING CHANNELS

Suppose that the bandwidth of the DS complex envelope $v(t)$ is $W/2$. This can be achieved, for example, by using spectral raised cosine pulse shaping. Since the low-pass signal $v(t)$ is band-limited to $|f| \leq W/2$, the sampling theorem can be invoked and $v(t)$ can be completely described by the set of samples $\{v(n/W)\}_{n=-\infty}^{\infty}$. The sampled version of $v(t)$ is

$$v_\delta(t) = \sum_{n=-\infty}^{\infty} v\left(\frac{n}{W}\right) \delta\left(t - \frac{n}{W}\right) \tag{8.47}$$

$$= v(t) \sum_{n=-\infty}^{\infty} \delta\left(t - \frac{n}{W}\right) . \tag{8.48}$$

Taking the Fourier transform of both sides of (8.48) gives

$$V_\delta(f) = V(f) * W \sum_{n=-\infty}^{\infty} \delta(f - nW)$$

$$= W \sum_{n=-\infty}^{\infty} V(f) * \delta(f - nW)$$

$$= W \sum_{n=-\infty}^{\infty} V(f - nW) . \tag{8.49}$$

From (8.49) we can see that

$$V(f) = \frac{1}{W} V_\delta(f) \qquad 0 \leq |f| \leq W/2 . \tag{8.50}$$

Another useful expression can be obtained by taking the Fourier transform of both sides of (8.47) giving

$$V_\delta(f) = \sum_{n=-\infty}^{\infty} v\left(\frac{n}{W}\right) e^{-j2\pi nf/W} \ . \tag{8.51}$$

Combining (8.50) and (8.51) gives

$$V(f) = \frac{1}{W} \sum_{n=-\infty}^{\infty} v\left(\frac{n}{W}\right) e^{-j\pi nf/W} \qquad 0 \le |f| \le W/2 \ . \tag{8.52}$$

If the low-pass signal $v(t)$ is transmitted over a multipath fading channel with time-variant transfer function $T(f, t)$, the received (noiseless) complex envelope is

$$w(t) = \int_{-\infty}^{\infty} V(f)T(f, t)e^{j2\pi ft} df \ . \tag{8.53}$$

Substituting $V(f)$ from (8.52) gives

$$
\begin{aligned}
w(t) &= \frac{1}{W} \sum_{n=-\infty}^{\infty} v\left(\frac{n}{W}\right) \int_{-\infty}^{\infty} T(f, t) e^{-j2\pi f(t-n/W)} df \\
&= \frac{1}{W} \sum_{n=-\infty}^{\infty} v\left(\frac{n}{W}\right) c\left(t - \frac{n}{W}, t\right) \\
&= \frac{1}{W} \sum_{n=-\infty}^{\infty} v\left(t - \frac{n}{W}\right) c\left(\frac{n}{W}, t\right)
\end{aligned}
\tag{8.54}
$$

where $c(\tau, t)$ is the time-variant impulse response of the channel. By defining

$$c_n(t) = \frac{1}{W} c\left(\frac{n}{W}, t\right) \tag{8.55}$$

the noiseless received complex envelope can be written as

$$w(t) = \sum_{n=-\infty}^{\infty} c_n(t) v\left(t - \frac{n}{W}\right) \tag{8.56}$$

and it follows that the complex low-pass impulse response of the channel is

$$c(\tau, t) = \sum_{n=-\infty}^{\infty} c_n(t) \delta\left(\tau - \frac{n}{W}\right) \ . \tag{8.57}$$

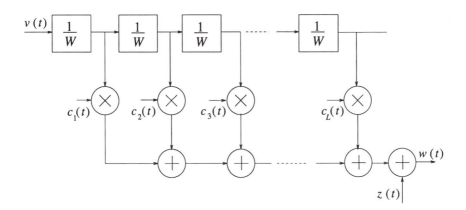

Figure 8.6 Tapped delay line model of a frequency selective fading channel, from [256].

For uncorrelated scattering channels, the $\{c_n(t)\}$ in (8.55) are independent complex Gaussian random processes. Suppose that the channel is causal with an impulse response that is nonzero over a time interval of duration $T_{\max}$. In this case, $c_n(t) = 0$, $n \leq 0, n > L$, where $L = \lfloor T_{\max}/W \rfloor + 1$ and $\lfloor x \rfloor$ is the smallest integer greater than x. It follows that the channel impulse response is

$$c(\tau, t) = \sum_{n=1}^{L} c_n(t)\delta\left(\tau - \frac{n}{W}\right) \quad . \qquad (8.58)$$

In conclusion, the frequency selective fading channel can be modeled by an L-tap $1/W$-spaced tapped delay line with tap vector $\mathbf{c} = (c_1, c_2, \ldots, c_L)$ as shown in Fig. 8.6. Note that the actual tap spacing is somewhat of a fuzzy concept because real pulses are time-limited and, therefore, they are only approximately bandlimited. Finally, it should be emphasized that the channel vector $\mathbf{c}$ is *not* the same as the channel vector $\mathbf{g}$ associated with the discrete-time white noise channel model.

If ideal Nyquist signaling is used with the chip amplitude shaping function $h_a(t) = S_a(\pi t/T_c)$, then $W = 1/T_c$ and the channel can be represented exactly as a T_c-spaced tapped delay line. Such a model is convenient because it leads to a simplified analysis. However, if any other pulse shape is used, such as a raised cosine pulse, then the equivalent tapped delay line channel model is not T_c-spaced, e.g., a raised cosine pulse with unity roll-off results in a $T_c/2$-spaced tapped delay line.

8.3.1   RAKE Receiver

A variety of receiver structures can be used with DS spread spectrum signaling
on frequency selective fading channels. One type of receiver simply uses the autocorrelation properties of the PN sequences to reject the multipath interference
[114], [113]. However, a much better strategy is to exploit the autocorrelation
properties of the spreading sequences by resolving and combining the multipath
components to obtain **multipath diversity** . This type of receiver is called a
RAKE receiver due to its similarity to an ordinary garden rake. [255]. Various
diversity combining techniques may be used to achieve the desired tradeoff between performance and complexity. The performance of DS spread spectrum
with RAKE reception has been studied by a very large number of authors [35],
[161, 162], [170, 171], [183, 184, 185], [235], [263], [274], [304, 305], [341].

To develop the RAKE receiver, suppose that DS/BPSK signaling is used on
the frequency-selective fading channel shown in Fig. 8.6[1]. During the nth bit
interval, one of two complex low-pass waveforms $v_1(t)$ or $v_2(t) = -v_1(t)$ is
transmitted over the channel, where

$$v_1(t) = \sum_{k=1}^{G} a_{nG+k} h_a(t - (nG+k)T_c) \, , \quad nT \le t \le (n+1)T \, . \quad (8.59)$$

For our purpose, the bit duration T is assumed to satisfy the condition $T \gg LT_c$
so as to neglect the ISI between data bits. If $v_1(t)$ is transmitted, then the
received complex envelope during the nth bit interval is

$$
\begin{aligned}
w(t) &= \sum_{\ell=1}^{L} c_\ell(t) v_1 \left(t - \frac{\ell}{W} \right) + z(t) \\
&= \hat{v}_1(t) + z(t) \, , \quad nT \le t \le (n+1)T
\end{aligned}
\quad (8.60)
$$

where $\hat{v}_1(t) = \sum_{\ell=1}^{L} c_\ell(t) v_1(t - \ell/W)$. As discussed in Section 5.2, the maximum
likelihood coherent receiver employs a filter that is matched to the received
pulse $\hat{v}_1(t)$ followed by a baud-rate sampler and decision device; alternatively,
a correlator can be used in place of the matched filter. In either case, the
decision variable at the sampler output is

$$
\begin{aligned}
\mu &= \text{Re} \left\{ \int_{nT}^{(n+1)T} w(t) \hat{v}_1^*(t) dt \right\} \\
&= \text{Re} \left\{ \int_{nT}^{(n+1)T} w(t) \sum_{\ell=1}^{L} c_\ell^*(t) v_1^*(t - \ell/W) dt \right\} \, .
\end{aligned}
\quad (8.61)
$$

[1]The switch from DS/QPSK to DS/BPSK is made for simplicity.

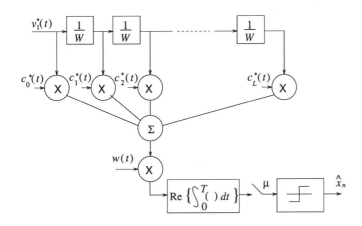

Figure 8.7   RAKE receiver for BPSK signals.

The decision variable μ is compared with a zero-threshold to generate bit decisions.

Notice that the coherent receiver described by (8.61) correlates the received complex envelope $w(t)$ with delayed versions of the waveform $v_1(t)$, followed by maximal ratio diversity combining. This leads to the receiver structure shown in Fig. 8.7. By changing the variable of integration in (8.61) an alternate form of the RAKE receiver can be obtained as shown in Fig. 8.8. In this case the waveform $v_1(t)$ is correlated with delayed versions of the received complex envelope $w(t)$.

Performance of the RAKE Receiver

Suppose that $v_1(t)$ is transmitted and substitute (8.59) into (8.61) to obtain[2]

$$\mu = \sum_{m=1}^{L}\sum_{\ell=1}^{L}\mathrm{Re}\left\{ c_m c_\ell^* \int_{nT}^{(n+1)T} v_1(t-m/W)v_1^*(t-\ell/W)dt \right\} + n_c \qquad (8.62)$$

where

$$n_c = \sum_{m=1}^{L}\mathrm{Re}\left\{ c_m^* \int_{nT}^{(n+1)T} z(t)v_1^*(t-m/W)dt \right\}$$

---

[2] It is assumed that the channel impulse response is randomly static during the transmission of each data bit, so that the dependency of $c_n(t)$ on time is not shown explicitly.

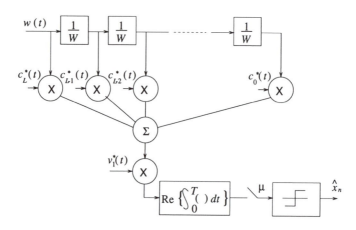

Figure 8.8   Alternate form of RAKE receiver for BPSK signals.

$$= \sum_{m=1}^{L} \alpha_m \mathrm{Re} \left\{ e^{-j\theta_m} \int_{-\infty}^{\infty} z(t) v_1^*(t - mT_c) dt \right\} . \qquad (8.63)$$

and $c_m \triangleq \alpha_m e^{j\phi_m}$. The random variable n_c is Gaussian with zero-mean and variance

$$\sigma_{n_c}^2 = 2EN_o \sum_{m=1}^{L} \alpha_m^2 . \qquad (8.64)$$

In general, the integral in (8.62) is a complicated function of the spreading sequence and chip amplitude shaping pulse that is used. However, certain cases will lead to some useful insight. For example, the ideal Nyquist pulse $h_a(t) = S_a(\pi t/T_c)$ with bandwidth $W = 1/T_c$ leads to[3]

$$\begin{aligned}
I &\equiv \int_{nT}^{(n+1)T} v_1(t - m/W) v_1^*(t - \ell/W) dt \\
&= \sum_{k=1}^{G} \sum_{j=1}^{G} a_k a_j \int_{nT}^{(n+1)T} h_a(t - (m+k)T_c) h_a(t - (\ell+j)T_c) dt \\
&= \sum_{k=1}^{G} a_k a_{m+k-\ell} \int_{-\infty}^{\infty} h_a^2(t) dt \\
&= 2E_c G \phi_{aa}(m - \ell) = 2E \phi_{aa}(m - \ell)
\end{aligned} \qquad (8.65)$$

---

[3] Since DS/BPSK signaling is used the spreading sequence a is real with autocorrelation function $\phi_{aa}(n) = \mathrm{E}[a_i a_{i+n}]$.

where the second last step follows under the assumption that the period of the spreading sequence is equal to the processing gain, i.e., $N = G$. Note that we also assume $G \gg L$ so that the effects of ISI can be neglected. Therefore, (8.62) becomes

$$\mu = 2\mathcal{E} \sum_{m=1}^{L} \alpha_m^2 + 2\mathcal{E} \sum_{m=1}^{L} \sum_{\substack{\ell=1 \\ \ell \neq m}}^{L} \mathrm{Re} \left\{ c_m c_\ell^* \right\} \phi_{aa}(m - \ell) + n_c \qquad (8.66)$$

The second term in the above expression is **self interference** that arises from the non-ideal autocorrelation properties of the spreading sequence.

To demonstrate the effect of the self interference, assume that the channel is Rayleigh faded and consider the random variable

$$\begin{aligned} Y_{m,\ell} &= \mathrm{Re} \left\{ c_m c_\ell^* \right\} \\ &= \phi_m \cos \phi_m \alpha_\ell \cos \phi_\ell + \alpha_m \sin \phi_m \alpha_\ell \sin \phi_\ell \qquad (8.67) \end{aligned}$$

Define the new random variables

$$X_k^R \stackrel{\Delta}{=} \alpha_k \cos \phi_k \qquad X_k^I \stackrel{\Delta}{=} \alpha_k \sin \phi_k \qquad (8.68)$$

Then

$$\alpha_k = \sqrt{(X_k^R)^2 + (X_k^I)^2} \qquad (8.69)$$

$$\phi_k = \mathrm{Tan}^{-1}(X_k^I / X_k^R) \qquad (8.70)$$

Therefore,

$$Y_{m,\ell} = X_m^R X_\ell^R + X_m^I X_\ell^I \qquad (8.71)$$

Since the X_k^R and X_k^I are independent zero-mean Gaussian random variables with variance σ_k^2, $Y_{m,\ell}$ has the Laplacian density

$$p_{Y_{m,\ell}}(y) = \frac{1}{2\sigma_m \sigma_\ell} e^{-\frac{|y|}{\sigma_m \sigma_\ell}} . \qquad (8.72)$$

Making the substitution for $Y_{m,\ell}$ and rearranging the sum in the second term in (8.66) gives

$$\mu = 2E \sum_{m=1}^{L} \alpha_m^2 + 2E \sum_{m=1}^{L} \sum_{\substack{\ell=1 \\ \ell \neq m}}^{L} Y_{m,\ell} \phi_{aa}(m - \ell) + n_c . \qquad (8.73)$$

It is difficult to evaluate the effect of the self-interference, because the $Y_{m,\ell}$ are non-Gaussian and correlated. However, the self interference due to multipath can be minimized by using spreading codes that have small autocorrelation sidelobes in the time intervals during which delayed signals with significant power are expected. For large delays, the stringent requirements on the auto-correlation function can be relaxed. For the reverse channel in cellular CDMA applications, the spreading codes still must have small crosscorrelation sidelobes over all delays, because the reverse channel transmissions are asynchronous. It is easy to find reasonably large sets of sequences that satisfy these properties. For example, a set of $2^m + 1$ Gold sequences can be generated of length $2^m - 1$. Of these $2^m + 1$ sequences, $2^{m-n+1} + 1$ will have their first autocorrelation off-peak ($t_m - 2$ or t_m) at least n chip durations from the main autocorrelation peak. Consequently, these $2^{m-n+1} + 1$ sequences will introduce negligible self interference if they are used on a channel having n or fewer paths.

If the spreading sequences have an ideal autocorrelation function, i.e., $\phi_{aa}(n - m) = \delta_{nm}$, then there is no self interference and (8.73) becomes

$$\mu = 2E \sum_{m=1}^{L} \alpha_m^2 + n_c \qquad (8.74)$$

For a fixed set of $\{\alpha_m\}$, the probability of bit error is

$$P_b(\gamma_b) = Q\left(\sqrt{2\gamma_b}\right) \qquad (8.75)$$

where γ_b is the received bit energy-to-noise ratio given by

$$\gamma_b = \frac{1}{\sigma_{n_c}^2}\left(2E \sum_{m=1}^{L} \alpha_m^2\right)^2 = \sum_{m=1}^{L} \gamma_m \qquad (8.76)$$

with

$$\gamma_m = \frac{E}{N_o}\alpha_m^2 \ . \qquad (8.77)$$

Each of the γ_m are exponentially distributed with density function

$$p(\gamma_m) = \frac{1}{\bar{\gamma}_m}e^{-\gamma_m/\bar{\gamma}_m} \qquad (8.78)$$

where $\bar{\gamma}_m$ is the average received bit energy-to-noise ratio for the kth channel path. To compute the density of γ_b, first note that the characteristic function of γ_m is

$$\psi_{\gamma_m}(jv) = \frac{1}{1 - jv\bar{\gamma}_m} \qquad (8.79)$$

so that the characteristic function of γ_b is

$$\psi_{\gamma_b}(jv) = \prod_{m=1}^{L} \frac{1}{1 - jv\bar{\gamma}_m} \tag{8.80}$$

By using a partial fraction expansion, the density of γ_b is

$$p_{\gamma_b}(x) = \sum_{m=1}^{L} \frac{A_m}{\bar{\gamma}_m} e^{-x/\bar{\gamma}_k} \tag{8.81}$$

where

$$A_m = \prod_{\substack{i=1 \\ i \neq m}}^{L} \frac{\bar{\gamma}_m}{\bar{\gamma}_m - \bar{\gamma}_i} \tag{8.82}$$

The average probability of bit error is

$$
\begin{aligned}
P_b &= \int_0^\infty Q\left(\sqrt{2x}\right) p_{\gamma_b}(x)\,dx \\
&= \frac{1}{2} \sum_{m=1}^{L} A_m \left[1 - \sqrt{\frac{\bar{\gamma}_m}{1 + \bar{\gamma}_m}}\right]
\end{aligned} \tag{8.83}
$$

In order to proceed further the $\bar{\gamma}_m$ must be specified. One plausible model is to assume an exponentially decaying power delay profile, e.g.,

$$\bar{\gamma}_m = Ce^{-k/\varepsilon} \tag{8.84}$$

where ε controls the delay spread and C is chosen to satisfy the constraint $\sum_{m=1}^{L} \bar{\gamma}_m = \bar{\gamma}_b$. Solving for C yields

$$\bar{\gamma}_m = \frac{(1 - e^{-1/\varepsilon})e^{-k/\varepsilon}}{e^{-1/\varepsilon} - e^{-(L+1)/\varepsilon}} \bar{\gamma}_b \tag{8.85}$$

The probability of bit error is plotted in Fig. 8.9 for $L = 4$ and various values of ε. For small ε, the channel is not dispersive and very little multipath diversity is obtained. However, as ε becomes large the channel becomes more dispersive and an Lth-order diversity gain is achieved. Finally, we note that the number of taps actually used in the RAKE receiver can be less than the channel length L with the sacrifice of some performance.

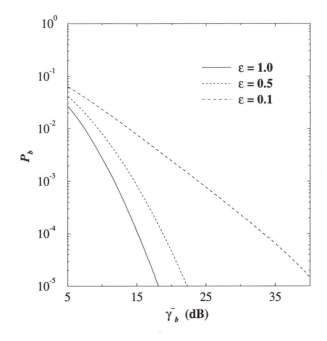

Figure 8.9   Bit error probability with a RAKE receiver for DS/BPSK signaling on a multipath fading channel. The channel has $L = 4$ taps and a small ε indicates a small delay spread.

8.4   ERROR PROBABILITY FOR DS CDMA ON AWGN CHANNELS

DS CDMA systems achieve multiple-access capability by assigning each user a unique spreading sequence. In general, however, the transmissions from the various users are not synchronized so that they arrive at each receiver with random delays and carrier phases as shown in Fig. 8.10. The true error probability depends on the particular spreading sequences that are employed and is a function of the random amplitudes, delays, and carrier phases of the signals that arrive at the intended receiver. Unfortunately, the exact error probability is difficult to evaluate so that a variety of upper and lower bounds, and Gaussian approximations have been derived in the literature.

Suppose that M users simultaneously access the channel. The transmitted complex envelope for the ith user is

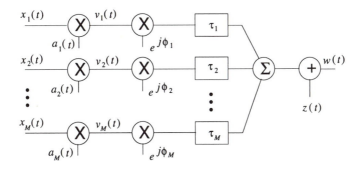

Figure 8.10 DS CDMA signaling on an AWGN channel.

$$v^{(i)}(t) = A \sum_n \sum_{k=1}^{G} x_n^{(i)} a_{nG+k}^{(i)} h_a(t - (nG + k)T_c) \qquad (8.86)$$

where $\mathbf{a}^{(i)} = \{a_n^{(i)}\}$ and $\mathbf{x}^{(i)} = \{x_n^{(i)}\}$ are the ith user's spreading and data sequences, respectively. The data sequence $\mathbf{x}^{(i)}$ is assumed to be a sequence of independent, identically distributed random variables with $P_r\{x_n^{(i)} = +1\} = P_r\{x_n^{(i)} = -1\} = 1/2$. In practice, the spreading sequences $\{a_n^{(i)}\}$ are carefully chosen to have good correlation properties, e.g., Gold sequences or Kasami sequences. However, many of the error probability approximations in the literature assume random spreading sequences, where the error probability is obtained by ensemble averaging over all possible combinations of spreading sequences, including those where multiple transmitters choose the same spreading sequence. For our purpose, we assume that each bit is spread by N chips and the spreading sequences have period N.

In general the signals from the various transmitters arrive at the intended receiver with different power levels. However, DS CDMA cellular systems such as IS-95 are power controlled with the objective of having all the signals arrive at the intended receiver with the same power level. Under the assumption of perfect power control and a frequency non-selective channel, the composite received complex envelope is[4]

$$w(t) = \sum_{i=1}^{M} e^{j\phi_i} v^{(i)}(t - \tau_i) + z(t) \qquad (8.87)$$

---

[4] Here we assume the normalization $\alpha = 1$.

where the $\{\tau_i\}$ and $\{\phi_i\}$ are the random delays and carrier phases of the received signals. In this section, we consider the performance with an ideal correlation receiver, where the composite received signal is multiplied by a synchronized replica of the spreading sequence of the intended transmission; since the two sequences cancel, the desired data sequence can be obtained at the output of the correlator. Because of symmetry, we only need to consider the receiver that is matched to the first transmitter. Furthermore, since only the relative delays and phases are important, we can set $\tau_1 = 0$ and $\phi_1 = 0$, and assume that ϕ_i is unformly distributed on $[0, 2\pi)$ and τ_i is uniformly distributed on $[0, T)$, for $i \neq 0$.

The decision variable at the output of the correlator in Fig. 8.1 has been derived by Lehnert and Pursley [184] and is given by

$$\mu_n = 2\mathcal{E}\left[x_n^{(1)} + \sum_{k=2}^{M} B_{k,1}(x_n, x_{n-1}, \tau_k) \cos\phi_k \right] + n_c \qquad (8.88)$$

where

$$B_{k,1}(x_n, x_{n-1}, \tau_k) = x_{n-1}^{(k)} R_{k,1}^p(\tau_k) + x_n^{(k)} \hat{R}_{k,1}^p(\tau_k) \qquad (8.89)$$

and $R_{k,m}^p(\tau)$ and $\hat{R}_{k,m}^p(\tau)$ are the continuous-time partial crosscorrelation functions of $\mathbf{a}^{(k)}$ and $\mathbf{a}^{(m)}$, defined by

$$R_{k,m}^p(\tau) = \frac{1}{T}\int_0^\tau a^{(k)}(t-\tau)a^{(m)^*}(t)dt \qquad (8.90)$$

$$\hat{R}_{k,m}^p(\tau) = \frac{1}{T}\int_\tau^T a^{(k)}(t-\tau)a^{(m)^*}(t)dt \ . \qquad (8.91)$$

The functions $R_{k,m}^p(\tau)$ and $\hat{R}_{k,m}^p(\tau)$ can be expressed in terms of the discrete aperiodic crosscorrelation function $\phi_{k,m}^a(n)$ and the continuous-time partial chip autocorrelation functions $R_h(\delta)$ and $\hat{R}_h(\delta)$ as

$$R_{k,m}^p(\tau) = \phi_{k,m}^a(\ell - N)\hat{R}_h(\delta) + \phi_{k,m}^a(\ell + 1 - N)R_h(\delta) \qquad (8.92)$$

$$\hat{R}_{k,m}^p(\tau) = \phi_{k,m}^a(\ell)\hat{R}_h(\delta) + \phi_{k,m}^a(\ell + 1)R_h(\delta) \qquad (8.93)$$

where $\ell = \lfloor \tau/T_c \rfloor$ and $\delta = \tau - \ell T_c$. Note that δ is uniform on $[0, T_c)$ and ℓ is uniform on the set $\{0, 1, \ldots, N-1\}$. Combining (8.89), (8.92), and (8.93) gives

$$\begin{aligned}
B_{k,1}(x_n, x_{n-1}, \tau_k) = &\left[x_{n-1}^{(k)}\phi_{k,1}^a(\ell_k - N) + x_n^{(k)}\phi_{k,1}^a(\ell_k) \right] \hat{R}_h(\delta_k) \\
&+ \left[x_{n-1}^{(k)}\phi_{k,1}^a(\ell_k + 1 - N) + x_n^{(k)}\phi_{k,1}^a(\ell_k + 1) \right] R_h(\delta_k) \ .
\end{aligned}$$
$$(8.94)$$

To proceed any further requires information about the aperiodic crosscorrelation functions of the spreading sequences being used, as well as the chip amplitude shaping function. For the special case of random spreading sequences and a rectangular shaping function $h_a(t) = u_{T_c}(t)$, Morrow and Lehnert [220] have shown that

$$B_{k,1}(x_n, x_{n-1}, \tau_k) = P_k \zeta_k + Q_k(1 - \zeta_k) + X_k + Y_k(1 - 2\zeta_k) . \qquad (8.95)$$

where $\zeta = R_h(\delta)$ is uniform on $[0, 1)$, P_k and Q_k are uniform on $\{-1, +1\}$, and X_k and Y_k have the probability distributions

$$p_{X_k}(i) = \frac{1}{2^{-A}}\binom{A}{\frac{1+A}{2}}, \quad i \in \{-A, -A+2, \ldots, A-2, A\} \qquad (8.96)$$

$$p_{Y_k}(i) = \frac{1}{2^{-B}}\binom{B}{\frac{i+B}{2}}, \quad i \in \{-B, -B+2, \ldots, B-2, B\} . \qquad (8.97)$$

The quantities A and B are related to

$$C \equiv N\phi_{1,1}^a(1) = N\sum_{j=0}^{N-2} a_j^{(1)} a_{j+1}^{(1)} \qquad (8.98)$$

by

$$A = \frac{N-1+C}{2} \qquad (8.99)$$

$$B = \frac{N-1-C}{2} \qquad (8.100)$$

where $\phi_{1,1}^a(1)$ is the aperiodic crosscorrelation of the spreading sequence of the first user as defined in (8.18). The parameter B is the number of chip boundaries in one period of the sequence $\mathbf{a}^{(1)}$ at which a transition to a different value occurs. For random spreading sequences, C has the probability distribution

$$p_C(i) = \frac{1}{2^{N-1}}\binom{N-1}{\frac{i+N-1}{2}}, \quad i \in \{-N+1, -N+3, \ldots, N-3, N-1\} . \qquad (8.101)$$

Standard Gaussian Approximation

The **standard Gaussian approximation** assumes that the multiple access interference

$$I = \sum_{k=2}^{M} W_k \qquad (8.102)$$

with

$$W_k = B_{k,1}(x_n, x_{n-1}, \tau_k) \cos \phi_k \qquad (8.103)$$

can be modeled as a Gaussian random variable with a distribution that is completely specified by its mean and variance. The approximation is obtained by conditioning the multiple access interference on the random set of parameters $\{\zeta_k, \phi_k, B\}$ followed by ensemble averaging. It is not difficult to show that

$$E[Z_k | C] = 0 \qquad (8.104)$$

where $Z_k \in \{P_k, Q_k, X_k, Y_k\}$ since the conditional density functions for P_k, Q_k, X_k, and Y_k are symmetrical about zero. Hence $E[W_k] = 0$ and finally $E[I] = 0$. The variance of the multiple access interference is

$$
\begin{aligned}
\sigma_I^2 &= \mathrm{E}\left[I^2 \, |\zeta_k, \phi_k, B \right] \\
&= \sum_{k=2}^{M} \mathrm{E}\left[W_k^2 \, |\zeta_k, B \right] \mathrm{E}\left[\cos^2 \phi_k \, |\phi_k \right] \\
&= \frac{1}{2} \sum_{k=2}^{M} (1 + \cos(2\phi_k)) \mathrm{E}\left[W_k^2 \, |\zeta_k, B \right] \qquad (8.105)
\end{aligned}
$$

Since all the Z_k are independent it follows that

$$\mathrm{E}\left[W_k^2 \, |\zeta_k, B \right] = 2(2B+1)(S_k^2 - S_k) + N \qquad (8.106)$$

so that

$$\sigma_I^2 = \frac{1}{2} \sum_{k=2}^{M} (1 + \cos(2\phi_k)) 2(2B+1)(S_k^2 - S_k) + N \quad . \qquad (8.107)$$

If the intended sequence is known, then B is known. For random sequences $E[B] = (N-1)/2$ giving

$$\sigma_I^2 = \sum_{k=2}^{M} (1 + \cos(2\phi_k)) N(S_k^2 - S_k + 1/2) \quad . \qquad (8.108)$$

Several possibilities can be examined from here, including the following two important cases.

Chip and Phase Asynchronous: The interfering signals are characterized by S uniform on $[0,0)$ and ϕ uniform on $[0, 2\pi)$ so that $E[S_k^2 - S_k] = -1/6$ and $E[\cos(2\phi)] = 0$. In this case, $\sigma_I^2 = (K-1)N/3$. Hence, the decision variable in (8.88) can be interpreted as Gaussian random variable with mean N and variance $(K-1)N/3$ leading to the probability of bit error

$$P_b = Q\left(\sqrt{\frac{3N}{K-1}}\right) . \tag{8.109}$$

The carrier to interference ratio C/I can be defined as the carrier power divided by the total noise power

$$\frac{C}{I} = \frac{1}{K-1} . \tag{8.110}$$

By comparing (8.109) with the probability of bit error for binary signaling on an AWGN channel, i.e., $P_b = Q(\sqrt{2\gamma_b})$ we see that the C/I and γ_b are simply related by

$$\gamma_b = \frac{2}{3}\frac{C}{I} . \tag{8.111}$$

Chip and Phase Synchronous: The interfering signals have $S_k = 0$ and $\phi_k = 0$ so that $\sigma_I^2 = (K-1)N$ and

$$P_b = Q\left(\sqrt{\frac{N}{K-1}}\right) . \tag{8.112}$$

For chip and phase synchronous signals C/I and γ_b are related by

$$\gamma_b = 2\frac{C}{I} . \tag{8.113}$$

Coherent addition of interfering signals yields worst case interference with random spreading sequences. The Orthogonal Walsh-Hadamard sequences are less random (secure) but yield zero correlation (better performance) under this condition.

The standard Gaussian approximation can be quite inaccurate when the number of simultaneous users K is small or the processing gain N is large. To circumvent this deficiency a number of improved approximations have been developed.

Improved Gaussian Approximation

An improved Gaussian approximation can be obtained by averaging the conditional probability of error over the variance of the multiple access interference.

That is

$$P_b = \int_0^\infty Q\left(\frac{N}{\sqrt{\sigma_I^2}}\right) p_{\sigma_I^2}(\psi)d\psi \qquad (8.114)$$

where $p_{\sigma_I^2}(\psi)$ is the pdf of the multiple-access variance σ_I^2. From (8.107)

$$\sigma_I^2 = \sum_{k=2}^M L_k \qquad (8.115)$$

where $L_k = U_k V_k$ and

$$U_k = (1 + \cos(2\phi_k)) \qquad (8.116)$$
$$V_k = (2B + 1)(S_k^2 - S_k) + N/2 . \qquad (8.117)$$

Note that the $\{U_k\}$ are independent and the $\{V_k\}$ are conditionally independent given B. By using the results in [172] and [97, pp. 79-82], [172, pg. 123,244] the conditional pdf of L_k is

$$p_{L|B}(\ell) = \frac{1}{2\pi\sqrt{\tilde{B}}z}\log\left|\frac{\sqrt{N-\ell} + \sqrt{\tilde{B}}}{\sqrt{N-\ell} - \sqrt{\tilde{B}}}\right| \qquad (8.118)$$

where $\tilde{B} = B + 1/2$. Since the L_k are independent and identically distributed, the density of σ_I^2 is obtained by taking the $(M-2)$-fold convolution and removing the condition on B, i.e.,

$$p_{\sigma_I^2}(\psi) = E_B\left[p_{L|B}(\ell) * \cdots * p_{L|B}(\ell)\right] . \qquad (8.119)$$

This improved Gaussian approximation has been shown to be much more accurate that the standard Gaussian approximation [220]. However, the $(M-2)$-fold convolution in (8.119) must be obtained numerically followed by an additional numerical integration for computing the probability of error. Hence, the utility of this improved Gaussian approximation is limited.

Simplified Improved Gaussian Approximation

Another simpler but still accurate Gaussian approximation has been derived by Holtzman [146]. To describe this method, let $P(\theta)$ be any function of a random variable θ having mean μ and variance σ^2. The using a Taylor series expansion about the mean μ gives

$$P(\theta) = P(\mu) + (\theta - \mu)P'(\mu) + \frac{1}{2}(\theta - \mu)^2 P''(\mu) + \cdots \qquad (8.120)$$

Taking expectations

$$E[P(\theta)] \approx P(\mu) + \frac{1}{2}P''(\mu)\sigma^2 \ . \tag{8.121}$$

Instead of using the Taylor series expansion, we can start with differences (Stirling's formula) and write

$$
\begin{aligned}
P(\theta) &= P(\mu) + (\theta - \mu)\frac{P(\mu + h) - P(\mu - h)}{2h} \\
&+ \frac{1}{2}(\theta - \mu)^2\frac{P(\mu + h) - 2P(\mu) + P(\mu - h)}{h^2} + \cdots
\end{aligned} \tag{8.122}
$$

Taking expectations

$$E[P(\theta)] \approx P(\mu) + \frac{1}{2}\frac{P(\mu + h) - 2P(\mu) + P(\mu - h)}{h^2}\sigma^2 \ . \tag{8.123}$$

Holtzman [146] has shown that $h = \sqrt{3}\sigma$ yields good results so that

$$E[P(\theta)] \approx \frac{2}{3}P(\mu) + \frac{1}{6}P(\mu + \sqrt{3}\sigma) + \frac{1}{6}P(\mu - \sqrt{3}\sigma) \ . \tag{8.124}$$

To apply the above result, we let μ and σ^2 be the mean and variance of σ_I^2 in (8.115). Then

$$
\begin{aligned}
\mu &= (K - 1)E[Z_k] \\
&= (K - 1)\left(\frac{N}{2} - \frac{E[B]}{3} - \frac{1}{6}\right) \\
&= (K - 1)N/3
\end{aligned} \tag{8.125}
$$

where the last line assumes ensemble averaging with random spreading sequences. The variance is

$$
\begin{aligned}
\sigma^2 &= (K - 1)\left(E[Z_k^2] - E^2[Z_k] + (K - 2)\mathrm{cov}(Z_j, Z_k)\right) \quad \text{for any } j \neq k. \\
&= (K - 1)\left(\frac{23}{360}N^2 + \left(\frac{1}{20} + \frac{K - 2}{36}\right)N - \frac{1}{20} - \frac{K - 2}{36}\right) \ . \tag{8.126}
\end{aligned}
$$

This yields

$$
\begin{aligned}
P_b &\approx \frac{2}{3}Q\left(\sqrt{\frac{3N}{K - 1}}\right) + \frac{1}{6}Q\left(\frac{N}{\sqrt{(K - 1)N/3 + \sqrt{3}\sigma}}\right) \\
&+ \frac{1}{6}Q\left(\frac{N}{\sqrt{(K - 1)N/3 - \sqrt{3}\sigma}}\right) \ . \tag{8.127}
\end{aligned}
$$

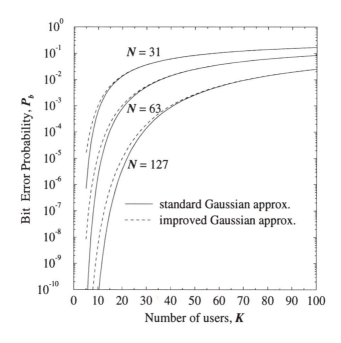

Figure 8.11   Bit error probability against the number of users and various processing gains. The standard Gaussian approximation is shown to under estimate the error probability for small numbers of users.

The above calculations are very simple and lead to quite accurate results for all values of K and N. Fig. 8.11 compares the standard and simplified improved Gaussian approximations for various processing gains and number of simultaneous users. Note that the standard Gaussian approximation under estimates the error probability for small numbers of users. In this case, the improved Gaussian approximation should be used. However, the accuracy of the standard Gaussian approximation improves when the number of simultaneous users increases.

Note that the above approximations assume an AWGN channel. For frequency selective fading channels, the approximations must be modified to account for the effects of self interference, multipath interference, and envelope fading. In this case the complex low-pass received signal is

$$w(t) = \sum_{i=1}^{M} \sum_{k=l}^{L_i} g_{i,k} v^{(i)}(t - \tau_i - k/W) + z(t) \qquad (8.128)$$

where $g_{i,k}$ is the complex gain associated with the ith user and the kth channel path. A variety of RAKE receiver structures can be employed to gain a diversity advantage including the maximal ratio scheme in Section 8.3.1.

8.5 PERFORMANCE ANALYSIS OF CELLULAR CDMA

CDMA is an attractive proposition for increasing cellular system capacity in dense urban areas, due to its many inherent benefits like the ability to mitigate multipath fading and interference, universal frequency reuse, soft handoff capability, and the ability to exploit voice activity detection. Numerous authors have investigated the capacity and performance of CDMA cellular systems for a propagation environment characterized by path loss and shadowing, including Gilhousen *et al.* [118], Kudoh and Matsumoto [167], and Newson and Heath [233]. Mokhtar and Gupta [215] considered reverse channel capacity on shadowed Nakagami fading channels, where the desired and interfering signals have the same fading statistical characteristics.

CDMA systems must use reverse channel power control; otherwise, the link performance will suffer from the **near-far effect** , a condition where the transmissions received from distant MSs experience excessive interference from nearby MSs. The IS-95 reverse link employs a fast closed-loop power control algorithm to combat variations in the received signal power due to path loss, shadowing, and fast envelope fading (at low Doppler frequencies). A large number of power control algorithms have been suggested in the literature. Ariyavistitakul and Chang [11] proposed a fast signal-to-interference ratio (SIR) based feedback power control algorithm that can mitigate both multipath fading and shadowing. For our purpose, we consider a simple closed-loop reverse channel power control scheme that equalizes the received power C from all MSs that are served by the same CS.

Power control is also useful on the forward channel of CDMA systems for combating the **corner effect**, a condition where a MS experiences a decrease in received signal strength and an increase in multiple-access interference as it exits a cell corner. Various "power balancing" schemes have been proposed to balance the BS transmit power for each [118] , [34]. Chang and Ren [34] have compared power balancing and mobile assisted SIR-based forward channel power control algorithms. They have shown the former to be better than the latter. Zorzi [356] has analyzed some simplified power control algorithms in the absence of shadowing. For our purpose, we will assume a slow open-loop power balancing algorithm.

8.5.1   Capacity of Cellular CDMA

CDMA cellular systems typically employ **universal frequency reuse** , where the bandwidth is shared by all the cells and transmissions are distinguished through the assignment of unique spreading sequences. For such systems, multiple-access interference from neighboring cells must be carefully accounted for. The propagation path loss associated with these interfering signals is relatively small compared to those found in narrow-band and mid-band TDMA systems that employ frequency reuse plans.

With cellular CDMA systems, any technique that reduces multiple-access interference translates into a capacity gain. Since cellular CDMA systems use speech coding, the multiple-access interference can be reduced by using voice activity detection along with variable rate speech transmission. This technique reduces the rate of the speech coder when silent periods are detected in the speech waveform. Voice activity detection has often been cited as an advantage of CMDA systems over TDMA systems. However, TDMA systems can also benefit from voice activity detection and discontinuous transmission, through a reduction in the level of co-channel interference.

Cell sectoring is another very effective method for reducing multiple-access interference, where each cell is sectored by using directional antennas. With 120° cell sectors, multiple-access interference on the reverse channel will only arise from MSs that are located in the shaded area of Fig. 8.12, where only the adjacent cells are shown. Likewise, multiple-access interference on the forward channel is generated by BSs that are transmitting to MSs located in the shaded regions of Fig. 8.13, where again only the adjacent cells are shown. In either case, 120° cell sectoring reduces the multiple-access interference by roughly a factor of three (on average); we say on average because the MSs are randomly distributed throughout the plane. Further improvements can be gained by using simple switched beam smart antenna systems with 30° or 15° sectors. A straight forward application of these antenna systems reduces the multiple access interference by a factor of 12 and 24, respectively, over a system using omni-directional antennas.

Our analysis of cellular CDMA starts with a cellular layout described by a uniform plane of hexagonal cells of radius R. Each cell contains a centrally

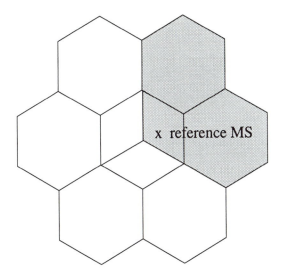

Figure 8.12 Reverse channel transmissions from MSs located in the shaded area will cause multiple-access interference with the reverse channel transmission from the reference MS.

located BS with 120° cell sectors. It is further assumed that the MSs are uniformly distributed throughout the system area with a density of K MSs per cell sector. For hexagonal cells of radius R, this yields a subscriber density of

$$\rho = \frac{2K}{3\sqrt{3}R^2} \text{ per unit area .} \tag{8.129}$$

The effects of voice activity detection can be modeled by assuming that each transmitter is independently active with probability p, so that the number of active transmitters in each cell has a (K, p) binomial distribution. The average number of active transmitters in a cell sector is Kp.

The standard Gaussian approximation in Section 8.10 has been extensively employed in the literature for the performance prediction of cellular CDMA systems. For random spreading sequences, we have seen that the standard Gaussian approximation for a power controlled chip and phase asynchronous reverse channel of a CDMA system predicts a bit signal-to-noise ratio of $\gamma_b = 3G/(2(K-1))$, where G is the processing gain and K is the number of simultaneously received signals. This assumes the use of a coherent correlation receiver with bit-by-bit decisions. If the signals are chip and phase synchronous as is the case in the forward channel of a CDMA cellular system, then the standard Gaussian approximation yields a bit signal-to-noise ratio of $\gamma_b = G/(2(K-1))$.

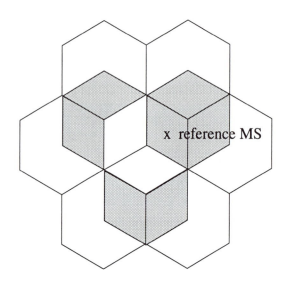

Figure 8.13 Forward channel transmissions to MSs located in the shaded areas will cause multiple-access interference with the forward channel transmission to the reference MS.

However, it is important to realize that this expression assumes random spreading sequences. If orthogonal spreading codes such as Walsh-Hadamard codes are used on the forward channel as is the case with the IS-95 system [79], then the multiple-access interference from the serving BS is effectively zero unless the channel delay spread destroys the orthogonality of the received signals. Under this condition out-of-cell interference dominates the performance.

8.5.2   Reverse Link Capacity

Perfect reverse channel power control maintains a constant received power C at the BS for all MSs. The jth MS located in cell i is denoted by MS_{ij}. The power transmitted by MS_{ij}, located at distance d_{ij} from its serving BS, BS_j, is P_{ij}. The received power at BS_j is

$$C = P_{ij} 10^{\frac{\varsigma_{ij}}{10}} \tag{8.130}$$

where ς_{ij} is a random variable due to shadowing and fading. MS_{ij} is also at distance d_{i0} to the reference BS, BS_0, and will produce an out-of-cell interference equal to

$$
\begin{aligned}
\frac{I_o(ij)}{C} &= 10^{\varsigma_{io}/10} \cdot \left(\frac{1}{10^{\varsigma_{ij}/10}}\right) \\
&= 10^{(\varsigma_{io}-\varsigma_{ij})/10} \le 1 \ .
\end{aligned} \tag{8.131}
$$

The first term is due to path loss and shadowing to BS_0, while the second term is the effect of the power control to compensate for the corresponding attenuation to BS_j. Note that $I_o(ij)/C$ is always less than unity; otherwise the MS would execute a handoff to the BS which makes it less than unity.

For our purpose, we assume a shadowed Nakagami fading channel, where the received signal power has the composite *Gamma*-log-normal pdf in (2.187). The composite pdf is approximated by a purely log-normal pdf with mean and standard deviation given by (2.188). Hence, the random variables ς_{ij} and ς_{io} are treated as Gaussian random variables with means and variances, respectively,

$$
\begin{aligned}
\mu_{ij} &= \xi[\psi(m_{ij}) - \ln(m_{ij})] + \mu_{ij}(\text{area}) \\
\mu_{io} &= \xi[\psi(m_{io}) - \ln(m_{io})] + \mu_{io}(\text{area}) \\
\sigma_{ij}^2 &= \xi^2\,\zeta(2, m_{ij}) + \sigma^2 \\
\sigma_{io}^2 &= \xi^2\,\zeta(2, m_{io}) + \sigma^2
\end{aligned} \tag{8.132}
$$

where $\xi = 10/\ln 10$, m_{ij} and m_{io} are the Nakagami shape factors, and σ is the shadow standard deviation. The difference $\varsigma_{io} - \varsigma_{ij}$ is a Gaussian random variable with variance $2\sigma^2$. The parameters $\mu_{ij}(\text{area})$ and $\mu_{io}(\text{area})$ are determined by the path loss. Assuming the simple path loss model in (1.6), yields their difference (in dB) as

$$
\mu_{io(\text{area})} - \mu_{ij(\text{area})} \ (\text{dB}) = 10\beta \log_{10}(d_{ij}/d_{io}) \ . \tag{8.133}
$$

The total out-of-cell interference-to-signal ratio is equal to

$$
\frac{I_o}{C} = \int \int \chi I_o(ij)\Phi_{0j}\rho\,dA \tag{8.134}
$$

where

$$
\Phi_{0j} = \begin{cases} 1, & \text{if } 10^{(\varsigma_{io}-\varsigma_{ij})/10} < 1 \\ 0, & \text{otherwise} \end{cases} \tag{8.135}
$$

ρ is user density over the area A, and χ is the voice activity variable

$$
\chi = \begin{cases} 1, & \text{with probability } p \\ 0, & \text{with probability } 1-p \end{cases} \ . \tag{8.136}
$$

The total out-cell interference I_o can be modeled as a Gaussian random variable by invoking the central limit theorem. The mean of the total out-of-cell interference-to-carrier ratio is

$$
\begin{aligned}
E[I_o/C] &= \int\int E[\chi 10^{(\varsigma_{i0}-\varsigma_{ij})/10}\Phi_{0j}]\rho dA \\
&= \int\int pE[10^{(\varsigma_{i0}-\varsigma_{ij})/10}\Phi_{0j}]\rho dA
\end{aligned}
\tag{8.137}
$$

Let $x = \varsigma_{i0} - \varsigma_{ij}$ and define

$$
\begin{aligned}
\mu_x &= \mu_{i0} - \mu_{ij} \\
\sigma_x^2 &= \sigma_{i0}^2 + \sigma_{ij}^2 \ .
\end{aligned}
\tag{8.138}
$$

Then the inner expectation in (8.137) is

$$
\begin{aligned}
E[10^{(\varsigma_{i0}-\varsigma_{ij})/10}\Phi_{0j}] &= \int_{-\infty}^{0} e^{x/\xi}\frac{1}{\sqrt{2\pi}\sigma_x}\exp\left\{-\frac{(x-\mu_x)^2}{2\sigma_x^2}\right\}dx \\
&= \exp\left\{\frac{\sigma_x^2}{2\xi^2}+\frac{\mu_x}{\xi}\right\}\frac{1}{\sqrt{2\pi}\sigma_x}\int_{-\infty}^{0} \\
&\quad \times \exp\left\{-\left(\frac{x}{\sqrt{2}\sigma_x}-\frac{\sigma_x}{\sqrt{2}\alpha}-\frac{\mu_x}{\sqrt{2}\sigma_x}\right)^2\right\}dx \\
&= \exp\left\{\frac{\sigma_x^2}{2\xi^2}+\frac{\mu_x}{\xi}\right\}\left[1-Q\left(-\frac{\sigma_x}{\xi}-\frac{\mu_x}{\sigma_x}\right)\right]
\end{aligned}
\tag{8.139}
$$

Therefore,

$$
E[I_o/C] = p\int\int \exp\left\{\frac{\sigma_x^2}{2\xi^2}+\frac{\mu_x}{\xi}\right\}\left[1-Q\left(-\frac{\sigma_x}{\xi}-\frac{\mu_x}{\sigma_x}\right)\right]\rho dA \ .
\tag{8.140}
$$

In a similar fashion,

$$
\begin{aligned}
E[(I_o/C)^2] &= \int\int E[\chi^2 10^{(\varsigma_{i0}-\varsigma_{ij})/5}\Phi_{0j}^2]\rho dA \\
&= p\int\int\int_{-\infty}^{0} e^{2x/\xi}\frac{1}{\sqrt{2\pi}\sigma_x}\exp\left\{-\frac{(x-\mu_x)^2}{2\sigma_x^2}\right\}dx\rho dA \\
&= \eta\int\int \exp\left\{\frac{2\sigma_x^2}{\xi^2}+\frac{2\mu_x}{\xi}\right\}\left[1-Q\left(-\frac{2\sigma_x}{\xi}-\frac{\mu_x}{\sigma_x}\right)\right]\rho dA
\end{aligned}
\tag{8.141}
$$

Finally, the variance of I_o/C is

$$
\begin{aligned}
\text{Var}[I_o/C] &= p\int\int \exp\left\{\frac{2\sigma_x^2}{\xi^2}+\frac{2\mu_x}{\xi}\right\}\left[1-Q\left(-\frac{2\sigma_x}{\xi}-\frac{\mu_x}{\sigma_x}\right)\right]\rho dA \\
&\quad -E^2[I_o/C] \ .
\end{aligned}
\tag{8.142}
$$

The integrals in (8.140) and (8.142) must be numerically evaluated over the random mobile locations in the area A, as defined by the set of shaded sectors in Fig. 8.12.

With perfect power control, the in-cell interference I_{in} is

$$I_{in} = C \sum_{i=1}^{K-1} \chi_i \qquad (8.143)$$

where χ_i is a Bernoulli random variable equal to 1 with probability p and 0 with probability $1 - p$. Let $I = I_o + I_{in}$ be the total interference. Then the probability that the received γ_b at a BS is below a required value, $\gamma_{b\ req}$, is

$$
\begin{aligned}
P_{out} &= \mathrm{Pr}(\gamma_b < \gamma_{b\ req}) \\
&= \mathrm{Pr}((I/C)|_{\gamma_b} > (I/C)_{\gamma_b\ req}) \\
&= \mathrm{Pr}\left(\sum_{i=1}^{K-1} \chi_i + I_o/S > I/C|_{\gamma_b\ req} \right) \\
&= \mathrm{Pr}\left(I_o/S > I/C|_{\gamma_b\ req} - \sum_{i=1}^{K-1} \chi_i \right) \\
&= \sum_{k=0}^{K-1} \mathrm{Pr}\left(I_o/S > I/C \Big|_{\gamma_b\ req} - k \Big| \sum_{i=1}^{K-1} \chi_i = k \right) \mathrm{Pr}\left(\sum_{i=1}^{K-1} \chi_i = k \right) \\
&= \sum_{k=0}^{K-1} \binom{K-1}{k} p^k (1-p)^{K-1-k} Q\left(\frac{(I/C)|_{\gamma_b\ req} - k - \mathrm{E}[I_o/C]}{\sqrt{\mathrm{Var}[I_o/C]}} \right)
\end{aligned}
$$

$$(8.144)$$

In all of our numerical results, we assume a chip rate of $R = 1.25$ Mchips/s and a source symbol rate of 8 kb/s, yielding a processing gain of $G = 156.25$. We further assume a voice activity factor of $p = 3/8$. Fig. 8.14 shows the reverse channel capacity for different $\gamma_{b\ req}$ and shadow standard deviations. The reverse channel capacity is greatly increased by a reduction in $\gamma_{b\ req}$ and slightly reduced when the shadow standard deviation is increased. Fig. 8.15 shows the reverse channel capacity with different Nakagami shape factors for the desired and interfering signals. Observe that a change in the Nakagami shape factor m_I of interfering signals has very little effect on the reverse channel capacity. Fig. 8.16 further illustrates the effect of fading and shadowing on the reverse channel capacity. As expected, shadowing and fading have relatively little impact on the reverse channel capacity, since these components of the received

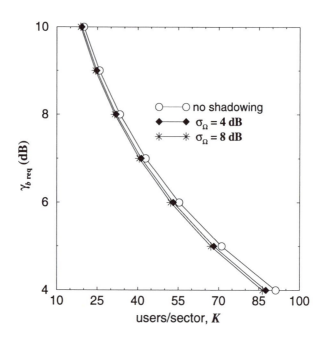

Figure 8.14   Reverse channel capacity with $P_{\text{out}} = 10^{-2}$ for different $\gamma_{b\,\text{req}}$; the Nakagami shape factors are $m_d = 1$ and $m_I = 1$ (Rayleigh fading), and σ_Ω is the shadow standard deviation.

signal are power controlled. Therefore, fading and shadowing variations only affect the out-of-cell interference.

The ratio of the mean out-of-cell interference to the mean in-cell interference ratio is

$$\theta = \frac{E[I_o]}{E[I_{\text{in}}]} = \frac{E[I_o/C]}{E[I_{\text{in}}/C]} = \frac{E[I_o/C]}{pK} \tag{8.145}$$

With a 4th-order path loss exponent, Newson and Heath [233] showed that $\theta = 0.5$ when no fading and shadowing are considered and $\theta \approx 0.66$ when shadowing is considered with $\sigma = 8$ dB. This translates into a frequency reuse efficiency f, defined as the ratio of mean in-cell interference to the total mean interference, of 0.66 and 0.38, respectively. Table 8.1 tabulates the corresponding values of θ and f for the CDMA cellular system under consideration for different propagation conditions. The calculations only consider the first tier of interfering cells. Observe that the frequency reuse efficiency decreases with the shadow standard deviation, σ, and slightly increases when m_d increases or m_I decreases.

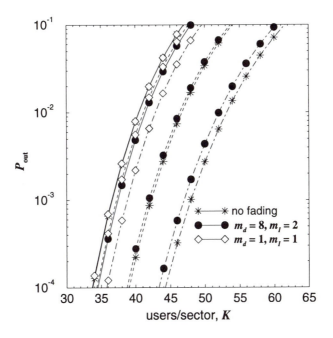

Figure 8.15 Reverse channel capacity for different propagation environments with $\gamma_{b\ \text{req}} = 8.76$ dB. Solid lines denote $\sigma_{\Omega}mega = 0$ dB, dotted lines denote $\sigma_{\Omega}mega = 4$ dB, and dot-dashed lines denote $\sigma_{\Omega} = 8$ dB.

To show that the values of θ and $\bar{f}$ in Table 8.1 do not depend on the number of users per cell, K, we derive the cdf of the out-of-cell interference to the in-cell interference $I_{\text{o}}/I_{\text{in}}$ for the reverse channel as

$$
\begin{aligned}
\Pr(I_{\text{o}}/I_{\text{in}} < z) &= \Pr\left(\frac{I_{\text{o}}/C}{I_{\text{in}}/C} < z\right) \\
&= 1.0 - \Pr(I_{\text{o}}/C > z\, I_{\text{in}}/C) \\
&= 1.0 - \left\{\sum_{k=0}^{K-1} \binom{K-1}{k} \eta^k (1-\eta)^{K-1-k} Q\left(\frac{kz - \mathrm{E}[I_{\text{o}}/C]}{\sqrt{\mathrm{Var}[I_{\text{o}}/C]}}\right)\right\}
\end{aligned}
$$

(8.146)

Fig. 8.17 plots the distribution of $I_{\text{o}}/I_{\text{in}}$ (in dB) with different shadow standard deviations. Although the distribution varies with K, the mean value $\mathrm{E}[I_{\text{o}}/I_{\text{in}}]$ remains almost the same, i.e., all the curves cross at the 50% point. This implies that the values of θ and f in Table 8.1 do not depend on K.

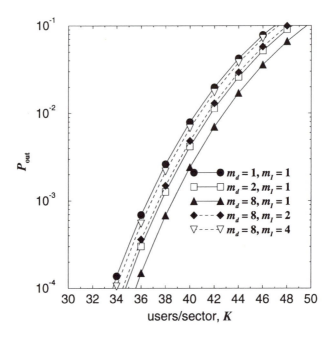

Figure 8.16 Reverse channel capacity for different Nakagami shape factors with $\gamma_{b\ req} = 8.76$ dB, $\sigma_\Omega = 8$ dB.

Extensions of the above results to include the effects of soft handoff have been provided in [323]. Soft handoff was shown to improve coverage by a factor of 2 to 2.5 in cell area, i.e., the number of BSs can be reduced by this factor. It was also shown to increase the reverse channel capacity by a factor better than 2.

8.5.3 Forward Link Capacity

For the forward channel, a pilot signal is transmitted from each BS. The pilot signal is a spread spectrum signal that causes interference in every cell, thereby reducing the capacity. However, this is offset by a decrease in $\gamma_{b\ req}$ due to coherent modulation. With forward channel balancing power control, the mobile measures the received signal and periodically transmits the measurement to its serving BS [118]. When the total power requested by mobiles is below the maximum allowable transmit power, the BS will reduce its transmit power, thereby reducing interference; otherwise, the BS will redistribute the power from the forward links with good quality to those with poor quality.

m_d	m_I	σ (dB)	$E\left[I_o/I_{in}\right]$	f
8	8	8	60.14%	62.45%
8	4	8	58.83%	62.96%
8	2	8	56.20%	64.02%
8	1	8	51.11%	66.1i8%
4	1	8	52.36%	65.63%
2	1	8	54.90%	64.56%
1	1	8	59.73%	62.61%
1	1	10	57.34%	63.56%
1	1	6	60.16%	62.44%
1	1	4	57.82%	63.36%
1	1	noshadowing	51.76%	65.89%
no fading	no fading	no shadowing	21.81%	82.10%
no fading	no fading	4	39.42%	71.73%
no fading	no fading	8	60.02%	62.49%
no fading	no fading	10	61.65%	61.86%

Table 8.1 Ratio of the mean of out-of-cell interference to the mean in-cell interference ratio, θ, and frequency reuse efficiency, f, under different propagation conditions.

In the worst case situation, each BS always transmits with the maximum allowable power $P_{\max}$. From (8.130), the γ_b at the ith mobile under this condition is

$$\gamma_{b,i} \;=\; \frac{3G}{2(I/C)_i} \;=\; \frac{3G}{2\left(\sum_{j=0}^{M} C_{T_j} - \delta\phi_i C_{T_0}\right)/\delta\phi_i C_{T_0}} \tag{8.147}$$

$$C_{T_j} \;=\; P_{\max} 10^{\frac{\varsigma_j}{10}} \tag{8.148}$$

where M is the number of surrounding BSs that are included in the calculation, C_{T_j} is the received power from BS_j, $1 - \delta$ is the fraction of the total power allocated to the pilot, and the weighting factor ϕ_i is the fraction of the remaining power allocated to the ith mobile. Note that (8.148) assumes the worst case condition of chip and phase asynchronous random spreading sequences. As in [118] our results assume that 20% of the total BS transmit power is allocated

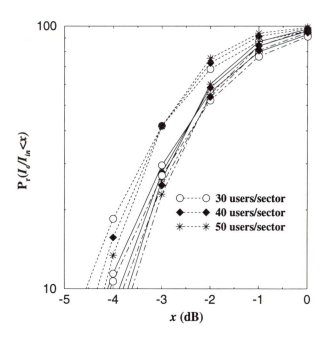

Figure 8.17 Distribution of out-of-cell interference to in-cell interference, $I_o/I_{\rm in}$, for the reverse channel. Solid lines denote $\sigma_\Omega = 0$ dB, dotted lines denote $\sigma_\Omega = 4$ dB, and dot-dashed lines denote $\sigma_\Omega = 8$ dB; $m_d = 1, m_I = 1$ (Rayleigh fading).

to the pilot signal. Once again, the ς_i are Gaussian random variables due to shadow and fading variations, with means and variances obtained from (2.188).

The BS distributes its transmit power proportionally according to the needs of each mobile within its cell. This is accomplished by first obtaining the required ϕ_i for each mobile, $(\phi_i)_{\rm req}$, by setting $\gamma_b = \gamma_{b\ {\rm req}}$ in (8.147). To account for the voice activity, we then calculate the modified weighting factor [167]

$$\overline{\phi}_i = \frac{(\phi_i)_{\rm req}}{\sum_{\substack{j=1 \\ j \neq i}}^{K} \chi_j (\phi_j)_{\rm req} + (\phi_i)_{\rm req}} \ . \tag{8.149}$$

The power balancing scheme in [34] does the same thing, except that the voice activity factors, χ_j, are not considered. The outage probability then becomes

$$
\begin{aligned}
P_{\text{out}} &= \Pr\left(\gamma_b < \gamma_{b \text{ req}}\right) \\
&= \Pr\left((I/C)|_{\gamma_b} > (I/C)|_{\gamma_{b \text{ req}}}\right) \\
&= \Pr\left(\left[1 + \sum_{j=1}^{K} C_{T_j}/C_{T_0} - \delta\overline{\phi_i}\right]\frac{1}{\delta\overline{\phi_i}} > \right. \\
&\qquad\qquad \left.\left[1 + \sum_{j=1}^{K} C_{T_j}/C_{T_0} - \delta(\phi_i)_{\text{req}}\right]\frac{1}{\delta(\phi_i)_{\text{req}}}\right) \\
&= \Pr\left(\overline{\phi_i} < (\phi_i)_{\text{req}}\right) \ .
\end{aligned}
\tag{8.150}
$$

Results must be obtained from the last equation in (8.150) by using Monte Carlo simulation techniques to account for the random user locations, and shadow and fading variations.

Fig. 8.18 shows how the forward channel capacity depends on $\gamma_{b \text{ req}}$ and the shadow standard deviation. Shadowing has a slightly stronger effect on forward channel capacity compared to the reverse channel. Fig. 8.19 shows the forward channel capacity for various Nakagami shape factors. The Nakagami shape factor also plays a significant role in forward channel capacity, and overly optimistic capacity estimates will be obtained if fading is neglected.

8.5.4   Imperfect Power Control

Any power control algorithm will inevitably be subject to some degree of error. It has been experimentally verified that the power control error (in dB) can be modeled as a zero-mean Gaussian random variable with variance σ_E^2 [167], [233]. In this case, (8.130) has the modified form

$$
C 10^{\varsigma_E/10} = P_{ij} 10^{\varsigma_d/10}
\tag{8.151}
$$

where ς_E is the power control error. Then the mean and variance of I_o/C with imperfect power control are similar to (8.138), but have the form

$$
\mu_x = \mu_{i0} - \mu_{ij} \qquad\qquad \sigma_x^2 = \sigma_{i0}^2 + \sigma_{ij}^2 + \sigma_E^2 \ .
\tag{8.152}
$$

With imperfect power control, the in-cell interference is

$$
I_{\text{in}} = C \sum_{\substack{j=1 \\ j \neq i}}^{K} 10^{\varsigma_{Ej}/10} \chi_j = C\kappa \ .
\tag{8.153}
$$

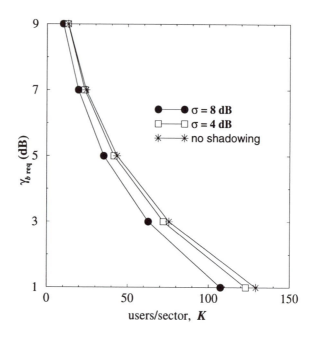

Figure 8.18 Forward channel capacity with $P_{\text{out}} = 10^{-2}$ for different $\gamma_{b\ \text{req}}$; the Nakagami shape factors are $m_d = 1$ and $m_I = 1$ (Rayleigh fading), and σ_Ω is the shadow standard deviation.

Then

$$
\begin{aligned}
P_{\text{out}} &= \Pr(\gamma_b < \gamma_{b\ \text{req}}) \\
&= \Pr(I_o/C + I_{\text{in}}/C > I/C|_{\gamma_b\ \text{req}}) \\
&= \Pr(I_o/S + \kappa > I/C|_{\gamma_b\ \text{req}}) \\
&= \Pr(I_o/S > I/C|_{\gamma_b\ \text{req}} - \kappa) \\
&= \sum_{k=0}^{K-1} \binom{K-1}{k} p^k (1-p)^{K-1-k} \\
&\quad \times \Pr\left(I_o/C > (I_o/C)|_{\gamma_b\ \text{req}} - \kappa|\kappa \right) \Pr\left(\kappa|k \right) \ . \qquad (8.154)
\end{aligned}
$$

Note that the conditional density of κ given k is also Gaussian. Observe from Fig. 8.20 that the reverse channel capacity is dramatically decreased as the power control error increases. For $P_{\text{out}} = 0.01$ and power control errors of $\sigma_E = 1$ dB, 2 dB, and 3 dB, the reverse channel capacity is decreased by 24%, 50%, and 68%, respectively. To consider the effect of power control error on

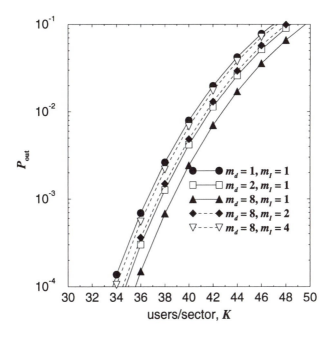

Figure 8.19 Forward channel capacity for different Nakagami shape factors with $\gamma_{b\ req} = 6.76$ dB, $\sigma_\Omega = 8$ dB.

the forward channel, (8.149) becomes

$$\overline{\phi_i} = \frac{\phi_i 10^{\varsigma E_i/10}}{\sum_{\substack{j=1 \\ j \neq i}}^{K} \chi_j \phi_j 10^{\varsigma E_j/10} + \phi_i 10^{\varsigma E_i/10}} \tag{8.155}$$

Fig. 8.20 shows that the forward channel capacity is reduced by 31%, 64%, and 83% for $\sigma_E = 1.0$ dB, 2.0 dB, and 3.0 dB, respectively. Note that imperfect power control has a more severe effect on the forward channel than the reverse channel for the same propagation conditions.

8.5.5 Error Probability with RAKE Reception

Suppose that the reference MS is located in CS_0. The instantaneous received bit energy-to-*total* noise ratio that is associated with the $\hat{j}$th MS and the ℓth

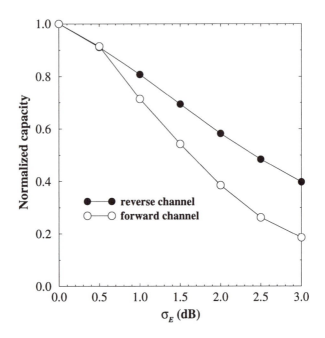

Figure 8.20   Forward and reverse channel capacity with imperfect power control. The capacity is normalized with respect to the capacity with perfect power control; σ_E is the standard deviation of the power control error, $m_d = 1$, $m_I = 1$, and $\sigma_\Omega = 8$ dB.

path is

$$R_{j,\ell} = \frac{3G\lambda_\ell^{\hat{j}(0)}}{2\sum_{c,i,n \in U} \lambda_n^{i(c)} + N_o} \tag{8.156}$$

where $\lambda_n^{i(c)} = |g_n^{i(c)}|^2 E_b$, E_b is the transmitted energy per data bit, G is the processing gain and $g_n^{i(c)}$ is the nth channel tap gain that is associated with MS_i in SC_c. Very often the thermal noise N_o can be neglected in deference to the typically dominant effect of the multiple-access interference. The set U in (8.156) is defined as

$$U \triangleq \bigcup_{c \in A} U_c \tag{8.157}$$

where

$$U_c = \{(c, i, n) : \chi_{ci} = 1, \ 1 \le i \le K, \ 1 \le n \le L, \ \text{and} \ (c, i, n) \ne (0, \hat{j}, n)\} \tag{8.158}$$

and $\chi_{ci} = 1$ if transmitter i in SC_c is active, and $\chi_{ci} = 0$ otherwise. In (8.158) the assumption is made that the spreading sequences are chosen so that the self-interference due to multipath can be neglected. Note that the multipath increases the level of multiple access by extending the size of the set U_c in (8.158).

The mean interference power that is received at the serving BS from all out-of-cell MSs is a constant. With the use of cell sectoring and voice activity gating, Kp MSs will be actively transmitting in a CS on average. Of course the number of active MSs is binominally distributed, but our simplified analysis will use the average number of active MSs. The ratio of the average received power received over path ℓ to the average received total noise power is approximately

$$\bar{\gamma}_\ell \approx \frac{3G\Gamma_\ell^{j(0)}}{\left(\frac{Kp}{3} - 1\right)\sum_{n=1}^{L}\Gamma_n^{i(0)} + \sum_{c\neq 0, i, n\in U}\Gamma_n^{i(c)}} \quad . \tag{8.159}$$

The second term in the denominator of (8.159) sums the multiple-access interference over all MS, multipaths, and interfering cells. It must be determined by careful study of the particular CDMA deployment. For uniform hexagonal cells, we have seen earlier that this term is approximately 50% of the first term in the denominator [117]. However, for microcells this term can be as large as 100% [143], depending on the cell layout, user spatial distribution and propagation path loss exponent. Assuming a fractional value of η_F for the out-of-cell interference (8.159) becomes

$$\bar{\gamma}_\ell \approx \frac{3G\Gamma_\ell^{j(0)}}{(1 + \eta_F)(Kp - 1)\sum_{n=1}^{L}\Gamma_n^{i(c)}} \quad . \tag{8.160}$$

The values of $\Gamma_\ell^{j(0)}$ and $\Gamma_n^{i(c)}$ depend, among other things, on the delay spread of the channel. If the delay spread exhibits an exponential decay, then it follows from (8.85) that

$$\Gamma_n^{i(0)} = \frac{(1 - \exp\{-1/\varepsilon\})\exp\{-n/\varepsilon\}}{\exp\{-1/\varepsilon\} - \exp\{-(L+1)/\varepsilon\}} \cdot \Gamma_t \tag{8.161}$$

where

$$\Gamma_t = \sum_{n=1}^{L}\Gamma_n^{i(c)} \tag{8.162}$$

is the total power received from each transmitter.

Since the received signals from the interfering users are Rayleigh faded, they can be treated as zero-mean complex Gaussian random processes. Furthermore, the multiple-access interference consists of a large number of terms so that it can be treated as a stationary Gaussian random process. However, we must still account for the random fluctuations in the envelope of the desired signal.

The error probability depends on the type of diversity combining and detection being used. We assume that an M-tap RAKE receiver with maximal ratio combining. In general, $M \neq L$, where L is the number of taps in the discrete-time channel model. The receivers use D-branch spatial diversity so there are DM replicas of the signal that are available for processing. The instantaneous received bit energy-to-*noise* ratio that is associated with path ℓ and antenna m, $\gamma_{m,\ell}$, is exponentially distributed with density

$$p_\gamma(x) = \frac{1}{\bar{\gamma}_\ell} e^{-x/\bar{\gamma}_\ell} \tag{8.163}$$

where we have assumed that identical antenna elements so that $\bar{\gamma}_{m,\ell} = \bar{\gamma}_\ell$ with $\bar{\gamma}_\ell$ given by (8.160).

Maximal Ratio Combining

This section presents a simplified performance analysis of cellular CDMA. Although there are many different performance aspects, we have chosen to focus on bit error probability of cellular CDMA with RAKE reception. While the multipath allows us to gain a diversity advantage by employing a RAKE receiver, it also has the undesirable effect of accentuating the effect of the multiple-access interference. Hence, we are interested in assessing the diversity gains that we can expect.

Following the treatment of maximal ratio combining in Section 5.5.3, the instantaneous *processed* bit energy-to-total noise ratio is defined as

$$\gamma_{\mathrm{p}} = \sum_{m=1}^{D} \sum_{k=1}^{M} \gamma_{m,k} \ . \tag{8.164}$$

With coherent BPSK or QPSK signaling, the bit error probability conditioned on γ_{p} is

$$P_b(\gamma_{\mathrm{p}}) = Q(\sqrt{2\gamma_{\mathrm{p}}}) \ . \tag{8.165}$$

Since the $\gamma_{m,k}$ are statistically identical with respect to the index m, it follows that the characteristic function of γ_{p} is

$$\psi_{\gamma_{\mathrm{p}}}(s) = \prod_{k=1}^{M} \frac{1}{(1 - s\bar{\gamma}_k)^D}$$

$$= \sum_{k=1}^{M} \sum_{j=1}^{D} \frac{A_{kj}}{(1 - s\bar{\gamma}_k)^j} \tag{8.166}$$

where

$$A_{kj} = \frac{1}{(D-j)!(-\bar{\gamma}_k)^{D-j}} \left\{ \frac{d^{D-j}}{ds^{D-j}} (1 - s\bar{\gamma}_k)^D \psi_{\gamma_{\mathrm{p}}}(s) \right\} \Bigg|_{s=1/\bar{\gamma}_k} . \tag{8.167}$$

It follows that the density of γ_{p} is

$$p_{\gamma_{\mathrm{p}}}(x) = \sum_{k=1}^{M} \sum_{j=1}^{D} A_{kj} \frac{1}{(j-1)!(\bar{\gamma}_k)^j} x^{j-1} \exp\{-x/\bar{\gamma}_k\} . \tag{8.168}$$

Therefore, the average bit error probability becomes

$$
\begin{aligned}
P_b &= \int_0^\infty P_b(x) p_{\gamma_{\mathrm{p}}}(x) dx \\
&= \sum_{k=1}^{M} \sum_{j=1}^{D} A_{kj} \int_0^\infty Q(\sqrt{2x}) \frac{1}{(j-1)!(\bar{\gamma}_k)^j} x^{j-1} \exp\{-x/\bar{\gamma}_k\} dx \\
&= \sum_{k=1}^{M} \sum_{j=1}^{D} A_{kj} \left(\frac{1-\mu_k}{2}\right)^j \sum_{n=0}^{j-1} \binom{j-1+n}{n} \left(\frac{1+\mu_k}{2}\right)^n
\end{aligned} \tag{8.169}
$$

where

$$\mu_k = \sqrt{\frac{\bar{\gamma}_k}{1 + \bar{\gamma}_k}} . \tag{8.170}$$

It is useful to express the performance as a function of the total average received bit energy-to-noise ratio per antenna branch, $\bar{\gamma}_c$, defined as

$$\bar{\gamma}_c \triangleq \sum_{k=1}^{L} \bar{\gamma}_k$$

$$= \frac{3G}{(1 + \eta_{\mathrm{F}})\left(\frac{Kp}{3} - 1\right)} \tag{8.171}$$

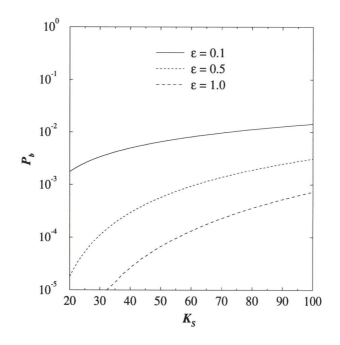

Figure 8.21 Bit error probability against the number of MS per cell $K_S = 3K$. A 4-tap RAKE receiver is used without antenna diversity on a 4-tap channel. The voice activity factor is $p = 0.5$ and the out-of-cell interference has the fractional value $\eta_F = 0.5$. Values of $\varepsilon = 1.0, 0.5$ and 0.1 are shown.

where the last step is obtained by using (8.160)–(8.162). Note that $\bar{\gamma}_p$ is always less than or equal to $\bar{\gamma}_c$. Then

$$\bar{\gamma}_k = \frac{3G}{(1 + \eta_F)\left(\frac{K_p}{3} - 1\right)} \cdot \frac{(1 - \exp\{-1/\varepsilon\})\exp\{-k/\varepsilon\}}{\exp\{-1/\varepsilon\} - \exp\{-(L+1)/\varepsilon\}} . \tag{8.172}$$

Fig. 8.21 shows the CDMA reverse channel performance with $D = 1$, $L = 4$, $M = 4$, $p = 0.5$, $\eta_F = 0.5$ and various ε. For cellular CDMA systems an error probability on the order of 10^{-2} to 10^{-3} is deemed acceptable. Note that the error probability increases as the channel becomes less dispersive (smaller ε) because the RAKE receiver cannot achieve a diversity advantage. In order to prevent poor performance in a nondispersive channel environment, antenna diversity can be used. Fig. 8.22 shows the performance under conditions identical to those in Fig. 8.21 except that 2-branch antenna diversity is used, with independently faded antenna branches. Note the scale change on the abssisa in Figs. 8.21 and 8.22.

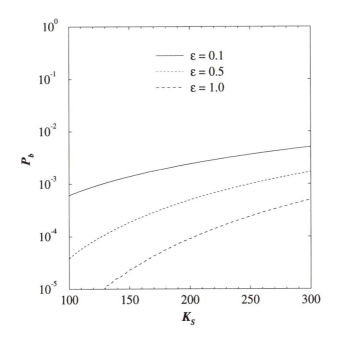

Figure 8.22 Bit error probability against the number of MS per cell $K_S = 3K$. A 4-tap RAKE receiver is used with 2-branch antenna diversity on a 4-tap channel. The voice activity factor is $p = 0.5$ and the out-of-cell interference has the fractional value $\eta_F = 0.5$. Values of $\varepsilon = 1.0, 0.5$ and 0.1 are shown.

Problems

8.1. Suppose that a DS/BPSK spread spectrum signal is corrupted by a single, phase-asynchronous, interfering tone at the carrier frequency. The received low-pass waveform is

$$w(t) = v(t) + i(t)$$

where $v(t)$ is defined in (8.4) and

$$i(t) = A_i e^{j\phi}$$

where ϕ is an arbitrary phase offset. Assume the use of a Gold code having a period equal to the bit duration. Compute the probability of bit error with a simple correlation detector.

8.2. The generator polynomials for constructing Gold code sequences of length $N = 7$ are

$$p_1(x) = 1 + p + p^3$$
$$p_2(x) = 1 + p^2 + p^3 \ .$$

Generate all the Gold codes of length 7 and determine the crosscorrelation functions on one sequence with each of the others.

8.3. Write a computer program to generate a set of Gold sequences of length 127.

a) Plot the mean and variance of the partial period autocorrelation as a function of the processing gain $10 \leq G \leq 20$ for this set of Gold codes.

b) Repeat part a) for the partial period crosscorrelation.

8.4. Plot the continuous-time partial autocorrelation functions of the chip waveform, $R_h(\delta)$ and $\hat{R}_h(\delta)$, as a function of the fractional chip delay δ for the following chip waveforms

$$h_a(t) = \begin{cases} u_{T_c}(t) & \text{non-return-to-zero} \\ \sin(\pi t/T_c)u_{T_c}(t) & \text{half-sinusoid} \\ 1 - 2|t - T_c/2|/T_c u_{T_c}(t) & \text{triangular} \end{cases}$$

8.5. Consider the set of Walsh-Hadamard sequences of length 16.

a) Determine full period autocorrelation $\phi_{k,k}(n)$ for this set of sequences. Tabulate your results in the matrix

$$\boldsymbol{\rho}_{k,k} = [\rho_{k,n}]$$

where $\rho_{k,n} = \phi_{k,k}(n)$.

b) Suppose that each Walsh-Hadamard sequence is supposed to represent a symbol from a 16-symbol alphabet. The rows of the Walsh-Hadamard matrix are orthogonal and, therefore, 16-ary orthogonal modulation is achieved. The channel consists of two equal strength T_c-spaced rays, i.e.,

$$c(t) = \delta(t) + \delta(t - kT_c) \ .$$

For each 16-ary symbol, determine the delay k that results in the largest probability of symbol error.

8.6. Suppose that the multipath intensity profile of a channel is given by

$$\phi_c(\tau) = \frac{A}{\mu_\tau} e^{-\tau/\mu_\tau} \ .$$

a) What is the average delay and delay spread of the channel?

b) Suppose further that DS spread spectrum is used with a two-tap RAKE receiver (assume ideal Nyquist pulses and maximal ratio combining). The tap spacing of the RAKE tapped delay line is equal to the chip duration T_c. Neglecting self-interference, write down an expression for the probability of error in terms of the average delay of the channel and the average received bit energy-to-noise ratio.

c) If the bit error probability for a nondispersive channel ($\mu_\tau = 0$) is 10^{-3}, what value of delay spread μ_τ will reduce the bit error probability from 10^{-3} to 10^{-4}?

8.7. A multipath fading channel has the multipath intensity profile

$$\phi_{cc}(\tau) = \frac{A}{\mu_\tau} e^{-\tau/\mu_\tau} \ .$$

Suppose that DS spread spectrum is used on this channel with a 3-tap, T_c-spaced, RAKE receiver. Assume ideal Nyquist pulses and spreading sequences having an ideal autocorrelation. Find the probability of error with selective combining in terms of the average received bit energy-to-noise ratio.

8.8. Consider the perfectly power controlled DS CDMA system that is analyzed in Section 8.4. Determine the probability of bit error with a standard Gaussian approximation for the following cases;

a) Chip synchronous and phase asynchronous.

b) Chip asynchronous and phase synchronous.

9

CELLULAR COVERAGE PLANNING

Many current cellular systems employ fixed channel assignment (FCA) schemes, where the cells are permanently assigned a subset of the available channels according to a frequency reuse plan. This chapter discusses the issues associated with radio coverage planning for cellular systems. Several schemes are suggested for improving the spectral efficiency in a cellular system including cell sectoring, cell splitting, reuse partitioning, switched beam smart antenna systems, selective base station (BS) diversity, and overlay/underlay schemes. Although the concepts are developed for systems that employ FCA, they will in many cases apply to those that use dynamic channel assignment (DCA).

Since spectral efficiency is inversely proportional to the cell area, microcellular systems are one popular approach for increasing the spectral efficiency in high density urban and suburban areas. However, as cell sizes decrease the problems associated with fast handoff are exasperated. Consequently, mixed cell architectures (microcells overlaid with macrocells) are preferred for handling both low density fast moving vehicular mobile stations (MSs) and high density low speed portable MSs [42]. For macro-macrocell and macro-microcell handoffs, a backward handoff protocol can be adopted since the required handoff times are less stringent. However, micro-macrocell and micro-microcell handoffs still need to be fast and a forward handoff protocol may be a better choice.

Rather than reducing the cell size, another approach for high spectral efficiency is to reduce the size of frequency reuse clusters. In rural areas, global roaming radio coverage is also important. The system deployment objective in rural areas with low density MSs is to provide radio coverage with the minimum number of base stations (BSs). For this application, radio coverage extension is an important consideration for reduced infrastructure cost.

467

Very recently the use of smart antennas has gained enormous interest [26], [110], [165], [289], [290], [296]. A marked improvement in spectrum efficiency, interference reduction and capacity can be obtained by using smart antennas for cellular systems. Smart antennas can not only improve the performance of TDMA cellular systems, but can also improve the performance of CDMA cellular systems [228], [303] and indoor wireless networks [288]. There are three basic approaches to the design a smart antenna system [165]: *adaptive null steering, phased array, and switched beam.* The realization of adaptive null steering smart antenna and phased array smart antenna systems requires an architecture capable of tracking any MS, or group of MSs, within the radio coverage area and a beamforming network capable of producing the appropriate multiple independent beams. These approaches have strong implementation cost and complexity constraints, because they cannot be integrated into an arbitrary cellular system in a straight forward manner [296], [343]. Furthermore, in macrocellular system with relatively high BS antennas, the signals are received at the BS with a narrow angle-of-arrival (AOA) spread [303]. In this case, the signals received from the MSs on the multiple antenna beams will be strongly correlated which limits the gains that can be achieved from adaptive null steering and phased array antennas. Hence, switched-beam smart antennas remain the most attractive because no complicated multi-beam beamforming (combining) is needed and no significant changes to the existing cellular systems are required.

9.1   CELL SECTORING

Consider a uniform deployment of hexagonal cells where the BSs employ omni-directional antennas. If the effects of shadowing are neglected and uniform path loss is assumed from all BSs, then the worst case forward channel co-channel interference scenario is illustrated in Fig. 9.1 In this case, the MS is located at the corner of a cell. There are six co-channel BSs, two each at approximate distances of $D - R$, D, and $D + R$. It follows that the worst case carrier-to-interference ratio is

$$
\begin{aligned}
\Lambda &= \frac{1}{2} \frac{R^{-\alpha}}{(D-R)^{-\alpha} + D^{-\alpha} + (D+R)^{-\alpha}} \\
&= \frac{1}{2} \frac{1}{\left(\frac{D}{R} - 1\right)^{-\alpha} + \left(\frac{D}{R}\right)^{-\alpha} + \left(\frac{D}{R} + 1\right)^{-\alpha}} \ .
\end{aligned}
\tag{9.1}
$$

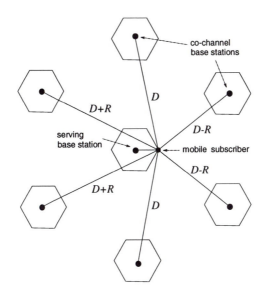

Figure 9.1 Worst case co-channel interference situation on the forward channel.

For a 7-cell reuse cluster $D/R = \sqrt{3N} = 4.6$, and with a path loss exponent of 4, the worst case Λ is 17 dB. Unfortunately, this Λ is may be too low to yield acceptable performance, especially if shadowing is present.

Cell sectoring is a very common method that is employed in macrocellular systems to improve the performance against co-channel interference, whereby each cell is divided into radial sectors with directional BS antennas. The carriers assigned to each cell are divided into disjoint sets with each set assigned to a cell sector. Current cellular systems are quite often deployed with 120° cell or 60° cell sectors. As shown in Fig. 9.2, cell sectoring reduces the number of primary co-channel interferers from six to two. With 120° cell sectoring the primary interferers are located at approximate distances D and $D+0.7R$. The worst case carrier-to-interference ratio in this case is

$$
\begin{aligned}
\Lambda &= \frac{R^{-\alpha}}{D^{-\alpha} + (D + 0.7R)^{-\alpha}} \\
&= \frac{1}{\left(\frac{D}{R}\right)^{-\alpha} + \left(\frac{D}{R} + 0.7\right)^{-\alpha}} \cdot
\end{aligned}
\qquad (9.2)
$$

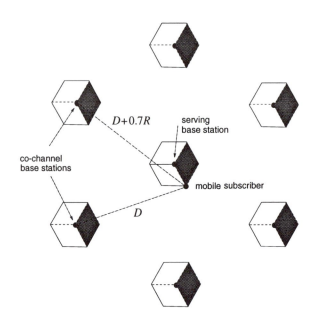

Figure 9.2 Worst case co-channel interference situation on the forward channel with 120° cell sectoring.

For $N = 7$, $D/R = 4.6$ resulting in $\Lambda = 24.5$ dB, a gain of 7.5 dB over omnidirectional cells. With 60° sectoring there is only one primary interferer at distance D, resulting in $\Lambda = 26.5$ dB, a gain of 9.5 dB.

9.2 SWITCHED-BEAM ANTENNAS

Switched-beam smart antennas are based on the retro-targeting concept. The selection of the activated receive beam is based on the (Received Signal Strength Indicator) RSSI and SAT tone[1]. Forward channel transmissions are over the best received beam, i.e., the same beam is used for both reception and transmission. Beam forming is accomplished by using physically directive elements to create aperture, and thus gain [26]. If the received carrier-to-interference ratio (CIR) falls below some preset level during a call, then the BS then switches to the best available beam for transmission and reception. The directive nature

---

[1] The detection of SAT tone can prevent *beam falsing* where the system is spoofed into thinking that the desired MS is located in a different beam when a strong co-channel interferer is present on another beam.

of the narrow-beam ensures that the mean interference experienced by any one user is much less than that experienced using conventional wide coverage BS antennas [295], thus offering substantial performance advantages [267].

Spatial diversity is typically not used when a smart antenna system is deployed at a BS, because the physical tower structures will prevent it. Angular diversity is a possibility, but it is not effective for macrocellular applications with their characteristically small AOA spreads. The pdf of the received signal power s is modeled by the Gamma distribution in (2.187). We have already seen that the Gamma distribution can be approximated by a log-normal distribution with mean and variance given by (2.188).

Comparisons will be made with an AMPS reference system, where the BSs use $120°$ sectoring with two-branch spatial diversity and selective combining. The received signal having the best quality (determined by RSSI) is selected for output. Assuming that the branches experience independent Rayleigh fading[2] and correlated shadowing, the pdf of the conditional received signal power with two branch microscopic selection diversity is

$$p_{s|\Omega_p}(x) = \frac{2}{\Omega_p} e^{-x/\Omega_p}(1 - e^{-x/\Omega_p}) \ . \tag{9.3}$$

By averaging over the log-normal shadowing, the pdf of the received signal power is

$$p_s(x) = \int_0^\infty \frac{2}{\Omega_p} e^{-x/\Omega_p}(1 - e^{-x/\Omega_p}) \frac{\xi}{\sqrt{2\pi}\sigma_\Omega\Omega_p} \exp\left\{-\frac{(10\log_{10}\Omega_p - \mu_{\Omega_p})^2}{2\sigma_\Omega^2}\right\} d\Omega_p \tag{9.4}$$

where $\xi = 10/\ln 10$. Appendix 9A shows that the pdf $p_s(x)$ can be approximated by a purely log-normal distribution with mean and variance given by

$$
\begin{aligned}
\mu_{(new)} &= \xi[\ln 2 - C] + \mu_{\Omega_p} \\
\sigma_{(new)}^2 &= \xi^2[\zeta(2,1) - 2(\ln 2)^2] + \sigma_\Omega^2 \ .
\end{aligned} \tag{9.5}
$$

The MS antennas are assumed to be omni-directional. With omni-directional BS antennas, there are six first-tier co-channel interferers for both the forward and reverse channels. The number of first-tier interferers is reduced to two with $120°$ sectoring. With a switched beam smart antenna, the number of first-tier co-channel interferers on the forward channel is a random variable

---

[2] Actually, a separation of 30 wavelengths will still result in a branch correlation of about 0.7, making the performance of the reference system optimistic [154].

ranging from 0 to 6, due to the narrow-beam directional antennas and the dependency of the activated beam on the MS location. If there are N_I co-channel interferers each with mean μ_i and variance σ_i^2 (in natural units), then the total interfering power is approximately log-normal. For our purpose, the mean and variance of the approximate log-normal distribution is obtained by using Fenton-Wilkinson method as described in Section 3.1.1. Finally, if the co-channel interference from the antenna sidelobes is ignored, there is at most one interferer on the reverse channel when the smart antenna beamwidth is less than 60°.

9.2.1 Trunkpool Techniques

In switched beam smart antenna systems, the narrow-beam directional antennas are analogous to cell sectoring that can reduce unnecessary spillage of radiation [204] and mitigate the effects of channel time dispersion [214]. Higher antenna gains also can be achieved because of narrow antenna beamwidth. However, switched beam smart antennas will have more frequent handoffs (inter-sector handoffs) that result in reduced trunking efficiency. To overcome the trunking efficiency degradation caused by narrow beam sectoring, **sector-trunkpool** and **omni-trunkpool** load sharing schemes are suggested.

Fig. 9.3 shows a switched-beam smart antenna with 4 azimuthal elements (beams) per 120° degree sector, i.e., 30° beam widths. With the sector-trunkpool technique, all the channels assigned to a 120° sector are shared by all four beams in that sector. Each sector antenna acts as a common aperture for one of four beams. No handoffs are needed unless the MS crosses sector or cell boundaries. In this case, the trunking efficiency will remain the same as the reference system, where each wide-beam sector has a unique channel assignment. This concept can be extended to the omni-trunkpool technique, where any of the channels assigned to a cell can be assigned to any one of the activated beams. In this case, no handoffs are needed unless a MS enters or exits the cell.

Usually, the trunking efficiency is measured by the *channel usage efficiency* (or loading factor) [296]

$$\eta_T = \rho(1 - P_B)/m \qquad \text{Erlangs/channel} \qquad (9.6)$$

where ρ is the offered traffic, P_B is the blocking probability, and m is the number of channels. From the Erlang-B formula under the blocked-calls-cleared

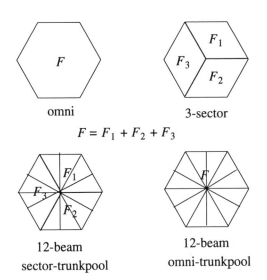

omni

3-sector

$$F = F_1 + F_2 + F_3$$

12-beam
sector-trunkpool

12-beam
omni-trunkpool

Figure 9.3   Trunkpool schemes for switched-beam antennas.

assumption, P_B can be shown to be[3]

$$P_B = \frac{\rho^m / m!}{\sum_{k=0}^{m} \rho^k / k!} \cdot \tag{9.7}$$

With AMPS, each service provider has 416 duplex channels[4]. Fig. 9.4 shows the channel usage efficiency for different trunkpool techniques. With the omni-trunkpool technique, the channel usage efficiency is increased 31.2% as compared to a 7-cell reuse reference system when the blocking probability is 0.01. In contrast, channel usage efficiency is increased only 17.4% when the frequency reuse cluster size is reduced from 7 to 4 cells. Therefore, the omni-trunkpool technique is helpful for increasing the trunking efficiency when smart antenna systems are employed.

---

[3] The Erlang B formula assumes an infinite subscriber population, and ignores handoff traffic. However, since we are interested in macrocellular systems, the handoff traffic is not expected to be substantial anyway.

[4] With AMPS, a total of 832 full duplex channels are divided in half for two competitive providers, each half consisting of 395 traffic channels and 21 control channels.

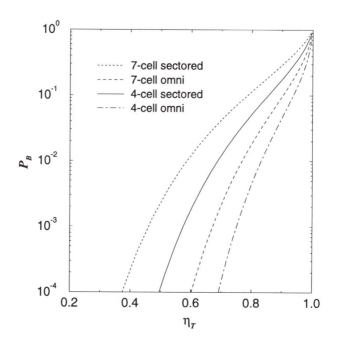

Figure 9.4 Channel usage efficiency with various trunkpool schemes.

9.2.2 Cellular Performance with Switched-beam Antennas

Our performance evaluation begins with the following assumptions:

1. Each cell shape is assumed to be circular under a hexagonal cell layout. The MSs are uniformly distributed within a cell.

2. Only the first tier of co-channel interferers is considered.

3. The system utilization is assumed to be 100% (worst case).

4. The 120° directional antennas used in the reference system have perfect directivity, i.e., there are no sidelobes. However, sidelobes adjacent to the main beam are considered for the switched-beam smart antenna system. The smart antenna front-to-back ratio is 30 dB, and the adjacent sidelobe attenuation is 12 dB.

Because power control is employed in the existing AMPS system, two different cases are studied here: power control and no power control. Practical power control algorithms usually react to the total received signal strength (C+I). However, for simplicity, the power control we consider here reacts to the desired signal strength C only[5]. The following notation will be used to distinguish between different frequency reuse factors, trunkpool techniques and antenna beamwidths, when switched-beam smart antennas are used:

sma - [*reuse cluster size*] - [*trunkpool type*] - [*antenna beamwidth*]

where

- *reuse cluster size:* 7-cell or 4-cell

- *trunkpool type:* omni or sectored

- *antenna beamwidth:* 30° (12 elements) or 15° (24 elements).

Reverse Channel

When no power control is employed, all MSs are assumed to transmit with the same power. The CIR at the serving BS is

$$\Lambda = \frac{10^{\frac{\varsigma_d}{10}}}{\sum_i 10^{\frac{\varsigma_i}{10}}} \tag{9.8}$$

where the subscripts d and i index the desired signals and interfering signals, respectively. The random variables ς_d and ς_i are Gaussian distributed, and their means and variances can be derived from (9.4) and (2.188) for the reference system and the smart antenna system, respectively. It is noted that the reference system has two-branch selection diversity in the reverse channel. The number of co-channel interferers depends on the position of the desired and interfering MSs. The CIR will vary as a function of the activated beams in the co-channel cells. When power control is employed, we assume that the power received at the serving BS from each MS is maintained at a constant C. The power transmitted by the ith mobile in the jth cell, MS_{ij}, at distance d_{ij} to its serving BS is P_{ij}. The power received at BS_j is

$$C = P_{ij} 10^{\frac{\varsigma_{ij}}{10}} . \tag{9.9}$$

---

[5]Since the required CIR is 17 dB in the AMPS system, the interference I can be safely neglected.

MS_{ij} is also at distance d_{i0} to the reference BS, BS_0, and will generate co-channel interference equal to

$$I_{i0} = P_{ij} 10^{\frac{\varsigma_{i0}}{10}} = C 10^{\frac{\varsigma_{i0}-\varsigma_{ij}}{10}} = C 10^{\frac{\varsigma_i}{10}} . \tag{9.10}$$

Then the CIR at BS_0 is

$$\Lambda = \frac{C}{C \sum_i 10^{\frac{\varsigma_i}{10}}} = \frac{1}{\sum_i 10^{\frac{\varsigma_i}{10}}} . \tag{9.11}$$

Assuming that the path loss follows a 4th law with distance, the mean and variance of ς_i are

$$\begin{aligned} \mu_i &= \mu_{i0} - \mu_{ij} = 10 \log_{10} \left(\frac{d_{i0}}{d_{ij}} \right)^4 \\ \sigma_i^2 &= \sigma_{i0}^2 + \sigma_{ij}^2 . \end{aligned} \tag{9.12}$$

Forward Channel

The major difference between the forward and reverse channels is the number of co-channel interferers. The calculation of the CIR is similar, but no antenna diversity is included for reference system. The means and variances of the log-normal random variables ς are calculated from (2.188).

Performance Criteria and Results

Two criteria will be used to evaluate the performance of the switched-beam smart antenna system.

Criterion 1: The area-averaged probability, P_1, that the received CIR exceeds a target value, Λ_{th}.

Criterion 2: The percentage of the cell area, P_2, where the received CIR exceeds a target value, Λ_{th}, 75% of the time.

Criterion 1 can be treated as a global performance measure. However, some bad locations will be masked from the area-averaged performance if there are some very good locations. This is particularly true when the performance is non-homogeneous over the cell area. In this case, Criterion 2 is useful. The performance with Criterion 1 is plotted in Figs. 9.5 and 9.6. A significant performance improvement is observed with switched-beam smart antenna systems, especially for the forward channel. For example, with a sma-7-omni-30° system, Criterion 1 yields an improvement of at least 5 dB compared to the reference

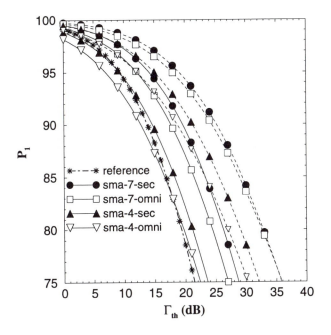

Figure 9.5 Area-averaged probability, P_1, for the reverse channel, with 30° antenna beamwidths (solid), and 15° antenna beamwidths (dashed); $\sigma = 8$ dB.

system. The major factors affecting the area-averaged CIR with switched beam smart antennas are the antenna beamwidth and the frequency re-use factor. The trunkpool techniques do not play an important role.

Figs. 9.7 and 9.8 show the performance with Criterion 2. In this case there is only about 2 dB improvement with respect to the reference system. Trunkpool techniques are shown have significant effect in the performance with Criterion 2. The performance with the sectored trunkpool is better than the omni-trunkpool. The sma-4-sec-30° system is worse than the reference system.

To explain the difference between performance Criterion 1 and 2 more clearly, Figs. 9.9 and 9.10 plot the points during the simulation where the the CIR is less than 17 dB more than 25% time. These points are called **bad points** [6]. With power control, bad points can occur anywhere within the cell and not just

---

[6] The location of these points are not fixed. Their locations vary with the locations of the co-channel interferers.

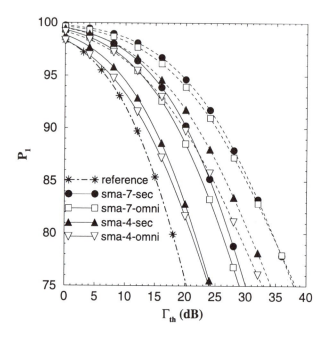

Figure 9.6   Area-averaged probability, P_1, for the forward channel, with 30° antenna beamwidths (solid), and 15° antenna beamwidths (dashed); $\sigma = 8$ dB.

on the cell boundary. This is because the power control algorithm only controls the power of the desired signal. When a MS is close to its serving BS the power levels are reduced. Hence, more interference will be experienced on both the forward and reverse channels when the co-channel interferers are close to this BS yet far from their own serving BSs. When switched-beam smart antennas are employed, the performance with Criterion 1 can be improved and yet the performance with Criterion 2 may not change much.

Another interesting phenomenon shown in Fig. 9.9 is that the reverse channel bad points with switched-beam smart antennas are concentrated in radial sectors called **bad areas** , i.e., the CIR improvement is not uniform within the cell area. We call this is the **cart-wheel** effect. However, it is not present in the forward channel. Fig. 9.11 replots Fig. 9.7 based only on the bad areas. As shown in Fig. 9.11, even the sma-4-sec-15° system is worse than the reference system. Bad areas will always exist in systems using switched-beam smart antennas no matter how narrow the antenna beamwidth is and regardless of whether or not power control is used. Of course, the number of bad points is reduced when the antenna beamwidth is decreased. To mitigate the cart-wheel

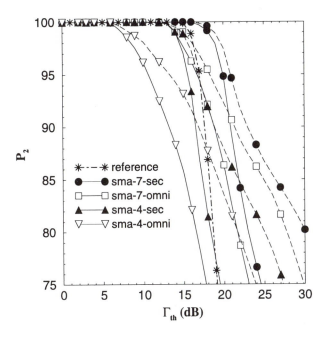

Figure 9.7 Percentage of the cell area, P_2, for the reverse channel, where the CIR exceeds the target value, Λ_{th}, 75% of the time with 30° antenna beamwidths (solid), and 15° antenna beamwidths (dashed); $\sigma = 8$ dB.

effect, one possibility is to rotate some switched beam cells with respect to others to distribute the bad points. However, this may be effective only when the rotation degree is larger than the AOA of the signals. Another approach is to use dynamic channel assignment to avoid using the same channel when a potential co-channel interferer is nearby.

9.3 CELL SPLITTING

Cell splitting refers to the process of splitting a cell into smaller cells. This is accomplished by establishing new and smaller cells at specific locations in the cellular system. By repeatedly splitting the cells, the cellular system can be tailored to meet traffic demands. Large cells can used in suburban and rural areas with low traffic loading, while smaller cells can be used in urban areas with high traffic loading. To demonstrate the concept of cell splitting, consider the uniform grid of hexagonal cells shown in Fig. 9.12. If heavy traffic loading

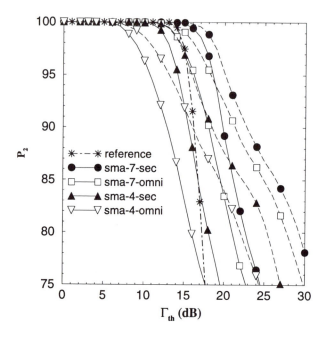

Figure 9.8 Percentage of the cell area, P_2, for the forward channel, where the CIR exceed the target value, Λ_{th}, 75% of the time with 30° antenna beamwidths (solid), and 15° antenna beamwidths (dashed); $\sigma = 8$ dB.

is experienced at the midpoint between two the cells labeled **1**, then a split cell labeled **1'** is introduced as shown. The area of the split cell is 1/4 of the area of the original cell. Additional split cells can be introduced to accommodate traffic loading in other locations throughout the system area. For example, the split cell **2'** can be located at the midpoint between the **2** cells.

After introducing the split cells, additional changes in the frequency assignments are required so that the split cells do not violate the reuse constraint. One possibility is to use **channel segmenting**, where the channel sets in the co-channel cells are subdivided into two groups; the split cells and inner co-channel cells are assigned one group of channels, while the outer co-channel cells use the remaining group of channels. Unfortunately, this arrangement decreases the trunking efficiency because the outer cells cannot use the channels assigned to the inner cells. Furthermore, if the larger cells are already near capacity, then segmenting the channels in these cells may result in their splitting as well, i.e., a propagation of splitting may occur throughout the system area.

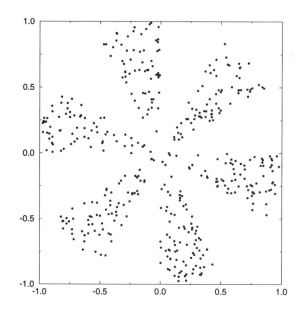

Figure 9.9 Reverse channel bad points for a 'sma-7-omni-30°' system with power control.

A more elegant solution for maintaining the co-channel reuse distance is shown in Fig. 9.13, where **overlaid** cells are introduced that are served by the same BS as the large cells. Once again, the available set of channels is divided into two groups. MSs located within the overlaid cells can reuse the same group of channels that are assigned to the split cells, while MSs located within the outer co-channel cells use the other group of channels. Whenever a MS moves between the inner and outer areas of a cell a hand-off is executed.

Because the split cells are smaller, the power transmitted to and from the MSs within the split cell can be reduced. To estimate the transmit power requirements, we note that the received power for a MS located at the corner of the original cell is

$$\Omega(R_o) = AP_o R_o^{-\beta} \tag{9.13}$$

while the received power at the boundary of the split cell is

$$\Omega(R_s) = AP_s R_s^{-\beta} \ . \tag{9.14}$$

where P_o and P_s, and R_o and R_s, are the transmit power and cell radius associated with the original cells and split cells, respectively. To keep the

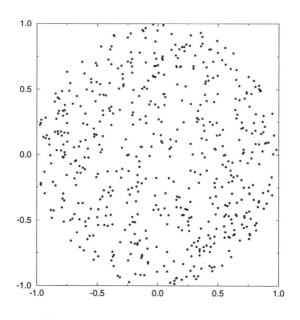

Figure 9.10   Forward channel bad points for a 'sma-7-omni-30°' system with power control.

received power associated with a MS located on the cell boundary constant, the required transmitter power can be obtained by using

$$P_s = P_o \left(\frac{R_s}{R_o} \right)^{-\alpha} .$$ (9.15)

If $\beta = 4$, then $P_s = P_o/16$, since $R_s = R_o/2$. Hence, the split cell can reduce the transmit power levels by 16 dB.

9.4   MICROCELLULAR SYSTEMS

Another approach for improving channel reuse efficiency is to deploy a microcellular system that uses multiple omni-directional antennas within each cell as shown in Fig. 9.14, a scheme proposed by Lee [177]. The cells are divided into zones with zone sites (antennas) at their centers . The zone sites are connected via fiber or microwave radio to a common BS. All three zones sites receive the

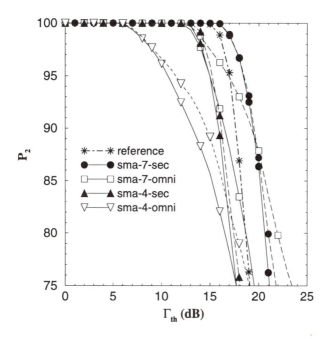

Figure 9.11 Percentage of the cell area, P_2, for the reverse channel, where the CIR exceeds the target value, Λ_{th}, 75% of the time with 30° antenna beamwidths (solid), and 15° antenna beamwidths (dashed); $\sigma = 8$ dB.

signal from a MS on the reverse channel; however, only the zone site that receives the largest C+I on the reverse channel will transmit to the MS on the forward channel. As a MS moves between the zones of a cell, a new zone site is selected to serve the MS on the forward channel, i.e., and intracell handoff is performed. Channel changes are still required, however, when the MSs move between cells.

The zones are arranged in a hexagonal grid as shown in Fig. 9.15. Note that certain hexagons are not assigned to any cell. However, because three zones sites are used, the coverage area will be roughly circular as shown in Fig. 9.15. Also note that a 3-cell cluster is used. Therefore, if the co-channel interference is acceptable, this arrangement will realize an improvement in channel reuse efficiency of $7/4 = 2.33$ over a system using a conventional 7-cell reuse cluster.

To evaluate the effect of co-channel interference, consider the situation where the MS is located at a zone-corner as shown in Fig. 9.15. In this case, the worst

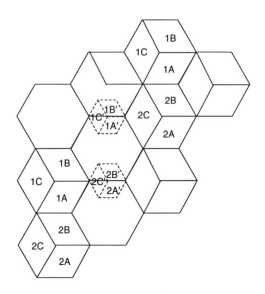

Figure 9.12 Cell splitting can be used to accommodate an increased traffic load by introducing smaller cells.

case carrier-to-interference ratio is

$$
\begin{aligned}
\Lambda &= \frac{R^{-\beta}}{(\sqrt{13}R)^{-\beta} + (4R)^{-\beta} + (\sqrt{28}R)^{-\beta} + (7R)^{-\beta} + (6R)^{-\beta}(\sqrt{31}R)^{-\beta}} \\
&= 81.380 = 19.1 \text{ dB} \tag{9.16}
\end{aligned}
$$

where the last line follows under the assumption that $\beta = 4$. Note that Λ is slightly larger than the 17 dB that is achieved with a 7-cell reuse cluster (with omni-directional antennas), but still less than that achieved with cell sectoring.

9.4.1   Edge Excited Zones

A cell with edge excited zones is shown in Fig. 9.16. Note that deployment of center-excited zones in Fig. 9.15 uses a 3-cell cluster, while deployment of edge-excited zones in Fig. 9.16 uses a 4-cell cluster. The worst case carrier-to-interference ratio for a 3-cell reuse cluster is

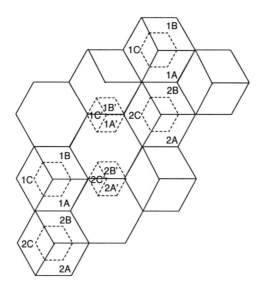

Figure 9.13 Overlaid inner cells can be used to maintain the frequency reuse constraint when cell splitting is used.

$$\Lambda = \frac{(2R)^{-\beta}}{3(8R)^{-\beta} + 3(\sqrt{52}R)^{-\beta}}$$
$$= 33.933 = 15.3 \text{ dB} \tag{9.17}$$

Hence, a 4-cell reuse cluster must be used, with the deployment in Fig. 9.17. The worst case Λ occurs when the MS is located in the center of a cell as shown in Fig. 9.16. If 120° directional antennas are used (actually any antenna with a beam width less than 146° will do), then the co-channel interferers are located as shown in Fig. 9.16. It follows that the worst case Λ is

$$\Lambda = \frac{(2R)^{-\beta}}{6(2\sqrt{19}R)^{-\beta}}$$
$$= 60.167 = 17.8 \text{ dB} \tag{9.18}$$

This Λ is comparable to that achieved with a 7-cell reuse cluster with omni-directional antennas. Provided that this level of co-channel interference is acceptable, the channel reuse efficiency will be increased by a factor of $7/4 = 1.75$ over that system.

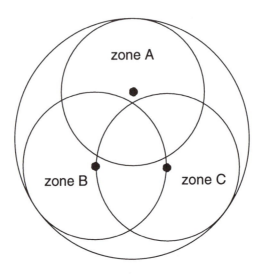

Figure 9.14   Microcell with center excited zones.

9.4.2   Street Microcell Deployment

Fig. 9.18 depicts a typical deployment of low elevation, low power BSs along city streets. Such a deployment is called a **Manhattan** microcell deployment, and for the particular deployment in Fig. 9.18 a BS is located at every intersection. Such a luxury in BS density virtually eliminates the street microcell corner effect. In a street microcell deployment, the affect of the non-line-of-sight (NLOS) co-channel interferers on the radio link performance can be neglected in deference to the dominant effect of the line-of-sight (LOS) co-channel interferers. The worst case Λ occurs when a MS is located on a cell edge as shown in Fig. 9.19. Using the two-slope path loss model in (2.212), the carrier-to-interference ratio is

$$
\begin{aligned}
\frac{C}{I} &= \frac{R^{-a}(1 + R/g)^{-b}}{(D_L - R)^{-a}(1 + (D_L - R)/g)^{-b} + (D_L + R)^{-a}(1 + (D_L + R)/g)^{-b}} \\
&= \left(\frac{g/R + 1}{g/R + D_L/R - 1} \right)^{-b} \frac{1}{\left(\frac{D_L}{R} - 1 \right)^{-a} + \left(\frac{D_L}{R} + 1 \right)^{-a}}
\end{aligned}
\tag{9.19}
$$

For the 2-cell reuse cluster in Fig. 9.19 $D_L/R = 4$. Using the parameters $R = 100$ m, $g = 150$ m, and $a = b = 2$, gives $\Lambda = 13.3$ dB. This may be too small. However, if $N = 4$-cell reuse cluster is used instead, then it can be shown that $D_L/R = 8$, resulting in $\Lambda = 25.48$ dB.

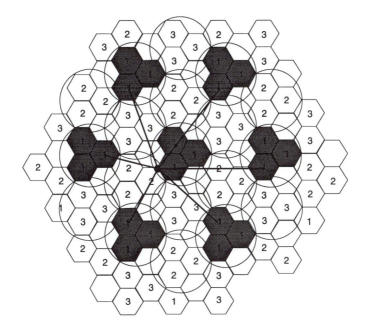

Figure 9.15 Cellular deployment with center excited zones. Also shown is the worst case forward channel co-channel interference scenario.

9.5 HIERARCHICAL ARCHITECTURES

It is well recognized that cellular PCS systems will most likely employ overlaid microcell/macrocell architectures [150]. A variety of microcell/macrocell architectures have been introduced in the literature [135], [151], [349], [107]. One method segregates the channel resources into two orthogonal groups, one for macrocells and one for microcells [135], [151]. In [349] another scheme is proposed where free macrocell channels of adjacent cells are borrowed by the underlaid microcells. Dynamic channel allocation (DCA) has been suggested as a method for selecting the available macrocell channels in [107]. Each of the above schemes has its drawbacks. Orthogonal sharing [135], [151] decreases the macrocell system capacity if all available channel resources have already been assigned to the macrocells. The overlaying structure with channel borrowing in [349] can only relieve hot spot traffic, but is ineffective if the traffic in neighboring cells is also heavy. The scheme in [107] requires forward and reverse link power control. The necessity of scanning all the macrocell channels to chose the available frequencies will cause serious delay in call setup [291].

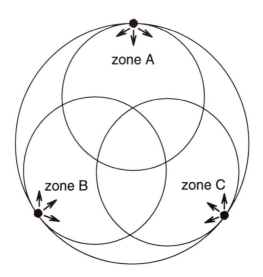

Figure 9.16   Cell with edge excited zones, from [177].

9.5.1   Reuse Partitioning

Halpern [135] has suggested an overlay/underlay scheme called **reuse partitioning** that increases the capacity of a cellular system. Reuse partitioning uses a deployment with multiple co-channel reuse factors. With this scheme an inner (micro)cell is created within each of the existing cells as shown in Fig. 9.20. Channels are assigned to the inner and outer cells with a 3-cell and 7-cell reuse plan, respectively. Channels assigned to the inner and outer cells are only used by MSs located within the inner and outer cells, respectively. Handoffs are required when a MS crosses the boundary of an inner and outer cell. The reduction in the cell radius for the inner cells lead to an increase in spatial reuse efficiency. To determine the achievable increase in spectral efficiency, we begin with the assumption that the link performance is determined by the co-channel reuse factor D/R, with all other things being equal. If the cell radius R is decreased, then the co-channel reuse distance D may also be decreased. Let

$$
\begin{aligned}
R_i &= \text{radius of the inner cells.} \\
R_o &= \text{radius of the outer cells.} \\
D_i &= \text{reuse distance for the inner cells.} \\
D_o &= \text{reuse distance for the outer cells.}
\end{aligned}
$$

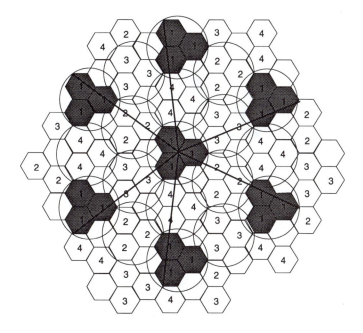

Figure 9.17 Cellular deployment with edge excited zones. Also shown is the worst case forward channel co-channel interference scenario.

Suppose that an acceptable link quality requires that $D_i/R_i = D_o/R_o = 4.6$. If a 7-cell and 3-cell reuse cluster is used for the outer and inner cells, respectively, then $D_i/R_o = 3$ and

$$\frac{D_i/R_i}{D_i/R_o} = \frac{4.6}{3} \ . \tag{9.20}$$

Hence, the inner and outer cell radii are related by $R_i = 0.65R_o$, i.e., the inner and outer cell areas are related by $A_i = (0.65)^2 A_o = 0.43A_o$. If a total of N_T channels are available, then $0.43N_T$ channels should be assigned to the inner cells and $0.57N_T$ channels assigned to the outer cells (assuming that the traffic is uniformly distributed throughout the system area). The channel reuse efficiency is

$$N_\mu = 0.57N_T/7 + 0.43N_T/3 = 0.225N_T \ \text{channels/cell} \ . \tag{9.21}$$

With a conventional 7-cell reuse plan

$$N_\mu = N_T/7 \ \text{channels/cell} \ . \tag{9.22}$$

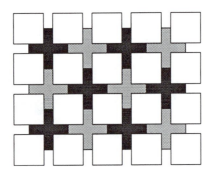

Figure 9.18   A typical Manhattan street microcell deployment.

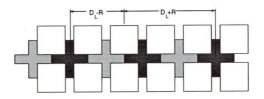

Figure 9.19   Worst case co-channel interference situation.

Hence, an improvement of 1.575 in channel reuse efficiency is realized. The cost associated with this scheme is an increase in number of handoffs that are required.

Cell Splitting with Reuse Partitioning

Cell splitting can also be used with reuse partitioning. An example is shown in Fig. 9.21 where a split cell is added between the large **B2** cells. The split cell also uses reuse partitioning. To maintain the co-channel interference at acceptable levels, some of the channels in the **B2** cells are moved to the inner cells and are denoted by **B2'**. Furthermore, the closest co-channel inner cells **A1** must have their channels partitioned in a similar fashion. Thus we see a drawback with the reuse partitioning scheme – the cells must be divide into many concentric rings of channels.

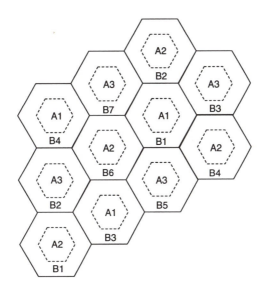

Figure 9.20 Reuse partitioning can be used to increase the channel reuse efficiency, from [135].

9.5.2 Cluster Planning

We now consider an innovative hierarchical architecture that applies the concept of **cluster planning**, to yield a sectoring arrangement that provides good shielding between the macrocells and microcells [327]. Cluster planning allows microcells and macrocells to reuse the same frequencies with complete coverage throughout the system area and without any decrease in the macrocell system capacity. The proposed architecture continues to use the existing macrocell

BSs thereby saving in infrastructure cost. Unlike the scheme in [349], the proposed architecture allows microcells to be deployed throughout the system area independent of the traffic loading of the macrocells. Moreover, the architecture does not require forward channel power control, unlike the scheme in [107].

System Architecture

A traditional cellular system with a 7-cell reuse cluster and 120° cell sectors is shown in Fig. 9.22. The channels are partitioned into 21 sets and each set is reused in a diamond-shaped sector with adequate distance of separation. Unfortunately. the interfering regions for each channel cover the whole service

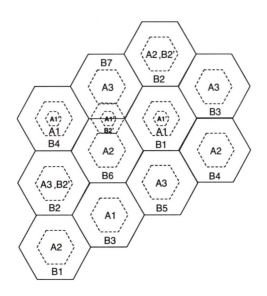

Figure 9.21   Cell splitting can be used in combination with reuse partitioning.

area. This widely distributed co-channel interference from the macrocells makes it impossible to reuse the same frequencies in the microcells.

Cluster planning can be used to change the conventional cell sectoring scheme into one having some areas of very low interference for a specified set of channels. The basic procedure for cluster planning is as follows.

1. The macrocell clusters are divided into three adjacent groups as shown in Fig. 9.23.

2. Each cell site retains the same channels assigned to the original macrocellular system.

3. Let the first group be the reference group

4. Rotate the channel sets of sectors of the second group cells 120° clockwise with respect to the first group;

5. In the third group, rotate the channel sets of sectors cells 120° counterclockwise with respect to the first group.

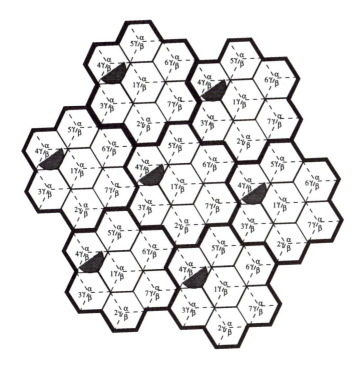

Figure 9.22 Traditional cellular system with a 7-cell, 3-sector, reuse cluster, from [327].

The only change to the existing 7-cell, 120° sectored system is the rotation of each cluster. Implementation of this scheme only requires a systematic shift of the original frequency sets in three sectors for each BS as shown by the example in Fig. 9.24. Observe that zone $A \sim F$ is a low interference zone for the channel set 4_β, since zone A is located in the back-lobe areas of the macrocell sectors using channel set 4_β. Thus the principle of cluster planning allows us to introduce a microcell in zone A that uses channel set 4_β. This zone is called a **micrco-area** and is defined by the back-lobe area of surrounding macrocell BSs for certain channel set.

We now show that micro-areas can be defined across the entire system area. From Fig. 9.24, observe that a micro-area consists of three macrocell sectors, each belonging to a different BS. For each micro-area, we define the **interference neighborhood** as the 18 neighboring macrocell sectors that surround the microcell as in Fig. 9.25. Because the interference neighborhood contains all the first-tier co-channel interferers, we only need to consider cells located

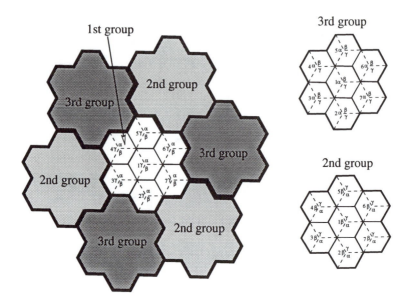

Figure 9.23   Proposed 7-cell, 3-sector, cellular system with cluster planning, from [327].

in the interference neighborhood when studying the effect of co-channel interference. For example, in Fig. 9.25, we see that micro-area A is located at the back-lobe area of the sectors using channel sets 4_β and 4_γ. Thus these two sets of frequencies can be used for micro-area A.

The following algorithm determines the macrocell channels that can be reused in each micro-area. Let c_i^j represent the channel set in sector $i = \alpha, \beta, \gamma$, of the cell site c, where $c = 1, \ldots, 7$. The superscript $j = 1, 2, 3$ in c_i^j is associated with the three kinds of rotated clusters.

- Given a desired micro-area and a corresponding interference neighborhood M, denote

$$\Theta = \left\{ c_i^j \in M \right\}$$

  as the union of channel sets c_i^j in the interference neighborhood M.

- From Θ, construct a 3×3 *indicator matrix* $\mathbf{B}_c = [a_{ij}]$ for BSs $c = 1, \ldots, 7$, where

$$a_{ij} = \begin{cases} 1 & \text{if the channel set } c_i^j \in M; \\ 0 & \text{otherwise} \end{cases}$$

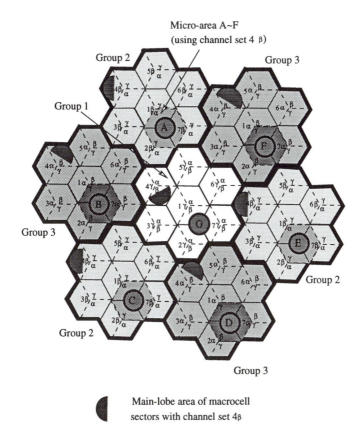

Figure 9.24 Microcells can reuse low-interference macrocell channels in the proposed hierarchical architecture. The macrocell channel set 4_β can be reused in the micro-area A $\sim$ F, from [327].

- If the indicator matrix $\mathbf{B}_c$ for some cell site c has full column rank, then the null row space of $\mathbf{B}_c$ defines the low-interference macrocell channel sets for the micro-area.

Example 7.1 _____According to Fig. 9.25, the interference neighborhood for micro-area A is

$$\Theta = \left\{1_\alpha^2, 1_\beta^2, 1_\gamma^2, 2_\alpha^2, 2_\beta^2, 2_\gamma^2, 3_\alpha^2, 3_\alpha^3, 3_\gamma^2, 4_\alpha^1, 4_\alpha^2, 4_\alpha^3, 5_\alpha^1, 5_\alpha^2, 5_\gamma^1, 6_\alpha^2, 6_\beta^2, 6_\beta^3, 7_\alpha^2, 7_\beta^2, 7_\gamma^2\right\}$$

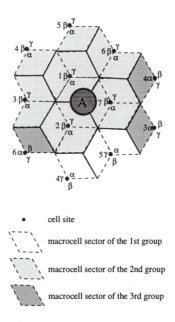

- •   cell site

  macrocell sector of the 1st group

  macrocell sector of the 2nd group

  macrocell sector of the 3rd group

Figure 9.25   Interference neighborhood for the micro-area A in Fig. 9.24, from [327].

Then the indicating matrices are

$$\mathbf{B}_1 = \begin{pmatrix} 0 & 1 & 0 \\ 0 & 1 & 0 \\ 0 & 1 & 0 \end{pmatrix}; \mathbf{B}_2 = \begin{pmatrix} 0 & 1 & 0 \\ 0 & 1 & 0 \\ 0 & 1 & 0 \end{pmatrix}; \mathbf{B}_3 = \begin{pmatrix} 0 & 1 & 1 \\ 0 & 0 & 0 \\ 0 & 1 & 0 \end{pmatrix}$$

$$\mathbf{B}_4 = \begin{pmatrix} 1 & 1 & 1 \\ 0 & 0 & 0 \\ 0 & 0 & 0 \end{pmatrix}; \mathbf{B}_5 = \begin{pmatrix} 1 & 1 & 0 \\ 0 & 0 & 0 \\ 1 & 0 & 0 \end{pmatrix}; \mathbf{B}_6 = \begin{pmatrix} 0 & 1 & 0 \\ 0 & 1 & 1 \\ 0 & 0 & 0 \end{pmatrix}$$

$$\mathbf{B}_7 = \begin{pmatrix} 0 & 1 & 0 \\ 0 & 1 & 0 \\ 0 & 1 & 0 \end{pmatrix}.$$

Examining the indicating matrix $\mathbf{B}_c, c = 1, \ldots, 7$, we find that only matrix $\mathbf{B}_4$ has full column rank. The null space of $\mathbf{B}_4$ consists of the second row and the third row. Based on the above algorithm, the low-interference macrocell channel sets for micro-area A are 4_β and 4_γ.

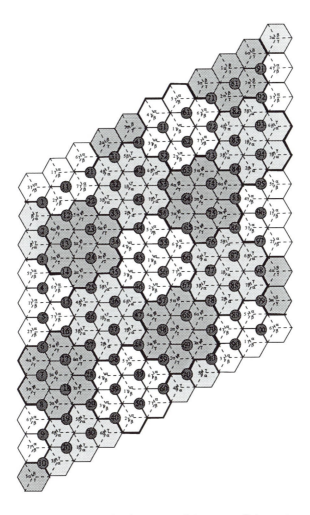

Figure 9.26 Channel reuse in the proposed 3-sector cellular system, where each micro-area consists of three sectors that belong to three different macro-cells, from [327].

To see if other micro-areas can be defined in the proposed system architecture, consider the system in Fig. 9.26 having 100 micro-areas defined over the service area. By applying the above algorithm, the available macrocell channel sets for each micro-area are listed in Table 9.1. Note that the micro-areas are capable of reusing two macrocell channel sets and they can be deployed throughout the whole service area. A detailed analysis of this system is beyond the scope of

zone	channel set	zone	channel set	zone	channel set	zone	channel set
1	$7_\alpha, 7_\beta$	26	$6_\alpha, 6_\gamma$	51	$6_\alpha, 6_\beta$	76	$3_\alpha, 3_\beta$
2	$5_\beta, 5_\gamma$	27	$7_\alpha, 7_\gamma$	52	$7_\alpha, 7_\beta$	77	$6_\alpha, 6_\gamma$
3	$1_\beta, 1_\gamma$	28	$5_\alpha, 5_\beta$	53	$5_\beta, 5_\gamma$	78	$7_\alpha, 7_\gamma$
4	$2_\beta, 2_\gamma$	29	$1_\alpha, 1_\beta$	54	$1_\beta, 1_\gamma$	79	$5_\alpha, 5_\beta$
5	$4_\alpha, 4_\gamma$	30	$2_\alpha, 2_\beta$	55	$2_\beta, 2_\gamma$	80	$1_\alpha, 1_\beta$
6	$3_\alpha, 3_\gamma$	31	$3_\alpha, 3_\beta$	56	$4_\alpha, 4_\gamma$	81	$4_\alpha, 4_\beta$
7	$6_\beta, 6_\gamma$	32	$6_\alpha, 6_\gamma$	57	$3_\alpha, 3_\gamma$	82	$3_\alpha, 3_\beta$
8	$7_\beta, 7_\gamma$	33	$7_\alpha, 7_\gamma$	58	$6_\beta, 6_\gamma$	83	$6_\alpha, 6_\gamma$
9	$5_\alpha, 5_\gamma$	34	$5_\alpha, 5_\beta$	59	$7_\beta, 7_\gamma$	84	$7_\alpha, 7_\gamma$
10	$1_\alpha, 1_\gamma$	35	$1_\alpha, 1_\beta$	60	$5_\alpha, 5_\gamma$	85	$5_\alpha, 5_\beta$
11	$4_\alpha, 4_\gamma$	36	$2_\alpha, 2_\beta$	61	$2_\beta, 2_\gamma$	86	$1_\alpha, 1_\beta$
12	$3_\alpha, 3_\gamma$	37	$4_\beta, 4_\gamma$	62	$4_\alpha, 4_\gamma$	87	$2_\alpha, 2_\beta$
13	$6_\beta, 6_\gamma$	38	$3_\beta, 3_\gamma$	63	$3_\alpha, 3_\gamma$	88	$4_\beta, 4_\gamma$
14	$7_\beta, 7_\gamma$	39	$6_\alpha, 6_\beta$	64	$6_\beta, 6_\gamma$	89	$3_\beta, 3_\gamma$
15	$5_\alpha, 5_\gamma$	40	$7_\alpha, 7_\beta$	65	$7_\beta, 7_\gamma$	90	$6_\alpha, 6_\beta$
16	$1_\alpha, 1_\gamma$	41	$1_\alpha, 1_\beta$	66	$5_\alpha, 5_\gamma$	91	$5_\alpha, 5_\beta$
17	$2_\alpha, 2_\gamma$	42	$2_\alpha, 2_\beta$	67	$1_\alpha, 1_\gamma$	92	$1_\alpha, 1_\beta$
18	$4_\alpha, 4_\beta$	43	$4_\beta, 4_\gamma$	68	$2_\alpha, 2_\gamma$	93	$2_\alpha, 2_\beta$
19	$3_\alpha, 3_\beta$	44	$3_\beta, 3_\gamma$	69	$4_\alpha, 4_\beta$	94	$4_\beta, 4_\gamma$
20	$6_\alpha, 6_\gamma$	45	$6_\alpha, 6_\beta$	70	$3_\alpha, 3_\beta$	95	$3_\beta, 3_\gamma$
21	$5_\alpha, 5_\gamma$	46	$7_\alpha, 7_\beta$	71	$7_\beta, 7_\gamma$	96	$6_\alpha, 6_\beta$
22	$1_\alpha, 1_\gamma$	47	$5_\beta, 5_\gamma$	72	$5_\alpha, 5_\beta$	97	$7_\alpha, 7_\beta$
23	$2_\alpha, 2_\gamma$	48	$1_\beta, 1_\gamma$	73	$1_\alpha, 1_\gamma$	98	$5_\beta, 5_\gamma$
24	$4_\alpha, 4_\beta$	49	$2_\beta, 2_\gamma$	74	$2_\alpha, 2_\gamma$	99	$1_\beta, 1_\gamma$
25	$3_\alpha, 3_\beta$	50	$4_\alpha, 4_\gamma$	75	$4_\alpha, 4_\beta$	100	$2_\beta, 2_\gamma$

Table 9.1 The macrocell channel sets that can be used in the underlaid microcells, from [327].

this text, but appears in [327]. However, Wang *et al.* have concluded that this hierarchical cellular architecture is feasible with AMPS.

Since a micro-area consists of 3 macrocell sectors, each cell has 5 channel sets – 3 assigned to macrocells and 2 assigned to microcells. Let C_μ represent the number of the microcell clusters in a micro-area. Since each microcell cluster can reuse these two sets of low-interference macrocell channels as shown in the above example, the capacity can be increased by factor of $1 + 2 \times C_\mu/3$ times.

Appendix 9A
DERIVATION OF EQUATIONS (9.4)

The conditional pdf of the received signal power is

$$p_{s|\Omega_p}(x) = \frac{2}{\Omega_p} e^{-x/\Omega_p}(1 - e^{-x/\Omega_p}) \ . \tag{9A.1}$$

Averaging over distribution of log-normal shadowing yields the pdf

$$p_s(x) = \int_0^\infty \frac{2}{\Omega_p} e^{-x/\Omega_p}(1-e^{-x/\Omega_p}) \frac{\xi}{\sqrt{2\pi}\sigma_\Omega \Omega_p} \exp\left\{ -\frac{(10\log_{10}\Omega_p - \mu_{\Omega_p})^2}{2\sigma_\Omega^2} \right\} d\Omega_p \tag{9A.2}$$

The mean of the approximate log-normal distribution is

$$
\begin{aligned}
E[10\log_{10}s] &= \int_0^\infty \int_0^\infty (10\log_{10}x) \frac{2}{\Omega_p} e^{-x/\Omega_p}(1 - e^{-x/\Omega_p}) \\
&\quad \times \frac{\xi}{\sqrt{2\pi}\sigma_\Omega \Omega_p} \exp\left\{ -\frac{(10\log_{10}\Omega_p - \mu_{\Omega_p})^2}{2\sigma_\Omega^2} \right\} d\Omega_p dx \\
&= \xi^2 \int_0^\infty \frac{1}{\sqrt{2\pi}\sigma_\Omega \Omega_p^2} \exp\left\{ -\frac{(10\log_{10}\Omega_p - \mu_{\Omega_p})^2}{2\sigma_\Omega^2} \right\} \\
&\quad \times \int_0^\infty 2\ln(x)e^{-x/\Omega_p}\left(1 - e^{-x/\Omega_p}\right) dx d\Omega_p \ .
\end{aligned}
\tag{9A.3}
$$

From [128, 4.352.1], the inner integral becomes

$$\int_0^\infty 2\ln(x)e^{-x/\Omega_p}(1 - e^{-x/\Omega_p})dx = \Omega_p[\ln(2) + \ln(\Omega_p) - C] \tag{9A.4}$$

where $C \simeq 0.5772$ is *Euler's constant*. Hence,

$$E[10\log_{10}s] = \xi[\ln(2) - C] + \mu_{\Omega_p} \tag{9A.5}$$

In a similar fashion, the mean square value is

$$
\begin{aligned}
E[(10\log_{10}(s))^2] &= \int_0^\infty \int_0^\infty (10\log_{10}x)^2 (\frac{2}{\Omega_p}) e^{-x/\Omega_p}(1 - e^{-x/\Omega_p}) \\
&\quad \times \frac{\xi}{\sqrt{2\pi}\sigma_\Omega \Omega_p} \exp\left\{ -\frac{(10\log_{10}\Omega_p - \mu_{\Omega_p})^2}{2\sigma_\Omega^2} \right\} d\Omega_p dx \\
&= \xi^3 \int_0^\infty \frac{1}{\sqrt{2\pi}\sigma_\Omega \Omega_p^2} \exp\left\{ -\frac{(10\log_{10}\Omega_p - \mu_{\Omega_p})^2}{2\sigma_\Omega^2} \right\} \\
&\quad \times \int_0^\infty 2(\ln x)^2 e^{-x/\Omega_p}(1 - e^{-x/\Omega_p})dx d\Omega_p \ .
\end{aligned}
\tag{9A.6}
$$

From [128, 4.358.2], the inner integrals become

$$\int_0^\infty 2(\ln x)^2 e^{-x/\Omega_p} dx = 2\Omega_p([\ln \Omega_p - C] + \zeta(2,1))$$

$$\int_0^\infty (\ln x)^2 e^{-2x/\Omega_p} dx = \Omega_p([\ln \Omega_p - \ln(2) - C]^2 + \zeta(2,1)) \quad (9A.7)$$

Finally, the variance is

$$\sigma^2_{(new)} = E[(10\log_{10}s)^2] - E^2[10\log_{10}s]$$
$$= \xi^2[\zeta(2,1) - 2(\ln 2)^2] + \sigma^2_\Omega . \quad (9A.8)$$

When m is an integer

$$\psi(m+1) = -C + \sum_{k=1}^{m} \frac{1}{k}$$

$$\zeta(2,m) = \sum_{k=0}^{\infty} \frac{1}{(m+k)^2} .$$

Problems

9.1. Consider a cellular system that uses a 7-cell reuse cluster without cell sectoring.

a) Show graphically the worst case co-channel interference situation for the reverse channel.

b) Ignoring envelope fading and shadowing and assuming the simple path loss model in (1.6), calculate the worst case carrier-to-interference ratio in terms of the co-channel reuse factor D/R.

c) Repeat parts a) and b) if 120° cell sectoring is used.

9.2. One method for improving the capacity of a cellular system employs a *two-channel bandwidth* scheme as suggested by Lee [179], where a hexagonal cell is divided into two concentric hexagons as shown in Fig. 9.27 below. The inner hexagon is serviced by 15 kHz channels, while the outer hexagon is serviced by 30 kHz channels. Suppose that the 30 kHz channels require $\Lambda = 18$ dB to maintain an acceptable radio link quality, while the 15 kHz channels require $\Lambda = 24$ dB.

Assume a fourth-law path loss model and suppose that the effects of envelope fading and shadowing can be ignored. Consider the mobile-to-base link and suppose that there are six co-channel interferers at distance D from the BS. For a 7-cell reuse cluster, it follows that the worst case carrier-to-interference ratio, Λ, when a mobile station (MS) is located at distance d from the BS is $\Lambda = (D/d)^4/6$. Hence, $\Lambda = 18$ dB requires $D/R_o = 4.6$, and $\Lambda = 24$ dB requires $D/R_i = 6.3$, where R_i and R_o are the radii of the inner and outer cells, respectively.

a) Use the values of D/R_i and D/R_o to determine the ratio of the inner and outer cell areas, A_i/A_o.

b) Let N_i and N_o be the number of channels that are allocated to the inner and outer areas of each cell, and assume that the channels are assigned such that $N_i/N_o = A_i/A_o$. Determine the increase in capacity (as measured in channels per cell) over a conventional *one-channel bandwidth* system that uses only 30 kHz channels.

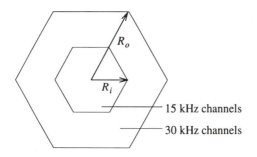

Figure 9.27 Cell division with two channel bandwidth scheme.

9.3. It has been suggested by [179] that the two-channel bandwidth scheme in Problem 9.2 can be combined with Halpern's reuse partitioning scheme. In this case, 15 kHz channels are used in the inner cells and 30 kHz channels are used in the outer cells. In order to have adequate performance in the inner or low bandwidth ring we must have $D_i/R_i = 6.3$, while the outer higher bandwidth ring can use $D_o/R_o = 4.6$.

Compute the increase in capacity (as measured in channels per cell) that will result from using this scheme, as compared to a conventional system using a 7-cell reuse cluster.

9.4. In Section 9.1 the worst case forward channel carrier-to-interference ratio was calculated by considering only the first tier of co-channel interferers. Calculate the amount of interference from the second tier of co-channel interferers. Is it reasonable to neglect this interference?

9.5. Microcells are characterized by very erratic propagation environments. This problem is intended to illustrate the imbalance in the forward and reverse channel carrier-to-interference ratio that could occur in a street microcell deployment. Consider the scenario shown in Fig. 9.28, that consists of two co-channel BSs, BS_1 and BS_2, communicating with two co-channel MSs, MS_1 and MS_2. Neglect the effects of shadowing and multipath, and assume that the NLOS corner path loss model in (2.214). Suppose that $a = 2$, $b = 4$, and $g = 150$ m. Plot Λ at BS_1, BS_2, MS_1, and MS_2 as MS_2 moves from A to C. When plotting your results, assume a received power level of 1 dBm at a distance of one meter.

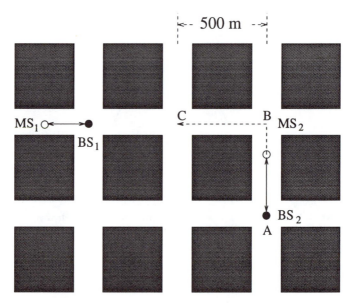

Figure 9.28   Microcellular propagation environment for Problem 9.5.

10

LINK QUALITY MEASUREMENT AND HANDOFF INITIATION

Cellular systems must have the ability to maintain a call while a mobile station (MS) moves throughout a cellular service area. This is accomplished by performing a handoff to an alternate base station (BS) whenever the link quality with the serving BS becomes unacceptable. A variety of parameters such as bit error rate (BER) [54], carrier-to-interference ratio (C/I) [103], distance [209], [85], traffic load, signal strength [209], [130], [132], [225], [41], [319], and various combinations of these fundamental schemes have been suggested for evaluating the link quality and deciding when a handoff should be performed. Of these, temporal averaging signal strength based handoff algorithms that measure the received carrier plus interference power (C+I) have received the most attention due to their simplicity and good performance in macrocellular systems. As a reflection of this trend, much of this chapter will be devoted to these types of handoff algorithms. However, spectrally efficient cellular systems are interference limited and a large C+I does not necessary imply a large C/I. Since radio link quality is more closely related to C/I than to C+I, it is apparent that C/I based handoff algorithms are highly desirable for microcellular systems with their characteristically erratic propagation environments. Unfortunately, C/I measurements can be quite difficult to obtain in practice [180], [348] and, hence, practical resource allocation algorithms often resort to received (C+I) measurements as a quality measure. Nevertheless, some discussion of C/I measurement techniques will be presented in this chapter.

Signal strength based handoff algorithms have been optimized by minimizing two conflicting design criteria; the handoff delay and the mean number of handoffs between BSs. It is important to keep the handoff delay small to prevent dropped calls and to prevent an increase in co-channel interference due to distortion of the cell boundaries. Likewise, it is important to keep the mean

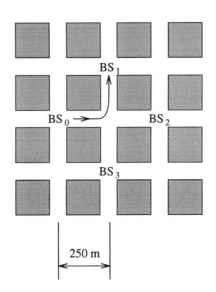

Figure 10.1 Typical NLOS handoff scenario. The MS rounds the corner, losing the LOS from BS_0 and gaining the LOS from BS_1. The frequency reuse plan is specified by using the different numbers at the BSs, from [15].

number of handoffs between BSs along a handoff route at a reasonably low value to prevent excessive loading and resource consumption on the network. Several authors [225], [319], [209], [132] have applied these (or similar) design criteria while adjusting two important design parameters; the required average signal strength difference, or **hysteresis** , between the BSs before a handoff is initiated, and the window length over which the signal strength measurements are averaged.

Murase [225] studied the tradeoff between the hysteresis and window length for line-of-sight (LOS) and non line-of-sight (NLOS) handoffs. For LOS handoffs, the MS always maintains a LOS with both the serving and target BS. This would be the case, for example, when a MS traverses along a route from BS_0 to BS_2 in Fig. 10.1. NLOS handoffs, on the other hand, arise when the MS suddenly loses the LOS component with the serving BS while gaining a LOS component with the target BS. This phenomenon is called the **corner effect** [225], [41] since it occurs while turning corners in urban microcellular settings like the one shown in Fig. 10.1 where the MS traverses along a route from BS_0 to BS_1. In this case, the average received signal strength with the serving BS can drop by 25-30 dB over distance as small as 10 m [225].

Corner effects may also cause link quality imbalances on the forward and reverse channels due to the following mechanism. Quite often the co-channel interference will arrive via a NLOS propagation path. Hence, as a MS rounds a corner, the received signal strength at the serving BS suffers a large decrease while the NLOS co-channel interference remains the same, i.e., the corner effect severely degrades the C/I on the reverse channel. Meanwhile, the corner will cause the same attenuation to both the desired and interfering signals that are received at the MS. Therefore, unless there are other sources of co-channel interference that become predominant as the MS rounds the corner, the C/I on the forward channel will remain about the same.

If the handoff requests from rapidly moving MSs in microcellular networks are not processed quickly, then excessive dropped calls will occur. Fast temporal based handoff algorithms can partially solve this problem, where short temporal averaging windows are used to detect large, sudden, drops in signal strength [225]. However, the shortness of a temporal window is relative to the MS velocity and, furthermore, a fixed time averaging interval makes the handoff performance sensitive to velocity with the best performance being achieved at only a particular velocity. Velocity adaptive handoff algorithms can overcome these problem, and are known to be robust to the severe propagation environments that are typical of urban microcellular networks [15].

The remainder of this chapter is organized as follows. Section 10.1 presents several different types of signal strength based handoff algorithms. This is followed by a detailed treatment of spatial signal strength averaging in Section 10.2. Guidelines are developed on the window averaging length that is needed so that Rician fading can be neglected in analog and sampled averaging. These guidelines are necessary for local mean and velocity estimation. Section 10.3 motivates the need for velocity adaptive handoff algorithms and presents three velocity estimators. The velocity estimators are compared in terms of their sensitivity to Rice factor, directivity, and additive Gaussian noise. In Section 10.4, the velocity estimators are incorporated into a velocity adaptive handoff algorithm. Section 10.5 provides an analytical treatment of conventional signal strength based handoff algorithms. In Section 10.6, methods are discussed for C/I measurements in TDMA cellular systems. Finally 10.7 wraps up with some concluding remarks.

10.1   SIGNAL STRENGTH BASED HANDOFF ALGORITHMS

Traditional mobile assisted handoff algorithms calculate signal strength *time* averages $< r_i(t) >$ from N neighboring BSs, BS_i, $i = 0, \cdots, N - 1$, and reconnect the MS to an alternate BS whenever the signal strength of the target BS exceeds that of the serving BS by at least H dB. For example, a handoff is performed between two BSs, BS_0 and BS_1, when

$$
\begin{aligned}
Y_1(n) &> Y_0(n) + H \quad \text{if the serving BS is } BS_0 \\
Y_0(n) &> Y_1(n) + H \quad \text{if the serving BS is } BS_1
\end{aligned}
\tag{10.1}
$$

where H denotes the hysteresis (in dB), and $Y_0(n)$ and $Y_1(n)$ are the estimated mean signal strengths (in dBm) of BS_0 and BS_1, given by

$$
Y_0(n) = \frac{1}{N} \sum_{k=n-N+1}^{n} |r_0(k)|^2_{(\text{dB})}
\tag{10.2}
$$

$$
Y_1(n) = \frac{1}{N} \sum_{k=n-N+1}^{n} |r_1(k)|^2_{(\text{dB})}
\tag{10.3}
$$

respectively, where $|r_i(k)|^2_{(\text{dB})}$ is the kth sample of the squared envelope (in dBm) and N is the window length.

Many other variations of signal strength based handoff algorithms have been suggested in the literature. In one variation, handoffs are also triggered when the measured signal strength of the serving BS drops below a threshold. For example, a handoff could be performed between BS_0 and BS_1 when

$$
\begin{aligned}
Y_1(n) &> Y_0(n) + H \text{ and } Y_0(n) > \Omega_L, \text{ if the serving BS is } BS_0 \\
Y_1(n) &> Y_0(n) \text{ and } Y_0(n) < \Omega_L, \text{ if the serving BS is } BS_0 \\
Y_0(n) &> Y_1(n) + H \text{ and } Y_1(n) > \Omega_L, \text{ if the serving BS is } BS_1 \\
Y_0(n) &> Y_1(n) \text{ and } Y_1(n) < \Omega_L, \text{ if the serving BS is } BS_1
\end{aligned}
\tag{10.4}
$$

This scheme encourages a handoff whenever the signal strength with the serving BS drop below the threshold Ω_L, thereby reducing the probability of dropped call.

Another variation discourages handoffs when the signal strength of the serving BS exceeds another threshold Ω_U. For example, a handoff is performed between BS_0 and BS_1 when

$$Y_1(n) \; > \; Y_0(n) + H \text{ and } \Omega_L < Y_0(n) < \Omega_U, \text{ if the serving BS is } BS_0$$
$$Y_1(n) \; > \; Y_0(n) \text{ and } Y_0(n) < \Omega_L, \text{ if the serving BS is } BS_0$$
$$Y_0(n) \; > \; Y_1(n) + H \text{ and } \Omega_L < Y_1(n) < \Omega_U, \text{ if the serving BS is } BS_1$$
$$Y_0(n) \; > \; Y_1(n) \text{ and } Y_1(n) < \Omega_L, \text{ if the serving BS is } BS_1 \qquad (10.5)$$

This scheme avoids unnecessary handoffs, thereby reducing the network load and network delay.

Direction biased handoff algorithms have also been suggested for improving the handoff performance in urban microcells [19]. These algorithm incorporate moving direction information into the handoff algorithm to encourage handoffs to BSs that the MS is approaching, and to discourage handoffs to BSs that the MS is moving away from. Let BS_{serv} denote the serving BS. A direction biased handoff algorithm can be defined by grouping all the BSs being considered as handoff candidates, including BS_{serv}, into two sets based on their direction information. Define

$$\mathcal{A} \; := \; \text{the set of BSs the MS is approaching} \qquad (10.6)$$
$$\mathcal{R} \; := \; \text{the set of BSs the MS is moving away from.} \qquad (10.7)$$

By introducing an encouraging hysteresis H_e, and a discouraging hysteresis H_d, a direction biased handoff algorithm requests a handoff to BS_j if $BS_j \in \mathcal{R}$ and

$$Y_j(n) \; > \; X_s(n) + H, \text{ if } BS_{serv} \in \mathcal{R}$$
$$Y_j(n) \; > \; X_s(n) + H_d, \text{ if } BS_{serv} \in \mathcal{A} \qquad (10.8)$$

or if $BS_j \in \mathcal{A}$ and

$$Y_j(n) \; > \; X_s(n) + H_e, \text{ if } BS_{serv} \in \mathcal{R}$$
$$Y_j(n) \; > \; X_s(n) + H, \text{ if } BS_{serv} \in \mathcal{A} \; . \qquad (10.9)$$

To encourage handoffs to BSs in $\mathcal{A}$ and discourage handoffs to BSs in $\mathcal{R}$, the hysteresis values should satisfy $H_e \leq H \leq H_d$. When equality holds, the algorithm reduces to the conventional method described in (10.1). Good values for H_e, H, and H_d depend on the propagation environment and BS layout. In general, a direction biased handoff algorithm can maintain a lower mea number of handoffs and handoff delay, and provide better cell membership properties.

10.2   SIGNAL STRENGTH AVERAGING

The squared envelopes $|r_i(t)|^2$ are affected by Rician fading, log-normal shadowing, and path loss attenuation. Here we assume a narrow-band system with flat fading, although the techniques that are described in the sequel can be extended to other systems with some modification. For middle-band TDMA systems, it is likely that the necessary signal strength information can be obtained from the adaptive equalizer or channel estimator. Likewise, for wide-band CDMA systems, the tap weightings in a RAKE receiver could be used to estimate the received signal strength.

Two Rician fading models are considered. The first model assumes that the received bandpass signal is

$$x(t) = r_I(t) \cos 2\pi f_c t - r_Q(t) \sin 2\pi f_c t \tag{10.10}$$

where f_c is the carrier frequency, and $r_I(t)$ and $r_Q(t)$ are independent Gaussian random processes with variance σ^2 and means m_I and m_Q, respectively. As discussed in Section 2.1.2 the envelope of $x(t)$, given by $|r(t)| = |r_I(t) + jr_Q(t)|$, is Rician distributed with Rice factor $K = s^2/(2\sigma^2)$, where $s^2 = m_I^2 + m_Q^2$.

The second Rician fading model has been proposed by Aulin [12], and is used when the autocorrelation and LOS effects of the received signal are important. With Aulin's model the means $m_I(t)$ and $m_Q(t)$ correspond to the in phase and quadrature components of the LOS signal and are given by

$$m_I(t) = s \cdot \cos(\omega_0 t + \theta_0) \tag{10.11}$$

$$m_Q(t) = s \cdot \sin(\omega_0 t + \theta_0) \tag{10.12}$$

where ω_0 and θ_0 are the Doppler shift and angle offset of the LOS signal, respectively. Once again, the envelope $|r(t)|$ is Rician distributed with Rice factor $K = s^2/(2\sigma^2)$. Both models are equivalent for Rayleigh fading ($K = 0$).

As suggested in Section 2.4.1, the spatial correlation of the log-normal shadowing can be effectively described by the negative exponential model

$$\phi_{\Omega_{(dB)}\Omega_{(dB)}}(d) = \sigma_\Omega^2 \exp(-|d|/D) \tag{10.13}$$

where σ_Ω is the shadow standard deviation (typically between 4 and 12 dB), d is the spatial distance, and D controls the spatial correlation of the shadowing.

For LOS propagation we assume the two-slope path loss model given by (2.212). For NLOS propagation we use the model in (2.214) yielding, for example, the signal strength profile in Fig. 10.2.

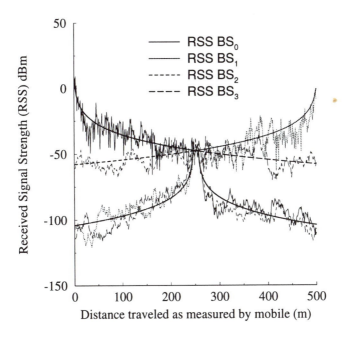

Figure 10.2 Typical average received signal strength (thick lines) and instantaneous received signal strength (thin lines) for a NLOS handoff scenario. For each particular BS, the received signal strength is shown when the MS is connected to that particular BS and the MS moves along the route in Fig. 10.1, from [15].

Time averaging $< |r_i(t)|^2 >$ and hysteresis H reduce the effect of fading and shadowing variations that would otherwise cause large numbers of unnecessary handoffs. Short spatial windows average over the fades while longer spatial windows average over the shadows as well. The effect of the spatial window length on handoffs is well documented in the literature [209], [132], [225], [319]. However, for the development of fast microcellular handoff algorithms, new guidelines must be developed so that spatial averaging can be used effectively for reducing the effects of fading in microcells.

10.2.1 Choosing the Proper Window Length

One method for determining the proper window length is to use analog averaging. The following development extends the original work of Lee [182] by incorporating Aulin's Rician fading model. With Lee's multiplicative model, the received signal strength at position y is, from (2.182),

$$|\hat{r}_c(y)|^2 = |r(y)|^2 \cdot \Omega_p(y) \qquad (10.14)$$

where $|\hat{r}_c(y)|^2$ is the composite squared envelope, $|r(y)|^2$ is a non-central chi-square random variable with 2 degrees of freedom (Rician fading), and $\Omega_p(y)$ is a log-normal random variable (log-normal shadowing). If the local mean is constant with distance, then $\Omega_p(y) = \Omega_p$. Assuming ergodicity, an integral spatial average of $|\hat{r}_c(y)|^2$ can be used to estimate the local mean Ω_p, i.e.,

$$\bar{\Omega}_p = \frac{1}{2L} \int_{x-L}^{x+L} |\hat{r}_c(y)|^2 dy = \frac{\Omega_p}{2L} \int_{x-L}^{x+L} |r(y)|^2 dy \qquad (10.15)$$

where the second equality holds since $\Omega_p(y)$ is constant over the spatial interval $(x-L, x+L)$. The accuracy of the estimate can be determined from the variance of (10.15), calculated as [175]

$$\sigma_{\bar{\Omega}_p}^2 = \frac{1}{L} \int_0^{2L} \left(1 - \frac{\ell}{2L}\right) \mu_{|r|^2|r|^2}(\ell) d\ell \qquad (10.16)$$

where $\mu_{|r|^2|r|^2}(\ell) = E[|r(y)|^2|r(y+\ell)|^2] - E[|r(y)|^2]E[|r(y+\ell)|^2]$ is the *spatial* autocovariance of the squared envelope, and $E[\,x\,]$ denotes the ensemble average of x. Aulin [12] derived $\mu_{|r|^2|r|^2}(\ell)$ as (see 2.70),

$$\mu_{|r|^2|r|^2}(\ell) = \left(\frac{\Omega_p}{K+1}\right)^2 \left[J_0^2(2\pi\ell/\lambda_c) + 2K J_0(2\pi\ell/\lambda_c)\cos\left(2\pi\ell\cos(\theta_0)/\lambda_c\right)\right] \qquad (10.17)$$

where $J_0(\,\cdot\,)$ is the zero-order Bessel function of the first kind, K is the Rice factor, λ_c is the carrier wavelength, and θ_0 is the angle that the specular component makes with the MS direction of motion. The spatial autocovariance of the squared envelope can be obtained directly from Fig. 2.13 by using the time-distance transformation $f_m\tau = \ell/\lambda_c$.

Substituting (10.17) into (10.16) yields

$$\begin{aligned} \sigma_{\bar{\Omega}_p}^2 &= \left(\frac{\Omega_p}{K+1}\right)^2 \frac{1}{L} \int_0^{2L} \left(1 - \frac{\ell}{2L}\right) \\ &\quad \times \left[J_0^2(2\pi\ell/\lambda_c) + 2K J_0(2\pi\ell/\lambda_c)\cos\left(2\pi\ell\cos(\theta_0)/\lambda\right)\right] d\ell \ . \end{aligned} \qquad (10.18)$$

As desired, $\sigma_{\bar{\Omega}_p}^2 \to 0$ as $L \to \infty$. If L is large, then $\bar{\Omega}_p$ can be considered Gaussian since it is the summation of many independent random variables. However if $\sigma_{\bar{\Omega}_p}^2$ is relatively large compared to Ω_p (due to small L or small Ω_p), then it is more appropriate to treat $\bar{\Omega}_p$ as a non-central chi-square random variable. In this case, it may be more appropriate to approximate $\bar{\Omega}_p$ as a log-normal random variable which has the same general shape as a non-central chi square distribution (i.e., zero at the origin with an infinitely long tail) [123], [124].

Proceeding under the assumption that $\bar{\Omega}_p$ is approximately Gaussian, the 1σ spread can be calculated to measure the accuracy of the estimator, where

$$1\sigma \text{ spread } = 10 \cdot \log_{10} \frac{\Omega_p + \sigma_{\bar{\Omega}_p}}{\Omega_p - \sigma_{\bar{\Omega}_p}} \text{ dB} \qquad (10.19)$$

with the interpretation that $\text{Prob}(|\bar{\Omega}_{p\ (\text{dB})} - \Omega_{p\ (\text{dB})}| \leq 1\ \sigma \text{ spread}) = 0.68.$[1] Observe from (10.18) and (10.19) that the accuracy of the local mean estimate depends on K, L, and θ. Fig. 10.3 shows the 1σ spread when $\theta = 60°$ for various values of K. In general, $\bar{\Omega}_p$ approaches Ω_p with increasing K. However, the angle θ also affects the accuracy as shown in Fig. 10.4. When $\theta = 90°$ the $1\ \sigma$ spread is minimized, resulting in the best estimate of the local mean. Conversely, the worst estimates occur for small θ (in the neighborhood of $10°$ in Fig. 10.4). The actual angle that the maximum occurs is a function of L, and it can easily be shown that the $1\ \sigma$ spread has a local minimum at $\theta = 0°$ and global minimum at $\theta = 90°$ for all L. In any case, the required spatial averaging distance for local mean estimation in microcells depends on K and θ.

10.2.2 Choosing the Proper Number of Samples to Average

Most practical signal strength estimators use samples of the signal strength rather than analog averaging. We must determine the required number and spacing of samples that should be used, to sufficiently mitigate the effects of fading. Consider the sampled signal

$$|r[i]|^2 \stackrel{\triangle}{=} |r(iS)|^2 \qquad (10.20)$$

where S is the spatial sampling period, and i is an integer. Then the unbiased estimate

$$\bar{\Omega}_p = \frac{1}{N} \sum_{i=0}^{N-1} |r[i]|^2 \qquad (10.21)$$

---

[1]The probability of lying within one standard deviation of the mean of a Gaussian random variable is 0.68.

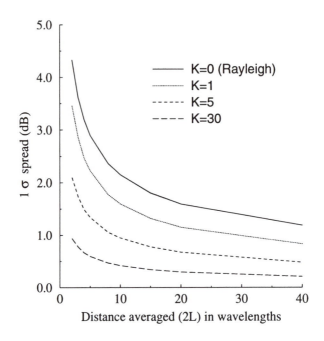

Figure 10.3 Dependency of the 1σ spread for squared-envelope samples on the averaging distance $(2L)$ and Rice factor K when $\theta = 60°$, from [15].

can be used to determine an estimate of the local mean squared-envelope Ω_p. As with analog averaging, the variance of this estimate can be used to measure its accuracy, where

$$\sigma_{\bar{\Omega}_p}^2 = \frac{1}{N^2} \sum_{i=0}^{N-1} \sum_{j=0}^{N-1} \mathrm{E}\left[|r[i]|^2 |r[j]|^2\right] - \left(\mathrm{E}\left[\bar{\Omega}_p\right]\right)^2 \ . \qquad (10.22)$$

By using (10.17) along with the symmetric properties of the autocovariance, (10.22) becomes

$$
\begin{aligned}
\sigma_{\bar{\Omega}_p}^2 &= \frac{\mu_{|r|^2|r|^2}(0)}{N} + 2 \sum_{j=1}^{N-1} \left(\frac{N-j}{N^2}\right) \mu_{|r|^2|r|^2}(Sj) \\
&= \frac{\Omega_p^2}{(K+1)^2} \left[\frac{1+2K}{N} + 2 \sum_{j=1}^{N-1} \left(\frac{N-j}{N^2}\right) J_0^2(2\pi Sj/\lambda_c) \right. \\
&\quad \left. + 2K J_0(2\pi Sj/\lambda_c) \cos\left(2\pi Sj \cos(\theta)/\lambda_c\right) \right] \qquad (10.23)
\end{aligned}
$$

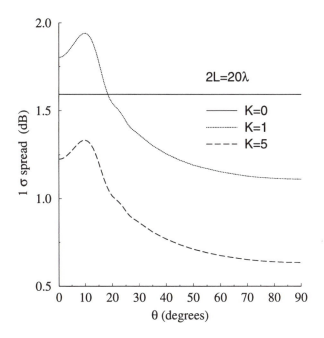

Figure 10.4 Dependency of the 1σ spread on the specular angle θ mod $90°$, from [15].

where S is measured in wavelengths (λ_c). Note that $\sigma_{\bar{\Omega}_p}$ depends on N, K, S, and θ. Fortunately, the effect of each parameter is nearly independent of the others. Fig. 10.5 illustrates the relationship between S and K for $\theta = 0°$, where $N = \lceil 20\lambda_c/S \rceil$ so that the averages are over $20\lambda_c$ (and $\lceil x \rceil$ denotes the smallest integer greater than or equal to x). Increasing N for a fixed S will increase the spatial averaging distance, thereby lowering the 1σ spread in a manner similar to analog averaging in Fig. 10.3. The discontinuities in Fig. 10.5 are due to the $\lceil x \rceil$ function. Observe that if $S < 0.5\lambda_c$ then the discrete local mean estimate is approximately equivalent to the estimate from analog averaging ($\theta = 0°$ in Fig. 10.4) over the same spatial distance. Similar to Fig. 10.4, we also observe that small Rice factors, e.g., $K = 0.1$ and $K = 1$, at $\theta = 0°$ increase the 1σ spread. The spikes at $0.5\lambda_c$ and $1\lambda_c$, correspond to the location of the first lobe of the autocovariance function given by (10.17) and plotted in Fig 2.13.

Although we often assume $\theta = 0°$ in our treatment, Fig. 10.6 shows the relationship between the 1σ spread and S, for $K = 1$, $N = \lceil 20\lambda_c/S \rceil$, and several values of θ. Increasing θ generally lowers the 1σ spread except for some small

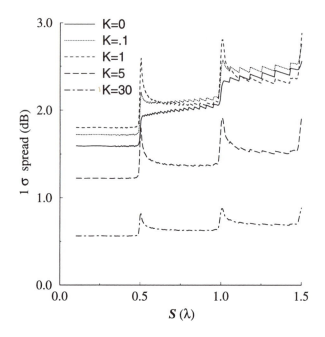

Figure 10.5 1 σ spread versus S for various K, with $\theta = 0°$, distance averaged $= 20\lambda_c$, from [15].

angles as shown in Fig. 10.4; it also shifts the spike at $0.5\lambda_c$ to the right, because the first sidelobe of (10.17) shifts as θ increases.

To summarize, the spatial averaging distance that is needed to sufficiently reduce the effects of fading depends on K and θ. If sample averaging is used, then the sample spacing should be less than $0.5\lambda_c$. As a rule of thumb, a spatial averaging distance of 20 to 40 λ_c should be sufficient for most applications.

10.3 VELOCITY ESTIMATION IN CELLULAR SYSTEMS

Temporal based handoff algorithms can yield poor handoff performance in microcells due to the diverse propagation environment and the wide range of MS velocities. Consider the NLOS handoff scenario shown in Fig. 10.1, where a MS traveling from BS_0 has a Rician faded log-normal shadowed LOS signal

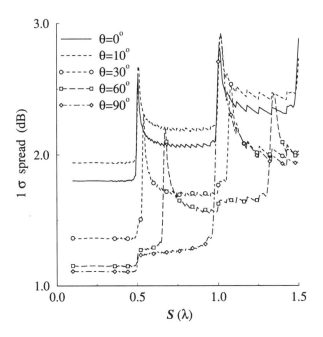

Figure 10.6 1 σ spread versus S (λ_c) for various θ, $K = 1$, distance averaged $= 20\lambda_c$, from [15].

from BS_0, and a Rayleigh faded log-normal shadowed NLOS signal from BS_1, until it rounds the corner where the situation is suddenly reversed. The loss (gain) of the LOS component causes a rapid decrease (increase) in the signal strength. Effective handoff algorithms for this scenario should use short temporal averaging window and a large hysteresis, so that rapid changes in the mean signal strength are detected and unnecessary handoffs are prevented [225]. Unfortunately, temporal averaging with a short fixed window length gives optimal handoff performance for only a single velocity. For example, consider again the handoff scenario in Fig. 10.1 along with the received signal strength profile in Fig. 10.2. Assume log-normal shadowing with $\sigma_\Omega=6$ dB and choose D so that $\phi_{\Omega_{(dB)}\Omega_{(dB)}}(d) = 0.1\sigma_\Omega^2$ at $d = 30$ m in (10.13). As a preview, the simulation of a 2.27 s non-overlapping temporal power averaging handoff algorithm with a hysteresis $H = 8$ dB has a handoff performance shown by the lines in Fig. 10.12.[2] The handoff performance is evaluated by the mean number of handoffs, averaged over 1000 runs, versus the distance from BS_0 where 50%

---

[2] A 2.27 s window corresponds to a $20\lambda_c$ spatial window at a velocity of 5 km/h, assuming a carrier frequency of 1.9 GHz. Section 10.4 further details the simulation.

(and 90%) of the MSs have made a handoff to BS_1, i.e., $Pr(BS_1) = 0.5$ and $Pr(BS_1) = 0.9$ at the abscissa. This distance gives a measure of the handoff delay, assuming that handoffs will occur between BS_0 and BS_1 only.

Fig 10.12 only shows the handoff request delay, while in a real system the network delay should also be included. However, the performance of a velocity adaptive handoff algorithm can still be evaluated without knowledge of the network delay. For example, suppose that the receiver threshold is -90 dBm. Also, assume that a good handoff algorithm should have at least 90 % of the MSs handed off before a distance d_{cutoff}, where d_{cutoff} is chosen as that distance where the mean signal strength is $2\sigma_\Omega$ above -90 dBm. If $\sigma_\Omega = 6$ dB and the data from Fig. 10.2 is used, then a signal strength of $-90 + 12 = -78$ dBm occurs at 283 m for BS_0. Hence, if the velocity adaptive handoff algorithm can adapt to the point at 5 km/h in Fig. 10.12, corresponding to handoff requests at a distance 262 m, and the maximum speed of a MS turning the corner is 40 km/h ($40/3.6$ m/s), then a maximum network delay of

$$\text{Max Network Delay} = \frac{3.6 \text{ s}}{40 \text{ m}}(283 - 262) \text{ m} = 1.89 \text{ s}$$

can be tolerated. For some of the newer cellular standards, e.g., GSM, this network delay is tight but acceptable, implying the usefulness of velocity adaptive handoff algorithms discussed here. In the above example the 5 km/h point on the curve in Fig. 10.12 was chosen as the desired operating point, because the best handoff performance occurs near the knee of the curve where the mean number of handoffs and handoff delay are jointly minimized. Other hysteresis and window lengths could possibly result in better performance. However, the settings used here ($H = 8$ dB and a $20\lambda_c$ spatial window) are adequate to illustrate the usefulness of velocity adaptive handoff algorithms.

Some cellular system proposals have suggested the deployment of microcells along with "umbrella" macrocells for accommodating high speed MSs. Velocity estimation will be necessary for these systems along with a macro-to-microcell and micro-to-macrocell handoff scheme. Alternatively, if a low network handoff delay can be achieved, then a velocity adaptive handoff algorithm can maintain good link quality without the need for umbrella macrocells.

10.3.1   Level Crossing Rate Estimators

It is well known that the level (or zero) crossing rates of the inphase $r_I(t)$ and quadrature $r_Q(t)$ low-pass components, or the envelope $|r(t)| = \sqrt{r_I^2(t) + r_Q^2(t)}$

of a received sinusoid in noise, are functions of the MS velocity as discussed in Section 2.1.4. The envelope **level crossing rate** (LCR) is defined as the average number of positive-going crossings per second, a signal makes of a predetermined level R. Likewise, the **zero crossing rate** (ZCR) is defined as the average number of positive going zero crossings a signal makes per second.

Assuming fading model in (10.10), the means m_I and m_Q can be subtracted from the inphase and quadrature components and the ZCR of the resulting signals be used to estimate the velocity. Rice gives the ZCR of $r_I(t) - m_I$ or $X_Q(t) - m_Q$ as [269]

$$L_{ZCR_1} = \frac{1}{\pi}\sqrt{\frac{b_2}{b_0}} \tag{10.24}$$

and the envelope LCR with respect to the level R as [269, 4.7]

$$L_R = \int_0^\infty \dot{r} p(R, \dot{r}) d\dot{r} = \frac{R(2\pi)^{-3/2}}{\sqrt{\mathcal{B}b_0}} \int_{-\pi}^{\pi} d\psi \int_0^\infty \dot{r} d\dot{r} \tag{10.25}$$

$$\exp\left\{-\frac{1}{2\mathcal{B}b_0}\left[\mathcal{B}\left(R^2 - 2Rs\cos\psi + s^2\right) + (b_0\dot{r} + b_1 s \sin\psi)^2\right]\right\}$$

where $p(R, \dot{r})$ is the joint probability density function of the envelope r (evaluated at $r = R$) and the slope of the envelope $\dot{r}$, and $\mathcal{B} = b_0 b_2 - b_1^2$. From (2.80), the b_n are equal to

$$b_n = (2\pi f_m)^n b_0 \int_0^{2\pi} \hat{p}(\alpha)\cos^n(\alpha) d\alpha + (2\pi)^n \int_{-B/2}^{B/2} \frac{N_o}{2} f^n df \tag{10.26}$$

where $b_0 = \sigma^2$, v is the velocity, λ_c is the carrier wavelength, $f_m = v/\lambda_c$ is the maximum Doppler frequency, and $\hat{p}(\alpha)$ is the spatial distribution of the *scatter* component of the arriving plane waves [154]. The second term in (10.26) is due to additive bandpass Gaussian noise, centered at f_c, with a two-sided power spectral density of $N_o/2$ watts/Hz and a noise bandwidth of B Hz, resulting in a total power of $N_o B$ watts. For two-dimensional isotropic scattering $\hat{p}(\alpha) = 1/(2\pi)$, $-\pi \leq \alpha \leq \pi$ and (10.26) can be written as

$$b_n = (2\pi)^n \frac{\sigma^2}{\pi} \int_{-f_m}^{f_m} \frac{f^n}{\sqrt{f_m^2 - f^2}} df + (2\pi)^n \int_{-B/2}^{B/2} \frac{N_0}{2} f^n df \ . \tag{10.27}$$

With Aulin's Rician fading in (10.12), the ZCR of $r_I(t)$ or $r_Q(t)$ is [269]

$$L_{ZCR_2} = L_{ZCR_1}\left[e^{-\gamma} I_0(\beta) + \frac{b^2}{2\gamma} I_e\left(\frac{\beta}{\gamma}, \gamma\right)\right] , \tag{10.28}$$

where $I_0(x)$ is the zero order modified Bessel function of the first kind, and

$$\gamma = \frac{a^2 + b^2}{4}, \quad \beta = \frac{a^2 - b^2}{4}, \quad a = \sqrt{2K}, \quad b = 2\pi f_m \cos\theta \sqrt{\frac{2Kb_0}{b_2}}$$

$$I_e(k, x) = \int_0^x e^{-u} I_0(ku) du .$$

(10.29)

For macrocells two-dimensional isotropic scattering is a reasonable assumption. However, in microcells the scattering is often non-isotropic. Nevertheless, one approach is to derive the velocity estimators under the assumption of isotropic scattering with no additive noise, and afterwards study the effects of the mismatch caused by nonisotropic scattering and noise. Using the definition for b_n in (10.27) and Rician fading model in (10.10) gives

$$L_{ZCR_1} = \sqrt{2}v/\lambda_c$$

(10.30)

and (2.84)

$$L_R = (v/\lambda_c)\sqrt{2\pi(K+1)}\rho e^{-K-(K+1)\rho^2} I_0\left(2\rho\sqrt{K(K+1)}\right)$$

(10.31)

where $\rho = R/R_{\text{rms}}$, where $R_{\text{rms}} = \sqrt{\Omega_p}$ is the rms signal level. Likewise, for $\theta = 0°$ and Rician fading model in (10.12) we have

$$\gamma = \frac{3K}{2}, \quad \beta = -\frac{K}{2}, \quad a = \sqrt{2K}, \quad b = 2\sqrt{K}$$

(10.32)

and L_{ZCR_2} reduces to

$$L_{ZCR_2} = (v/\lambda_c)\sqrt{2}\left[e^{-3K/2}I_0\left(-\frac{K}{2}\right) + \frac{4}{3}I_e\left(-\frac{1}{3}, \frac{3K}{2}\right)\right] .$$

(10.33)

Clearly, each of the above level crossing rate estimators is proportional to the velocity v and, hence, can be used as a velocity estimator. However, it remains to be seen if they are robust to K, non-isotropic scattering, additive noise, and other factors. We first consider the robustness with respect to K and treat the other factors afterwards.

L_{ZCR_1} is not affected by K. Fig. 2.14 in compared the level crossing rate L_R for different K with the conclusion that the LCR around $\rho = 0$ dB is roughly independent of K. This attractive property suggests that the level crossing rate can be used to provide a velocity estimate that is robust to K. Consequently, the steps for using the LCR (or ZCR) of the envelope (or $r_I(t)$ or $r_Q(t)$), for

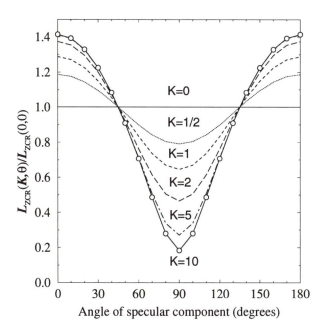

Figure 10.7 Normalized zero crossing rate versus K and θ, from [15].

velocity estimation are; determine R_{rms} (or m_I or m_Q), estimate the number of crossings per second $\hat{L}_{R_{\mathrm{rms}}}$ (or $\hat{L}_{\mathrm{ZCR}_1}$), and use (10.31) to solve for v, with $\rho = 1$ and $K = 0$ (or 10.30 for ZCR_1). Thus, the following velocity estimators are robust with respect to K assuming the Rician fading model in (10.10):

$$\hat{v}_{\mathrm{ZCR}_1} \approx \frac{\lambda_c \hat{L}_{\mathrm{ZCR}}}{\sqrt{2}}, \quad \hat{v}_{\mathrm{LCR}} \approx \frac{\lambda_c \hat{L}_{R_{\mathrm{rms}}}}{\sqrt{2\pi e^{-1}}} . \tag{10.34}$$

Fig. 10.7 shows the effect of K and θ on L_{ZCR_2}. Notice that if the angle of the specular component is $\theta = 0°$ or $180°$, then L_{ZCR_2} can have up to 40% relative error. Consequently, a non-zero value of K should be chosen as default to minimize the effect of K. Choosing $K \approx 0.61$ yields a maximum error of at most 20% which is quite acceptable for urban LOS velocity adaptive handoff applications. In this case, the velocity estimate from (10.33) becomes

$$\hat{v}_{\mathrm{ZCR}_2} \approx \frac{\lambda_c \hat{L}_{\mathrm{ZCR}_2}}{1.2\sqrt{2}} . \tag{10.35}$$

10.3.2   Covariance Approximation Methods

Recently, a velocity estimator has been proposed by Holtzman and Sampath that relies upon an estimate of the autocovariance between faded samples $r[i]$, where the $r[i]$ can be envelope, squared-envelope, or log-envelope samples [147], [273]. With this method, referred to here as the **covariance method** (COV), the statistic

$$V = \frac{1}{N} \sum_{k=1}^{N} (r[k+\tau] - r[k])^2 \tag{10.36}$$

is calculated. If N is large and ergodicity applies, then V can be replaced by the ensemble average

$$E[V] = 2\mu_{rr}(0) - 2\mu_{rr}(\tau) \tag{10.37}$$

where $\mu_{rr}(\tau)$ denotes the autocovariance of $r[k]$. The general form for $\mu_{rr}(\tau)$, assuming squared-envelope samples, can be derived from [12] and [273] as

$$\mu_{rr}(\tau) = 4a(\tau)\left[a(\tau) + s^2 \cdot \cos(\omega\tau\cos(\theta))\right] + 4c^2(\tau) + \frac{2N_o a(\tau)\sin(B\pi\tau)}{\pi\tau}$$
$$+ \frac{4KN_o\sigma^2\cos(\omega\tau\cos(\theta))\sin(B\pi\tau)}{\pi\tau} + \frac{N_o{}^2\sin(B\pi\tau)^2}{\pi^2\tau^2} \tag{10.38}$$

where [12]

$$a(\tau) = b_0 \int_0^{2\pi} \hat{p}(\alpha)\cos\left(2\pi f_m\tau\cos(\alpha)\right) d\alpha \tag{10.39}$$

$$c(\tau) = b_0 \int_0^{2\pi} \hat{p}(\alpha)\sin\left(2\pi f_m\tau\cos(\alpha)\right) d\alpha \ . \tag{10.40}$$

This estimator depends on $\hat{p}(\alpha)$ and, hence, is also a function of the scattering environment. Like the LCR estimator, we first assume isotropic scattering without additive noise to derive a velocity estimator and afterwards evaluate the effect of non-isotropic scattering and noise.[3] If the channel is characterized by isotropic scattering and squared-envelope samples are used, then using (10.17) gives

$$E[V] = \overline{V} = 2 \cdot \left(\frac{\Omega_p}{K+1}\right)^2 [(1+2K) \tag{10.41}$$
$$- (J_0^2(2\pi\tau/\lambda_c) + 2KJ_0(2\pi\tau/\lambda_c)\cos(2\pi\cos(\theta)\tau/\lambda_c))]$$

---

[3] Only isotropic scattering was considered in [147], [273]

which is dependent on K and θ. If $\mu_{rr}(0)$ is known exactly, then the bias with respect to K can be eliminated for small τ by the normalization [273]

$$\frac{\overline{V}}{\mu_{rr}(0)} \approx (2\pi v \tau_t/\lambda_c)^2 \frac{1 + 2K + K\cos(2\theta)}{(1 + 2K)} \tag{10.42}$$

so that [273]

$$\hat{v}_{\text{COV}} \approx \frac{\lambda_c}{2\pi\tau_t} \sqrt{\frac{\overline{V}}{\mu_{rr}(0)}} \tag{10.43}$$

where τ_t is the sample spacing in seconds/sample.

In large co-channel interference situations it may be preferable to modify the above scheme since the empirical average in (10.36), and in particular $\mu_{rr}(0)$, is sensitive to co-channel interference as shown in [166]. Consequently, defining,

$$U(\tau) = \frac{1}{N} \sum_{k=1}^{N} r[k + \tau]r[k] - \left(\frac{1}{N} \sum_{k=1}^{N} r[k] \right)^2 \tag{10.44}$$

and $V_2 = 2U(\tau_1) - 2U(\tau_2)$, yields

$$E[V_2] = 2\mu_{rr}(\tau_1) - 2\mu_{rr}(\tau_2) \tag{10.45}$$

so that $E[V_2]/\mu_{rr}(0)$ is equal to (10.42) with $\tau_t = \tau_2^2 - \tau_1^2$, and a result similar to (10.43) follows.

Whether V or V_2 is used, $\mu_{rr}(0)$ is never known exactly and must be estimated by the MS in the same way that m_I, m_Q, and R_{rms} must be estimated in the ZCR and LCR methods, respectively. Consequently, to actually use (10.43) it must be shown or verified that

$$v \propto E\left[\sqrt{\frac{\overline{V}}{\mu_{rr}(0)}} \right]. \tag{10.46}$$

This is analytically difficult, but simulation results in Section 10.4 suggest that (10.43) is a useful approximation to (10.46).

It is also shown in Appendix 10A that

$$\lim_{\tau \to 0} \hat{v}_{\text{COV}} = \lim_{\tau \to 0} \frac{\lambda_c}{2\pi\tau} \sqrt{\frac{\overline{V}}{\mu_{rr}(0)}} = v \sqrt{\frac{1 + 2K + K\cos(2\theta)}{1 + 2K}}. \tag{10.47}$$

It follows from (10.43) and (10.47) that K and θ cause at most 20% error in v [273], thus providing a velocity estimator that is reasonably robust with respect to K.

10.3.3   Velocity Estimator Sensitivity

To illustrate the sensitivity of the velocity estimators, the ratio of the corrupted velocity estimate to the ideal velocity estimate is used. For the LCR and ZCR velocity estimators with the fading model in (10.10) we have

$$\frac{\tilde{v}}{v} = \frac{\tilde{L}_{R_{rms}}(\tilde{b}_0, \tilde{b}_1, \tilde{b}_2)}{L_{R_{rms}}(b_0, b_1, b_2)} \tag{10.48}$$

and

$$\frac{\tilde{v}}{v} = \frac{\tilde{L}_{ZCR_1}}{L_{ZCR_1}} = \sqrt{\frac{\tilde{b}_2}{\tilde{b}_0} \cdot \frac{b_0}{b_2}} \tag{10.49}$$

where $\tilde{v}$ denotes the corrupted velocity estimate, and $\tilde{L}_{R_{rms}}(\tilde{b}_0, \tilde{b}_1, \tilde{b}_2)$ and $L_{R_{rms}}(b_0, b_1, b_2)$ are given by (10.25) with the appropriate values of $\tilde{b}_n$ and b_n, respectively. Little simplification results for the LCR method in general. However, when $K = 0$ (10.49) simplifies to [154]

$$\frac{\tilde{v}}{v} = \sqrt{\frac{\frac{\tilde{b}_2}{\tilde{b}_0} - \frac{\tilde{b}_1^2}{\tilde{b}_0^2}}{\frac{b_2}{b_0} - \frac{b_1^2}{b_0^2}}} \ . \tag{10.50}$$

For Aulin's fading model in (10.12) the sensitivity of the ZCR is

$$\frac{\tilde{v}}{v} = \frac{\tilde{L}_{ZCR_2}}{L_{ZCR_2}} = \frac{\tilde{L}_{ZCR_1}}{L_{ZCR_1}} \frac{\left[e^{-\tilde{\gamma}} I_0(\tilde{\beta}) + \frac{\tilde{b}^2}{2\tilde{\gamma}} I_e\left(\frac{\tilde{\beta}}{\tilde{\gamma}}, \tilde{\gamma}\right) \right]}{\left[e^{-\gamma} I_0(\beta) + \frac{b^2}{2\gamma} I_e\left(\frac{\beta}{\gamma}, \gamma\right) \right]} \tag{10.51}$$

where $\tilde{\beta}$, $\tilde{\gamma}$, and $\tilde{b}$ are given by (10.29) using (10.26) where appropriate. Likewise, for the covariance method using squared envelope samples we have

$$\frac{\tilde{v}}{v} = \frac{\frac{\lambda_c}{2\pi\tau} \sqrt{\frac{2\tilde{\mu}_{rr}(0) - 2\tilde{\mu}_{rr}(\tau)}{\tilde{\mu}_{rr}(0)}}}{\frac{\lambda_c}{2\pi\tau} \sqrt{\frac{2\mu_{rr}(0) - 2\mu_{rr}(\tau)}{\mu_{rr}(0)}}} \ . \tag{10.52}$$

Effect of the Scattering Distribution

Here we study the sensitivity of the velocity estimators to the scattering distribution by using four different non-isotropic scattering models. With the first model S1, plane waves arrive from one direction only with a varying degree of directivity as might happen when signals are channeled along a city street. The probability density of the scatter component of the arriving plane waves as a function of angle of arrival has the form in (2.36), where the vehicle motion is in the direction of $\alpha = 0°$, and α_m determines the directivity of the incoming plane waves. Fig. 2.7 shows a polar plot of $\hat{p}(\alpha)$ for $\alpha_m = 30°, 60°$ and $90°$. The second model S2, assumes that the plane waves can arrive from either the front ($\alpha = 0°$) or back ($\alpha = 180°$), which may be typical for city streets that dead end at another street. In this case $\frac{\hat{p}(\alpha)}{2}$ and $\frac{\hat{p}(\alpha-\pi)}{2}$ are combined to form the distribution versus angle of arrival. The resulting density is similar to Fig. 2.7 but with lobes extending in both the $0°$ and $180°$ directions. The third and fourth models S3 and S4, respectively, are similar to S1 and S2 except that the distributions are rotated by $90°$, so that the plane waves tend to arrive perpendicular to the direction of travel. This may occur when a MS passes through a street intersection. The effect of the scattering distribution is determined for the cases when the velocity estimator has been designed for i) isotropic scattering and, ii) scattering model S1 with $\alpha_m = 90°$. The scattering model that the velocity estimator has been designed for will determine the values of b_0, b_1 and b_2 in the denominators of (10.48)–(10.51), while the values of $\tilde{b}_0$, $\tilde{b}_1$ and $\tilde{b}_2$ depend on the scattering environment that is actually present. The effect of non-isotropic scattering on the COV estimate (10.52) can be found from the results in Appendix 10A with $N_o = 0$, or by using small values of τ in (10.52). Here we chose the latter with $\tau = 1/50$.

Fig. 10.8 shows the effect of the scattering distribution on each of the velocity estimators. Due to the very large number of possible scenarios, only the most significant results are plotted in Fig. 10.8 and curves similar (but not equal) to the plotted curves are simply asterisked in the accompanying table. Velocity estimators with the subscript "d" in Fig. 10.8 correspond to those that are designed for scattering model S1 with $\alpha_m = 90°$. By using Fig. 10.8 the relative robustness of the various velocity estimators to the scattering distribution has been summarized by the ranking in Table 10.1.

Curve	Rank	$\tilde{v}/v$ at $\alpha_m = 10°$	$\tilde{v}/v$ at $\alpha_m = 90°$
d	Excellent	1.06	1.0
e , c	Very Good	.66 , 1.52	.85 , 1.24
b , f	Good	.34 , .107	2.6 , .82
a , g , h	Poor	4.5 , .014 , .004	3.6 , 1.0 , .32

Table 10.1   Robustness to the scattering distribution.

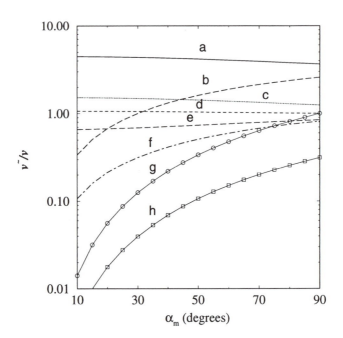

Figure 10.8   Non-isotropic scattering effects. — corresponds to a curve that had insufficient precision to be reported. Superscript * denotes that the curve is approximately equal the letter curve, from [15].

In urban situations, robustness with respect to scattering models S1 and S2 is important. The LCR and COV methods are very sensitive to the directivity in scattering model S1 when $K = 0$ as shown by curve "h". This sensitivity can be partially mitigated by using the velocity estimators LCR_d and COV_d that have been designed for scattering model S1 with $\alpha_m = 90°$ as shown by curve "g." However, the price for increased robustness to scattering model S1 is the increased sensitivity of LCR_d to scattering models S2, S3, and S4 when $K = 0$. Fortunately, the presence of even a small specular component $(K = 1)$ reduces the sensitivity as seen in COV $(K = 1)$ and LCR_d. In contrast a specular component does not reduce the sensitivity of the LCR estimator in scattering models S2, S3, and S4, because $b_1 = 0$, and therefore the ratio of the crossing rates in (10.50) depends on $\tilde{b}_2$ and b_2 and is independent of K. Results are not shown for LCR or LCR_d with $K = 1$ scattering model S1, due to numerical difficulties in calculating (10.25) for small α_m. For large $\alpha_m > 80°$ the results obtained were very close to curve "d."

The ZCR velocity estimator is generally more robust than the LCR and COV methods. The presence of a small specular component improves robustness to the scattering distribution as seen in ZCR_2, $\mathrm{ZCR}_{2,d}$ (ZCR_1 and ZCR_{1_d} are independent of K). Also, velocity estimators that have been designed for scattering model S1 with $\alpha_m = 90°$ perform slightly better than those designed for isotropic scattering. However, the improvement obtained by using these velocity estimators must be weighed against the relative error that will be introduced if the scattering is actually isotropic. For LCR_d and COV_d, $\tilde{v}/v = .316$ and for ZCR_d $\tilde{v}/v = 1.15$. Since all the velocity estimators seem to have some sensitivity to the scattering distribution, and sensitivity is greatly reduced when $K > 0$, we conclude that those designed for isotropic scattering should be adequate.

In summary, for very directive situations where the plane waves arrive from either the front or back but not both, the ZCR, COV_d, or LCR_d methods are the most robust. If the plane waves arrive from both the front and back, then all the velocity estimators with the exception of LCR_d are reasonably robust. The sensitivity to directivity is reduced when a specular component is present. In the unlikely event that $K = 0$ and plane waves arrive from the perpendicular direction with high directivity, all methods will have a significant bias. Finally, another method for overcoming the sensitivity to the scattering distribution is to obtain velocity estimates from signals arriving from a distant cell or an umbrella cell, since they will experience isotropic scattering.

Effects of Additive Gaussian Noise

Since the effect of the scattering distribution has already been established, the sensitivity to additive white Gaussian noise (AWGN) is determined by using (10.48) – (10.52) with isotropic scattering. With AWGN the rms value of the received signal is $\tilde{R} = \sqrt{s^2 + 2\sigma^2 + N_o B}$, and the values of $\tilde{b}_n$ and b_n in (10.48) are

$$\tilde{b}_0 = \sigma^2 + \frac{N_o B}{2} \qquad\qquad b_0 = \sigma^2$$
$$\tilde{b}_2 = 2\left(\pi\sigma f_m\right)^2 + \frac{N_o B^3 \pi^2}{6} \qquad b_2 = 2\left(\pi\sigma f_m\right)^2 \qquad (10.53)$$

For the LCR velocity estimator (10.48) with (10.25) become, after considerable algebra,

$$\frac{\tilde{v}}{v} = \left(1 + \frac{K+1}{6\gamma_S}\left(\frac{B}{f_m}\right)^2\right)^{\frac{1}{2}} \frac{\sqrt{\gamma_S(\gamma_S + 1)}}{\gamma_S + K + 1}$$
$$\times \frac{I_0\left(\frac{2\sqrt{\gamma_S(\gamma_S+1)K(K+1)}}{\gamma_S+K+1}\right)}{I_0\left(2\sqrt{K(K+1)}\right)}$$
$$\times \exp\left\{2K + 1 - \frac{\gamma_S(2K+1) + K + 1}{\gamma_S + K + 1}\right\}, \qquad (10.54)$$

where

$$\gamma_S \triangleq \frac{s^2 + 2\sigma^2}{N_o B} = \frac{2\sigma^2(K+1)}{N_o B} \qquad (10.55)$$

is defined as the signal-to-noise ratio. Likewise, for the ZCR velocity estimator ZCR$_1$ (10.49) becomes

$$\frac{\tilde{v}}{v} = \sqrt{\frac{\gamma_S + \left(\frac{B}{f_m}\right)^2 \frac{K+1}{6}}{\gamma_S + K + 1}} . \qquad (10.56)$$

For Aulin's fading model in (10.12), the effect of AWGN on ZCR$_2$ can be obtained from (10.51) with $\tilde{L}_{\text{ZCR}_1}/L_{\text{ZCR}_1}$ in (10.56),

$$\tilde{\beta} = \frac{K}{2}\left(1 - 2\cos^2(\theta)\left[\frac{\gamma_S + K + 1}{\gamma_S + \frac{K+1}{6}\left(\frac{B}{f_m}\right)^2}\right]\right) \qquad (10.57)$$

$$\tilde{\gamma} = \frac{K}{2}\left(1 + 2\cos^2(\theta)\left[\frac{\gamma_S + K + 1}{\gamma_S + \frac{K+1}{6}\left(\frac{B}{f_m}\right)^2}\right]\right) \qquad (10.58)$$

$$\tilde{b} = 2\sqrt{K}\cos(\theta)\sqrt{\frac{\gamma_S + K + 1}{\gamma_S + \frac{K+1}{6}\left(\frac{B}{f_m}\right)^2}} \qquad (10.59)$$

and a, b, γ, and β given by (10.32).

In [273], the effect of AWGN on the COV velocity estimator has been derived as a function of $\tau > 0$. Here we provide a closed form analytic result for the effect of AWGN on the COV velocity estimate for the limiting case when $\tau \to 0$. The limiting case is important for comparisons to AWGN effects on level crossing rate estimators, and since (10.42) is only valid for small τ. Consequently, the $\lim_{\tau \to 0} \tilde{v}/v$ in (10.52) is found, and afterwards, the effect of $\tau > 0$ in (10.52) is compared. It is shown in Appendix 10A that

$$\lim_{\tau \to 0} \tilde{v}/v = \frac{\sqrt{\zeta}}{\sqrt{\frac{(1+2K+K\cos(2\theta))}{(1+2K)}}} \tag{10.60}$$

where ζ is given by (10A.4), with $a(0) = \sigma^2, a'(0) = c(0) = c'(0) = c''(0) = 0$ and $a''(0) = \sigma^2 \omega/2$ for isotropic scattering.

It is apparent from (10.54), and (10.56)–(10.60) that the effect of AWGN depends on K, B, γ_S, v and θ. For a practical system, the bandwidth B can be chosen as the maximum expected Doppler frequency over the range of velocities. However, a smaller B in reference to the actual maximum Doppler frequency f_m will result in velocity estimates that are less sensitive to noise. Therefore, a better approach is to use the velocity estimate $\hat{v}$ to continuously adjust B to be just greater than the current maximum Doppler frequency, i.e., $B \gtrsim \hat{v}/\lambda_c$. Fig. 10.9 shows the effect of AWGN on each of the velocity estimators with respect to K, γ_S, and v, assuming $\theta = 0°$ (head-on LOS specular component). A value of $B = 357$ Hz is chosen which allows speeds up to 100 km/h at $f_c = 1.9$ GHz. For $K = 0$, AWGN has the same effect on all the velocity estimators. For larger velocities, e.g., 20 km/h, the bias becomes insignificant because B/f_m is small. However, for small velocities, e.g., 1 km/h, a very large B/f_m results in a significant bias. As mentioned above, this slow speed bias can be reduced by adapting the filter bandwidth B. It must also be remembered that Fig. 10.9 shows the worst case performance of the COV method as $\tau \to 0$. Any $\tau > 0$ will reduce the bias of the COV method due to AWGN. For example, if $2\pi v \tau/\lambda_c = 0.5$ in (10.52) then a large reduction in the effect of AWGN is realized, as shown by the curves labeled COV(.5) in Fig. 10.9. However, the accuracy of the COV velocity estimate itself improves with smaller τ, so that increasing τ for reduced noise sensitivity must be weighed against the reduced accuracy of the velocity estimate itself. This will be discussed further in the next section.

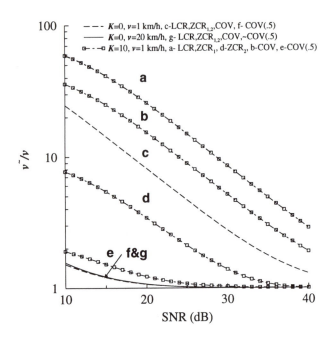

Figure 10.9 The effect of AWGN on the velocity estimates. COV(.5)⇒ $2\pi v \tau / \lambda_c = .5$, from [15].

10.4 VELOCITY ADAPTIVE HANDOFF ALGORITHMS

To study other velocity adaptive handoff issues we now assume $K = 0$, isotropic scattering, and no AWGN.

A velocity adaptive handoff algorithm must adapt the temporal window over which the mean signal strength estimates are taken by either keeping the sampling period constant and adjusting the number of samples per window, or vice versa. Here, we assume the latter. To reduce the variance in the velocity estimate, a sum of weighted past velocity estimates is performed using an exponential average of past estimates, i.e.,

$$\tilde{v}(n) = a\tilde{v}(n-1) + (1-a)\hat{v}(n) \tag{10.61}$$

where a controls the weighting of past estimates used in the average, and $\hat{v}(n)$ is the current velocity estimate. The accuracy of the velocity estimates will be affected by the window length W_l used to obtain the velocity estimates (not to be confused with the window length over which the signal strengths are averaged), and the number of samples per wavelength N_λ.

To show the effect of parameters a, N_λ, and W_l, simulation of the NLOS handoff scenario shown in Figs. 10.1 and 10.2 was performed. The path loss was assumed to follow the two-slope model in Section 10.2 with $a = 2$, $b = 2$, and $g = 150$ m in (2.212). Drastic path loss at the corner was assumed to take effect 5 m into the corner, so that the MS moving from BS_0 to BS_1 would experience the corner effect at 255 m from BS_0. The corner effect was modeled by choosing the average received signal strength at 255 m as the initial signal strength in a new LOS path loss model with $a = 2$, $b = 2$ and $g = 150$ m as before. Correlated log-normal shadows were used having a standard deviation of $\sigma_\Omega = 6$ dB and D in (10.13) set so that shadows decorrelated to $0.1\sigma_\Omega^2$ at 30 m. The instantaneous signal strength samples were affected by Rayleigh fading using the Jakes' simulator presented in Section 2.3.2. Samples were taken of the log-envelope and appropriately converted to envelope or squared envelope samples for the velocity estimator under study. Two-branch antenna diversity was assumed, so that the $\hat{v}(n)$ in (10.61) represent the average estimate out of the diversity branches at position n.

As mentioned previously, Fig. 10.12 shows the performance of a temporal handoff algorithm with $H = 8$ dB, signal strength averaging over 2.27 s, and overlapping windows by $2.27/2 = 1.135$ s. Slightly better temporal handoff performance can probably be obtained by fine tuning these values. However, for purposes of studying the velocity adaptive algorithms it is sufficient to maintain $H = 8$ dB and adapt to some point near the knee of the performance curve. Consequently, the velocity estimators were designed to adapt to the 5 km/h operating point which corresponds to signal strength window averages over $\approx 20\lambda_c$ with a window overlap of $\approx 10\lambda_c$.

A total of 1000 runs were made from BS_0 to BS_1, and the 95% confidence intervals were calculated for i) the velocity at 100 m, ii) the corner at 255 m, and iii) the probability of being assigned to BS_0 at 255 m. This resulted in a 95% confidence interval spread of $\hat{v} \pm .5$ km/h and $\Pr(BS_0) \pm .025$. Likewise, the mean number of handoff values had a 95% confidence interval spread of approximately .05 (mean number of handoffs $\pm .025$).

10.4.1   Effect of N_λ

To examine the effect of N_λ, assume that $a = .1$ and $W_l = 10\lambda_c$ for the LCR, ZCR and COV velocity estimators, and assume that the MS traverses the NLOS handoff route in Fig. 10.1 at 30 km/h. Furthermore, assume that the velocity estimators are initialized to 5 km/h, and that the MS is measuring signals from BS_0 and BS_1 only. Fig. 10.10 shows the effect of N_λ on the velocity estimate, in the first 90 m of the call as the MS moves from BS_0 to BS_1, in terms of the response time and final velocity estimate. The LCR velocity estimator requires a higher sampling density than the COV or ZCR methods and its final velocity and response time to an incorrect startup value (5 km/h) improve dramatically when N_λ is increased from 10 to 30 samples/wavelength. For $N_\lambda = 30$ the COV method shows a slight overshoot in the initial convergence, a characteristic seen with all the velocity estimators as the sampling density is increased. It is interesting to note that for $N_\lambda = 10$ samples/wavelength $2\pi(v\tau)/\lambda_c = 2\pi0.1\lambda_c/\lambda_c = .628$ and the final COV velocity estimate is close to the actual 30 km/h with a reasonable response time. This fact, along with the results of the Section 10.1 where $2\pi v\tau/\lambda_c = .5$ confirm that the effects of AWGN can be mitigated by using a larger sample spacing without drastically affecting the velocity estimate. We also note that the simulations used an estimate of the rms value $R_{\rm rms}$ in the LCR method and an estimate of the variance $\mu_{rr}(0)$ in the COV method. Thus the practicality of the velocity estimators that have been derived assuming perfect knowledge of these values is confirmed. Although not shown here, the Rice factor K was also found to have little effect thus confirming the claimed robustness of the estimators. Over the 1000 runs, the ZCR had the smallest velocity variance followed by the COV and LCR methods, respectively.

10.4.2   Corner Effects and Sensitivity to a and W_l

The sharp downward spike at the corner (255 m) for the LCR velocity estimate in Fig. 10.11 is typical of the corner effects on the velocity estimators. The effect is caused by a sudden change in path loss which lowers the local mean estimate in the LCR method thus yielding fewer level crossings per second. This corner effect is apparent, although less acute in the ZCR and COV methods due to their quick adaptability. The LCR and ZCR methods may exhibit a drop in estimated velocity when the average signal strength changes abruptly. Although not shown here, the COV method has an upward bias with an abrupt increase in the average signal strength, and a downward bias when the opposite

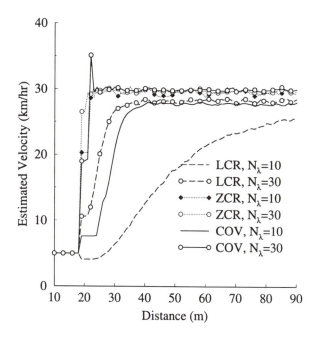

Figure 10.10   The effect of N_λ on the mean response time to a change in velocity; $a = .1$, $W_l = 10\lambda_c$; the true MS velocity is 30 km/h, from [15].

occurs. These corner effect properties could possibly be exploited to provide a combined corner detecting velocity adaptive handoff algorithm [14].

Larger values of a reduce the variance of the velocity estimate while sacrificing response time. Smaller values of a provide faster startup convergence and more sensitivity to corner effects.

Although a velocity window length W_l less than $20\lambda_c$ will increase the variance of the velocity estimates, it is beneficial for reducing the corner effect on the velocity estimator, as shown for the LCR method. Although not shown, the same is true for the ZCR and COV methods. The ZCR curve with $W_l = 20\lambda_c$ and $a = 0.5$ shows an overshoot in the initial convergence. This arises because the $W_l = 20\ \lambda_c$ windows that are used to obtain the velocity estimates overlap by $10\lambda_c$. Hence, part of the velocity estimate is derived from the previous window which may have a different sampling period due to adaptation. Note that we have used overlapped windows because they result in less handoff delay. Thus, it is probably better for initial startup to derive velocity estimates from the non-overlapped portions of the signal strength windows.

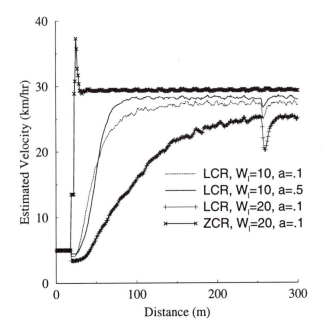

Figure 10.11   The effect of a and W_l on the mean response and corner effects; the true MS velocity is 30 km/h, from [15].

10.4.3   Velocity Adaptive Handoff Performance

Now that the effect of each parameter has been determined, the performance of the velocity adapted handoff algorithm is shown by the various symbols in Fig. 10.12 for a MS traveling at 30 km/h. The estimators, were selected to adapt to the 5 km/h operating point, the algorithm parameters were chosen as $a = .1$, $W_l = 10\lambda_c$ with an initial startup velocity of 5 km/h. The mean number of handoffs were found to have a 95% confidence interval with a span of about 0.05 (mean number of handoffs ±0.025) about the mean that is plotted. The velocity adaptive handoff algorithm performs very well by maintaining the desired operating point near the 5 km/h point.

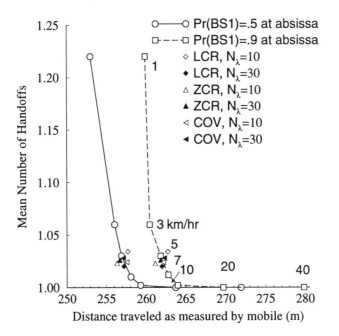

Figure 10.12 Handoff performance of a 2.27 second averaging handoff algorithm in comparison with a velocity adaptive handoff algorithm using the LCR, ZCR or covariance method for velocity control. $H = 8$ dB, $\sigma_\Omega = 6$ dB, from [15].

10.5 HANDOFF ANALYSIS

The classical signal strength based handoff algorithm compares signal strength averages measured over a time interval (T), and executes a hand-off if the average signal strength of the target BS is at least H (dB) larger than that of the serving BS [132], [15], [318], [317]. The analytical computation of the handoff characteristics for this classical signal strength based handoff algorithm is generally intractable. However, for the case when the average signal strength decays smoothly along a handoff route and the handoff hysteresis H is not too small compared to the shadow standard deviation, Vijayan and Holtzman [318], [317] have developed an analytical method to characterize the performance of the classical signal strength based handoff algorithm. They have also extended their results to include handoff algorithms that use absolute measurements [352], similar to the one in (10.4), and to soft handoff analysis for CDMA systems in [353].

Consider the case of a MS moving at a constant velocity along a straight line between two BSs, BS_0 and BS_1, that are separated by a distance of D meters. We neglect envelope fading under the assumption that the received signal strength estimates are averaged by using a window with an appropriate length as explained in Section 10.2. In any case, however, the signal strength estimates will respond to path loss and shadowing variations. Considering the effects of path loss and shadowing, the signal levels $\Omega_0 \ _{(\mathrm{dB})}(d)$ and $\Omega_1 \ _{(\mathrm{dB})}(d)$ that are received from BS_0 and BS_1, respectively, are (1.4)

$$\Omega_0 \ _{(\mathrm{dB})}(d) = \Omega_{(\mathrm{dB})}(d_0) - 10\beta\log_{10}(d/d_0) + \epsilon_0 \ _{(\mathrm{dB})} \qquad (10.62)$$

$$\Omega_1 \ _{(\mathrm{dB})}(d) = \Omega_{(\mathrm{dB})}(d_0) - 10\beta\log_{10}((D-d)/d_0) + \epsilon_1 \ _{(\mathrm{dB})} \quad (10.63)$$

where d is the distance between BS_0 and the MS. The parameters $\epsilon_0 \ _{(\mathrm{dB})}$ and $\epsilon_1 \ _{(\mathrm{dB})}$ are independent zero-mean Gaussian random processes with variance σ_Ω^2, reflecting a log-normal shadowing model. The signal strength measurements are assumed to be averaged by using an exponential averaging window with parameter d_{av} so that the averaged signal levels from the two BSs are, respectively,

$$\overline{\Omega}_0 \ _{(\mathrm{dB})}(d) = \frac{1}{d_{\mathrm{av}}} \int_0^d \Omega_0 \ _{(\mathrm{dB})}(d-x)e^{-x/d_{\mathrm{av}}} dx \qquad (10.64)$$

$$\overline{\Omega}_1 \ _{(\mathrm{dB})}(d) = \frac{1}{d_{\mathrm{av}}} \int_0^d \Omega_1 \ _{(\mathrm{dB})}(d-x)e^{-x/d_{\mathrm{av}}} dx \ . \qquad (10.65)$$

To describe the signal strength based handoff algorithm, let

$$x(d) = \overline{\Omega}_0 \ _{(\mathrm{dB})}(d) - \overline{\Omega}_1 \ _{(\mathrm{dB})}(d) \qquad (10.66)$$

denote the difference between the averaged signal strength estimates for BS_0 and BS_1. Consider the crossings of $x(d)$ with respect to the hysteresis levels $\pm H$ (dB) as illustrated in Fig. 10.13. A handoff is triggered if $x(d)$ has a down-crossing at $-H$ (dB) given that the last level crossing was an up-crossing at H (dB), or if $x(d)$ has an up-crossing at H (dB) given that the last level crossing was a down-crossing at $-H$ (dB). Vijayan and Holtzman verified that the two point processes, up-crossings of H (dB) and down-crossings of $-H$ (dB), can be modeled as independent Poisson processes under the assumption that $x(d)$ is a stationary *zero-mean* Gaussian random process, i.e., changes in the mean are ignored and the MS is moving along the boundary between two cells [318]. This result also applies when $x(d)$ has non-zero mean, but in this case the up-crossing and down-crossing rates are not equal. The Poisson assumption is asymptotically true for large H, but has been shown to hold true for H values

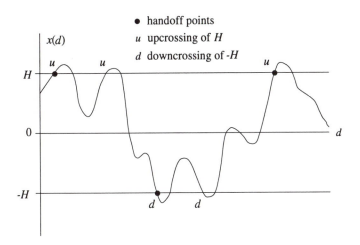

Figure 10.13 Handoff initiation points with their associated hysteresis level crossings. The MS is moving from BS_0 to BS_1 and assumed to be communicating with BS_1 at the beginning of the interval shown, from [318].

of practical interest, i.e., those on the order of the shadow standard deviation σ_Ω [318].

The handoff analysis proceeds by dividing up a handoff route into small spatial intervals of length d_s, such that only one level crossing is likely to occur within each interval. The probability of handoff at distance $d = nd_s$ is [318]

$$p_{ho}(n) = p_d(n)p_{lu}(n) + p_u(n)(1 - p_{lu}(n)) \qquad (10.67)$$

where $p_u(n)$ and $p_d(n)$ is the probability of an up-crossing or down-crossing in the nth interval, and $p_{lu}(n)$ is the probability that the last event was an up-crossing. In other words, the MS was assigned to BS_0 at the beginning of the nth interval. This can happen in one of two mutually exclusive ways; i) there is an up-crossing but no down-crossing in the $(n-1)$th interval, and ii) there are no crossings in the the $(n-1)$th interval, and the last event before the $(n-1)$th interval was an up-crossing. By assuming $p_{lu}(1) = 1$, the following recursive equation for $p_{lu}(n)$ can be derived as a function of $p_u(n-1)$, $p_d(n-1)$ and $p_{lu}(n-1)$ [318]

$$p_{lu}(n) = p_u(n-1)(1 - p_d(n-1)) + (1 - p_u(n-1))(1 - p_d(n-1))p_{lu}(n-1) . \qquad (10.68)$$

As detailed in [318], the probabilities $p_d(n)$ and $p_u(n)$ are functions of the mean $\mu_x(d)$, variance $\sigma_x^2(d)$, and variance of the derivative of $x(d)$, $\sigma_{\dot{x}}^2(d)$. These in turn are functions of the statistics of $\Omega_{0\ (\mathrm{dB})}(d)$ and $\Omega_{1\ (\mathrm{dB})}(d)$, which depend on the path loss and shadowing. Austin and Stüber have shown how these statistics depend on the co-channel interference [18]. We will first evaluate the statistics of $\Omega_{0\ (\mathrm{dB})}(d)$ and $\Omega_{1\ (\mathrm{dB})}(d)$ and afterwards derive the appropriate expressions for $p_d(n)$ and $p_u(n)$.

As discussed in Chapter 3, co-channel interference is usually assumed to add on a power basis [278], [249]. Hence, in the presence of N co-channel interferers the signals received from BS_0 and BS_1 are, respectively,

$$\Omega_{0\ (\mathrm{dB})}(d) = 10\log_{10}\left(\sum_{k=0}^{N}\Omega_{0,k(\mathrm{dB})}(d)\right) \tag{10.69}$$

$$\Omega_{1\ (\mathrm{dB})}(d) = 10\log_{10}\left(\sum_{k=0}^{N}\Omega_{1,k(\mathrm{dB})}(d)\right) \tag{10.70}$$

where $\Omega_{0,0(\mathrm{dB})}(d)$ and $\Omega_{1,0(\mathrm{dB})}(d)$ are the power of the desired signals from BS_0 and BS_1, respectively, and $\Omega_{0,k(\mathrm{dB})}(d)$ and $\Omega_{1,k(\mathrm{dB})}(d)$ $k = 1, \ldots, N_I$ are the powers of the interfering co-channel signals received at the same BSs. Once again, the $\Omega_{0,k(\mathrm{dB})}(d)$ and $\Omega_{1,k(\mathrm{dB})}(d)$ are log-normally distributed. As discussed in Section 3.1, the sum of log-normal random variables can be approximated by another log-normal random variable and, hence, $\Omega_{0\ (\mathrm{dB})}(d)$ and $\Omega_{1\ (\mathrm{dB})}(d)$ remain Gaussian. Here we consider the approximations suggested by Fenton [278], [249], and Schwartz and Yeh [278].

Following the notation in Section 3.1, define $\hat{\Omega} = \xi\Omega_{(\mathrm{dB})}$, where $\xi = (\ln 10)/10 = 0.23026$. If the N_I interferers for BS_0 have means $\mu_{\hat{\Omega}_{0,k}}(d)$ and variance $\sigma_{\hat{\Omega}}^2$, then the mean and variance of $\hat{\Omega}_0(d)$ using the Fenton-Wilkinson approach are

$$\mu_{\hat{\Omega}_0}(d) = \frac{\sigma_{\hat{\Omega}}^2 - \sigma_{\hat{\Omega}_0}^2(d)}{2} + \ln\left(\sum_{k=0}^{N_I}e^{\mu_{\hat{\Omega}_{0,k}}(d)}\right) \tag{10.71}$$

$$\sigma_{\hat{\Omega}_0}^2(d) = \ln\left((e^{\sigma_{\hat{\Omega}}^2} - 1)\frac{\sum_{k=0}^{N_I}e^{2\mu_{\hat{\Omega}_{0,k}}(d)}}{\left(\sum_{k=0}^{N_I}e^{\mu_{\hat{\Omega}_{0,k}}(d)}\right)^2} + 1\right) \tag{10.72}$$

where the conversion of $\mu_{\hat{\Omega}_0}(d)$ and $\sigma_{\hat{\Omega}_0}^2(d)$ to units of decibels is $\mu_{\Omega_0}(d) = 1/\xi\mu_{\hat{\Omega}_0}(d)$, and $\sigma_{\Omega_0}^2(d) = (1/\xi)^2\sigma_{\hat{\Omega}_0}^2(d)$, respectively. Schwartz and Yeh's approach is an recursive technique that combines only two log-normal variates at a time. For example, combining $\hat{\Omega}_{0,0}(d)$ and $\hat{\Omega}_{0,1}(d)$ gives the intermediate result

$$\mu_{\hat{\Omega}_0}(d) = \mu_{\hat{\Omega}_{0,0}}(d) + G_1 \tag{10.73}$$

$$\sigma^2_{\hat{\Omega}_0}(d) = \sigma^2_{\hat{\Omega}} - G_1^2 - 2\sigma^2_{\hat{\Omega}}G_3 + G_2 \tag{10.74}$$

where G_1, G_2, and G_3 are defined by (3.20), (3.23), and (3.24), respectively. The final values of $\mu_{\hat{\Omega}_0}(d)$ and $\sigma^2_{\hat{\Omega}_0}(d)$ are obtained by recursion.

By using either approach the mean $\mu_x(d)$ can be determined. Since $x(d)$ is modeled as a Gaussian random process, the probabilities $p_d(n)$ and $p_u(n)$ can be computed by using the same procedure used to determine the envelope level crossing rates in Section 2.1.4. In particular,

$$p_u(n) = d_s \int_0^\infty \dot{x} p(H, \dot{x}) d\dot{x}$$

$$p_d(n) = d_s \int_{-\infty}^0 |\dot{x}| p(H, \dot{x}) d\dot{x} \tag{10.75}$$

where $p(H, \dot{x})$ is the joint density function of $x(kd_s)$ and its derivative $\dot{x}(kd_s)$. Since $x(kd_s)$ and $\dot{x}(kd_s)$ are independent Gaussian random variables

$$p_u(kd_s) = \frac{d_s}{\sqrt{2\pi b_0}} \exp\left\{-\frac{(H - \mu_x(kd_s))^2}{2b_0}\right\} \tag{10.76}$$

$$\times \left[\mu_{\dot{x}}(kd_s)Q\left(-\frac{\mu_{\dot{x}}(kd_s)}{\sqrt{b_2}}\right) + \sqrt{\frac{b_2}{2\pi}}\exp\left\{-\frac{\mu_{\dot{x}}^2(kd_s)}{2b_2}\right\}\right]$$

where, from (2.79)

$$b_0 = \sigma_x^2(d) = 2\int_0^\infty \hat{S}_{xx}(f)df \tag{10.77}$$

$$b_2 = \sigma_{\dot{x}}^2(d) = 2(2\pi)^2 \int_0^\infty f^2 \hat{S}_{xx}(f)df \tag{10.78}$$

and $\hat{S}_{xx}(f)$ is the power spectrum of $x(d)$ that includes the effect of co-channel interference. Likewise,

$$p_d(kd_s) = \frac{d_s}{\sqrt{2\pi b_0}} \exp\left\{-\frac{(H + \mu_x(kd_s))^2}{2b_0}\right\} \tag{10.79}$$

$$\times \left[-\mu_{\dot{x}}(kd_s)Q\left(\frac{\mu_{\dot{x}}(kd_s)}{\sqrt{b_2}}\right) + \sqrt{\frac{b_2}{2\pi}}\exp\left\{-\frac{\mu_{\dot{x}}^2(kd_s)}{2b_2}\right\}\right]$$

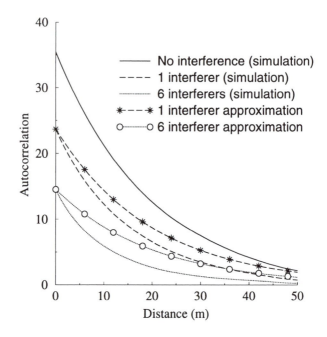

Figure 10.14   Shadow autocorrelation with and without co-channel interference, from [18].

The autocovariance of $\Omega_{0\ (dB)}(d)$ or $\Omega_{1\ (dB)}(d)$ (equal to the shadow autocorrelation) *without* co-channel interference is modeled by

$$\mu_{\Omega_{(dB)}\Omega_{(dB)}}(d) = \sigma_\Omega^2 \exp(-|d_1 - d_2|/d_0) \qquad (10.80)$$

where $d = d_1 - d_2$, and d_0 controls the decorrelation with distance. Let $\tilde{\mu}_{\Omega_{(dB)}\Omega_{(dB)}}(d)$ denote the same function when co-channel interference is present. The value $\tilde{\mu}_{\Omega_{(dB)}\Omega_{(dB)}}(0)$ can be accurately approximated by using either (10.72) or (10.74)). An approximation of $\tilde{\phi}_{\Omega_{(dB)}\Omega_{(dB)}}(d)$ for $d > 0$ can be obtained by substituting σ_Ω^2 in (10.80) with the value obtained in (10.72) or (10.74). The accuracy of this approximation was tested through the simulation of mutually uncorrelated log-normal interferers, each having the shadow autocovariance in (10.80) with $\sigma_\Omega = 6$ dB and $d_0 = 20$ m. Fig. 10.14 shows the results and verifies that the proposed approximation of $\tilde{\mu}_{\Omega_{(dB)}\Omega_{(dB)}}(d)$ is fairly accurate. Also, very accurate modeling of $\tilde{\mu}_{\Omega_{(dB)}\Omega_{(dB)}}(d)$ is not essential in handoff analysis [317].

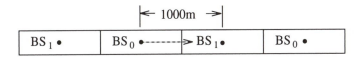

Figure 10.15 Base station layout, MS route (dotted line), and location of co-channel interferers.

Using the above approximation gives

$$\hat{S}_{xx}(f) = \frac{2(\sigma_{\Omega_0}^2(d) + \sigma_{\Omega_1}^2(d))d_0}{(1 + d_0^2(2\pi f)^2)(1 + d_{av}^2(2\pi f)^2)} \tag{10.81}$$

so that

$$\sigma_x^2(d) = \frac{(\sigma_{\Omega_0}^2(d) + \sigma_{\Omega_1}^2(d))d_0}{d_0 + d_{av}}, \quad \sigma_{\dot{x}}^2(d) = \frac{\sigma_x^2(d)}{d_{av}\,d_0} \ . \tag{10.82}$$

10.5.1 Simulation Results

Consider a MS traversing from BS_0 to BS_1 separated by 1000 m with two co-channel interferers as shown in Fig. 10.15. Assume a square-law path loss with distance (used here to accentuate the co-channel interference effects), $d_{av} = 10$ m, $d_0 = 20$ m, and choose $\sigma_\Omega = 4$ dB so that both the Fenton and Schwartz and Yeh log-normal approximations are accurate. Fig. 10.16 compares analytical and simulation results for the handoff probabilities in the presence and absence of co-channel interference. Note that the presence of co-channel interference actually lowers the probability of handoff. Schwartz and Yeh's method leads to an accurate prediction of the handoff probabilities while Fenton's method does not lead to as much accuracy. Finally, the accuracy of the prediction of handoff probabilities leads us to conclude that the assumptions made for $\tilde{\phi}_{\Omega_{(dB)}\Omega_{(dB)}}(d)$ were reasonable.

10.6 CIR-BASED LINK QUALITY MEASUREMENTS

Cellular radio resource allocation algorithms have been developed for handoffs [103], dynamic channel assignment [232], [126], and power control [10], [11], under the assumption that the MSs and/or BSs have access to real time measurements of the received carrier-to-interference plus noise ratio C/(I+N). However, very little literature has appeared on methods for measuring C/(I+N).

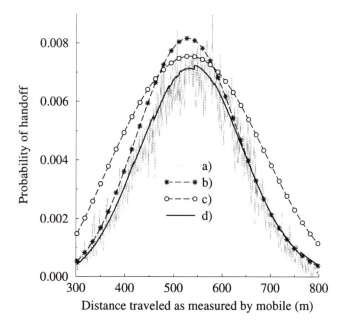

Figure 10.16 Simulation vs. analytical model performance. a) Simulation of LOS handoff with co-channel interference, b) handoff analysis model in the absence of co-channel interference, c) handoff analysis model including co-channel interference and using the Fenton-Wilkinson log-normal approximation, d) handoff analysis model including co-channel interference and using the Schwartz and Yeh log-normal approximation, from [18].

Kozono [166] suggested a method for measuring co-channel interference (CCI) in AMPS, by separating two terms at different frequencies which are known functions of the signal and interference. Yoshida [348] suggested a method for in-service monitoring of multipath delay spread and CCI for a QPSK signal. He reported that the CCI can be monitored provided that the delay spread is negligible compared to the symbol duration. Sollenberger [287] used the eye-opening as a measure of signal quality.

In this section we present a technique for estimating (S+I+N) and S/(I+N) that could be used in signal quality based resource allocation algorithms in TDMA cellular systems [13], [16]. Section 10.6.1 presents the discrete channel model. Estimation methods for the received (I+N) and C/(I+N) are then derived in Section 10.6.1 whose accuracy is only a function of the symbol error statistics. These estimators are evaluated by software simulation for an IS-54 frame structure in Section 10.6.2.

10.6.1 Discrete-Time Model for Signal Quality Estimation

As shown in Section 6.3.1, the overall channel consisting of the transmit filter, channel, matched filter, sampler (and noise whitening filter) can be modeled by a T-spaced, $L+1$-tap, transversal filter[4]. The overall discrete-time is described by the channel vector $\mathbf{g} = [g_0, g_1, \ldots, g_L]^T$, where T denotes transposition. Let $\mathbf{v} = [v_0, \ldots, v_M]^T$ denote the received signal vector consisting of M samples, where $v_k = \sum_{i=0}^{L} g_i x_{k-i}$. Assuming that the channel does not change significantly over a block of $L + M + 1$ symbols, the received vector $\mathbf{v}$ can be written as

$$\mathbf{v} = \mathbf{X}\mathbf{g} + \mathbf{w} \tag{10.83}$$

where $\mathbf{X}$ is an $(M + 1) \times (L + 1)$ Toeplitz matrix consisting of the transmitted symbols of the form

$$\mathbf{X} = [x_{i,j}] = \begin{bmatrix} x_0 & x_{-1} & \cdots & x_{-L} \\ x_1 & x_0 & \cdots & x_{1-L} \\ \vdots & \vdots & \cdots & \vdots \\ x_M & x_{M-1} & \cdots & x_{M-L} \end{bmatrix} \tag{10.84}$$

and $\mathbf{w} = [w_0, \ldots, w_M]^T$ is a vector consisting of the samples of the received interference plus noise.

Estimation of (I+N)

An (I+N) or C/(I+N) estimator requires a method for separating $\mathbf{g}$ and $\mathbf{w}$ from the observation of $\mathbf{v}$. Consider the situation where $L > M$, so that $\mathbf{A}$ has more rows than columns. Then, there exists a vector $\mathbf{c} = [c_0, \ldots, c_M]^T$ in the *null space* of $\mathbf{X}$ such that $\mathbf{c}^T \mathbf{X} = 0$. If $\mathbf{X}$ is known, then $\mathbf{c}$ can be easily determined. Then

$$\mathbf{c}^T \mathbf{v} = 0 + \mathbf{c}^T \mathbf{w} \tag{10.85}$$

and, therefore, $\mathbf{g}$ and $\mathbf{w}$ are completely separated from the observation $\mathbf{v}$. However, with the exception of the training and perhaps the color code sequences, $\mathbf{X}$

---

[4] If rate $2/T$ sampling is used, then the overall channel is a $t/2$-spaced, $2L+1$-tap, transversal filter.

is not known exactly because the data symbols comprising $\mathbf{X}$ must be obtained from decisions. Therefore the matrix of decisions $\hat{\mathbf{X}}$ must be used instead, where $\hat{\mathbf{A}} = \mathbf{A} + \boldsymbol{\Delta}$ and $\boldsymbol{\Delta} = [\delta_{i,j}]$ is the matrix of symbol errors. Nevertheless, a vector $\hat{\mathbf{c}}$ can still be found in the *null space* of $\hat{\mathbf{X}}$ so that

$$\hat{\mathbf{c}}^T \mathbf{v} = \hat{\mathbf{c}}^T \mathbf{X} \mathbf{g} + \hat{\mathbf{c}}^T \mathbf{w} \ . \tag{10.86}$$

Hence, an (I+N) estimate can be obtained from

$$
\begin{aligned}
\hat{\sigma}^2_{I+N} &= \ \mathrm{E}\left[\frac{\mathbf{v}^H \hat{\mathbf{c}}^* \hat{\mathbf{c}}^T \mathbf{v}}{||\hat{\mathbf{c}}||^2}\right] \\
&= \ \mathrm{E}\left[\frac{\mathbf{g}^H \boldsymbol{\Delta}^H \hat{\mathbf{c}}^* \hat{\mathbf{c}}^T \boldsymbol{\Delta} \mathbf{g}}{||\hat{\mathbf{c}}||^2}\right] + \mathrm{E}\left[\frac{\mathbf{w}^H \hat{\mathbf{c}}^* \hat{\mathbf{c}}^T \mathbf{w}}{||\hat{\mathbf{c}}||^2}\right] \\
&= \ \left(\sum_{i=0}^{L}\sum_{j=0}^{M}\sum_{k=0}^{L}\sum_{\ell=0}^{M} \mathrm{E}\left[\frac{\delta_{i,j}^* f_j^* \hat{c}_i^* \delta_{k,\ell} f_\ell \hat{c}_k}{||\hat{\mathbf{c}}||^2}\right]\right. \\
&\quad \left. +\sum_{i=0}^{M}\sum_{j=0}^{M} \mathrm{E}\left[\frac{\hat{c}_i \hat{c}_j^*}{||\hat{\mathbf{c}}||^2}\right] \mathrm{E}[w_i w_j^*]\right)
\end{aligned}
\tag{10.87}
$$

where H is the Hermitian transpose, and where the second equality is obtained by using $\hat{\mathbf{A}} = \mathbf{A} + \boldsymbol{\Delta}$ along with the reasonable assumption that $\mathbf{w}$ has zero mean and is uncorrelated with $\mathbf{A}$ and $\mathbf{g}$. Although the errors will appear in bursts we invoke the assumption that the symbol errors are independent with a constant variance, i.e.,

$$\mathrm{E}[|\delta_{j,i}|^2] = \sigma^2_\Delta \ . \tag{10.88}$$

We then have

$$\hat{\sigma}^2_{I+N} = \sigma^2_F \sigma^2_\Delta + \sum_{i=0}^{M}\sum_{j=0}^{M} \mathrm{E}\left[\frac{\hat{c}_i \hat{c}_j^*}{||\hat{\mathbf{c}}||^2}\right] \mathrm{E}[w_i w_j^*] \tag{10.89}$$

where $\sigma^2_F = \sum_{i=0}^{L} \sigma^2_f(i)$ and $\sigma^2_f(i)$ is the variance of the ith channel tap.

To determine $\mathrm{E}[w_i w_j^*]$, define the vector $\mathbf{w}$ as

$$\mathbf{w} = \sum_{k=1}^{N_I} \mathbf{B}(k)\mathbf{g}(k) + \mathbf{n} \tag{10.90}$$

where $\mathbf{B}(k)$ is an $(M+1) \times (L+1)$ matrix consisting of the symbols from the kth interferer with associated channel tap vector $\mathbf{g}(k)$, N is the number of interferers, and $\mathbf{n}$ is the vector of additive white Gaussian noise samples. The elements of $\mathbf{w}$ are

$$w_i = \sum_{k=1}^{N_I} \sum_{\ell=0}^{L} b(k)_{i,\ell}\, g(k)_\ell + n_i\ , \quad i = 0\ ,\ldots,\ M \tag{10.91}$$

where $\mathbf{B}(k) = [b(k)_{i,\ell}]$ and $\mathbf{g}(k) = [g(k)_\ell]$. We now assume that the data symbols have zero mean, the data sequences comprising the $\mathbf{B}(k)$ matrices for the interferers are both uncorrelated and mutually uncorrelated, and the n_i are independent zero mean Gaussian random variables with variance σ_n^2. Then $E[w_i w_j^*] = 0$ for $i \neq j$ and

$$\begin{aligned}
\sigma_w^2 = E[|w_i|^2] &= E\left[\sum_{k=1}^{N_I} \sum_{\ell=0}^{M} |b(k)_{i,\ell}|^2\, |g(k)_\ell|^2 + |n_i|^2\right] \\
&= \sum_{k=1}^{N_I} \sigma_b^2 \sum_{\ell=0}^{L} \sigma_{g(k)_\ell}^2 + \sigma_n^2 \\
&= \sigma_I^2 + \sigma_n^2 \\
&= \sigma_{I+N}^2
\end{aligned} \tag{10.92}$$

where σ_b^2 denotes the symbol variance of the interferers, $\sigma_{g(k)_\ell}^2$ denotes the variance of the ℓth channel tap gain associated with the kth interferer, and σ_I^2 denotes the total interference power. Using this result, (10.89) becomes

$$\hat{\sigma}_{I+N}^2 = \sigma_F^2 \sigma_\Delta^2 + \sigma_w^2 = \sigma_F^2 \sigma_\Delta^2 + \sigma_{I+N}^2\ . \tag{10.93}$$

In practice, the ensemble averaging in (10.87) must be replaced by an empirical average over P independent output vectors $\mathbf{v}_i$ so as to provide the unbiased estimate

$$\hat{\sigma}_{I+N}^2 = \frac{1}{P} \sum_{i=1}^{P} \frac{\mathbf{v}_i^H \hat{\mathbf{c}}_i^* \hat{\mathbf{c}}_i^T \mathbf{v}_i}{\|\hat{\mathbf{c}}_i\|^2}\ . \tag{10.94}$$

Estimation of C/(I+N)

A C/(I+N) estimator can be formed by using $\hat{\sigma}_{I+N}^2$, and one possibility is as follows. The total received signal power from the desired signal, interfering signals, and noise is

$$\sigma^2_{C+I+N} = \frac{1}{L+1} E\left[\mathbf{v}^H \mathbf{v} \right]$$

$$= \frac{1}{L+1} E\left[\mathbf{g}^H \mathbf{A}^H \mathbf{A} \mathbf{g} + \mathbf{w}^H \mathbf{w} \right]$$

$$= \frac{1}{L+1} \left(\sum_{j=0}^{L} \sigma^2_f(j) \sum_{i=0}^{M} |a_{i,j}|^2 + (L+1)\sigma^2_w \right) \qquad (10.95)$$

where the second equality follows from the assumption that $\mathbf{w}$ has zero mean, and the third equality requires that either the elements of the data sequence comprising the $\mathbf{A}$ matrix or the channel taps are uncorrelated. Once again, when $|a_{i,j}|^2 = \sigma^2_a$ (a constant) then

$$\sigma^2_{C+I+N} = \sigma^2_a \sigma^2_F + \sigma^2_w = \sigma^2_C + \sigma^2_w \ . \qquad (10.96)$$

Using (10.93) and assuming that σ^2_Δ is small yields the C/(I+N) estimate

$$\widehat{CIR} = \left(\frac{\sigma^2_{C+I+N}}{\hat{\sigma}^2_{I+N}} - 1 \right) \approx \frac{\sigma^2_C}{\sigma^2_{I+N}} \ . \qquad (10.97)$$

The above approximation becomes exact when $\mathbf{A}$ is known exactly. Finally, by replacing ensemble averages with empirical averages we obtain the C/(I+N) estimate

$$\widehat{CIR} = \frac{\frac{1}{L+1} \sum_{i=1}^{P} \mathbf{v}_i^H \mathbf{y}_i}{\sum_{i=1}^{P} \frac{\mathbf{v}_i^H \hat{\mathbf{c}}_i^* \hat{\mathbf{c}}_i^T \mathbf{v}_i}{\|\hat{\mathbf{c}}_i\|^2}} - 1 \ . \qquad (10.98)$$

10.6.2 Training Sequence Based C/(I+N) Estimation

The bursts in TDMA cellular systems contain known training and color code sequences. These sequences are used for BS and sector identification, sample timing, symbol synchronization, and channel estimation. As mentioned in the previous section, the (I+N) and C/(I+N) estimators will only work well when σ^2_Δ is small. Fortunately, if the (I+N) and C/(I+N) estimators are used during the training and color code sequences[5], $\sigma^2_\Delta = 0$.

The (I+N) and C/(I+N) estimators of the previous section were evaluated through the software simulation of an IS-54 [78] system . The baud rate is

---

[5]The color code is known provided the MS has correctly determined its serving BS.

24,300 symbols/s and each frame is composed of 6 bursts of 162 symbols so that the frame rate is 25 frames/s. The MS is assumed to have correctly determined the serving BS, i.e., the color code is known, and is monitoring its half rate channel (one burst per frame). Therefore, the known symbols within a burst consist of the 14 symbol training sequence at the beginning of the burst, and a 6 symbol color code sequence in the middle of the burst as shown in Fig. 1.5. For simulation purposes, a two-equal-ray T-spaced Rayleigh fading channel was chosen. The channel taps were assumed to be uncorrelated, although tap correlation will not affect the proposed algorithms because the various estimates depend only on the sum of the tap variances $\sigma_F^2 = \sum_{i=1}^{M} \sigma_f^2(i)$. Shadowing is assumed to remain constant over the estimates and is therefore neglected. Finally, it is assumed that the receiver can correctly synchronize onto each of the received bursts, i.e., perfect sample timing is assumed.

Four consecutive symbols were used to form a 3×2 Toeplitz non-symmetric matrix $\mathbf{A}$. Let $\{y_1(i), \ldots, y_{14}(i)\}$ denote the 14 received symbols corresponding to the training sequence of the ith frame, and $\{y_{15}(i), \ldots, y_{21}(i)\}$ the 6 received symbols in the color code. From the training sequence 4 estimates of (I+N) and C/(I+N) were formed by using the following 4 sets

$$\{\{y_1(i), \quad \ldots, y_4(i)\}, \{y_5(i), \ldots, y_8(i)\},$$
$$\{y_9(i), \ldots, y_{12}(i)\}, \{y_{11}(i), \ldots, y_{14}(i)\}\}$$

$$(10.99)$$

where the fourth set shares two symbols with the third set. Likewise, 2 estimates of (I+N) and C/(I+N) were formed from the 6 symbol color code sequence by using the 2 sets

$$\{\{y_{15}(i), \ldots, y_{18}(i)\}, \{y_{17}(i), \ldots, y_{21}(i)\}\} \qquad (10.100)$$

which share two common symbols. Although the (I+N) and C/(I+N) estimators in (10.94) and (10.98) assume independent $\mathbf{v}_j(i)$, $j = 1, \ldots, 21$, the additional estimates of (I+N) and C/(I+N) which use overlapped symbols at the ends of the training and color code sequences was found to improve the I+N and C/(I+N) estimates. The channel tap gains associated with the interferers were assumed to be constant during known symbols. Additive white Gaussian noise at 20 dB below the interference power was also included.

To evaluate the performance of the (I+N) estimator, we define the average absolute percentage error between the (I+N) estimate and the true interference plus noise power as

$$\frac{|\hat{\sigma}_{I+N}^2 - \sigma_{I+N}^2|}{\sigma_{I+N}^2} \times 100 \ . \qquad (10.101)$$

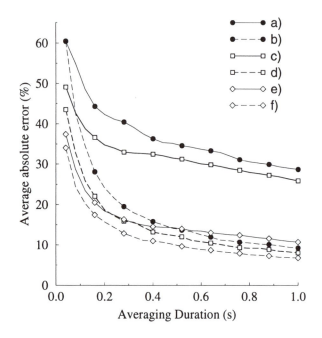

Figure 10.17 Average absolute percent error of the (I+N) estimator against the averaging time, from [13], [16]. The frame duration is 40 ms. Legend: a) $N_I = 1$, $v = 5$ km/h, b) $N_I = 1$, $v = 100$ km/h, c) $N_I = 2$, $v = 5$ km/h, d) $N_I = 2$, $v = 100$ km/h, e) $N_I = 6$, $v = 5$ km/h, f) $N_I = 6$, $v = 100$ km/h.

Fig. 10.17 depicts the average absolute percentage error over 500 independent averages for a specified averaging time (s), MS velocity (v), and number of interferers (N_I). Since the interference plus noise estimator is compared against σ^2_{I+N} under the assumption that the fading has been averaged out, it is natural to expect the estimator to perform worse for lower MS velocities when the averaging length is short, as Fig. 10.17 illustrates. Nevertheless, the presence of multiple interferers can improve the estimate, since with multiple interferers it is less likely that the total interference power will be small due to fading.

Likewise, Fig. 10.18 depicts the average absolute percentage error between the (C+I+N) estimate, $\hat{\sigma}^2_{C+I+N}$, and the true total received power, σ^2_{C+I+N}. As before, the MS velocity has a large effect on the estimator performance. Also, the C/I has a minor effect. However, in contrast to the (I+N) estimator, the number of interferers has little effect for C/I between 5 and 20 dB and, hence, variations in the number of interferers are not shown in Fig 10.18.

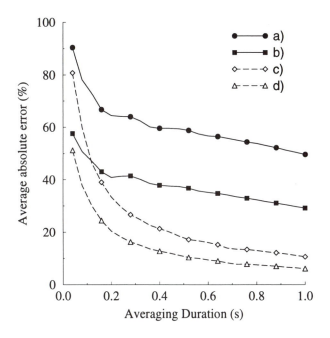

Figure 10.18 Average absolute percent error of the (C+I+N) estimator against the averaging time, from [13], [16]. The frame duration is 40 ms. Legend: a) $N_I = 1$, $v = 5$ km/h, C/I = 5 dB, b) $N_I = 1$, $v = 5$ km/h, C/I = 20 dB, c) $N_I = 1$, $v = 100$ km/h, C/I = 5 dB, d) $N_I = 1$, $v = 100$ km/h, C/I = 20 dB.

Fig. 10.19 depicts performance of the C/(I+N) estimator for an actual C/I of 5 dB. Only the performance with C/I = 5 dB is shown, since the estimator was found insensitive to C/I variations when the actual C/I was between 5 and 20 dB. For high speed MS, C/(I+N) can be estimated to within 2 dB in less than a second. A slight improvement is also obtained when the MS uses two slots per frame (a full rate channel) as shown in Fig. 10.20.

10.7 SUMMARY

This chapter has provided a detailed discussion of local mean estimation in microcells and presented three velocity estimators that can be used for adaptive signal strength window averaging. The accuracy of local mean estimation in microcells was shown to depend on the Rice factor, the angle of the specular

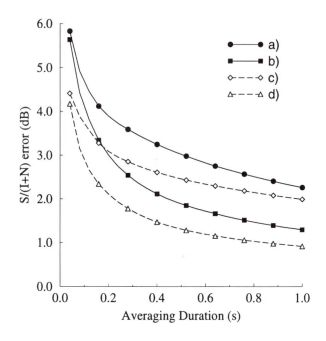

Figure 10.19   Average error of the C/(I+N) estimator against the averaging time, from [13], [16]. The frame duration is 40 ms. Legend: a) $N_I = 1$, $v = 5$ km/h, C/I = 5 dB, b) $N_I = 1$, $v = 100$ km/h, C/I = 5 dB, c) $N_I = 6$, $v = 5$ km/h, C/I = 5 dB, d) $N_I = 6$, $v = 100$ km/h, C/I = 5 dB.

component, and the averaging length. For sample averaging, sample spacings less than $.5\lambda_c$ should be used. All three velocity estimators are relatively insensitive to the Rice factor under isotropic scattering. The LCR and COV velocity estimators are highly sensitive to non-isotropic scattering, whereas the ZCR estimator is reasonably robust. However, as is likely in urban microcells, the presence of a specular component can significantly reduce non-isotropic scattering biases. When $K = 0$, AWGN has the same effect on each of the three methods. However, when $K \neq 0$ and infinitely small sample spacing is used, the best performance is achieved with the ZCR, COV, and LCR methods, in that order. With larger sample spacings, the COV method is able to show a greatly reduced sensitivity. To reduce AWGN effects, an adaptive filter bandwidth with respect to the maximum Doppler frequency and/or increasing the sampling period should be used. Increasing the sampling density reduces the bias in the final velocity estimate and improves the rate of convergence to changes in velocity or propagations effects such as the corner effect. The ZCR method has the fastest mean response time followed by the COV and LCR

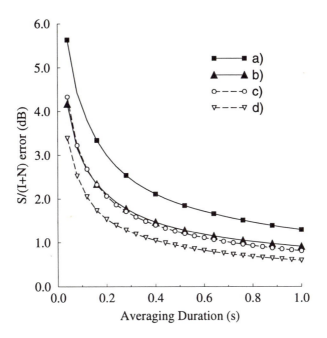

Figure 10.20 Average error of the C/(I+N) estimator for half rate and full rate channels against the averaging time, from [13], [16]. The frame duration is 40 ms. Legend: a) $N_I = 1$, $v = 100$ km/h, C/I = 5 dB, half rate channel, b) $N_I = 6$, $v = 100$ km/h, C/I = 5 dB, half rate channel, c) $N_I = 1$, $v = 100$ km/h, C/I = 5 dB, full rate channel, d) $N_I = 6$, $v = 100$ km/h, C/I = 5 dB, full rate channel.

methods. All the velocity estimators are biased by the corner effect. Averaging the velocity estimates over several windows gives a slower initial convergence but reduces prolonged biases in the velocity estimates when a MS turns a corner. Shorter window lengths can also be used for faster adaptation. Each of the velocity estimators can successfully maintain good handoff performance over a wide range of MS velocities in a typical NLOS handoff scenario.

An analytical technique has been discussed for evaluating handoff performance, where the handoff rates can be studied in terms of the level crossings of the averaged signal level process.

Signal quality estimation techniques were examined for multipath fading channels having co-channel interference and additive Gaussian noise in TDMA cellular systems. Estimators for (C+I+N), (I+N) and C/(I+N) have been developed whose accuracy is only a function of the symbol error statistics. These estimators have been applied to a cellular TDMA system, where knowledge of the training and color code sequences is used to form the estimates. Simulation results show that C/(I+N) can be estimated to within 2 dB in less than a second for high speed MSs.

Appendix 10A
DERIVATION OF EQUATIONS (10.47) AND (10.60)

The limit in (10.60) can be written as

$$
\lim_{\tau \to 0} \frac{\frac{\lambda}{2\pi\tau}\sqrt{\frac{2\tilde{\mu}_{rr}(0)-2\tilde{\mu}_{rr}(\tau)}{\tilde{\mu}_{rr}(0)}}}{\frac{\lambda}{2\pi\tau}\sqrt{\frac{2\mu_{rr}(0)-2\mu_{rr}(\tau)}{\mu_{rr}(0)}}} = \frac{\lim_{\tau \to 0} \frac{\lambda}{2\pi\tau}\sqrt{\frac{2\tilde{\mu}_{rr}(0)-2\tilde{\mu}_{rr}(\tau)}{\tilde{\mu}_{rr}(0)}}}{\lim_{\tau \to 0} \frac{\lambda}{2\pi\tau}\sqrt{\frac{2\mu_{rr}(0)-2\tilde{\mu}_{rr}(\tau)}{\mu_{rr}(0)}}} . \tag{10A.1}
$$

Note that the limit of the denominator gives (10.47) and is a special case of the numerator limit with $N_o = 0$. To find the numerator limit the following theorem can be used [297]

Theorem 1 *If a function $f(\tau)$ has a limit as τ approaches a, then*

$$
\lim_{\tau \to a} \sqrt[n]{f(t)} = \sqrt[n]{\lim_{\tau \to a} f(t)} \tag{10A.2}
$$

provided either τ is an odd positive integer or n is an even positive integer and $\lim_{\tau \to a} f(\tau) > 0$.

Therefore, if the limit

$$
\zeta = \lim_{\tau \to 0} f^2(\tau) = \lim_{\tau \to 0} \frac{\lambda^2}{(2\pi\tau)^2} \frac{2\tilde{\mu}_{rr}(0) - 2\tilde{\mu}_{rr}(\tau)}{\tilde{\mu}_{rr}(0)} \tag{10A.3}
$$

exists and is positive, the solution to (10A.1) will be readily determined. It is apparent that L'Hôpital's Rule should be applied to determine the limit in (10A.3). After substituting $\tilde{\mu}_{rr}(\tau)$ from (10.38) and applying L'Hôpital's Rule four times, the limit is found as

$$\zeta = \frac{\lambda^2 \left(B^4 N_o{}^2 \pi^2 + 3BK N_o \omega^2 \sigma^2 + 2B^3 K N_o \pi^2 \sigma^2 + 2B^3 N_o \pi^2 a(0) \right)}{6\pi^2 \left(B N_o + 2a(0) \right) \left(B N_o + 4K\sigma^2 + 2a(0) \right)}$$

$$+ \frac{\lambda^2 \left(6K\omega^2 \sigma^2 a(0) + 3BK N_o \omega^2 \sigma^2 \cos(2\theta) + 6K\omega^2 \sigma^2 a(0) \cos(2\theta) \right)}{6\pi^2 \left(B N_o + 2a(0) \right) \left(B N_o + 4K\sigma^2 + 2a(0) \right)}$$

$$+ \frac{\lambda^2 \left(-12a'(0)^2 - 12c'(0)^2 - 6B N_o a''(0) - 12K\sigma^2 a''(0) \right)}{6\pi^2 \left(B N_o + 2a(0) \right) \left(B N_o + 4K\sigma^2 + 2a(0) \right)}$$

$$+ \frac{\lambda^2 \left(-12a(0)a''(0) - 12c(0)c''(0) \right)}{6\pi^2 \left(B N_o + 2a(0) \right) \left(B N_o + 4K\sigma^2 + 2a(0) \right)} \tag{10A.4}$$

where $a(\tau)$ and $c(\tau)$ are given by (10.40) and $x'(0)$ denotes the derivative of $x(t)$ evaluated at 0. Consequently, $a(0) = b_0 = \sigma^2$, $a'(0) = c(0) = 0$, and

$$a''(0) = \sigma^2 \omega^2 \int_0^{2\pi} \hat{p}(\alpha) \cos^2(\alpha) d\alpha$$

$$c'(0) = \sigma^2 \omega \int_0^{2\pi} \hat{p}(\alpha) \cos(\alpha) d\alpha \ . \tag{10A.5}$$

Using these, and the fact that $(1 + \cos(2\theta))/2 = \cos^2(\theta)$, (10A.4) is positive for all θ under all scattering scenarios mentioned here. Consequently, applying theorem 1, the limit of the numerator of (10A.1) is the square root of (10A.4), which if desired can easily be put in terms of the signal-to-noise ratio γ_S using

$$\sigma = \sqrt{\frac{\gamma_S N_o B}{2(K+1)}} \tag{10A.6}$$

from (10.55). The denominator of (10A.1), which is also (10.47) is obtained by assuming isotropic scattering and no noise, so that $a(0) = \sigma^2$, $a'(0) = c(0) = c'(0) = c''(0) = 0$, $a''(0) = \sigma^2 \omega^2/2$, and $N_o = 0$ in (10A.4). After taking the square root, the result is

$$\lim_{\tau \to 0} \frac{\lambda}{2\pi\tau} \sqrt{\frac{2\mu_{rr}(0) - 2\mu_{rr}(\tau)}{\mu_{rr}(0)}} = v \sqrt{\frac{(1 + 2K + K\cos(2\theta))}{(1 + 2K)}} \ . \tag{10A.7}$$

Problems

10.1. Suppose that a MS is traveling along a straight line from BS_1 to BS_2, as shown in Fig. 10.21. The BSs are separated by distance D, and the MS is at distance r from BS_1 and distance $D - r$ from BS_2. Ignore the effects of fading and assume that the signals from the two BSs experience independent log-normal shadowing. The received signal power (in decibels) at the MS from each BS has the Gaussian density in (1.5), where the propagation path loss is described by

$$\mu_{\Omega_k} = \Omega_{(\text{dB})}(d_o) - 10\beta \log_{10}(d/d_o) \ .$$

- A handoff from BS_1 to BS_2, or vice versa, can *never* occur if $|\Omega_{1\ (\text{dB})} - \Omega_{2\ (\text{dB})}| < H$ but *may or may not* occur otherwise.

- A handoff from BS_1 to BS_2 will occur if the MS is currently assigned to BS_1 *and* $\Omega_{2\ (\text{dB})} \geq \Omega_{1\ (\text{dB})} + H$.

a) Find an expression for the probability that a handoff can never occur from BS_1 to BS_2, or vice versa.

b) Given that the MS is currently assigned to BS_1 what is the probability that a handoff will occur from BS_1 to BS_2.

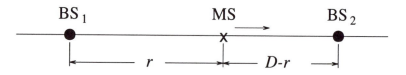

Figure 10.21   MS traversing from BS_0 to BS_1 along a handoff route.

10.2. A freeway with a speed limit of 120 km/h passes through a metropolitan area. If the average call duration is 120 s

a) What will be the average number of handoffs in a cellular system that uses omnidirectional cells having a 10 km radius.

b) Repeat part a) for a cellular system that uses 120° sectored cells having a 1 km radius.

10.3. Derive equation (10.23).

10.4. Derive equation (10.25).

10.5. Derive equation (10.38).

10.6. Derive equation (10.54).

11

CHANNEL ASSIGNMENT TECHNIQUES

First generation macrocellular systems typically use fixed channel assignment (FCA), where disjoint subsets of the available channels are permanently allocated to the cells in advance according to their *estimated* traffic loads. The cells are arranged in tessellating reuse clusters whose size is determined by the co-channel reuse constraint. For example, the North American AMPS system typically uses a 7-cell reuse cluster with 120° sectoring. The 12.5 MHz bandwidth allocation for AMPS can support a total of 416 two-way channels, 21 of which are control channels (one for each sector in a cluster), leaving a total of 395 traffic channels. This yields an allocation of 56 channels/cell with uniform FCA.

FCA provides adequate capacity performance in macrocellular systems that are characterized by stationary and homogeneous traffic, and a predictable propagation environment. In this case the channel resources can be allocated statically, since the call blocking probabilities can be predicted with reasonable certainty. Under conditions of nonstationary and nonhomogeneous traffic, however, FCA is spectrally inefficient because the channels are literally fixed to the cells. A new call or handoff arrival that finds all channels busy in a cell will be blocked even though there may be several idle channels in the adjacent cells that could service the call.

In microcellular systems the propagation environment is highly erratic, and the traffic is characterized by spatial and temporal variations. Furthermore, the decreased cell sizes imply an increase in handoff traffic, since a call may be handed off several times before its natural completion. Because of these properties, the channel assignment problem in microcellular and macrocellular networks is fundamentally different. The uneven nature of the traffic and the larger volume of handoff attempts in microcellular networks demand careful attention. Furthermore, a microcellular channel assignment strategy has to be fast, because the handoffs must be serviced quickly due to the small cell sizes and propagation anomalies such as the street corner effect.

One well known solution to the microcellular channel assignment problem is *distributed* dynamic channel assignment (DCA) , where the dynamic nature of the strategy permits adaptation to spatial and temporal traffic variations while the distribution of control reduces the required computation and communication among base stations (BSs), thereby reducing system latencies. DCA schemes have no exclusive relationship between cells and channels, and in their most general they allow any cell to use any channel that does not violate the co-channel reuse constraint. DCA schemes are known to outperform FCA under conditions of light nonstationary traffic. However, under conditions of heavy traffic FCA usually provides better performance, because the DCA schemes often yield an inefficient arrangement of channels. Although DCA has clear benefits, the cost can be quite high because it not only requires increased computation and communication among BSs but also an increased number of radio ports at the BSs; in the extreme case each BS must have the ability to use all channels simultaneously.

Practical DCA schemes differ in degree of network planning and the required communication among BSs. Centralized DCA schemes require centralized control with system-wide channel information. The extreme example is maximum packing (MP) [87], where a new call or handoff arrival is blocked only if there is no global rearrangement of calls to channels that will accommodate the service need. Unfortunately, the enormous computation and communication among cells render centralized DCA schemes impractical. In fact, the number of channel rearrangements required between two subsequent arrivals in a two dimensional network with MP can increase without bound with the number of cells in the network [265].

Fully decentralized DCA schemes require no network planning or communication among BSs. These scheme often rely upon the passive monitoring of idle channels at each BS, allowing the cells to acquire any idle channel that is deemed to provide a sufficient carrier-to-interference ratio (C/I). Decentralized DCA schemes require limited communication among the BSs. One such scheme is dynamic resource acquisition (DRA) [230], a DCA scheme where channels are acquired and released according to a reward/penalty function. The calculation of the reward/penalty function requires that each cell have channel usage information from a finite set of surrounding cells called the DRA neighborhood. Decentralized and fully decentralized schemes are not without their problems, including service interruption, deadlock, and instability. A service interruption occurs when a channel allocation causes an existing link to fall below the threshold C/I. The interrupted mobile station (MS) then tries to find an alternate link and if unsuccessful a service termination occurs. This is known as deadlock . A sequence of successive interrupts, or rippling effect, caused by channel allocations is called an instability .

DCA schemes also have the advantage of assigning the same channel to a MS moving from one cell to another provided that the level of co-channel interference is tolerable, while FCA must conduct a handoff with a channel change because the same channel is not available in adjacent cells. Handoffs without channel changes are attractive because they can eliminate the need for channel searching and ultimately relieve the BSs from extra computation. More important, this mechanism is essential for supporting macrodiversity TDMA cellular architectures where the signal from a MS can be simultaneously received by two or more BS yielding a diversity improvement against shadow (and fading) variations. Such architectures provide the same benefit as soft handoff in CDMA systems.

This chapter is intended to introduce the various approaches to cellular channel assignment. Unfortunately, most channel assignment schemes are quite detailed and founded largely on ad hoc principles. Furthermore, the channel assignment schemes are almost always evaluated by using detailed simulations with a variety of assumptions concerning the mobile radio environment, e.g., cellular topology and reuse factors, traffic patterns, propagation factors, mobility, etc.. The combination of these factors makes a systematic comparison of the various DCA schemes largely infeasible and a true consensus of the best scheme cannot be attained. Therefore, we will briefly outline some of the many different DCA schemes, followed by a detailed evaluation of a few specific schemes that serve to illustrate the basic concepts.

Throughout the chapter various performance measures will be used to evaluate the channel assignment schemes, including the following

- Probability of new call blocking , P_b, defined as

$$P_b = \frac{\text{number of new calls blocked}}{\text{number of new call arrivals}} \ .$$

- Probability of forced termination , P_f, defined as

$$P_f = \frac{\text{number of handoff calls blocked}}{\text{number of handoff attempts}} \ .$$

- Grade of service , GOS, defined as

$$\text{GOS} = \frac{P_b R_N}{(R_N + R_H)} + \frac{P_f R_H}{(R_N + R_H)}$$

  where R_N and R_H are the new call and handoff arrival rates, respectively.

- Channel changing rate , R_C, defined as

$$R_C = \frac{\text{number of channel changes}}{\text{number of handoffs}} \ .$$

The remainder of this chapter begins with an overview of some important DCA schemes. These include the fully centralized Maximum Packing (MP) and MAXMIN DCA strategies in Section 11.1. Decentralized DCA strategies, such as First Available (FA), Nearest Neighbor (NN), and Dynamic Resource Acquisition (DRA) are discussed in Section 11.2. Fully decentralized DCA schemes are the topic of Section 11.3, including Channel Segregation (CS) and Minimum Interference (MI), along with aggressive and timid strategies. Hybrid FCA/DCA schemes are the subject of Section 11.4. The important class of borrowing schemes are the topic of Section 11.5, including Borrowing with Channel Ordering (BCO), Borrowing with DireCtional Locking (BDCL), and Compact Pattern based DCA (CPDCA). Finally, our overview of DCA schemes wraps up with a treatment of Directed Retry (DR) and Directed Handoff (DH), Moving Direction (MD) strategies, reduced transceiver coverage, reuse partitioning, and handoff priority.

Following our results in [332], Section 11.10 provides some detailed and instructive examples of distributed DCA schemes for TDMA microcellular systems. In particular, two DCA strategies are presented that accommodate handoff queueing. An aggressive DCA policy with handoff queueing is also considered where a cell may be forced to terminate calls in progress in order to accommodate handoff requests in neighboring cells. The conditions for forced termination are carefully determined to ensure a performance improvement over a timid policy.

11.1 CENTRALIZED DCA

Centralized DCA schemes require system-wide information and control for making channel assignments. As expected, centralized DCA schemes can theoretically provide the best performance. However, the enormous amount of computation and communication among BSs leads to excessive system latencies

Figure 11.1 Five cell deployment with MP

and renders centralized DCA schemes impractical. Nevertheless, centralized DCA schemes often provide a useful benchmark to compare the more practical decentralized DCA schemes.

11.1.1 Maximum Packing (MP)

The Maximum Packing (MP) algorithm was originally presented by Everitt and Macfadyen in 1983 [88]. With the MP policy a call is blocked only if there is no global rearrangement of calls to channels that will accommodate the call. Accomplishing this task requires a controller with system-wide information along with the ability to perform call rearrangements. The MP policy has the ability to serve all calls in a network with the minimum number of channels. Equipped with the capability, MP can yield the lowest new call blocking and forced termination probabilities of any DCA scheme under any traffic conditions.

Kelly [164] presented an interesting and enlightening analytical approach to MP DCA, by modeling the MP policy as a circuit-switched network. This allows some very powerful and well known network analysis tools to be applied to the analysis of MP DCA. The analysis ignores situations where the MS is moving from one cell to another or out of the service area, i.e., the handoff and roaming problem. Upon a call arrival in a particular cell, the MP policy checks to see if all reuse clusters that contain that cell have at least one channel available. If so, then the call can be accommodated through channel rearrangements; otherwise, the call is blocked. For example, consider the simple system consisting of five cells shown in Fig. 11.1. In this example, co-channel cells must be separated by at least two cells so there are three reuse clusters; $CL_1 = (1, 2, 3)$, $CL_2 = (2, 3, 4)$, and $CL_3 = (3, 4, 5)$. When a call arrives in cell 2, it can be accommodated if there is at least one channel available in clusters CL_1 and CL_2.

The stochastic model for MP uses the following definitions:

$$
\begin{aligned}
\mathcal{R} &= \text{set of cells in the system.} \\
K &= |\mathcal{R}| = \text{number of cells in the system.} \\
N_T &= \text{total number of channels available.} \\
n_i &= \text{number of calls in progress in cell } i. \\
\mathbf{n} &= (n_i, i \in \mathcal{R}) = \text{state vector.} \\
\mathcal{S} &= \text{set of admissible states.} \\
\rho_i &= \text{traffic load in cell } i.
\end{aligned}
$$

The set of admissible states depends on the particular cell layout. Let J be the number of complete or partial reuse clusters $\mathrm{CL}_i, i = 1, \ldots, J$ that can be defined such that i) each reuse cluster differs by at least one cell, i.e., they are not totally overlapping, and ii) all cells are contained in at least one such cluster. For the example in Fig. 11.1, $J = 3$. Now let $\mathbf{A} = [a_{ij}]_{J \times K}$ be the **demand matrix**, where $a_{ij} = 1$, if $i = j$ and if cell j is in the same cluster as cell i; otherwise, $a_{ij} = 0$. For the example in Fig. 11.1

$$
\mathbf{A} = \begin{bmatrix} 1 & 1 & 1 & 0 & 0 \\ 0 & 1 & 1 & 1 & 0 \\ 0 & 0 & 1 & 1 & 1 \end{bmatrix} . \tag{11.1}
$$

Matrix $\mathbf{A}$ tabulates the channel requirements for servicing calls that arrive in each of the cells. For example, a call arrival in cell 2 requires that a channel be available in CL_1 and CL_2 but not in CL_3 and, therefore, $a_{1,2} = a_{2,2} = 1$ and $a_{3,2} = 0$. Finally, let $N_i, i = 1, \ldots, J$ be the number of channels that are available in CL_i, $N_i \leq N_T$, and $\mathbf{N} = (N_1, \ldots, N_J)$. Then the set of admissible states is then given by

$$
\mathcal{S} = \{\mathbf{n} : \mathbf{An}^T \leq \mathbf{N}\} . \tag{11.2}
$$

It is well known (e.g., [163]) that $\mathbf{n}$ has the steady-state distribution

$$
\pi(\mathbf{n}) = G(\mathbf{N}) \prod_{i \in \mathcal{R}} \frac{\rho_i^{n_i}}{n_i!} , \quad \mathbf{n} \in \mathcal{S} \tag{11.3}
$$

where $G(\mathbf{N})$ is the normalizing constant

$$
G(\mathbf{N}) = \left(\sum_{\mathbf{n} \in \mathcal{S}} \prod_{i \in \mathcal{R}} \frac{\rho_i^{n_i}}{n_i!} \right)^{-1} . \tag{11.4}
$$

Then the steady-state probability that a call arrival in cell i is blocked is

$$B_i = 1 - \frac{G(\mathbf{N})}{G(\mathbf{C} - \mathbf{A}\mathbf{e}_i^T)} \tag{11.5}$$

where $\mathbf{e}_i$ is a length K vector with a '1' at position i and '0' elsewhere. Even though B_i appears to have a compact closed form expression, the computation of $G(\mathbf{N})$ is prohibitive except for very simple cases. Therefore, approximate methods are usually employed. One approximation assumes that the availability of channels in the clusters CL_i are *independent* events. This independence assumption leads to

$$B_i \approx 1 - \prod_{j \in \mathrm{CL}_i} (1 - E_j) \tag{11.6}$$

where the $E_j, j = 1, \ldots, J$ solve the nonlinear equations

$$E_j = E\left(\sum_{r \in \mathcal{R}} a_{jr} \rho_r \prod_{i \in r - \{j\}} (1 - E_i), \ N_j \right), \quad j = 1, \ldots, J \tag{11.7}$$

with

$$E(\rho, N) = \frac{\rho^N}{N!} \left(\sum_{n=0}^{N} \frac{\rho^n}{n!} \right)^{-1} \tag{11.8}$$

being the Erlang-B formula. Kelly [164] showed that there is an unique solution to the above nonlinear equations. The intuitive notion behind this approximation is that when $a_{jr} = 1$ the call arrivals of rate ρ_r in cell r are thinned by a factor of $1 - E_i$ by each cluster $\mathrm{CL}_i, i \in r - \{j\}$ before being offered to CL_j.

11.1.2 MAXMIN Scheme

The MAXMIN scheme was introduced by Goodman *et al.* [126]. With the MAXMIN scheme, a MS is assigned a channel that maximizes the minimum C/I that any MS will experience in the system at the time of assignment. Assuming that the link quality depends on the average received C/I, the C/I of MS_i at its serving BS is

$$\Lambda(\mathbf{d})_{(\mathrm{dB})} = \Omega(d_i)_{(\mathrm{dB})} - 10 \log_{10} \sum_{k \in I} 10^{\Omega(d_k)_{(\mathrm{dB})}/10} \ . \tag{11.9}$$

where the $\Omega(d_j)_{(\mathrm{dB})}$ are independent Gaussian random variables with the density in (1.5) and (1.6), and d_j is the distance between MS_j and the BS for MS_i.

The set I consists of all MSs other than MS_i that are using the same channel. A MS that requires service is assigned the channel j that gives

$$\max_{j \in C} \min_{i \in S} \{\Lambda_i\} \qquad (11.10)$$

where i and j index the set of MSs and channels, respectively, C is the set of channels that are available at the BS corresponding to the MS that requires service, Λ_i is the C/I of MS_i at its BS, and S is the set of all MS in service including the MS that requires service. We have already seen methods for C/I monitoring in Section 10.6.

11.2   DECENTRALIZED DCA

11.2.1   First Available (FA) and Nearest Neighbor (NN)

In 1972, Cox and Reudnik [58] proposed four basic decentralized DCA algorithms and compared them to FCA for the case of linear highway macrocells: **First Available (FA)**, **Nearest Neighbor (NN)**, **Nearest Neighbor+1 (NN+1)** , and **Mean Square (MSQ)** . All four schemes allow a BS to acquire any idle channel that is not being used in its **interference neighborhood**, defined as the set of surrounding cells that can interfere with the BS. The schemes differ in the way that the channel selected should more than one channel be available for acquisition. The FA scheme acquires the first available channel found in the search. Assuming that a channel can be reused D_N cells away without causing excessive co-channel interference, the NN policy acquires the channel that is being used by the nearest BS at distance D_N or greater. The NN+1 policy acquires the channel that is in use at the nearest BS at distance $D_N + 1$ or greater with the goal of allowing more MSs to retain the same channel as they cross cell boundaries. The MSQ policy seeks to assign the available channel that minimizes the mean square of the distances among all BSs using the same channel. The DCA schemes were shown to outperform the FCA schemes in terms of probabilities of new call blocking and forced termination, except under conditions of heavy traffic. Of these four DCA schemes, the NN policy performs the best.

11.2.2   Dynamic Resource Acquisition (DRA)

Nanda and Goodman [230], have proposed a distributed DCA strategy called Dynamic Resource Acquisition (DRA). When a channel must be selected for acquisition or release by a BS, DRA calculates a reward/cost function for each channel. The reward associated with a channel release is the number of cells in the interference neighborhood of the BS that could acquire the channel after it is released. When a channel is released, the busy channel giving the largest reward is selected. Channel rearrangements may be required to do this. The cost associated with a channel acquisition is the number of cells in the interference neighborhood of the BS that would be deprived from using the acquired channel. When a channel is acquired, the available channel having the smallest cost is selected. In the event of a tie in the reward/cost function, the released/acquired channel is chosen randomly.

As described in [230], the calculation of the reward/cost function requires channel usage information from all the cells within the **DRA neighborhood** of a BS. The DRA neighborhood of a BS is the set of cells whose interference neighborhoods overlap with the interference neighborhood of that BS. Any cell outside the DRA neighborhood of a BS will not affect the calculation of the reward/cost function associated with that BS. Fig. 11.2 illustrates the reward/cost functions associated with three carriers for a 2-D grid of square cells. The cell under consideration is shaded black. In case of a carrier acquisition, Channel 2 would be selected by the given BS since it has the smallest cost. Channel 3 could not be selected, because it would violate the co-channel reuse constraint. If Channels 1 and 2 are active in the given cell and a carrier is to be released, then Channel 1 would be selected since it gives the largest reward.

11.3   FULLY DECENTRALIZED DCA

11.3.1   Channel Segregation (CS)

Akaiwa and Andoh [6] proposed a distributed adaptive self-organizing DCA strategy whereby the BSs use Channel Segregation (CS) to develop *favorite* channels through an evolutionary process that is based on the criteria of eliminating unnecessary interference. Their scheme has been developed for TDMA systems with the assumption that each BS can access any channel by tuning a carrier frequency and selecting a time slot. CS also accounts for the effect

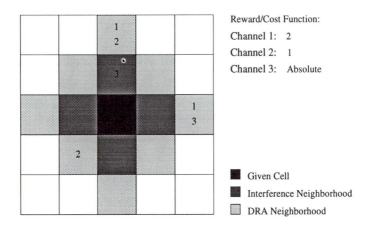

Figure 11.2 The DRA (dark shaded) and interference (light shaded) neighborhoods of a cell under consideration (shaded black). The numbers within
each cell indicate active channels.

of unaccessible channels where a call can be blocked in a cell even when there
are idle channels because of the restriction placed on the number of different
carrier frequencies that may be simulatneously used, i.e., the BS has a finite
number of radio ports each of which can be tuned to only one frequency.

A flowchart of the CS algorithm is shown in Fig. 11.3. Each BS ranks the
channels according to a priority function $P(i)$, where a large $P(i)$ corresponds
to a high priority, e.g., in [6] $P(i) = N_s/N_t$, where N_s is the number of successful
uses of the channel plus the number of accesses to the channel when it is idle but
unaccessible, and N_t is the total number of trials for the channel. When a call
arrives, the BS senses the highest priority channel from the list of channels it is
not currently using. If the channel is sensed idle, then the channel is checked for
accessibility. If accessible, it is acquired and its priority is increased; otherwise,
its prioity is increased and the BS recursively senses the next highest priority
channel that it is not currently using. If all channels are sensed busy, then the
call is blocked. Akaiwa and Andoh [6] demonstrated by simulation that the CS
policy outperforms FCA and the FA DCA algorithm.

The steps within the dashed box in Fig. 11.3 are a modification so that the original CS algorithm developed by Akaiwa [5] for FDMA systems can be applied
to TDMA systems. Simulation results show that this modification achieves the
goal of gathering channels with the highest priorities onto the same carrier frequency, thus reducing the probability of call blocking due to the unavailability
of a BS transceiver.

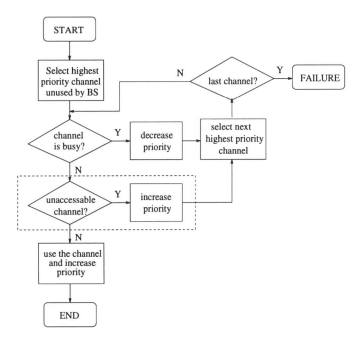

Figure 11.3 Channel segregation (CS) algorithm.

11.3.2 Channel Segregation with Variable Threshold

Another channel segregation scheme has been proposed by Hanabe *et al.* [138] that uses prioritized orderings with a variable interference threshold. The channels are ranked from highest to lowest according to their priority values. Each BS measures the interference levels of its currently unused channels. For each channel, the priority value is decreased if the interference level is higher than a predefined threshold and the threshold for that channel is decreased. Likewise, the priority value is increased if the interference level is lower than a predefined threshold and the threshold for that channel is increased. Hanabe *et al.* do not clearly define their priority function. However, it is likely that the priority function is defined as the ratio of number of times that the interference level of a channel is less than the threshold to the total number of times the channel is sensed. The interference threshold is varied depending on the ranking of the channel in the priority list. For example, the particular threshold that Hanabe *et al.* propose is as follows

$$T(k) = \begin{cases} T_o + 15 + S, & k = 1, 2, \ldots N_1 \\ T_o + 10 + S, & k = 9, 10, \ldots N_2 \\ T_o + 5 + S, & k = 17, 18, \ldots N_3 \end{cases} \quad . \quad (11.11)$$

For the example in [138], $N_1 = 8$, $N_2 = 16$, and $N_3 = 72$, where N_3 is the total number of channels in the system, $T(k)$ is the threshold, T_o is the minimum required C/I, and S is a constant margin. Upon call arrival, the highest priority idle channel that meets the C/I threshold is chosen. If no suitable channels are available, the call is blocked. If the C/I drops below the required level during a call, a handoff procedure is initiated. In Hanabe *et al.'s* scheme, handoffs are not prioritized and are treated the same as new call arrivals.

The rationale for using a variable threshold $T(k)$ for each channel in (11.11) can be answered by examining the case where the thresholds are fixed. Allocation of a high priority channel with a fixed threshold is more likely to cause interference since the C/I thresholds for all channels are the same. The reason is that a higher priority channel will be acquired over a lower priority channel if both channels exceed the C/I threshold, regardless of whether or not a lower priority channel would cause less interference to neighboring cells. Thus the allocation of the higher priority channel may cause service interruptions, deadlocks, and instability. For the case of a variable threshold, a higher threshold is assigned to higher priority channels to reduce the probability of co-channel interference, and a lower threshold is assigned to lower priority channels to decrease the probability of blocking.

11.3.3   Minimum Interference (MI) Schemes

Schemes based on Minimum Interference (MI) have been presented by Goodman *et al.* [126]. The basic MI scheme has been incorporated into the CT-2 and DECT systems. With these schemes, the MS signals the BS with the strongest paging signal for a channel. The BS measures the interference level on all channels that it is not already using. The MS is then assigned the channel with the minimum interference. This policy coupled with mobile controlled handoff (MCHO) guarantees good performance. Variations of the MI scheme have been proposed that differ in the order in which MSs are assigned channels. These include **Random Minimum Interference** (RMI) , **RMI with Reassignment** (RMIR) , and **Sequential Minimum Interference** (SMI) . The RMI scheme call requests in the order that they arrive. The RMIR scheme serves the call requests according to the RMI scheme, but afterwards each MS is reassigned according to the MI policy. The order of reassignments is random. Those MSs initially denied service try again to acquire a channel. The procedure is repeated a fixed number of times. The SMI algorithm assigns channels according to the MI scheme but in a sequential order. In [126] linear microcells are considered and the sequence that is followed is to serve a MS only after all MS to its left have had a chance to be served. This, however, requires some co-ordination between BSs and the extension to 2-D schemes is not obvious. Goodman *et al.* [126] showed that the probability of blocking decreases with FCA, RMI, RMIR, SMI, in that order.

11.3.4   Aggressive and Timid DCA Strategies

Distributed self organizing DCA algorithms that use aggressive and timid strategies were first introduced by Cimini and Foshini [43]. These simple autonomous DCA algorithms can self-organize with little loss in capacity compared to the best globally coordinated channel selection algorithm. In their paper, two classes of algorithms were studied; **timid** where a MS acquires a channel only if the channel is free of interference, and **aggressive** where a MS can acquire a channel even if it is not free of interference. The studies in [43] showed that a linear array of cells could self organize its placement of a single channel to **saturate** the array from random starting arrangements. An array is saturated when no additional cells can use a channel without violating the co-channel reuse constraint. Channel usage in the array organizes itself according to the DCA policy. The performance of the algorithm is measured in terms of the **saturation density**, defined as the ratio of the number of cells using a particular channel to the number of cells in the arrary. Timid algorithms which require no call rearrangements have been shown to have saturation densities that compare favorably with FCA, while the aggressive algorithms have higher saturation densities at the expense of some instability. This is due to a **simulated annealing** mechanism where an instability perturbs a system so as to escape a local optimum in an attempt to reach the global optimum.

The saturation densities can be derived for the case of linear and hexagonal planar cells with R-cell buffering[1]. For linear cells, the maximum and minimum saturation densities are $C_{\max} = 1/(R+1)$ and $C_{\min} = 1/(2R+1)$. The saturation density can also be obtained for the random placement of a channel in a linear array. In this case, cells sequentially acquire the channel; the next cell

---

[1]The reuse factor N is related to the number of buffer rings R as follows. For linear cells $N = R+1$. For hexagonal planar cells, $N = i^2 + ij + j^2$, where for R odd $i = j = (R+1)/2$, and for R even $i = R/2$ and $j = R/2+1$.

to acquire the channel is chosen uniformly from those cells not already using the channel that could use the channel without violating the co-channel reuse constraint. The derivation of the saturation density in this case is quite lengthy but leads to the result [45]

$$C_{\text{ran}} = \int_0^1 \exp\left\{2\sum_{i=0}^{R-1} \frac{(v^{i+1} - 1)}{i+1}\right\} dv \ . \tag{11.12}$$

The saturation density can also be obtained as a function of the traffic load ρ as [45]

$$C(\rho)\left(1 - C(\rho)R\right)^R = \rho\left(1 - (R+1)C(\rho)\right)^{R+1} \tag{11.13}$$

which has a unique solution in the interval $0 < C(\rho) < 1/(R+1)$. For hexagonal planar cells the minimum and maximum saturation densities are

$$C_{\text{min}} = \frac{1}{1 + 3R(R+1)} \tag{11.14}$$

$$C_{\text{max}} = \begin{cases} \frac{4}{3(R+1)^2} & , \quad R \text{ even} \\ \frac{4}{1+3(R+1)^2} & , \quad R \text{ odd} \end{cases} \ . \tag{11.15}$$

However, expressions for C_{ran} and $C(\rho)$ for the hexagonal planar array are unknown.

For the case of a single channel the blocking probability has the exact form [45]

$$P_b = 1 - C(\rho)/\rho \ . \tag{11.16}$$

For the case of multiple channels, Cimini *et al.* have derived a very accurate approximation for the call blocking performance of timid algorithms. They also derived lower bounds on the call blocking performance of aggressive algorithms [44]. If a total of N_T channels are available, the effective number of channels available for use in a reuse cluster of size N is δM, where δ is called the **normalized channel utilization** defined as the saturation density that is achieved with a particular algorithm C to the maximum possible saturation density C_{max}. Values for δ are tabulated in Table 11.1. For FCA, each cell has $m = N_T/N$ available channels and the blocking probability can be obtained from the Erlang-B formula $P_b = E(\rho, m)$ in (11.8), where ρ is the traffic load per cell. For the case of the timed algorithm a call is blocked if all channels are use in the interference neighborhood. To approximate the blocking probability for the timid algorithm, we replace ρ by $N\rho$ and m by δN_T, i.e., $P_b = E(N\rho, \delta N_T)$. To lower bound the blocking probability with an aggressive algorithm, we replace ρ by $N\rho$ as before, and m by N_T ($\delta = 1$), so that $P_b > E(N\rho, N_T)$. The

R	Linear	Planar hexagonal
1	0.864	0.693
2	0.825	0.658
3	0.804	0.627

Table 11.1 Normalized channel utilizations $\delta = C_{ran}/C_{max}$ for the timid DCA algorithm, from [44].

performance of a practical aggressive algorithm will lie somewhere between the timid algorithm and the aggressive bound. Finally, we note that the blocking probability with an aggressive algorithm includes the calls that are blocked and the calls that are dropped because the aggressive algorithm has taken the channel and another suitable channel can be found.

11.4 HYBRID FCA/DCA SCHEMES

DCA schemes perform very well under light non-stationary non-homogeneous traffic. However, under conditions of uniformly heavy traffic FCA outperforms most of the DCA schemes, except perhaps MP. As a result of this behavior efforts have been directed toward hybrid FCA/DCA schemes that are intended to provide a compromise between FCA and DCA. Cox and Reudink [59] introduced a hybrid scheme, called **Dynamic Channel Reassignment** (DCR) where each cell is assigned number of fixed channels, while the remaining channels are available for DCA. Fixed channels are used first to accommodate call requests. Calls that cannot be serviced by the fixed channels are offered to the dynamically assigned channels. The dynamic channel that is selected can be obtained by using any of the elementary schemes such as FA, NN, NN+1, etc. Upon a call completion on a fixed channel, DCR executes a search to determine if a call nominally assigned to a dynamic channel can be reassigned to the newly released fixed channel.

11.5 BORROWING SCHEMES

Engel and Peritsky [84] introduced an FCA scheme with **borrowing**. The channels that are assigned to each BS are divided into two sets, fixed and borrowable. The fixed channels can only be used by the BS they are assigned to, while the remaining channels can be borrowed by a neighboring BS if necessary. Calls are serviced by using the fixed channels whenever possible. If necessary a channel is borrowed from a neighboring cell to service the call provided that the use of the borrowed channel does not violate the co-channel reuse constraint. The channel is borrowed from the neighboring BS having the largest number of available channels for borrowing. Improvements on this scheme were also proposed by Engel and Peritsky, where a call being serviced by a borrowed channel is transferred to a fixed channel whenever a fixed channel becomes available. The same idea was proposed by Anderson [9]. Scheduled and predictive channel assignment schemes have also been proposed, where the ratio owned to borrowable channels is dynamically varied according to the traffic conditions.

11.5.1   Borrowing with Channel Ordering (BCO)

Elnoubi *et. al.* [83] proposed a channel borrowing strategy that makes use channel orderings, call Borrowing with Channel Ordering (BCO). A group of channels is initially assigned to each cell according to a fixed channel assignment; these channels are called nominal channels and are arranged in an ordered list. The call arrival policy for BCO is illustrated by the flow chart in Fig. 11.4. Upon a call arrival in a cell, the BS searches for an available nominal channel nearest to the beginning of the channel ordering. If a nominal channel is available it is assigned to the call; otherwise, the BS attempts to borrow a channel from the adjacent cell having the largest number of channels available for borrowing. A channel is available for borrowing if it is unused in the adjacent cell and the other two co-channel cells. To illustrate this point, refer to Fig. 11.6. If cell B1 borrows a channel c from cell $A1$, then cells A1, A2, and A3 are **locked** from using channel c since their use of channel c will violate the co-channel reuse constraint. Being blocked, channel c can neither be used to service a call in these three cells nor borrowed from these three cells. Finally, when a channel is borrowed from an adjacent cell, the available channel appearing nearest to the end of the channel ordering of the adjacent is selected. If no channels are available for borrowing, the call is blocked.

The call departure policy for BCO is illustrated in Fig. 11.5. When a call terminates on a borrowed channel, the borrowed channel is released in the three cells where it is locked. When a call terminates on a nominal channel and there are calls in progress with the same BS on borrowed channels, then the channel that is borrowed from the adjacent cell with the largest number of

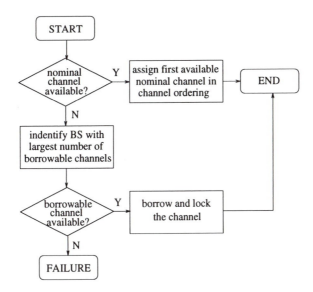

Figure 11.4 Call arrival policy for the BCO algorithm.

lent channels is released in the three cells where it is locked and its associated call is reassigned to the newly idle nominal channel. Finally, if a call completes on a nominal channel and there are no calls in progress with the same BS on borrowed channels, the call occupying the nominal channel nearest to the end of the channel ordering is reassigned to the newly idle nominal channel.

Kuek and Wong [168] introduced a DCA scheme called **Ordered Dynamic Channel Assignment/Reassignment** (ODCAR) that also combines channel ordering with channel rearrangements. The differences between the BCO and ODCAR schemes are very minor and quite subtle. BCO borrows a channel from the adjacent cell having the largest number of available *channels for borrowing*, while ODCAR borrows a channel from the adjacent cell having the largest number of available *nominal channels* that it could use to service its own calls. When a call completes on a nominal channel and there are calls in progress with the same BS on borrowed channels, then BCO releases the channel that is borrowed from the adjacent cell with the *largest number of lent channels* while ODCAR releases the channel that is borrowed from the adjacent cell with the *fewest number of nominal channels*. Finally, when a call completes on a borrowed channel BCO simply releases the channel in the three cells where it is locked, while ODCAR again releases the borrowed channel from the adjacent cell with the fewest number of nominal channels.

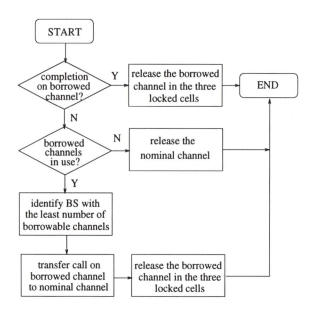

Figure 11.5 Call departure policy for the BCO algorithm.

11.5.2 Borrowing with Directional Locking

Zhang and Yum [351] introduced a new scheme called Borrowing with DireC-tional Locking (BDCL) and compared it with borrowing with channel ordering (BCO). Referring to the $N = 7$ cell reuse pattern in Fig. 11.6, the BCO scheme operates as follows. If cell B1 borrows a channel c from cell $A1$, then cells A1, A2, and A3 are **locked** from using channel c since there use of channel c would violate the co-channel reuse constraint. In the BDCL scheme, instead of locking channel c in cell A3 in all directions, channel c only needs to be locked in directions 1, 2, and 3. Cells that lie in the other three directions from A3, say B2, can freely borrow channel c from cell A3 without violating the co-channel reuse constraint. Whether or not channel c may be borrowed from A3 depends, however, on its locking conditions in A4, A5, and A6. Should the channel happen to be locked in A4, A5, or A6 but the cell locking is beyond B2's interference neighborhood, then B2 could still borrow channel c. This scheme increases the number of channels available for borrowing over the straight BCO scheme. Furthermore, the BDCL scheme uses channel rearrangements similar to the channel ordering scheme proposed by Elnoubi *et al.* [83], except that the directional locking mechanism is accounted for. Zhang and Yum [351] con-cluded that the BDCL scheme outperforms the BCO and FCA schemes in terms

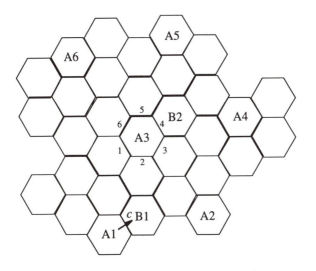

Figure 11.6 Principle of borrowing with directional channel locking (BDCL).

of blocking probabilities when the cells have nonuniform but stationary traffic loads.

11.5.3 Borrowing without Locking

A borrowing scheme, Channel Borrowing Without Locking (CBWL), has been proposed by Jiang and Rappaport [155] that does not require channel locking by using borrowed channels with a reduced power level to limit interference with co-channel cells. This allows the channel to be reused in all cells except the cell from which it has been borrowed. However, it also implies that channels can only be accessed in part of the borrowing cell. To determine if a channel can be borrowed with enough signal strength, the BS broadcast a **borrowed channel sensing signal** with the same reduced power of a borrowed channel.

The CBWL scheme divides the channels into six groups that can be lent to the neighboring cells, such that channels in the ith group can only be lent to the ith adjacent cell. This principle of **directional lending** is illustrated in Fig. 11.7, where channels in the group A1 can be borrowed by MSs in all of the B cells. Because of the reduced power level of borrowed channels, the MSs in the B cells that borrow the group A1 channels will be concentrated along the A-B cell boundaries. The CBWL scheme reduces the BS complexity because

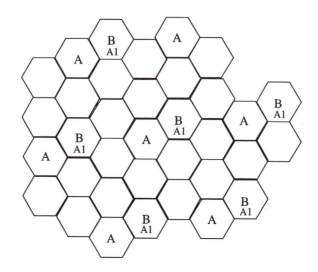

Figure 11.7   Principle of channel borrowing without locking (CBWL).

each BS does not need to have the capability of transmitting and receiving on all the channel assigned to its neighboring cells, but only a fraction of them in each cell. Furthermore, the division of borrowable channels into six groups limits co-channel interference so that locking is not required.

Various forms of channel rearrangements can be used enhance the scheme. For example, if cell B wishes to borrow a channel from cell A, the call is blocked if all the channels in group A1 of cell A are busy. However, it may be possible for cell A to transfer one of the calls to another group, say A2, to accommodate the borrow request. If this is not possible, cell A could itself borrow a channel from an adjacent cell to free up a channel to lend to cell B. Many other types of rearrangement policies are also possible.

11.5.4   Compact Pattern Based DCA

Yeung and Yum [347] introduced Compact Pattern based DCA (CPDCA), that attempts to dynamically keep the co-channel cells of any channel to a **compact pattern**, where a compact pattern of a network is the channel allocation pattern with the minimum average distance between co-channel cells. In other words, CPDCA attempts to increase spectral efficiency by keeping all channels at their minimum co-channel reuse distance. CPDCA accomplishes this task in two stages; i) channel acquisitions where an optimal idle channel is assigned to the MS, and ii) channel packing for the restoration of the compact patterns upon the release of a compact channel. Channel packing is achieved by reassigning at most one call per channel release.

Channels are assigned by using system-wide call arrival rate information to assign a channel that has a compact pattern that will yield the largest reduction in the overall system blocking probability. If a compact pattern is not available, the most optimal non-compact pattern is selected. If a call completes on a compact channel, CPDCA attempts to reassign a call on a noncompact channel to the newly idle compact channel. If no such call exists, CPDCA reassigns a call on the compact pattern that is least utilized to the newly idle channel. The first step minimizes the number of noncompact channels being used, while the second step packs the ongoing calls onto complete compact patterns. Yeung and Yun have shown their CPDCA scheme to outperform BDCL.

11.6 DIRECTED RETRY AND DIRECTED HANDOFF

Everitt [86] introduced the **directed retry** (DR) and **directed handoff** (DH) channel assignment algorithms. If a BS does not have an idle channel available to service a call with the DR policy, the MS tries to acquire an idle channel in any other cell that can provide a satisfactory signal quality. DR exploits the overlapping nature of cells in a practical cellular system, where some percentage of MSs can establish a suitable link with more than one BS. DH also exploits the overlapping nature of cells to direct some of the ongoing calls in a heavily loaded cell to an adjacent cell that is carrying a relatively light load. Both the DR and DH schemes can be used in conjunction with either FCA or DCA, and Everitt concluded that FCA and maximum packing DCA in conjunction with the combination of DR and DH offer about the same performance. Therefore, FCA in conjunction with DR and DH is the preferred scheme, since an improvement over FCA can be gained without the added complexity of DCA.

11.7 MOVING DIRECTION STRATEGIES

The moving direction (MD) strategy, proposed by Okada and Kubota, exploits information about the MS movement to reduce forced terminations and channel changes [237], [238]. The strategy attempts to assign an available channel from among those channels already assigned to MSs elsewhere in the service area that are moving in the same direction as the MS under consideration. Sets of MSs moving in the same direction are formed. When a MS enters a cell, a MS from the same set is probably leaving a cell. This allows both MSs to retain the same channel, thus reducing both the number of changes changes and probability of forced termination. This method is particularly useful for highway microcell deployment, where the traffic direction is highly predictable. Okada and Kubota compared the MD strategy with Cox and Reudnick's FA, NN, and NN+1 strategies [238]. The MD strategy was shown to offer the lowest channel changing rate and the lowest probability of forced termination. However, the NN strategy provided a slightly lower probability of new call blocking.

A variation of the MD scheme considers both **Speed and Moving Direction** (SMD) [236]. MS are divided into two classes; high speed MS (HSMS) who are traveling at 50 to 60 km/h and low speed MS (LSMS) who are traveling at 0 to 4 km/h. To reduce the probability of forced termination and channel changing rate, SMD uses the MD policy for its HSMS call requests. For the LSMS, the NN strategy is employed since LSMSs do not experience forced terminations or channel changes as frequently as HSMSs. Again, the SMD scheme was shown to outperform the FA, NN, and NN+1 policies in terms of channel changing rates and the probability of forced termination. Finally, we mention that a variety of velocity estimation techniques are available as discussed in Section 10.3. Moving direction information can be obtained by using the past signal strength history in LOS environments [19] or the sign of the Doppler.

11.8   REDUCED TRANSCEIVER COVERAGE

Takeo *et al.* [299] proposed a scheme where nonuniform traffic is handled by adjusting the BS transmit power level of the control channel according to the traffic variance for every control period. Since the MS uses the control channel to determine which BS to connect to, the effective cell size is dynamically varied. Highly loaded cells decrease the transmit power to shrink the cell sizes, while lightly loaded cells increase the transmit power to enlarge the cell sizes. This scheme may cause some unwanted side effects, for example, handoffs can occur even for stationary MSs. The experimental results in [299] suggest that the call blocking probability increases in proportion to a decrease in the control period and, therefore, frequent updating of the control channel power should be avoided. Takeo *et al.* [299] did not address the problem when many adjacent cells that are heavily loaded, a potentially deleterious situation since it may result in coverage gaps within a particular reuse cluster.

11.8.1 Reuse Partitioning

Reuse partitioning employs a two-level cell plan where clusters of size M are overlaid on clusters of size N, $N > M$. Fig. 9.20 shows a FCA scheme using reuse partitioning with $M = 3$ and $N = 9$. As discussed in Section 9.5.1, reuse partitioning divides the available channels into two sets; one set can be used by the inner cells only, while the other set can be used by both the inner and outer cells. Reuse partitioning uses rearrangements so that whenever possible MSs in the inner cells are assigned channels allocated for use in the inner cells only.

An **autonomous reuse partitioning** (ARP) scheme has been suggested by Kanai [157]. With this scheme an *identical* ordering of channels is given to all BSs. Upon call arrival, the channels are checked in order and the first one exceeding a C/I threshold for both the forward and reverse links is acquired. If no channels are available the call is blocked. The advantage of using a fixed ordering is that the channels higher in the ordering are used more frequently and, hence, have higher interference levels. This enables each BS to acquire channels with minimum C/I margins without the need for sorting channels according to their interference levels. The algorithm is self organizing in the sense that channels high in the ordering (with high interference levels) are allocated to MS that are close to a BS (with strong received signal levels). Channels low in the ordering tend to be allocated to MSs that are far from a BS with weak received signal levels.

Another scheme, **self-organized reuse partitioning** (SORP) has been proposed by Furukawa and Yoshihiko [108]. The BSs allocate channels by measuring the power levels transmitted from the MSs. This method relies upon a table at each BS that contains, for each channel, the average transmit power for MSs using the channel in its cell and all the surrounding cells. The table is updated with each call arrival and the update information is shared among the BSs. When a call arrives, the BS obtains the output power of the calling MS and assigns that channel with the corresponding average transmit power that is closest to that of the calling MS. The channel is acquired if available; otherwise the second closest candidate is examined, and so on. As a result of this procedure in each BS, channels that correspond to the same power are grouped autonomously for self-organizing reuse partitioning. The SORP scheme was shown to offer about the same blocking probability as the ARP scheme, but SORP requires less time to search for a channel and generally provides a higher C/I.

11.9   HANDOFF PRIORITY

Since the forced termination of a call in progress is worse than blocking of a new call, it is important to consider **handoff priority** in the design of a channel assignment strategy. This is especially important in microcellular systems with their increased number of handoffs. Two possible methods of achieving handoff priority are to use **guard channels** where a fraction of the channels are reserved for handoff requests only [148], and **handoff queueing** where a handoff request from a MS is placed in a queue with the target BS while the MS maintains a radio link with its serving BS [148], [109]. Both methods are known to decrease the probability of dropped calls. However, queueing does this with a smaller increase in the probability of new call blocking.

Handoff queueing exploits the time interval that the MS spends in the handoff region, i.e., between the time when the handoff request is generated and the time when the call will be dropped due to a degradation in link quality. The simplest queueing scheme uses a first in first out (FIFO) policy. More elaborate queueing schemes use measurement based priority, where the queue is ranked according to the measured link quality of the MSs in the queue [300]. MSs with the lowest link quality are placed in the highest priority class, and the handoff queue is sorted continuously according to the priority classes.

11.10   EXAMPLE DCA SCHEMES FOR TDMA SYSTEMS

We assume a TDMA system with **carrier groupings**, , where the calls are packed into TDMA carriers such that each cell acquires the minimum number of carriers required to carry the calls. This packing may require channel re-arrangements when the channels are released. A benefit of carrier groups is a reduction in the computation required to make decisions regarding acquisitions and releases. This reduction in complexity reduces the time required to select a channel, thus lowering the probability of dropped call.

Whenever a channel is needed a TDMA DCA scheme follows a strategy which, if necessary, selects a carrier for acquisition according to a **carrier acquisition criterion** . Likewise, when a channel is released another strategy is followed which, if necessary, selects a carrier to be released according to a **carrier release criterion** . The flow charts in Figs. 11.8 and 11.9 illustrate the general procedure for acquiring and releasing channels and carriers. The shaded blocks are steps that support handoff queueing and will be discussed later in the chapter.

ACQUIRE CARRIER Policy (non-queueing case) As Fig. 11.8 shows, the following policy is executed upon a new call or handoff arrival:

1. If at least one idle channel is available among the already acquired carriers, then assign an idle channel to the call; otherwise attempt to acquire a new carrier according to the carrier acquisition criterion.

    (a) If the carrier acquisition is successful, then assign one channel of the newly acquired carrier to the call; otherwise block the call.

RELEASE CARRIER Policy (non-queueing case) As Fig. 11.9 shows, the following policy is executed upon a call completion or a handoff departure:

1. If the channel release will not yield an idle carrier, then no carrier is released; otherwise a carrier is selected for release according to the carrier release criterion.

    (a) The call that occupies the carrier selected for release is reassigned to the newly idle channel, and the selected carrier is released.

11.10.1   The Simple DCA (SDCA) Strategy

Elnoubi *et al.* [83] proposed the BCO strategy that makes use of different channel orderings in each cell. Here, we consider a channel assignment strategy that uses **carrier orderings** rather than channel orderings [332]. However, unlike the BCO strategy, the carriers are not explicitly divided into nominal

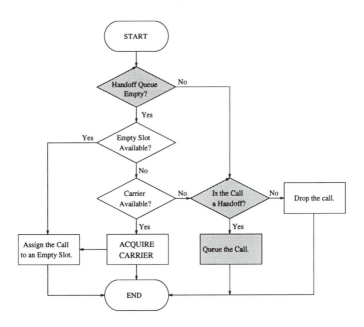

Figure 11.8 ACQUIRE CARRIER policy that is executed upon a new call or handoff arrival.

and borrowed sets with a specified rule for borrowing carriers. In our scheme, each cell has its own carrier ordering, and no two cells separated less than the frequency reuse distance have the same carrier ordering. The orderings are designed so that carriers occurring near the beginning of a cell's carrier ordering occur near the end of the carrier orderings of the cells in its interference neighborhood. For example, suppose there are 9 available carriers with a 3-cell reuse cluster. Three different carrier orderings are necessary to ensure that cells within the frequency reuse distance have distinct carrier orderings. For example, the following carrier orderings will do.

$$
\begin{array}{lll}
A: & \{1,4,7,8,9,5,6,2,3\} & \\
B: & \{2,5,8,9,7,6,4,3,1\} & (11.17) \\
C: & \{3,6,9,7,8,4,5,1,2\} &
\end{array}
$$

These carrier orderings are obtained by first listing the 9 available carriers column-wise until they are all assigned. Then columns 4 and 5 are permutations of the 3rd column, columns 6 and 7 are permutations of the 2nd column, and columns 8 and 9 are permutations of the 1st column. Notice that Carriers 1, 2, and 3 each appear first in one of the orderings and appear near the end of the other two orderings.

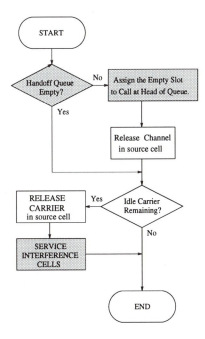

Figure 11.9 RELEASE CARRIER policy that is executed upon a call completion or handoff departure.

The carrier selection criterion is as follows. When a carrier is needed in a cell, the available carrier occurring nearest to the beginning of the cell's carrier ordering is selected. If there are no available carriers, then the carrier acquisition fails. When a carrier is released in a cell, the busy carrier occurring nearest to the end of the cell's carrier ordering is selected. This may require a rearrangement of calls within a cell to carriers that are closer to the beginning of the cell's carrier ordering. The above strategy is hereafter referred to as the **simple dynamic channel assignment** (SDCA) strategy, because of the simplicity of the carrier selection criterion.

Note that the SDCA scheme does not need an exchange of information within the interference neighborhoods. The busy/idle status of carriers can be determined by passive non-intrusive monitoring at each BS.

11.10.2   A Queueing DCA Strategy

Forced terminations of calls in progress are worse than blocking of new calls. Forced terminations or handoff blocking occurs when an active call crosses a cell boundary, and the target cell cannot accommodate the additional call. As described in [148], [109], one way to establish **handoff priority** is to queue the handoff attempts . If the target cell is momentarily unable to accommodate the additional call, the MS maintains its link with the source cell and enters a queue in the target cell. A queue failure occurs when either the signal level drops below some threshold before the call can be serviced by the target cell, the time spent in the queue exceeds a time-out interval, or the queue overflows. A queue success occurs when a channel becomes available and the queue is non-empty. The newly available channel is then assigned to the call at the head of the queue, and a channel within the source cell is released. Here we combine handoff queueing with DCA.

In a DCA strategy, there are two ways for a channel to become available in a cell. Either a call terminates (due to a handoff or completion) or a carrier is released somewhere in the interference neighborhood thus allowing the carrier to be acquired by the cell. When a cell releases a carrier, there may be multiple cells in its interference neighborhood that could acquire the released carrier to service their queued calls. However, the frequency reuse constraint will be violated if all these cells acquire the carrier. To determine which cells may acquire the carrier, we may assume that each cell has a subset of carriers designated as **owned carriers**. The owned carriers are a subset at the beginning of the carrier orderings. Owned carriers are distributed so that no two cells separated less than the frequency reuse distance share any owned carriers. The remaining carriers are designated as **borrowed carriers**. Considering the previous example in (11.17) where 9 carriers were distributed among three carrier orderings, the owned and borrowed carrier orderings are

$$
\begin{array}{ccc}
 & \text{Owned} & \text{Borrowed} \\
A: & \{1,4,7\} & \{8,9,5,6,2,3\} \\
B: & \{2,5,8\} & \{9,7,6,4,3,1\} \\
C: & \{3,6,9\} & \{7,8,4,5,1,2\}
\end{array}
\qquad (11.18)
$$

Cells tend to use their owned carriers before borrowing carriers from other cells. When a cell releases a borrowed carrier, the cells in the interference neighborhood that own the released carrier are given the first opportunity to service their handoff queues. If any of these cells have queued calls, then they can acquire the carrier without violating the frequency reuse constraint. If none of the owner cells in the interference neighborhood acquire the released carrier, then some of the remaining cells in the interference neighborhood may acquire the carrier to service their handoff queues.

A strategy combines DCA and handoff queueing is described below along with the flow charts in Figs. 11.8, 11.9 , 11.10, and 11.11.

ACQUIRE CARRIER Policy (queueing case) Referring to Fig. 11.8, the following policy is executed upon a new call arrival or handoff attempt:

1. If the handoff queue is not empty, then either queue the handoff call or drop the new call; otherwise

   (a) If there is at least one idle channel, then assign an idle channel to the call; otherwise try to acquire a according to the carrier acquisition criterion.

      i. If the carrier acquisition is successful, then assign one channel of the newly acquired carrier to the call; otherwise either queue the handoff call or drop the new call.

RELEASE CARRIER Policy (queueing case) Referring to Fig. 11.9, the following policy is executed upon a call completion, a handoff, or a failure from the handoff queue of an adjacent cell:

1. If the handoff queue is not empty then assign the newly available channel to the call at the head of the handoff queue. The channel currently serving the call is released according to the RELEASE CARRIER policy.

2. If the channel release will yield an idle carrier, then a carrier is selected for release according to the carrier release criterion. The call that occupies the carrier selected for release is reassigned to the newly idle channel, the selected carrier is released, and the SERVICE INTERFERENCE CELLS policy is executed.

SERVICE INTERFERENCE CELLS Policy Referring to Fig. 11.10, the following policy is executed whenever a carrier is released:

1. If a borrowed carrier is released then any owner cell in the interference neighborhood that has a non-empty handoff queue and can acquire the

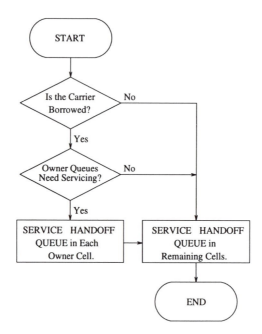

Figure 11.10 SERVICE INTERFERENCE CELLS Policy.

released carrier without violating the reuse constraint, will acquire the carrier and service its handoff queue according to the SERVICE HANDOFF QUEUE policy.

2. After the owner cells are given the opportunity to service their queues, the remaining cells in the interference neighborhood are given the opportunity to service their handoff queues by using the SERVICE HANDOFF QUEUE policy.

SERVICE HANDOFF QUEUE Policy

Referring to Fig. 11.11, whenever a carrier is acquired in a cell having a non-empty handoff queue, the following policy is executed:

1. An empty slot is assigned to the call at the head of the handoff queue, and the channel currently serving the call is released by using the RELEASE CARRIER policy.

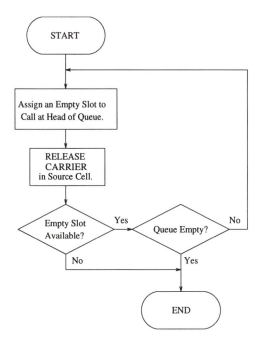

Figure 11.11   SERVICE HANDOFF QUEUE Policy.

(a) Step 1. is executed until either all of the available slots are filled or the handoff queue is empty.

11.10.3   An Aggressive DCA Strategy

DCA strategies increase trunking efficiency by assigning channels to cells as they are needed. Care must be taken to avoid a poor allocation of channels; otherwise capacity will suffer. With SDCA The carriers are acquired and released according to a carrier acquisition and release criteria that attempts to maximize capacity by favoring tightly packed arrangements of *co-carrier* cells. This strategy is similar to the 2-D RING strategy in [154] and suffers from the same problem; when a carrier is selected for acquisition, multiple carriers must be available for the carrier acquisition criteria to yield any advantage. The more carriers available for each selection process the better. At high traffic loads, very few carriers may be available for acquisition. In fact there may be only one or none, in which case there is no choice. Under such conditions, carriers tend to be assigned where they can be, rather than where they should be, and capacity suffers [87]. Under such conditions DCA strategies usually perform *worse* than FCA.

As discussed in Section 11.3.4 the performance at high traffic loads can be improved by using an aggressive policy where, under certain conditions, a cell that cannot acquire a carrier may force a surrounding cell to give up a carrier so that it may service one or more calls. Thus, a cell can actually *take* a carrier from its neighbors if none are otherwise available, according to the following TAKE CARRIER policy.

TAKE CARRIER Policy   Referring to Fig. 11.12, the following policy is executed when a call is in jeopardy due to a handoff failure (in the no queueing case), a queue failure, or the execution of the TAKE CARRIER policy in another cell:

1. The entire set of carriers is examined. If all carriers are being used, then the TAKE CARRIER policy fails; otherwise, all unused carriers are examined, and the number of calls within the interference neighborhood that will be placed in jeopardy by taking each carrier is calculated. Note that the number of jeopardized calls must be determined at each cell in the interference neighborhood. The carrier that will place the fewest number of calls in jeopardy is selected. In event of a tie, the carrier appearing earliest in the carrier ordering is selected. Let C_j be the number of calls that will be placed in jeopardy by taking the selected carrier.

2. The number of calls that will be serviced by taking the selected carrier, C_s is calculated. For a handoff attempt and no queueing, $C_s = 1$; for a queue failure, C_s is the minimum of the number of queued calls C_q and the number of slots per carrier N_s; for a carrier that is lost to another cell executing the TAKE CARRIER policy, C_s ranges from 1 to N_s.

3. If the selected carrier is owned and $C_j > C_s$ or if the selected carrier is not owned and $C_j \geq C_s$, then the TAKE CARRIER policy fails; otherwise all cells in the interference neighborhood that are currently using the selected carrier are told to release it.[2]

4. Each cell in the interference neighborhood releases the selected channel.

5. The selected channel is taken. For a handoff attempt (no queueing), the handoff is completed. For a queue failure, the SERVICE HANDOFF

---

[2] When there is no queueing $C_s = 1$ and, therefore, only an owned carrier can be taken that will not place more than one call in jeopardy.

QUEUE policy is executed. For a carrier that is lost to another cell executing the TAKE CARRIER policy, the slots of the taken carrier are assigned to the calls in jeopardy.

6. Each cell that was forced to release the selected carrier executes the SERVICE INTERFERENCE CELLS policy.

7. Each cell that was forced to release the selected channel and still has calls in jeopardy after the cell taking the carrier services its queue, executes the ACQUIRE CHANNEL policy.

   (a) If the carrier acquisition is successful, then the cell executes the SERVICE HANDOFF QUEUE policy; otherwise it executes the TAKE CARRIER policy.

   (b) If a cell from which a carrier was taken cannot obtain a new carrier, it must drop some of its calls. Queued calls are dropped first because they are in greater danger of being dropped than active calls. If more calls must be dropped after dropping the queued calls, then active calls are dropped until there are no excess calls.

Note that the TAKE CARRIER policy is only executed if the SCDA carrier acquisition criteria fails to acquire a carrier. Unlike SCDA, the TAKE CARRIER policy acquires carriers that place the fewest number of calls in jeopardy. From a practical standpoint it is important to note that the aggressive SCDA strategy requires communication among BSs in the interference neighborhood to execute the TAKE CARRIER policy.

11.10.4   Simulation Model, Results, and Discussion

Consider a microcellular environment consisting of a rectangular grid of intersecting streets, as shown in Fig. 11.13. It is assumed that MS traffic flowing off an edge of the grid wraps around to the opposite edge. However, the interference neighborhoods of each cell do not wrap around. If two cells are on opposite edges of the grid, such that MSs leaving one cell enter the other, they may simultaneously use the same carrier since they are not spatially adjacent.

Line-of-sight co-channel cells must be separated by at least 3 cells. There are no reuse constraints on non line-of-sight co-channel cells, due to the corner

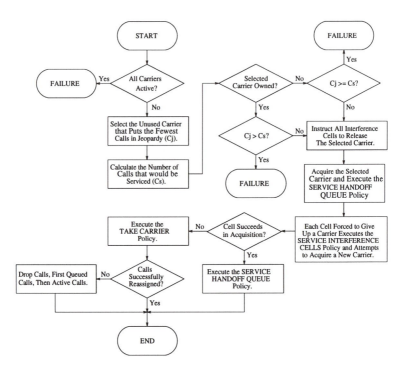

Figure 11.12 TAKE CARRIER Policy.

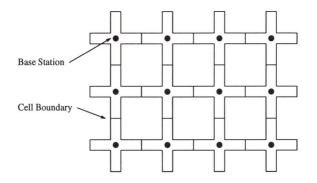

Figure 11.13 Cells and BSs in an urban microcellular environment.

effect. The frequency reuse factor is 4, meaning that the set of carriers must be divided into 4 subsets for FCA, and for SDCA there must be 4 different carrier orderings. The interference neighborhood and cell reuse pattern is shown in Fig. 11.14.

1	2	3	4	1	2	3	4	1
4	1	2	3	4	1	2	3	4
3	4	1	2	3	4	1	2	3
2	3	4	1	2	3	4	1	2
1	2	3	4	1	2	3	4	1
4	1	2	3	4	1	2	3	4
3	4	1	2	3	4	1	2	3
2	3	4	1	2	3	4	1	2
1	2	3	4	1	2	3	4	1

Figure 11.14 Interference neighborhood and cell reuse pattern.

To account for the uneven distribution of teletraffic in the microcellular environment the identical active-dormant Markov model from [230] is used, but modified to account for handoff queueing. The model is Markovian so that all events occur with exponentially distributed interarrival times. However, the parameters of the distributions change with time to reflect the time-varying nature of the model. The state of cell i at any time can be described by the following parameters

New call arrival rate: λ_i
Number of active calls: $N_{\text{active},i}$
Number of queued calls: $N_{\text{queued},i}$

New Call Arrivals: Call arrivals in cell i are Poisson with rate λ_i. This parameter is binary valued, where $\lambda \in \{\lambda_{ACT}, \lambda_{DOR}\}$. These two new call arrival rates correspond to two different cell modes, active and dormant. The arrivals of new calls in different cells are assumed to be independent, so the global call arrival rate is

$$\Lambda = \sum_i \lambda_i \ . \tag{11.19}$$

Call Completions: The duration of each call is exponentially distributed with mean μ. In cell i there are $N_{\text{active},i}$ active calls and $N_{\text{queued},i}$ queued calls, any of which could be completed at any time. These calls are assumed to be independent, so the call completion rate in cell i is

$$r_{c,i} = \frac{N_{\text{active},i} + N_{\text{queued},i}}{\mu} \quad . \tag{11.20}$$

The completion of calls in different cells are assumed to be independent. Therefore, the global call completion rate is:

$$r_c = \sum_i r_{c,i} \quad . \tag{11.21}$$

Handoff Attempts A handoff is attempted whenever an active call crosses a cell boundary and needs to be serviced by the target cell. To determine the handoff rate, it is assumed that each call is handed off an average of h times over its duration. Since the traffic flows wrap around the grid edges, the handoff calls are uniformly distributed to one of the four neighboring cells. Queued calls can be safely assumed to never cross a cell boundary, because the time required to traverse a cell will be much longer than the maximum time allowed in the handoff queue. Therefore, queued calls do not contribute to the handoff rate. The handoff rate in cell i is

$$r_{h,i} = \frac{h N_{\text{active},i}}{\mu} \quad . \tag{11.22}$$

Call handoffs in different cells are assumed to be independent, so the global handoff attempt rate is

$$r_h = \sum_i r_{h,i} \quad . \tag{11.23}$$

Mode Transitions Each cell remains in its current mode for duration D, where D is exponentially distributed with mean $1/\overline{D}$. If the cell is in active mode, then $\overline{D} = \overline{D}_{ACT}$, and if the cell is in dormant mode, then $\overline{D} = \overline{D}_{DOR}$. If there are N_{ACT} active cells and N_{DOR} dormant cells, then the global active-to-dormant and dormant-to-active transition rates are, respectively,

$$r_{ACT \to DOR} = \frac{N_{ACT}}{\overline{D}_{ACT}} \tag{11.24}$$

$$r_{DOR \to ACT} = \frac{N_{DOR}}{\overline{D}_{DOR}} \tag{11.25}$$

The probability of a cell being in the active mode is

$$P_{ACT} = \frac{\overline{D}_{ACT}}{\overline{D}_{ACT} + \overline{D}_{DOR}} \quad . \tag{11.26}$$

As the simulation progresses, five types of events are generated: new call arrivals, call completions, handoff attempts, active-to-dormant mode transitions, and dormant-to-active mode transitions. All events occur independently. Therefore, five random times are generated and the next event corresponds to the one with the minimum time. Once an event is selected, the event must be randomly assigned to a cell. The probability of cell i being selected for each type of event is

New call arrival: λ_i/Λ

Call completion: $r_{c,i}/r_c$

Handoff Attempt: $r_{h,i}/r_h$

ACT $\rightarrow$ DOR transition: $\begin{cases} 1/N_{ACT} & \text{, if cell } i \text{ is active} \\ 0 & \text{, if cell } i \text{ is dormant} \end{cases}$ The active to

DOR $\rightarrow$ ACT transition: $\begin{cases} 1/N_{DOR} & \text{, if cell } i \text{ is dormant} \\ 0 & \text{, if cell } i \text{ is active} \end{cases}$

dormant traffic ratio $R_{ACT/DOR} = \lambda_{ACT}/\lambda_{DOR}$ specifies the ratio of the new call arrival rates in the active and dormant cells. To complete the model, we specify the offered traffic per cell, ρ. Then the active and dormant call arrival rates are:

$$\lambda_{ACT} = \frac{\rho R_{ACT/DOR}}{\mu(1 + P_{ACT}(R_{ACT/DOR} - 1))} \tag{11.27}$$

$$\lambda_{DOR} = \frac{\rho}{\mu(1 + P_{ACT}(R_{ACT/DOR} - 1))} \tag{11.28}$$

where P_{ACT} is as defined in (11.26). The parameters used in the simulations are

Number of cells:	144 (12 x 12 square grid)
Total number of carriers:	40
Number of slots per carrier:	3
Number of channels per cell (FCA):	30
Number of owned carriers per cell (DCA):	10
Average call duration:	120 s
Average number of handoffs per call:	3
Average duration of the ACTIVE mode:	60 s
Average duration of the DORMANT mode:	600 s
ACTIVE to DORMANT traffic ratio:	5
Offered traffic:	0-50 Erlangs
Queue size:	10

as follows

Figs. 11.15 through 11.16 compare the probability of new call blocking and the probability of forced termination for the FCA, SDCA, and aggressive SDCA

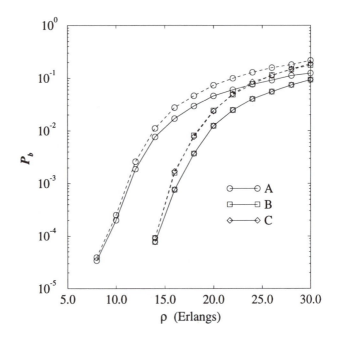

Figure 11.15   Probability of new call blocking versus offered traffic without queueing (solid) and 5-second handoff queue (dashed). Legend: A = FCA, B = SDCA, C = Aggressive SDCA, from [332].

strategies. Results are shown without handoff queueing and with a 5-second handoff queue. Observe from Fig. 11.15 that a substantial reduction in the probability of new call blocking is achieved by using SDCA as compared to FCA. Handoff queueing causes a sight increase in the probability of new call blocking because handoff calls are given priority over new calls when a channel has been released and is available for acquisition. Aggressive SDCA also causes a very slight increase in the probability of new call blocking over non-aggressive SDCA. Fig. 11.16 shows that both handoff queueing and SDCA significantly lower the probability of forced termination. Aggressive SDCA tends to be more effective than the non-aggressive SCDA when these schemes are combined with handoff queueing.

Carrier Acquisitions   It is useful to determine the increase in the rate of carrier acquisitions that results from using aggressive SDCA. Fig. 11.17 plots the cell carrier acquisition rate for aggressive and non-aggressive SDCA. Notice that the carrier acquisition rates at lower traffic loadings are almost the same. At higher traffic loadings hand-off queueing has the largest effect on the carrier

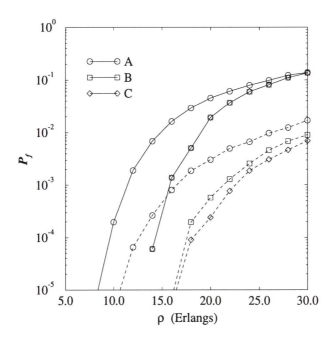

Figure 11.16 Probability of forced termination versus offered traffic without queueing (solid) and a 5-second handoff queue (dashed). Legend: A = FCA, B = SDCA, C = Aggressive SDCA, from [332].

acquisition rate. However, aggressive SDCA causes only a very slight increase in the carrier acquisition rate over non-aggressive SDCA.

The results presented here have been obtained under the assumption that the interference (and DRA) neighborhoods are symmetrical (cell A interferes biconditionally with cell B) and the average traffic loading is identical for all cells. This is not true of a practical system and, therefore, preassigned carrier orderings should not be used. In an actual microcellular system an adaptive, self-organizing algorithm for ordering of carriers and the selection of owned carriers is preferable. Also, an adaptive aggressive strategy may be employed that uses current performance (e.g., the current new call blocking and forced termination probabilities) and perhaps forward-looking strategies [87] to make a more informed decision when to *take* a carrier. It is expected that some performance deterioration will result over the ideal symmetrized case presented here, because of the aforementioned network asymmetries and the finite convergence rate of the adaptive algorithms.

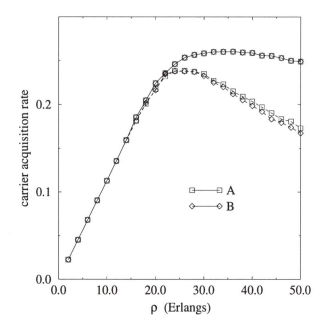

Figure 11.17 Carrier acquisitions per second (per cell) versus offered traffic without queueing (solid) and a 5-second handoff queue (dashed). Legend: A = SDCA, B = DRA, C = Aggressive SDCA, D = Aggressive DRA,from [332].

Finally, the channel assignment strategies as described do not take into consideration the arrangement of calls on the carriers. In reality, the C/I is not the same for each channel. A more effective strategy also arranges the calls in order to combat the unpredictable signal and interference variations present in microcells. Finally, the use of hand-off queueing will exaggerate the cell boundaries thereby causing increased co-channel interference. Unlike the purely statistical model that is used here, the study of these issues will require explicit models for the mobility of MSs and the radio propagation environment.

11.11 CONCLUDING REMARKS

Although it is very difficult to arrive at a consensus as to what the best channel assignment algorithm is, an effective DCA algorithm should possess distributed control mechanisms, handoff prioritization, high channel utilization, and stability. Unfortunately, there is no single DCA algorithm that combines all these features and the *best* solution is sure to depend on the service area characteristics. For example, cordless phones require a fully decentralized algorithm while urban microcells should allow some limited communication among BSs. Although some very interesting DCA schemes have been proposed in the literature, not all the issues have been sufficiently addressed to make them practical. Many of the current systems either require too much computation and communication among BSs, yield low channel utilization, or exhibit instability. As a result, DCA will be an active area of research for some time.

In general, the analytical treatment of DCA algorithms is quite difficult and few results have appeared in the literature. Most of DCA algorithms are derived on an ad hoc basis and evaluated by computer simulation. The development of new analytical tools is important for systematic development and will yield valuable insight into the performance of new DCA algorithms.

Much of the existing literature has separated the handoff problem from the channel assignment problem. However, these two problems linked and it is desirable that they receive unified treatment. For example, one performance measure for a handoff algorithm is the mean number of handoffs against the handoff delay. However, such an analysis usually proceeds under the assumption that a channel will always be available for a handoff. Clearly, this is not the case in practice. Most channel assignment schemes are designed for a single application, i.e., voice services. However, future systems will have to support a variety of multimedia applications that have different GOS requirements and require different types and amounts of network resources including channel resources, delay, etc. The channel assignment problem for multimedia services is an open area for research.

Problems

11.1. Suppose that the maximum packing (MP) policy is used with the system shown in Fig. 11.1. Suppose that 10 channels are available for use within each of the three reuse clusters $CL_i, i = 1, 2, 3$.

a) Compute the number of admissible states $|\mathcal{S}|$.

b) By using the approximation in (11.6) compute the approximate blocking probabilities for each cell assuming a traffic load of $\rho = 2$ Erlangs in each cell.

c) Compare the blocking probabilities in part b) with FCA for the same traffic load.

11.2. Show that the maximum and minimum saturation densities for a linear array of cells is $C_{\max} = (R + 1)^{-1}$ and $C_{\min} = (2R + 1)^{-1}$.

11.3. Show that the maximum and minimum saturation densities for a planar array of cells are

$$C_{\min} = \frac{1}{1 + 3R(R + 1)}$$

$$C_{\max} = \begin{cases} \frac{4}{3(R+1)^2} & , \quad R \text{ even} \\ \frac{4}{1+3(R+1)^2} & , \quad R \text{ odd} \end{cases} .$$

11.4. Consider a linear array of cells with $R = 1$ and a total of $N_T = 24$ channels. Plot the blocking probability, P_b, against the offered traffic per cell, ρ, with FCA, timid DCA, and aggressive DCA. What conclusions can you make?

11.5. Derive equation (11.12).

PROBABILITY AND RANDOM PROCESSES

The theory of probability and random processes is essential in the design and performance analysis of communication systems. This Appendix presents a brief review of the basic concepts of probability theory and random processes. It is intended that most readers have already had some exposure to probability and random processes, so that this Appendix is intended to provide a brief overview. A very thorough treatment of this subject is available in a large number of textbooks, including [242], [186].

A.1 CONDITIONAL PROBABILITY AND BAYES' THEOREM

Let A and B be two events in a sample space S. The **conditional probability** of A given B is

$$\mathrm{P_r}(A|B) = \frac{\mathrm{P_r}(A \bigcap B)}{\mathrm{P_r}(B)} \tag{A.1}$$

provided that $\mathrm{P_r}(B) \neq 0$. If $\mathrm{P_r}(B) = 0$, then $\mathrm{P_r}(A|B)$ is undefined.

There are several special cases.

- If $A \bigcap B = \emptyset$, then events A and B are **mutually exclusive**, i.e., if B occurs then A could not have occurred and $\mathrm{P_r}(A|B) = 0$.

- If $B \subset A$, then knowledge that event B has occurred implies that event A has occurred and so $\mathrm{P_r}(A|B) = 1$.

- If A and B are **statistically independent**, then $P_r(A \cap B) = P_r(A)P_r(B)$ and so $P_r(A|B) = P_r(A)$.

There is a strong connection between mutually exclusive and independent events. It may seem that mutually exclusive events are independent, but just the exact opposite is true. Consider two events A and B with $P_r(A) > 0$ and $P_r(B) > 0$. If A and B are mutually exclusive, then $A \cap B = 0$ and $P_r(A \cap B) = 0 \neq P_r(A)P_r(B)$. Therefore, mutually exclusive events with non-zero probability cannot be independent. Thus disjointness of events is a property of the events themselves, while independence is a property of their probabilities.

In general, the events $A_i, i = 1, \ldots, n$, are independent if and only if for all collections of k distinct integers $(i_1, i_2, \ldots, i_k)$ chosen from the set $(1, 2, \ldots, n)$, we have

$$P_r \left(A_{i_1} \cap A_{i_2} \cap \cdots \cap A_{i_k} \right) = P_r(A_{i_1})P_r(A_{i_2}) \cdots P_r(A_{i_k})$$

for $2 \leq k \leq n$.

In summary

- If $A_i, i = 1, \ldots, n$ is a sequence of mutually exclusive events, then

$$P_r \left(\bigcup_{i=1}^{n} \right) = \sum_{i=1}^{n} P_r(A_i) \ . \tag{A.2}$$

- If $A_i, i = 1, \ldots, n$ is a sequence of independent events, then

$$P_r \left(\bigcap_{i=1}^{n} \right) = \prod_{i=1}^{n} P_r(A_i) \ . \tag{A.3}$$

Total Probability

The collection of sets $\{B_i\}$, $i = 1, \ldots, n$ forms a *partition* of the sample space S if $B_i \cap B_j = \emptyset$, $i \neq j$ and $\bigcup_{i=1}^{n} B_i = S$. For any event $A \subset S$ we can write

$$A = \bigcup_{i=1}^{n} (A \cap B_i) \ . \tag{A.4}$$

That is, every element of A is contained in one and only one B_i. Since $(A \cap B_i) \cap (A \cap B_j) = \emptyset$, $i \neq j$, the sets $A \cap B_i$ are mutually exclusive. Therefore,

$$
\begin{aligned}
\Pr(A) &= \sum_{i=1}^{n} \Pr(A \cap B_i) \\
&= \sum_{i=1}^{n} \Pr(A|B_i)\Pr(B_i) \ .
\end{aligned}
\tag{A.5}
$$

This last equation is often referred to as total probability.

Bayes' Theorem

Let the events B_i, $i = 1, \ldots, n$ be mutually exclusive such that $\bigcup_{i=1}^{n} B_i = S$, where S is the sample space. Let A be an event with nonzero probability. Then as a result of conditional probability and total probability:

$$
\begin{aligned}
\Pr(B_i|A) &= \frac{\Pr(B_i \cap A)}{\Pr(A)} \\
&= \frac{\Pr(A|B_i)\Pr(B_i)}{\sum_{i=1}^{n} \Pr(A|B_i)\Pr(B_i)} \ .
\end{aligned}
$$

a result known as Bayes' theorem.

A.2 MEANS, MOMENTS, AND MOMENT GENERATING FUNCTIONS

The kth **moment** of a random variable, $E[X^k]$, is defined as

$$
E[X^k] \triangleq \begin{cases} \sum_{x_i \in R_X} x_i^k p_X(x_i) & \text{if } X \text{ is discrete} \\ \\ \int_{R_X} x^k p_X(x)\mathrm{d}x & \text{if } X \text{ is continuous} \end{cases}
\tag{A.6}
$$

where $p_X(x_i) \triangleq \Pr(X = x_i)$ is the **probability distribution function** of X, and $p_X(x) \triangleq \Pr(X = x)$ is the **probability density function** (pdf) of X. The kth **central moment** of the random variable X is $E[(X - E[X])^k]$. The **variance** is the second central moment.

The **moment generating function** or **characteristic function** of a random variable X is

$$\psi_X(jv) \triangleq E[e^{jvX}] = \begin{cases} \sum_{x_i \in R_X} e^{jvx_i} p_X(x_i) & \text{if } X \text{ is discrete} \\ \int_{R_X} e^{jvx} p_X(x) dx & \text{if } X \text{ is continuous} \end{cases} \quad (A.7)$$

where $j = \sqrt{-1}$. Note that the continuous version is a Fourier transform, except for the sign in the exponent. Likewise, the discrete version is a z-transform, except for the sign in the exponent.

The probability distribution and probability density functions of discrete and continuous random variables, respectively, can be obtained by taking the inverse transforms of the characteristic functions, i.e.,

$$p_X(x) = \frac{1}{2\pi} \int_{-\infty}^{\infty} \psi_X(jv) e^{-jvx} dv \quad (A.8)$$

and

$$p_X(x_k) = \frac{1}{2\pi} \oint_C \psi_X(jv) e^{-jvx_k} dv \quad . \quad (A.9)$$

The **cumulative distribution function** (cdf) of a random variable X is defined as

$$F_X(x) \triangleq P_r(X \le x) = \begin{cases} \sum_{x_i \le x} p_X(x_i) & \text{if } X \text{ is discrete} \\ \int_{-\infty}^{x} p_X(x) dx & \text{if } X \text{ is continuous} \end{cases} \quad . \quad (A.10)$$

The **complementary distribution function** (cdfc) is defined as

$$F_X^c(x) \triangleq 1 - F_X(x) \quad . \quad (A.11)$$

A.3 SOME USEFUL PROBABILITY DISTRIBUTIONS

A.3.1 Discrete Distributions

Binomial Distribution

Let X be a **Bernoulli random variable** such that $X = 0$ with probability $1 - p$ and $X = 1$ with probability p. Although X is a discrete random random variable with an associated probability distribution function, it is possible to treat X as a continuous random variable with a pdf by using dirac delta functions. In this case, the pdf of X has the form

$$p_X(x) = (1-p)\delta(x) + p\delta(x-1) . \tag{A.12}$$

Let $Y = \sum_{i=1}^n X_i$, where the X_i are independent and identically distributed with density $p_X(x)$. Then the random variable Y is an integer from the set $\{0, 1, \ldots, n\}$ and the probability distribution of Y is

$$p_Y(k) = P(Y = k) = \binom{n}{k} p^k (1-p)^{n-k}, \quad k = 0, 1, \ldots, n . \tag{A.13}$$

The random variable Y also has the pdf

$$p_Y(y) = \sum_{k=0}^n \binom{n}{k} p^k (1-p)^{n-k} \delta(y-k) . \tag{A.14}$$

Poisson Distribution

The random variable X has a **Poisson distribution** if

$$p_X(k) = \frac{\lambda^k e^{-\lambda}}{k!}, \quad k = 0, 1, \ldots, \infty \tag{A.15}$$

Geometric Distribution

The random variable X has a **geometric distribution** if

$$p_X(k) = (1-p)^{k-1} p, \quad k = 1, 2, \ldots, \infty . \tag{A.16}$$

A.3.2 Continuous Distributions

Many communication systems are affected by additive Gaussian noise. Therefore, the Gaussian distribution and various functions of Gaussian distributions play a central role in the characterization and analysis of communication systems.

Gaussian Distribution

A Gaussian random variable has the pdf

$$p_X(x) = \frac{1}{\sqrt{2\pi}\sigma} e^{-\frac{(x-\mu)^2}{2\sigma^2}} \tag{A.17}$$

where $\mu = E[X]$ is the mean and $\sigma^2 = E[(X - \mu)^2]$ is the variance. Sometimes we use the shorthand notation $X \sim N(\mu, \sigma^2)$ meaning that X is a Gaussian random variable with mean μ and variance σ^2. The random variable X is said to have a standard normal distribution if $X \sim N(0, 1)$.

The cumulative distribution function (cdf) of X is

$$F_X(x) = \int_{-\infty}^{x} \frac{1}{\sqrt{2\pi}\sigma} e^{-\frac{(y-\mu)^2}{2\sigma^2}} dy \ . \tag{A.18}$$

The cdf of a standard normal distribution defines the Q function

$$Q(x) \triangleq \int_{x}^{\infty} \frac{1}{\sqrt{2\pi}} e^{-y^2/2} dy \tag{A.19}$$

and the cdfc defines the Φ function

$$\Phi(x) \triangleq 1 - Q(x) \ . \tag{A.20}$$

If X is a non-standard normal random variable, $X \sim N(\mu, \sigma^2)$, then

$$F_X(x) \ = \ \Phi\left(\frac{x - \mu}{\sigma}\right) \tag{A.21}$$

$$F_X^c(x) \ = \ Q\left(\frac{x - \mu}{\sigma}\right) \ . \tag{A.22}$$

Quite often the cumulative distribution function of a Gaussian random variable is described in terms of the **error function** $\mathrm{erf}(x)$ and the **complementary error function** $\mathrm{erfc}(x)$, defined by

$$\mathrm{erfc}(x) \triangleq \frac{2}{\sqrt{\pi}} \int_{x}^{\infty} e^{-y^2} dy \triangleq 1 - \mathrm{erf}(x) \ . \tag{A.23}$$

The complementary error function and Q function are related as follows

$$\mathrm{erfc}(x) \ = \ 2Q(\sqrt{2}(x)) \tag{A.24}$$

$$Q(x) \ = \ \frac{1}{2}\mathrm{erfc}\left(\frac{x}{\sqrt{2}}\right) \ . \tag{A.25}$$

Rayleigh Distribution

Let $X_1 \sim N(0, \sigma^2)$ and $X_2 \sim N(0, \sigma^2)$ be independent normal random variables. The random variable $R = \sqrt{X_1^2 + X_2^2}$ is said to be Rayleigh distributed.

To find the pdf and cdf of R first define

$$V = \text{Tan}^{-1}\left(\frac{X_2}{X_1}\right) .$$

Then

$$X_1 = R\cos V$$
$$X_2 = R\sin V .$$

By using a bivariate transformation of random variables

$$p_{RV}(r, v) = p_{X_1 X_2}(r\cos v, r\sin v)\,|J(r, v)|$$

where

$$J(r, v) = \begin{vmatrix} \frac{\partial x_1}{\partial r} & \frac{\partial x_1}{\partial v} \\ \frac{\partial x_2}{\partial r} & \frac{\partial x_2}{\partial v} \end{vmatrix} = \begin{vmatrix} \cos v & r\sin v \\ \sin v & r\cos v \end{vmatrix} = r(\cos^2 v + \sin^2 v) = r$$

Since

$$p_{X_1 X_2}(x_1, x_2) = \frac{1}{2\pi\sigma^2}e^{-\frac{x_1^2 + x_2^2}{2\sigma^2}}$$

we have

$$p_{RV}(r, v) = \frac{r}{2\pi\sigma^2}e^{-\frac{r^2}{2\sigma^2}} . \tag{A.26}$$

It follows that the marginal pdf of R is

$$\begin{aligned} p_R(r) &= \int_0^{2\pi} p_{RV}(r, v)dv \\ &= \frac{r}{\sigma^2}e^{-\frac{r^2}{2\sigma^2}} \qquad r \geq 0 . \end{aligned} \tag{A.27}$$

The cdf is

$$F_R(r) = 1 - e^{-\frac{r^2}{2\sigma^2}} \quad r \geq 0 . \tag{A.28}$$

Rice Distribution

Let $X_1 \sim N(\mu_1, \sigma^2)$ and $X_2 \sim N(\mu_2, \sigma^2)$ be independent normal random variables with non-zero means. The random variable $R = \sqrt{X_1^2 + X_2^2}$ has a Rice distribution or is said to be Ricean distributed. To find the pdf and cdf of R again define $V = \text{Tan}^{-1}(X_2/X_1)$. Then by using a bivariate transformation $J(r, v) = r$ and

$$p_{RV}(r, v) = r \cdot p_{X_1 X_2}(r\cos v, r\sin v) . \tag{A.29}$$

However,

$$p_{X_1 X_2}(x_1, x_2) = \frac{1}{2\pi\sigma^2} \exp\left\{ -\frac{(x_1 - \mu_1)^2 + (x_2 - \mu_2)^2}{2\sigma^2} \right\}$$

$$= \frac{1}{2\pi\sigma^2} \exp\left\{ -\frac{x_1^2 + x_2^2 + \mu_1^2 + \mu_2^2 - 2(x_1\mu_1 + x_2\mu_2)}{2\sigma^2} \right\} .$$

Hence,

$$p_{RV}(r, v) = \frac{1}{2\pi\sigma^2} \exp\left\{ -\frac{r^2 + \mu_1^2 + \mu_2^2 - 2r(\mu_1 \cos v + \mu_2 \sin v)}{2\sigma^2} \right\} .$$

Now define $s \triangleq \sqrt{\mu_1^2 + \mu_2^2}$ and $t \triangleq \operatorname{Tan}^{-1}\mu_2/\mu_1$, $0 \le t \le 2\pi$, so that $\mu_1 = s \cos t$ and $\mu_2 = s \sin t$. Then

$$p_{RV}(r, v) = \frac{1}{2\pi\sigma^2} \exp\left\{ -\frac{r^2 + s^2 - 2rs(\cos t \cos v + \sin t \sin v)}{2\sigma^2} \right\}$$

$$= \frac{1}{2\pi\sigma^2} \exp\left\{ -\frac{r^2 + s^2 - 2rs\cos(v - t)}{2\sigma^2} \right\} .$$

The marginal pdf of R is

$$P_R(r) = \frac{r}{\sigma^2} e^{-\frac{(r^2+s^2)}{2\sigma^2}} \cdot \frac{1}{2\pi} \int_0^{2\pi} e^{-\frac{rs}{\sigma^2}\cos(v-t)} dv . \tag{A.30}$$

The zero order modified Bessel function of the first kind is defined as

$$I_0(x) \triangleq \frac{1}{2\pi} \int_0^{2\pi} e^{-x \cos\theta} d\theta . \tag{A.31}$$

Therefore,

$$P_R(r) = \frac{r}{\sigma^2} e^{-\frac{(r^2+s^2)}{2\sigma^2}} \cdot \frac{1}{2\pi} I_0\left(\frac{rs}{\sigma^2}\right), \qquad r \ge 0 . \tag{A.32}$$

The cdf of R is

$$F_R(r) = \int_0^r p_R(r) dr$$

$$= 1 - Q\left(\frac{s}{\sigma}, \frac{r}{\sigma}\right)$$

where $Q(a, b)$ is called the Marcum Q-function.

Central Chi-Square Distribution

Let $X \sim N(0, \sigma^2)$ and $Y = X^2$. Then it can be shown that

$$p_Y(y) = \frac{p_X(\sqrt{y}) + p_X(-\sqrt{y})}{2\sqrt{y}}$$

$$= \frac{1}{\sqrt{2\pi y}\sigma} e^{-\frac{y}{2\sigma^2}} \qquad y \geq 0 \ .$$

The characteristic function of Y is

$$\psi_Y(jv) = \int_{-\infty}^{\infty} e^{jvy} p_Y(y) dy$$

$$= \frac{1}{\sqrt{1 - j2v\sigma^2}} \ . \qquad\qquad (A.33)$$

Now define the random variable $Y = \sum_{i=1}^{n} X_i^2$, where the X_i are independent and $X_i \sim N(0, \sigma^2)$. Then

$$\psi_Y(jv) = \frac{1}{(1 - j2v\sigma^2)^{n/2}} \ . \qquad\qquad (A.34)$$

Taking the inverse transform gives

$$p_Y(y) = \frac{1}{2\pi} \int_{-\infty}^{\infty} \psi_Y(jv) e^{-jvy} dv$$

$$= \frac{1}{(2\sigma^2)^{n/2}\Gamma(n/2)} y^{n/2-1} e^{-\frac{y}{2\sigma^2}}, \quad y \geq 0 \ .$$

where $\Gamma(k)$ is the Gamma function and

$$\Gamma(k) = \int_0^{\infty} u^{k-1} e^{-u} du = (k-1)!$$

if k is a positive integer. If n is even (which is usually the case in practice) and we define $m = n/2$, then the pdf of Y defines the central chi-square distribution

$$p_Y(y) = \frac{1}{(2\sigma^2)^m (m-1)!} y^{m-1} e^{-\frac{y^2}{2\sigma^2}} \qquad y \geq 0 \qquad (A.35)$$

and the cdf of Y is

$$F_Y(y) = 1 - e^{-\frac{y}{2\sigma^2}} \sum_{k=0}^{m-1} \frac{1}{k!} \left(\frac{y}{2\sigma^2}\right)^k \qquad y \geq 0 \ . \qquad (A.36)$$

The **exponential distribution** is a special case of the central chi-square distribution when $m = 1$. In this case

$$p_Y(y) = \frac{1}{2\sigma^2} e^{-\frac{y}{2\sigma^2}}$$

$$F_Y(y) = 1 - e^{-\frac{y}{2\sigma^2}} . \tag{A.37}$$

Non-Central Chi-Square Distribution

Let $X \sim N(\mu, \sigma^2)$ and $Y = X^2$. Then

$$p_Y(y) = \frac{p_X(\sqrt{y}) + p_X(-\sqrt{y})}{2\sqrt{y}}$$

$$= \frac{1}{\sqrt{2\pi y}\sigma} e^{-\frac{(y+\mu^2)}{2\sigma^2}} \cosh\left(\frac{\sqrt{y}\mu}{\sigma^2}\right), \quad y \geq 0 .$$

The characteristic function of Y is

$$\psi_Y(jv) = \int_{-\infty}^{\infty} e^{jvy} p_Y(y) dy$$

$$= \frac{1}{\sqrt{1 - j2v\sigma^2}} \exp\left\{\frac{j\mu^2 v}{1 - j2v\sigma^2}\right\} .$$

Now define the random variable $Y = \sum_{i=1}^{n} X_i^2$, where the X_i are independent normal random variables and $X_i \sim N(\mu_i, \sigma^2)$. Then

$$\psi_Y(jv) = \frac{1}{(1 - j2v\sigma^2)^{n/2}} \exp\left\{\frac{jv \sum_{i=1}^{n} \mu_i^2}{1 - j2v\sigma^2}\right\} .$$

Taking the inverse transform gives

$$p_Y(y) = \frac{1}{2\sigma^2} \left(\frac{y}{s^2}\right)^{\frac{n-2}{4}} e^{-\frac{(s^2+y)}{2\sigma^2}} I_{n/2-1}\left(\sqrt{y}\frac{s}{\sigma^2}\right) \quad y \geq 0$$

where

$$s = \sum_{i=1}^{n} \mu_i^2$$

and $I_k(x)$ is the modified Bessel function of the first kind and order k, defined by

$$I_k(x) \triangleq \frac{1}{2\pi} \int_0^{2\pi} e^{x\cos\theta} \cos(n\theta) d\theta .$$

If n is even (which is usually the case in practice) and we define $m = n/2$, then the pdf of Y is

$$p_Y(y) = \frac{1}{2\sigma^2}\left(\frac{y}{s^2}\right)^{\frac{m-1}{2}} e^{-\frac{(s^2+y)}{2\sigma^2}} I_{m-1}\left(\sqrt{y}\frac{s}{\sigma^2}\right), \quad y \geq 0 \tag{A.38}$$

and the cdf of Y is

$$F_Y(y) = 1 - Q_m\left(\frac{s}{\sigma}, \frac{\sqrt{y}}{\sigma}\right) \tag{A.39}$$

where $Q_m(a, b)$ is called the generalized Q-function.

Multivariate Gaussian Distribution

Let $X_i \sim N(\mu_i, \sigma_i^2), i = 1, \ldots, n$, be correlated Gaussian random variables covariances

$$\begin{aligned}\mu_{X_i X_j} &= E\left[(X_i - \mu_i)(X_j - \mu_j)\right] \\ &= E[X_i X_j] - \mu_i \mu_j, \quad 1 \leq i, j \leq n .\end{aligned}$$

Let

$$\begin{aligned}\mathbf{X} &= (X_1, X_2, \ldots, X_n)^T \\ \boldsymbol{\mu}_X &= (\mu_1, \mu_2, \ldots, \mu_n)^T \\ \boldsymbol{\Lambda} &= \begin{bmatrix} \mu_{X_1 X_1} & \cdot & \cdot & \cdot & \mu_{X_1 X_n} \\ \vdots & & & & \vdots \\ \mu_{X_n X_1} & \cdot & \cdot & \cdot & \mu_{X_n X_n} \end{bmatrix}\end{aligned}$$

where $\mathbf{X}^T$ is the transpose of $\mathbf{X}$. Then the joint pdf of $\mathbf{X}$ defines the multivariate Gaussian distribution

$$p_{\mathbf{X}}(\mathbf{x}) = \frac{1}{(2\pi)^{n/2}|\Lambda|^{1/2}} \exp\left\{-\frac{1}{2}(\mathbf{x} - \boldsymbol{\mu}_X)^T \Lambda^{-1}(\mathbf{x} - \boldsymbol{\mu}_X)\right\} \tag{A.40}$$

where $|\Lambda|$ is the determinant of Λ.

A.4 UPPER BOUND ON THE CDFC

Several different approaches can be used to upper bound the area under the tails of a probability density function including the Chebyshev and Chernoff bounds.

Chebyshev Bound

The Chebyshev bound is derived as follows. Let X be a random variable with mean μ_X, variance σ_X^2, and pdf $p_X(x)$. Then the variance of X is

$$
\begin{aligned}
\sigma_X^2 &= \int_{-\infty}^{\infty} (x - \mu_X)^2 p_X(x) dx \\
&\geq \int_{|x - \mu_X| \geq \delta} (x - \mu_X)^2 p_X(x) dx \\
&\geq \delta^2 \int_{|x - \mu_X| \geq \delta} p_X(x) dx \\
&= \delta^2 P(|X - \mu_X| \geq \delta) .
\end{aligned}
$$

Hence,

$$
P_r(|X - \mu_X| \geq \delta) \leq \frac{\sigma_X^2}{\delta^2} . \tag{A.41}
$$

The Chebyshev bound is straightforward to apply but it tends to be quite loose.

Chernoff Bound

The Chernoff bound is more difficult to compute but is much tighter than the Chebyshev bound. To derive the Chernoff bound we use the following inequality

$$
u(x) \leq e^{\lambda x}, \qquad \forall \, x \text{ and } \forall \, \lambda \geq 0
$$

where $u(x)$ is the unit step function. Then,

$$
\begin{aligned}
P(X \geq 0) &= \int_0^{\infty} p_X(x) dx \\
&= \int_{-\infty}^{\infty} u(x) p_X(x) dx \\
&\leq \int_{-\infty}^{\infty} e^{\lambda x} p_X(x) dx \\
&= E[e^{\lambda X}] .
\end{aligned}
$$

The **Chernoff bound parameter** $\lambda > 0$ can be optimized to give the tightest upper bound. This can be accomplished by setting the derivative to zero

$$
\frac{d}{d\lambda} E[e^{\lambda X}] = E\left[\frac{d}{d\lambda} e^{\lambda X} \right] = E[X e^{\lambda X}] = 0 .
$$

Let $\lambda^* = \arg\min_{\lambda \geq 0} \mathrm{E}[e^{\lambda X}]$ be the solution to the above equation. Then

$$P(X \geq 0) \leq \mathrm{E}[e^{\lambda^* X}] . \tag{A.42}$$

Example A.1

Let X_i, $i = 1, \ldots, n$ be independent and identically distributed random variables with density

$$p_X(x) = p\delta(x - 1) + (1 - p)\delta(x + 1) .$$

Let

$$Y = \sum_{i=1}^{n} X_i .$$

Then

$$
\begin{aligned}
\mathrm{P_r}(Y \geq 0) &= \mathrm{P_r}\left(\lceil n/2 \rceil \text{ or more of the } X_i \text{ are ones }\right) \\
&= \sum_{k=\lceil n/2 \rceil}^{n} \binom{n}{k} p^k (1 - p)^{n-k} .
\end{aligned}
$$

For $n = 10$ and $p = 0.1$

$$\mathrm{P_r}(Y \geq 0) = 0.0016349 . \tag{A.43}$$

Chebyshev Bound

To compute the Chebyshev bound we first determine the mean and variance of Y.

$$
\begin{aligned}
\mu_Y &= n\mathrm{E}[X_i] \\
&= n[p - 1 + p] \\
&= n(2p - 1) .
\end{aligned}
$$

$$
\begin{aligned}
\sigma_Y^2 &= n\sigma_X^2 \\
&= n\left(\mathrm{E}[X_i^2] - \mathrm{E}^2[X_i]\right) \\
&= n\left(1 - (2p - 1)^2\right) \\
&= n\left(1 - 4p^2 + 4p - 1\right) \\
&= 4np(1 - p) .
\end{aligned}
$$

Hence,

$$\Pr(|Y - \mu_Y| \geq \mu_Y) \leq \frac{\sigma_Y^2}{\mu_Y^2} = \frac{4np(1-p)}{n^2(2p-1)^2} \ .$$

Then by symmetry

$$\Pr(Y \geq 0) = \frac{1}{2} P(|Y - \mu_Y| \geq \mu_Y)$$
$$\leq \frac{2p(1-p)}{n(2p-1)^2} \ .$$

For $n = 10$ and $p = 0.1$

$$P(Y \geq 0) \leq 0.028125 \ . \tag{A.44}$$

Chernoff Bound

The Chernoff bound is given by

$$\Pr(Y \geq 0) \leq \mathrm{E}[e^{\lambda y}]$$
$$= \left(\mathrm{E}[e^{\lambda x}]\right)^n \ .$$

However,

$$\mathrm{E}[e^{\lambda x}] = pe^\lambda - (1-p)e^{-\lambda} \ .$$

To find the optimal Chernoff bound parameter we solve

$$\frac{d}{d\lambda} \mathrm{E}[e^{\lambda x}] = pe^\lambda - (1-p)e^{-\lambda} = 0$$

giving

$$\lambda^* = \ln\left(\sqrt{\frac{1-p}{p}}\right) \ .$$

Hence,

$$\Pr(Y \geq 0) \leq \left(\mathrm{E}[e^{\lambda^* x}]\right)^n$$
$$= [4p(1-p)]^{n/2} \ .$$

For $n = 10$ and $p = 0.1$

$$P(Y \geq 0) \leq 0.0060466 \ .$$

Notice that the Chernoff bound is much tighter that the Chebyshev bound in this case.

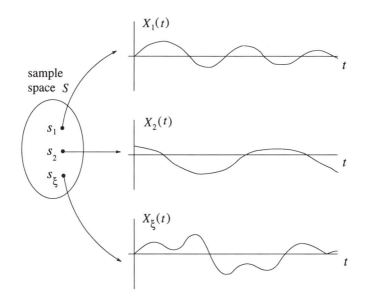

Figure A.1 Ensemble of sample functions for a random process.

A.5 RANDOM PROCESSES

A random process, or stochastic process, $X(t)$, is an ensemble of sample functions $\{X_1(t), X_2(t), \ldots, X_\xi(t)\}$ together with a probability rule which assigns a probability to any event associated with the observation of these functions. Consider the set of sample functions shown in Fig. A.1. The sample function x_i corresponds to the sample point s_1 in the sample space and, occurs with probability $P_r(s_1)$. The number of sample functions, ξ, in the ensemble may be finite or infinite. The function $X_i(t)$ is deterministic once the index i is known. Sample functions may be defined at discrete or continuous instants in time, and their values (parameters) at these time instants may be discrete or continuous in time also.

Suppose that we observe all the sample functions at some time instant t_1. and their values form the set of numbers $\{X_i(t_1)\}, i = 1, 2, \ldots, \xi$. Since $X_i(t_1)$ occurs with probability $P_r(s_i)$, the collection of numbers $\{X_i(t_1)\}, i = 1, 2, \ldots, \xi$ forms a random variable, denoted by $X(t_1)$. By observing the set of waveforms at another time instant t_2 we obtain a different random variable $X(t_2)$. A collection of n such random variables, $X(t_1)$, $X(t_2)$, \ldots, $X(t_n)$, has the joint cdf

$$F_{X(t_1),\dots,\ X(t_n)}(x_1,\dots,\ x_n) = \Pr(X(t_1) < x_1,\dots,X(t_n) < t_n)\ .$$

A more compact notation can be obtained by defining the vectors

$$\mathbf{x} = (x_1, x_2, \dots, x_n)$$
$$\mathbf{X}(t) = (X(t_1), X(t_2), \dots, X(t_n))\ .$$

Then the joint cdf and joint pdf are, respectively,

$$F_{\mathbf{X}(t)}(\mathbf{x}) = P(\mathbf{X}(t) \le \mathbf{x}) \tag{A.45}$$

$$p_{\mathbf{X}(t)}(\mathbf{x}) = \frac{\partial^n F_{\mathbf{X}(t)}(\mathbf{x})}{\partial x_1 \partial x_2 \cdots \partial x_n}\ . \tag{A.46}$$

A random process is **strictly stationary** if and only if the joint density function $p_{\mathbf{X}(t)}(\mathbf{x})$ is invariant under shifts of the time origin. In this case, the equality

$$p_{\mathbf{X}(t)}(\mathbf{x}) = p_{\mathbf{X}(t+\tau)}(\mathbf{x}) \tag{A.47}$$

holds for all sets of time instants $\{t_1, t_2, \dots, t_n\}$ and all time shifts τ. Many important random processes that are encountered in practice are strictly stationary.

A.5.1   Moments and Correlation Functions

To describe the moments and correlation functions of a random process, it is useful to define the following two operators

$$\mathrm{E}[\ \cdot\ ] \triangleq \text{ ensemble average}$$
$$<\ \cdot\ > \triangleq \text{ time average }.$$

The ensemble average of a random process at time t is

$$\mu_X(t) \equiv \mathrm{E}[X(t)] = \int_{-\infty}^{\infty} x p_{X(t)}(x) dx\ . \tag{A.48}$$

The time average of a random process is

$$<X(t)> = \lim_{T \to \infty} \frac{1}{2T} \int_{-T}^{T} X(t) dt\ . \tag{A.49}$$

In general, the time average $< X(t) >$ is also a random variable, because it depends on the particular sample function that is selected for time averaging.

The **autocorrelation** of a random process $X(t)$ is defined as

$$\phi_{XX}(t_1, t_2) = E\left[X(t_1)X(t_2)\right] \ . \tag{A.50}$$

The **autocovariance** of a random process $X(t)$ is defined as

$$
\begin{aligned}
\mu_{XX}(t_1, t_2) &= E\left[(X(t_1) - \mu_X(t_1))\left(X(t_2) - \mu_X(t_2)\right)\right] \\
&= \phi_{XX}(t_1, t_2) - \mu_X(t_1)\mu_X(t_2) \ .
\end{aligned} \tag{A.51}
$$

A random process is strictly stationary only if

$$E[X^n(t)] = E[X^n] \quad \forall \, t, n \ .$$

Hence, for a strictly stationary random process

$$
\begin{aligned}
\mu_X(t) &= \mu \\
\sigma_X^2(t) &= \sigma_X^2 \\
\phi_{XX}(t_1, t_2) &= \phi_{XX}(t_1 - t_2) \equiv \phi_{XX}(\tau) \\
\mu_{XX}(t_1, t_2) &= \mu_{XX}(t_1 - t_2) \equiv \mu_{XX}(\tau)
\end{aligned}
$$

where $\tau = t_1 - t_2$.

If a random process satisfies the following conditions

$$
\begin{aligned}
\mu_X(t) &= \mu_X \\
\mu_{XX}(t_1, t_2) &= \mu_{XX}(\tau)
\end{aligned}
$$

then it is said to be **wide sense stationary** . Note that a strictly stationary random process is always wide sense stationary, but the converse may not be true. However, a Gaussian random process is strictly stationary if it is wide sense stationary. The reason is that a joint Gaussian density of the vector $\mathbf{X}(t) = (X(t_1), X(t_2), \ldots, X(t_n))$ is completely described by the means and covariances of the $X(t_i)$.

Properties of $\phi_{XX}(\tau)$

The autocorrelation function, $\phi_{XX}(\tau)$, of a stationary random process satisfies the following properties.

1. $\phi_{XX}(0) = E[X^2(t)]$

2. $\phi_{XX}(\tau) = \phi_{XX}(-\tau)$

3. $|\phi_{XX}(\tau)| \leq \phi_{XX}(0)$, the Cauchy-Schwartz inequality.

4. $\phi_{XX}(\infty) = E^2[X(t)] = \mu_X^2$

5. If $X(t) = X(t + \tau)$, then $\phi_{XX}(\tau) = \phi_{XX}(\tau + T)$, i.e., if $X(t)$ is periodic, then $\phi_{XX}(\tau)$ is periodic.

A random process is **ergodic** if for all $g(\mathbf{X})$ and $\mathbf{X}$

$$
\begin{aligned}
E[g(\mathbf{X})] &= \int_{-\infty}^{\infty} g(\mathbf{X}) p_{\mathbf{X}(t)}(\mathbf{x}) d\mathbf{x} \\
&= \lim_{T \to \infty} \frac{1}{2T} \int_{-T}^{T} g[\mathbf{X}(t)] dt \\
&= < g[\mathbf{X}(t)] > .
\end{aligned}
\tag{A.52}
$$

For a random process to be ergodic, it must be strictly stationary. However, not all strictly stationary random processes are ergodic. A random process is **ergodic in the mean** if

$$< X(t) >= \mu_X$$

and **ergodic in the autocorrelation** if

$$< X(t + \tau)X(t) >= \phi_{XX}(\tau) .$$

Example A.2_____

Consider the random process

$$X(t) = A \cos(2\pi f_c t + \Theta)$$

where A and f_c are constants, and

$$
p_\Theta(\theta) = \begin{cases} 1/(2\pi) , & 0 \leq \theta \leq 2\pi \\ 0 , & \text{elsewhere} \end{cases} .
$$

The mean of $X(t)$ is

$$\mu_X(t) = E_\Theta[A \cos(2\pi f_c t + \theta)] = 0 = \mu_X$$

and autocorrelation of $X(t)$ is

$$
\begin{aligned}
\phi_{XX}(t_1, t_2) &= \mathrm{E}_{\Theta}[X(t_1)X(t_2)] \\
&= \mathrm{E}_{\Theta}[A^2 \cos(2\pi f_c t_1 + \theta) \cos(2\pi f_c t_2 + \theta)] \\
&= \frac{A^2}{2}\mathrm{E}_{\Theta}[\cos(2\pi f_c t_1 + 2\pi f_c t_2 + 2\theta)] + \frac{A^2}{2}\mathrm{E}_{\Theta}[\cos(2\pi f_c(t_1 - t_2)] \\
&= \frac{A^2}{2}\cos 2\pi f_c(t_1 - t_2) \\
&= \frac{A^2}{2}\cos 2\pi f_c \tau, \quad \tau = t_1 - t_2 \ .
\end{aligned}
$$

It is clear that this random process is wide sense stationary.

The time-average mean of $X(t)$ is

$$
< X(t) >= \lim_{T \to \infty} \frac{1}{2T} \int_{-T}^{T} A \cos(2\pi f_c t + \theta) dt = 0
$$

and the time average autocorrelation of $X(t)$ is

$$
\begin{aligned}
< X(t + \tau)X(t) > &= \lim_{T \to \infty} \frac{1}{2T} \int_{-T}^{T} A^2 \cos(2\pi f_c t + 2\pi f_c \tau + \theta) \cos(2\pi f_c t + \theta) dt \\
&= \lim_{T \to \infty} \frac{A^2}{4T} \int_{-T}^{T} A^2 [\cos(2\pi f_c \tau) \cos(4\pi f_c t + 2\pi f_c \tau + \theta)] \, dt \\
&= \frac{A^2}{2} \cos(2\pi f_c \tau) \ .
\end{aligned}
$$

By comparing the ensemble and time average mean and autocorrelation, we can conclude this random process is ergodic in the mean and ergodic in the autocorrelation.

Example A.3──

In this example we show that $|\phi_{XX}(\tau)| \leq \phi_{XX}(0)$. This inequality can be established through the following steps.

$$
\begin{aligned}
0 &\leq \mathrm{E}[(X(t + \tau) \pm X(t))^2] \\
&= \mathrm{E}[X^2(t) + X^2(t + \tau) \pm X(t + \tau)X(t)] \\
&= \mathrm{E}[X^2(t)] + \mathrm{E}[X^2(t + \tau)] \pm \mathrm{E}[X(t + \tau)X(t)] \\
&= 2\mathrm{E}[X^2(t)] \pm \mathrm{E}[X(t + \tau)X(t)] \\
&= 2\phi_{XX}(0) \pm 2\phi_{XX}(\tau) \ .
\end{aligned}
$$

Therefore,

$$\pm\phi_{XX}(\tau) \leq \phi_{XX}(0)$$
$$|\phi_{XX}(\tau)| \leq \phi_{XX}(0) \ .$$

Example A.4

Consider the random process

$$Y(t) = X \cos t, \quad X \sim N(0,1) \ .$$

In this example we will find the probability density function of $Y(0)$, the joint probability density function of $Y(0)$ and $Y(\pi)$, and determine whether or not $Y(t)$ is stationary.

1. To find the probability density function of $Y(0)$, note that

$$Y(0) = X \cos 0 = X \ .$$

   Therefore,

$$p_{Y(0)}(y_0) = \frac{1}{\sqrt{2\pi}} e^{-\frac{y_0^2}{2}} \ .$$

2. To find the joint density of $Y(0)$ and $Y(\pi)$, note that

$$Y(0) = X = -Y(\pi) \ .$$

   Therefore

$$p_{Y(0)|Y(\pi)}(y_0|y_\pi) = \delta(y_0 + y_\pi)$$

   and

$$
\begin{aligned}
p_{Y(0)Y(\pi)}(y_0, y_\pi) &= p_{Y(0)|Y(\pi)}(y_0|y_\pi) p_{Y(\pi)}(y_\pi) \\
&= \frac{1}{\sqrt{2\pi}} e^{-\frac{y_\pi^2}{2}} \delta(y_0 + y_\pi) \ .
\end{aligned}
$$

3. To determine whether or not $Y(t)$ is stationary, note that

$$
\begin{aligned}
E[Y(t)] &= E[X] \cos t = 0 \\
E[Y^2(t)] &= E[X^2] \cos^2 t \ .
\end{aligned}
$$

   Since the second moment varies with time, this random process is not stationary.

A.5.2 Crosscorrelation and Crosscovariance

Consider two random processes $X(t)$ and $Y(t)$. The **crosscorrelation** of $X(t)$ and $Y(t)$ is

$$\phi_{XY}(t_1, t_2) = E[X(t_1)Y(t_2)] \quad\quad (A.53)$$
$$\phi_{YX}(t_1, t_2) = E[Y(t_1)X(t_2)] \ . \quad\quad (A.54)$$

The **correlation matrix** of $X(t)$ and $Y(t)$ is

$$\phi(t_1, t_2) = \left[\begin{array}{cc} \phi_{XX}(t_1, t_2) & \phi_{XY}(t_1, t_2) \\ \phi_{YX}(t_1, t_2) & \phi_{YY}(t_1, t_2) \end{array} \right] \ . \quad\quad (A.55)$$

The **crosscovariance** of $X(t)$ and $Y(t)$ is

$$\begin{aligned} \mu_{XY}(t_1, t_2) &= E\left[(X(t_1) - \mu_X(t_1))(X(t_2) - \mu_X(t_2))\right] \\ &= \phi_{XY}(t_1, t_2) - \mu_X(t_1)\mu_X(t_2) \ . \end{aligned} \quad\quad (A.56)$$

The **covariance matrix** of $X(t)$ and $Y(t)$ is

$$\mu(t_1, t_2) = \left[\begin{array}{cc} \mu_{XX}(t_1, t_2) & \mu_{XY}(t_1, t_2) \\ \mu_{YX}(t_1, t_2) & \mu_{YY}(t_1, t_2) \end{array} \right] \ . \quad\quad (A.57)$$

If $X(t)$ and $Y(t)$ are each wide sense stationary and jointly wide sense stationary, then

$$\phi(t_1, t_2) = \phi(t_1 - t_2) = \phi(\tau) \quad\quad (A.58)$$
$$\mu(t_1, t_2) = \mu(t_1 - t_2) = \mu(\tau) \quad\quad (A.59)$$

where $\tau = t_1 - t_2$.

Properties of $\phi_{XY}(\tau)$

The crosscorrelation function $\phi_{XY}(\tau)$ has the following properties.

1. $\phi_{XY}(\tau) = \phi_{YX}(-\tau)$
2. $|\phi_{XY}(\tau)| \leq \frac{1}{2}[\phi_{XX}(0) + \phi_{YY}(0)]$
3. $|\phi_{XY}(\tau)|^2 \leq \phi_{XX}(0)\phi_{YY}(0)$ if $X(t)$ and $Y(t)$ have zero mean.

Classifications of Random Processes

Two random processes $X(t)$ and $Y(t)$ are said to be

- **uncorrelated** if and only if $\mu_{XY}(\tau) = 0$.

- **orthogonal** if and only if $\phi_{XY}(\tau) = 0$.

- **statistically independent** if and only if

$$p_{\mathbf{X}(t)\mathbf{Y}(t+\tau)}(\mathbf{x}, \mathbf{y}) = p_{\mathbf{X}(t)}(\mathbf{x})p_{\mathbf{Y}(t+\tau)}(\mathbf{y}) \ .$$

Furthermore, if $\mu_X = 0$ or $\mu_Y = 0$, then

$$
\begin{aligned}
\text{uncorrelated} \quad &\longleftrightarrow \quad \text{orthogonal} \\
\text{statistically independent} \quad &\longrightarrow \quad \text{uncorrelated} \ .
\end{aligned}
$$

Example A.5

Find the autocorrelation function of the random process

$$Z(t) = X(t) + Y(t)$$

where $X(t)$ and $Y(t)$ are wide sense stationary random processes.

The autocorrelation function is

$$
\begin{aligned}
\phi_{XX}(\tau) &= \mathrm{E}[Z(t+\tau)Z(t)] \\
&= \mathrm{E}\left[(X(t+\tau) + Y(t+\tau))(X(t) + Y(t))\right] \\
&= \phi_{XX}(\tau) + \phi_{YX}(\tau) + \phi_{XY}(\tau) + \phi_{YY}(\tau) \ .
\end{aligned}
$$

If $X(t)$ and $Y(t)$ are uncorrelated, then

$$\phi_{YX}(\tau) = \phi_{XY}(\tau) = \mu_X \mu_Y$$

and

$$\phi_{ZZ}(\tau) = \phi_{XX}(\tau) + \phi_{YY}(\tau) + 2\mu_X \mu_Y \ .$$

If $X(t)$ and $Y(t)$ are uncorrelated and zero-mean, then

$$\phi_{ZZ}(\tau) = \phi_{XX}(\tau) + \phi_{YY}(\tau) \ .$$

Example A.6_____

Can the following be a correlation matrix for two jointly wide sense stationary zero-mean random processes?

$$\phi(\tau) = \begin{bmatrix} \phi_{XX}(\tau) & \mu_{XY}(\tau) \\ \mu_{YX}(\tau) & \mu_{YY}(\tau) \end{bmatrix} = \begin{bmatrix} A^2 \cos\tau & 2A^2 \cos(3\tau/2) \\ 2A^2 \cos(3\tau/2) & A^2 \sin 2\tau \end{bmatrix} .$$

Note that the following two conditions are violated.

1. $|\phi_{XY}(\tau)| \le \frac{1}{2}[\phi_{XX}(0) + \phi_{YY}(0)]$

2. $|\phi_{XY}(\tau)|^2 \le \phi_{XX}(0)\phi_{YY}(0)$ if $X(t)$ and $Y(t)$ have zero mean.

Therefore, $\phi(\tau)$ is not a valid correlation matrix.

_____

A.5.3 Complex-Valued Random Processes

A complex-valued random process is given by

$$Z(t) = X(t) + jY(t)$$

where $X(t)$ and $Y(t)$ are real random processes.

Autocorrelation Function

The autocorrelation function of a complex-valued random process is

$$\begin{aligned} \phi_{ZZ}(t_1, t_2) &= \frac{1}{2}\mathrm{E}[Z(t_1)Z^*(t_2)] \\ &= \frac{1}{2}\mathrm{E}\left[(X(t_1) + jY(t_1))(X(t_2) - jY(t_2))\right] \\ &= \frac{1}{2}\{\phi_{XX}(t_1, t_2) + \phi_{YY}(t_1, t_2) \\ &\quad + \{j[\phi_{YX}(t_1, t_2) - \phi_{XY}(t_1, t_2)]\} . \end{aligned} \tag{A.60}$$

If $Z(t)$ is wide sense stationary, then

$$\phi_{ZZ}(t_1, t_2) = \phi_{ZZ}(t_1 - t_2) = \phi_{ZZ}(\tau) .$$

Crosscorrelation Function

Consider two complex-valued random processes

$$Z(t) = X(t) + jY(t)$$
$$W(t) = U(t) + jV(t) .$$

The crosscorrelation function of $Z(t)$ and $W(t)$ is

$$\phi_{ZW}(t_1, t_2) = \frac{1}{2} E[Z(t_1) W^*(t_2)]$$
$$= \frac{1}{2} \{ \phi_{XV}(t_1, t_2) + \phi_{YV}(t_1, t_2)$$
$$+ j[\phi_{YU}(t_1, t_2) - \phi_{XV}(t_1, t_2)] \} . \qquad (A.61)$$

If $X(t)$, $Y(t)$, $U(t)$, and $V(t)$ are pairwise wide sense stationary random processes, then

$$\phi_{ZW}(t_1, t_2) = \phi_{ZW}(t_1 - t_2) = \phi_{ZW}(\tau) . \qquad (A.62)$$

The crosscorrelation of a complex wide sense stationary random process satisfies the following property

$$\phi_{ZW}^*(\tau) = \frac{1}{2} E[Z^*(t_1) W(t_1 - \tau)]$$
$$= \frac{1}{2} E[Z^*(\hat{t}_1 + \tau) W(\hat{t}_1)]$$
$$= \frac{1}{2} E[W(\hat{t}_1) Z^*(\hat{t}_1 + \tau)]$$
$$= \phi_{WZ}(-\tau) . \qquad (A.63)$$

It also follows that

$$\phi_{XX}^*(\tau) = \phi_{XX}(-\tau) \qquad (A.64)$$
$$\phi_{XX}(\tau) = \phi_{XX}^*(-\tau) . \qquad (A.65)$$

A.5.4 Power Spectral Density

The power spectral density (psd) of a random process $X(t)$ is the Fourier transform of the autocorrelation function, i.e.,

$$\Phi_{XX}(f) = = \int_{-\infty}^{\infty} \phi_{XX}(\tau) e^{-j2\pi f\tau} d\tau \qquad (A.66)$$
$$\phi_{XX}(\tau) = \int_{-\infty}^{\infty} \Phi_{XX}(f) e^{j2\pi f\tau} df . \qquad (A.67)$$

If $X(t)$ is real, then $\phi_{XX}(\tau)$ is real and even. Therefore, $\Phi_{XX}(-f) = \Phi_{XX}(f)$ meaning that $\Phi_{XX}(f)$ is also real and even. If $X(t)$ is complex, then $\phi_{XX}(\tau) = \phi_{XX}^*(-\tau)$, and $\Phi_{XX}^*(f) = \Phi_{XX}(f)$ meaning that $\Phi_{XX}(f)$ is real but not necessarily even.

The power, P, in a random process $X(t)$ is

$$\begin{aligned} P &= \mathrm{E}[X^2(t)] \\ &= \phi_{XX}(0) \\ &= \int_{-\infty}^{\infty} \Phi_{XX}(f)df \end{aligned}$$

a result known as **Parseval's theorem** .

The **cross power spectral density** between two random processes $X(t)$ and $Y(t)$ is

$$\Phi_{XY}(f) = \int_{-\infty}^{\infty} \phi_{XY}(\tau)e^{j2\pi f\tau}d\tau \ . \tag{A.68}$$

If $X(t)$ and $Y(t)$ are both real random processes, then

$$\phi_{XY}(\tau) = \phi_{YX}(\tau)$$

and

$$\Phi_{XY}(f) = \Phi_{YX}(-f) \ .$$

If $X(t)$ and $Y(t)$ are complex random processes, then

$$\phi_{XY}^*(\tau) = \phi_{YX}(-\tau)$$

and

$$\Phi_{XY}^*(f) = \Phi_{YX}(f) \ .$$

A.5.5 Random Processes Filtered by Linear Systems

Consider the linear system with impulse response $h(t)$, shown in Fig. A.2. Suppose that the input to the linear system is a wide sense stationary random process $X(t)$, with mean μ_X and autocorrelation $\phi_{XX}(\tau)$. The input and output are related by the convolution integral

$$Y(t) = \int_{-\infty}^{\infty} h(\tau)X(t-\tau)d\tau \ .$$

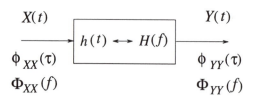

Figure A.2   Random process through a linear system.

Hence,

$$Y(f) = H(f)X(f) .$$

The output mean is

$$\mu_Y = \int_{-\infty}^{\infty} h(\tau)\mathrm{E}[X(t-\tau)]d\tau = \mu_X \int_{-\infty}^{\infty} h(t\tau)d\tau = \mu_X H(0) .$$

The output autocorrelation is

$$
\begin{aligned}
\phi_{YY}(\tau) &= \mathrm{E}[Y(t+\tau)Y(t)] \\
&= \mathrm{E}\left[\int_{-\infty}^{\infty} h(\alpha)X(t+\tau+\alpha)\int_{-\infty}^{\infty} h(\beta)X(t-\beta)d\beta d\alpha\right] \\
&= \int_{-\infty}^{\infty} \infty_{-\infty}^{\infty} h(\alpha)h(\beta)\phi_{XX}(\tau-\alpha+\beta)d\beta d\alpha \\
&= \int_{-\infty}^{\infty} h(\beta)\int_{-\infty}^{\infty} h(\alpha)\phi_{XX}(\tau+\beta-\alpha)d\alpha d\beta \\
&= \left\{\int_{-\infty}^{\infty} h(\beta)\phi_{XX}(\tau+\beta)d\beta\right\} * h(\tau) \\
&= h(-\tau) * \phi_{XX}(\tau) * h(\tau) .
\end{aligned}
$$

Taking transforms, the output psd is

$$
\begin{aligned}
\Phi_{YY}(f) &= H(f)H^*(f)\Phi_{XX}(f) \\
&= |H(f)|^2 \Phi_{XX}(f) .
\end{aligned}
$$

Example A.7_____

Consider the linear system shown in Fig. A.2. In this example we will find the crosscorrelation between the input and output random processes, $X(t)$ and

$Y(t)$, respectively. The crosscorrelation $\phi_{YX}(\tau)$ is given by

$$
\begin{aligned}
\phi_{YX}(\tau) &= \mathrm{E}[Y(t+\tau)X(t)] \\
&= \mathrm{E}\left[\int_{-\infty}^{\infty} h(\alpha)X(t+\tau-\alpha)X(t)d\alpha\right] \\
&= \int_{-\infty}^{\infty} h(\alpha)\mathrm{E}[X(t+\tau-\alpha)X(t)]d\alpha \\
&= \int_{-\infty}^{\infty} h(\alpha)\phi_{XX}(t-\alpha)d\alpha \\
&= h(\tau) * \phi_{XX}(\tau) \ .
\end{aligned}
$$

Also,

$$
\Phi_{YX}(f) = H(f)\Phi_{XX}(f) \ .
$$

Example A.8_____

Suppose that $X(t)$ is a Gaussian random process with mean μ_X and covariance function $\mu_{XX}(\tau)$. In this example we find the joint density of $X_1 \equiv X(t_1)$ and $X_2 \equiv Y(t_2)$. If a Gaussian random process is passed through a linear filter, then the output process is also Gaussian. Hence, X_1 and X_2 have joint Gaussian density function as defined in (A.40) that is completely described in terms of their means and covariances.

Step 1: Obtain the mean and covariance matrix of X_1 and X_2.

The crosscovariance of X_1 and X_2 is

$$
\begin{aligned}
\mu_{X_2 X_1}(\tau) &= \mathrm{E}\left[(Y(t+\tau) - \mu_Y)(X(t) - \mu_X)\right] \\
&= \mathrm{E}\left[Y(t+\tau)X(t)\right] - \mu_Y \mu_X \ .
\end{aligned}
$$

Now $\mu_Y = H(0)\mu_X$. Also, from the previous example

$$
\begin{aligned}
\mathrm{E}[Y(t+\tau)X(t)] &= \int_{-\infty}^{\infty} h(\alpha)\phi_{XX}(\tau-\alpha)d\alpha \\
&= \int_{-\infty}^{\infty} h(\alpha)[\mu_{XX}(\tau-\alpha) + \mu_X^2]d\alpha \\
&= \int_{-\infty}^{\infty} h(\alpha)\mu_{XX}(\tau-\alpha)d\alpha + H(0)\mu_X^2 \ .
\end{aligned}
$$

Therefore,

$$\mu_{X_2 X_1}(\tau) = \int_{-\infty}^{\infty} h(\alpha)\mu_{XX}(\tau - \alpha)d\alpha = h^*(\tau) * \mu_{XX}(\tau) \ .$$

Also

$$
\begin{aligned}
\mu_{X_1 X_2}(\tau) &= \mu_{Z_2 Z_1}(-\tau) \\
\mu_{X_1 X_1}(\tau) &= \mu_{XX}(\tau) \\
\mu_{X_2 X_2}(\tau) &= h(\tau) * h(-\tau) * \mu_{XX}(\tau) \ .
\end{aligned}
$$

Hence, the covariance matrix is

$$\Lambda = \begin{bmatrix} \mu_{X_1 X_1}(0) & \mu_{X_1 X_2}(\tau) \\ \mu_{X_2 X_1}(\tau) & \mu_{X_2 X_2}(0) \end{bmatrix} = \begin{bmatrix} \mu_{XX}(0) & h(\tau) * \mu_{XX}(-\tau) \\ h(\tau)\mu_{XX}(\tau) & h(\tau) * h(-\tau) * \mu_{XX}(\tau) \end{bmatrix}$$

Step 2: Write the joint density function of X_1 and X_2.

Let

$$
\begin{aligned}
\mathbf{X} &= (X_1, X_2)^T \\
\boldsymbol{\mu}_X &= (\mu_X, \mu_Y)^T = (\mu_X, H(0)\mu_X)^T \ .
\end{aligned}
$$

Then

$$P_{\mathbf{X}}(\mathbf{x}) = \frac{1}{2\pi|\Lambda|^{1/2}} \exp\left\{ -\frac{1}{2}(\mathbf{z} - \boldsymbol{\mu}_X)'\Lambda^{-1}(\mathbf{z} - \boldsymbol{\mu}_X) \right\} \ .$$

A.5.6 Discrete-time Random Processes

Let $X_n \equiv X(n)$, where n is an integer time variable, be a discrete-time random process. Then the mth moment of X_n is

$$E[X_n^m] = \int_{-\infty}^{\infty} x_n^m p_X(x_n)dx_n \ . \tag{A.69}$$

The autocorrelation of X_n is

$$\phi(n, k) = \frac{1}{2}E[X_n X_k^*] = \frac{1}{2} \int_{-\infty}^{\infty} \int_{-\infty}^{\infty} x_n x_k^* p_X(x_n)p_X(x_k)dx_m dx_k \tag{A.70}$$

and the autocovariance is

$$\mu(n, k) = \phi(n, k) - \mathrm{E}[X_n]\mathrm{E}[X_k] \ . \tag{A.71}$$

If X_n is a wide sense stationary random process, then

$$\phi(n, k) \equiv \phi(n - k) \tag{A.72}$$
$$\mu(n, k) \equiv \mu(n - k) = \phi(n - k) - \mu_X^2 \ . \tag{A.73}$$

The psd of a discrete random process is

$$\Phi(f) = \sum_{n=-\infty}^{\infty} \phi(n)e^{-j2\pi f n} \tag{A.74}$$

where

$$\phi(n) = \int_{-1/2}^{1/2} \Phi(f)e^{j2\pi f n}\, df \ . \tag{A.75}$$

Note that $\Phi(f) = \Phi(f + k)$ for any integer k.

Consider a discrete-time linear time-invariant system with impulse response $h_n \equiv h(n)$. The input, X_n, and output, Y_n, are related by the convolution sum

$$Y_n = \sum_{-\infty}^{\infty} h_k X_{n-k} \ . \tag{A.76}$$

The output mean is

$$\mu_Y = \mathrm{E}[Y_n] = \sum_{k=-\infty}^{\infty} h_k \mathrm{E}[X_{n-k}] = \mu_X \sum_{k=-\infty}^{\infty} h_k = \mu_X H(0)$$

and the output autocorrelation is

$$\begin{aligned}
\phi_{YY}(k) &= \frac{1}{2}\mathrm{E}[Y(n+k)Y^*(n)] \\
&= \frac{1}{2}\mathrm{E}\left[\sum_{\ell=-\infty}^{\infty} h_\ell X^*(n-\ell) \sum_{m=-\infty}^{\infty} h_m X(n+k-m) \right] \\
&= \frac{1}{2} \sum_{\ell=-\infty}^{\infty} \sum_{m=-\infty}^{\infty} h_\ell h_m \mathrm{E}[X(n+k-m)X^*(n-\ell)] \\
&= \sum_{\ell=-\infty}^{\infty} \sum_{m=-\infty}^{\infty} h_\ell h_m \phi_{XX}(k+\ell-m) \ .
\end{aligned}$$

The output psd is

$$
\begin{aligned}
\Phi_{YY} &= \sum_{k=-\infty}^{\infty} \phi_{YY}(k) e^{-j2\pi fk} \\
&= \sum_{k=-\infty}^{\infty} \sum_{\ell=-\infty}^{\infty} \sum_{m=-\infty}^{\infty} h_\ell h_m \phi_{XX}(k+\ell-m) e^{j2\pi fk} \\
&= \sum_{n=-\infty}^{\infty} \sum_{\ell=-\infty}^{\infty} \sum_{m=-\infty}^{\infty} h_\ell h_m \phi_{XX}(n) e^{-j2\pi f(n-\ell+m)} \\
&= \sum_{\ell=-\infty}^{\infty} h_\ell e^{j2\pi f\ell} \sum_{m=-\infty}^{\infty} e^{-j2\pi fm} \\
&\quad \times \sum_{n=-\infty}^{\infty} \phi_{XX}(n) e^{-j2\pi fn} \\
&= H^*(f) H(f) \Phi_{XX}(f) \\
&= |H(f)|^2 \Phi_{XX}(f) \ .
\end{aligned}
$$

A.5.7 Cyclostationary Random Processes

Consider the random process

$$
X(t) = \sum_{n=-\infty}^{\infty} a_n \psi(t - nT)
$$

where $\{a_n\}$ is a sequence of complex random variables with mean μ_a and auto-correlation $\phi_{aa}(n)$, and $\psi(t)$, $0 \le t \le T$ is a real deterministic shaping function. Note that the mean of $X(t)$

$$
\mu_X = \mu_a \sum_{n=-\infty}^{\infty} \psi(t - nT)
$$

is periodic. The autocorrelation of $X(t)$ is

$$
\begin{aligned}
\phi_{XX}(t+\tau, t) &= \frac{1}{2} E[X(t+\tau) X^*(t)] \\
&= \frac{1}{2} E\left[\sum_{-\infty}^{\infty} a_n \psi(t+\tau-nT) \sum_{-\infty}^{\infty} a_m^* \psi(t-mT) \right] \\
&= \sum_{-\infty}^{\infty} \sum_{-\infty}^{\infty} \phi_{aa}(n-m) \psi(t-mT) \psi(t+\tau-nT) \ .
\end{aligned}
$$

It is easy to show that

$$\phi_{XX}(t + \tau + kT, t + kT) = \phi_{XX}(t + \tau, t) \ .$$

Therefore, $\phi_{XX}(t + \tau, t)$ is periodic in t with period T.

The *time-averaged* psd of $X(t)$ can be computed by first determining the time-average autocorrelation

$$\phi_{XX}(\tau) = \frac{1}{T} \int_{-T/2}^{T/2} \phi_{XX}(t + \tau, t) dt$$

and then taking the Fourier transform in (A.66).

REFERENCES

[1] M. Abramowitz and I. A. S. (ed.), *Handbook of Mathematical Functions.* New York, NY: Dover, 1965.

[2] A. Abu-Dayya and N. Beaulieu, "Outage probabilities of cellular mobile radio systems with multiple Nakagami interferers," *IEEE Trans. Veh. Technol.,* Vol. 40, pp. 757–768, November 1991.

[3] A. S. Acampora, "Maximum likelihood decoding of binary convolutional codes on band-limited satellite channels," *IEEE Trans. Commun.,* Vol. 26, pp. 766–776, June 1978.

[4] A. S. Acampora, "Analysis of maximum likelihood sequence estimation performance for quadrature amplitude modulation," *Bell System Tech. J.,* Vol. 60, pp. 865–885, July 1981.

[5] Y. Akaiwa, "A conceptual design of microcellular radio communication system," in *IEEE Veh. Technol. Conf.,* Orlando, FL, pp. 156–160, May 1990.

[6] Y. Akaiwa and H. Andoh, "Channel segregation - a self organized dynamic channel allocation method: application to TDMA/FDMA microcellular systems," *IEEE J. Selec. Areas Commun.,* Vol. 11, pp. 949–954, August 1993.

[7] J. B. Andersen, T. Rappaport, and S. Yoshida, "Propagation measurements and models for wireless communications channels," *IEEE Commun. Mag.,* Vol. 33, pp. 42–49, January 1995.

[8] J. B. Anderson, T. Aulin, and C. E. Sundberg, *Digital Phase Modulation.* New York, NY: Plenum, 1986.

[9] L. Anderson, "A simulation study of some dynamic channel assignment algorithms in a high capacity mobile telecommunications system," *IEEE Trans. Veh. Technol.,* Vol. 22, pp. 210–217, November 1973.

[10] S. Ariyavisitakul, "SIR-based power control in a CDMA system," in *IEEE Global Commun. Conf.,* Orlando, FL, pp. 868–873, December 1992.

[11] S. Ariyavisitakul and L. F. Chang, "Signal and interference statistics of a CMDA system with feedback power control," *IEEE Trans. Commun.*, Vol. 41, pp. 1626–1634, November 1993.

[12] T. Aulin, "A modified model for the fading signal at a mobile radio channel," *IEEE Trans. Veh. Technol.*, Vol. 28, pp. 182–203, August 1979.

[13] M. D. Austin and G. L. Stüber, "In-service signal quality estimation for TDMA cellular systems," to appear in Kluwer J. Wireless Personal Commun.

[14] M. D. Austin and G. L. Stüber, "Velocity adaptive handoff algorithms for microcellular systems," in *IEEE Conf. Universal Personal Commun.*, Ottawa, Canada, pp. 793–797, October 1993.

[15] M. D. Austin and G. L. Stüber, "Velocity adaptive handoff algorithms for microcellular systems," *IEEE Trans. Veh. Technol.*, Vol. 43, pp. 549–561, August 1994.

[16] M. D. Austin and G. L. Stüber, "In-service signal quality estimation for TDMA cellular systems," in *IEEE Int. Symp. on Pers., Indoor and Mobile Radio Commun.*, Toronto, Canada, September 1995.

[17] M. E. Austin, "Decision-feedback equalization for digital communication over dispersive channels," Tech. Rep., MIT Lincoln Lab., Lexington, MA., August 1967.

[18] M. D. Austin and G. L. Stüber, "Co-channel interference modeling for signal strength based handoff analysis," *Electronics Letters*, Vol. 30, pp. 1914–1915, November 1994.

[19] M. D. Austin and G. L. Stüber, "Direction biased handoff algorithms for urban microcells," in *IEEE Veh. Technol. Conf.*, Stockholm, Sweden, pp. 101–105, June 1994.

[20] M. D. Austin and G. L. Stüber, "Exact co-channel interference analysis for log-normal shadowed Rician fading channels," *Electronic Letters*, Vol. 30, pp. 748–749, May 1994.

[21] C. T. Beare, "The choice of the desired impulse response in combined linear-Viterbi algorithm equalizer," *IEEE Trans. Commun.*, Vol. 26, pp. 1301–1327, August 1978.

[22] N. C. Beaulieu, A. A. Abu-Dayya, and P. J. McLane, "Comparison of methods of computing lognormal sum distributions and outages for digital wireless applications," in *IEEE Int. Conf. on Commun.*, New Orleans, LA, pp. 1270–1275, May 1994.

[23] C. A. Belfiore and J. J.H. Park, "Decision-feedback equalization," in *Proc. IEEE*, pp. 1143–1156, August 1979.

[24] P. Bello, "Characterization of random time-variant linear channels," *IEEE Trans. Commun.*, Vol. 11, pp. 360–393, December 1963.

[25] J-E. Berg, R. Bownds, and F. Lotse, "Path loss and fading models for microcells at 900 MHz," in *IEEE Veh. Technol. Conf.*, Denver, CO, pp. 666–671, May 1992.

[26] R. C. Bernhardt, "The use of multiple-beam directional antennas in wireless messaging systems," in *IEEE Veh. Technol. Conf.*, pp. 858–861, 1995.

[27] E. Biglieri, "High-level modulation and coding for nonlinear satellite channels," *IEEE Trans. Commun.*, Vol. 32, pp. 616–626, May 1984.

[28] E. Biglieri, D. Divsalar, P. McLane, and M. Simon, *Introduction to Trellis-Coded Modulation with Applications.* New York, NY: McMillan, 1991.

[29] M. A. Birchler and S. C. Jasper, "A 64 kbps digital land mobile radio system employing M-16QAM," in *Proc. of 5th Nordic Sem. Land Mobile Radio*, Helsinki, Finland, pp. 237–241, December 1992.

[30] D. M. Brady, "An adaptive coherent diversity receiver for data transmission through dispersive media," in *IEEE Int. Conf. on Commun.*, San Francisco, CA, pp. 21.35–21.40, June 1970.

[31] A. R. Calderbank and J. Mazo, "A new description of trellis codes," *IEEE Trans. Inform. Theory*, Vol. 30, pp. 784–791, November 1984.

[32] A. R. Calderbank and N. J. Sloane, "New trellis codes based on lattices and cosets," *IEEE Trans. Inform. Theory*, Vol. 33, pp. 177–195, March 1987.

[33] J. Cavers and P. Ho, "Analysis of the error performance of trellis-coded modulation in Rayleigh-fading channels," *IEEE Trans. Commun.*, Vol. 40, pp. 74–83, January 1992.

[34] C. J. Chang and F. C. Ren, "Downlink power control in DS/CDMA cellular mobile radio networks," in *IEEE Conf. Universal Personal Commun.*, pp. 89–93, 1994.

[35] I-K. Chang, G. L. Stüber, and A. M. Bush, "Performance of diversity combining techniques for DS/DPSK signaling over a pulse jammed multipath fading channel," *IEEE Trans. Commun.*, Vol. 38, pp. 1823–1834, October 1990.

[36] U. Charash, "Reception through Nakagami fading multipath channels with random delays," *IEEE Trans. Commun.*, Vol. 27, pp. 657–670, April 1979.

[37] P. Chevillat and J. D. J. Costello, "A multiple stack algorithm for erasure free decoding of convolutional codes," *IEEE Trans. Commun.*, Vol. 25, pp. 1460–1470, December 1977.

[38] P. Chevillat and E. Eleftheriou, "Decoding of trellis-encoded signal in the presence of intersymbol interference and noise," *IEEE Trans. Commun.*, Vol. 37, pp. 669–676, July 1989.

[39] P. Chevillat and E. Eleftheriou, "Decoding of trellis-encoded signals in the presence of intersymbol interference and noise," *IEEE Trans. Commun.*, Vol. 37, pp. 669–676, July 1989.

[40] S. Chia, R. Steele, E. Green, and A. Baran, "Propagation and bit-error ratio measurements for a microcellular system," *J. Inst. Electron. Radio Eng. (UK)*, Vol. 57, pp. 255–266, November 1987.

[41] S. T. S. Chia and R. J. Warburton, "Handover criteria for city microcellular systems," in *IEEE Veh. Technol. Conf.*, Orlando, FL, pp. 276–281, May 1990.

[42] S. Chia, "The universal mobile telecommunication system," *IEEE Commun. Mag.*, Vol. 30, pp. 54–62, December 1992.

[43] L.J. Cimini and G. Foschini, "Distributed algorithms for dynamic channel allocation in microcellular systems," in *IEEE Veh. Technol. Conf.*, Denver, CO, pp. 641–644, May 1992.

[44] L. J. Cimini, G. J. Foschini, C.-L. I, and Z. Miljanic, "Call blocking performance of distributed algorithms for dynamic channel allocation in microcells," *IEEE Trans. Commun.*, Vol. 42, pp. 2600–2607, August 1994.

[45] L. J. Cimini, G. J. Foschini, and L. Shepp, "Single-channel user-capacity calculations for self-organizing cellular systems," *IEEE Trans. Commun.*, Vol. 42, pp. 3137–3143, December 1994.

[46] M. Cioffi and T. Kailath, "Fast, recursive-least-squares, transversal filters for adaptive filtering," *IEEE Trans. Acoutics, Speech and Signal Proc.*, Vol. 32, pp. 304–337, April 1984.

[47] A. P. Clark, ed., *Advanced Data Transmission Systems*. London, UK: Pentech Press, 1977.

[48] A. P. Clark, S. N. Abdullah, S. J. Jayasinghe, and K. Sun, "Pseudobinary and pseudoquaternary detection processes for linearly distorted multilevel QAM signals," *IEEE Trans. Commun.*, Vol. 33, pp. 639–645, July 1985.

[49] A. P. Clark and M. Clayden, "Pseudobinary Viterbi detector," *IEE Proc. Part F*, Vol. 131, pp. 208–218, April 1984.

[50] A. P. Clark and R. Harun, "Assessment of Kalman-filter channel estimator for an HF radio link," *IEE Proc.*, Vol. 133, pp. 513–521, October 1986.

[51] G. C. Clark, Jr. and J. B. Cain, *Error-Correction Coding for Digital Communications*. New York, NY: Plenum, 1981.

[52] R. Clarke, "A statistical theory of mobile radio reception," *Bell System Tech. J.*, Vol. 47, pp. 957–1000, 1968.

[53] G. R. Cooper and R. W. Nettleton, "A spread-spectrum technique for high-capacity mobile communications," *IEEE Trans. Veh. Technol.*, Vol. 27, pp. 264 – 275, November 1978.

[54] K. G. Cornett and S. B. Wicker, "Bit error rate estimation techniques for digital land mobile radios," in *IEEE Veh. Technol. Conf.*, Saint Louis, MO, pp. 543–548, May 1991.

[55] COST 207 TD(86)51-REV 3 (WG1):, "Proposal on channel transfer functions to be used in GSM tests late 1986," September 1986.

[56] COST 231 TD(91)109, "1800 MHz mobile net planning based on 900 MHz measurements," 1991.

[57] COST 231 TD(973)119-REV 2 (WG2):, "Urban transmission loss models for mobile radio in the 900- and 1,800-MHz bands," September 1991.

[58] D. C. Cox and D. O. Reudnik, "A comparison of some channel assignment strategies in large-scale mobile communication systems," *IEEE Trans. Commun.*, Vol. 20, pp. 190–195, February 1972.

[59] D. C. Cox and D. O. Reudnik, "Increasing channel occupancy in large-scale mobile radio systems: dynaic channel reassignment," *IEEE Trans. Veh. Technol.*, Vol. 22, pp. 218–222, November 1973.

[60] D. C. Cox, "Cochannel interference considerations in frequency reuse small-coverage-area radio systems," *IEEE Trans. Commun.*, Vol. 30, pp. 135–142, January 1982.

[61] D. C. Cox, "Portable digital radio communications – an approach to tetherless access," *IEEE Commun. Mag.*, Vol. 27, pp. 30–40, July 1989.

[62] D. C. Cox, "Wireless network access for personal communications," *IEEE Commun. Mag.*, Vol. 30, pp. 96–115, December 1992.

[63] K. Daikoku and H. Ohdate, "Optimal channel reuse in cellular land mobile radio systems," *IEEE Trans. Veh. Technol.*, Vol. 32, pp. 217–224, August 1983.

[64] G. D'aria and V. Zingarelli, "Results on fast-Kalman and Viterbi adaptive equalizers for mobile radio with CEPT/GSM system characteristics," in *IEEE Global Commun. Conf.*, Hollywood, FL, pp. 815–819, December 1988.

[65] G. D'Avella, L. Moreno, and M. Sant'Agostino, "An adaptive MLSE receiver for TDMA digital mobile radio," *IEEE J. Selec. Areas Commun.*, Vol. 7, pp. 122–129, January 1989.

[66] W. B. Davenport and W. L. Root, *An Introduction to the Theory of Random Signals and Noise.* New York, NY: McGraw-Hill, 1958.

[67] G. W. Davidson, D. D. Falconer, and A. U. H. Sheikh, "An investigation of block adaptive decision feedback equalization for frequency selective fading channels," in *IEEE Int. Conf. on Commun.*, Philadelphia, PA, pp. 360–365, June 1988.

[68] P. Dent, G. E. Bottomley, and T. Croft, "Jakes fading model revisited," *Electronic Letters*, Vol. 7, pp. 1162–1163, June 1993.

[69] D. Divsalar and M. Simon, "Trellis coded modulation for 4800-9600 bit/s transmission over a fading mobile satellite channel," *IEEE J. Selec. Areas Commun.*, Vol. 5, pp. 162–174, February 1987.

[70] D. Divsalar and M. Simon, "The design of trellis coded MPSK for fading channels: performance criteria," *IEEE Trans. Commun.*, Vol. 36, pp. 1004–1012, September 1988.

[71] D. Divsalar and M. Simon, "The design of trellis coded MPSK for fading channels: set partitioning for optimum code design," *IEEE Trans. Commun.*, Vol. 36, pp. 1013–1022, September 1988.

[72] D. Divsalar, M. Simon, and J. Yuen, "Trellis coding with asymmetric modulations," *IEEE Trans. Commun.*, Vol. 35, pp. 130–141, February 1987.

[73] R. C. Dixon, *Spread Spectrum Techniques*. New York, NY: IEEE Press, 1976.

[74] J. Driscoll and N. Karia, "Detection process for V32 modems using trellis coding," *Proc. IEEE*, Vol. 135, pp. 143–154, April 1988.

[75] A. Duel-Hallen, *Detection for Channels with Intersymbol Interference*. Ph. D. thesis, Cornell University, Ithica, NY, 1987.

[76] A. Duel-Hallen and C. Heegard, "Delayed decision feedback sequence estimation," *IEEE Trans. Commun.*, Vol. 37, pp. 428–436, May 1989.

[77] F. Edbauer, "Performance of interleaved trellis-coded differential 8-PSK modulation over fading channels," *IEEE J. Selec. Areas Commun.*, Vol. 7, pp. 1340–1346, December 1989.

[78] EIA/TIA IS-54, "Cellular system dual-mode mobile station - base station compatibility standard,".

[79] EIA/TIA IS-95, "Mobile station – base station compatability standard for dual-mode wideband spread spectrum cellular system,".

[80] E. Eleftheriou and D. D. Falconer, "Restart methods for stabilizing FRLS adaptive equalizers in digital HF transmission," in *IEEE Global Commun. Conf.*, Atlanta, GA, pp. 1558–1562, November 1984.

[81] E. Eleftheriou and D. D. Falconer, "Tracking properties and steady-state performance of RLS adaptive filter algorithms," *IEEE Trans. Acoutics, Speech and Signal Proc.*, Vol. 34, pp. 1097–1110, October 1986.

[82] E. Eleftheriou and D. D. Falconer, "Adaptive equalization techniques for HF channels," *IEEE J. Selec. Areas Commun.*, Vol. 5, pp. 238–247, February 1987.

[83] S. Elnoubi, R. Singh, and S. Gupta, "A new frequency channel assignment algorithm in high capacity mobile communications systems," *IEEE Trans. Veh. Technol.*, Vol. 31, pp. 125–131, August 1982.

[84] J. Engel and M. Peritsky, "Statistically-optimum dynamic server assignment in systems with interfering servers," *IEEE Trans. Veh. Technol.*, Vol. 22, pp. 203–209, November 1973.

[85] ETSI – European Telecommunications Standards Institute, *GSM Recommendation 05.08*. January 1991.

[86] D. Everitt, "Traffic capacity of cellular mobile communications systems," *Computer Networks and ISDN Systems*, Vol. 20, pp. 447–454, December 1990.

[87] D. Everitt and D. Manfield, "Performance analysis of cellular mobile communication systems with dynamic channel assignment," *IEEE J. Selec. Areas Commun.*, Vol. 7, pp. 1172–1179, October 1989.

[88] D. E. Everitt and N. W. MacFadyen, "Analysis of multicellular mobile radio-telephone systems: a model and evaluation," *British Telecom Tech. J.*, Vol. 1, pp. 37–45, 1983.

[89] V. M. Eyuboğlu, "Detection of coded modulation signals on linear, severely distorted channels using decision-feedback noise prediction with interleaving," *IEEE Trans. Commun.*, Vol. 36, pp. 401–409, April 1988.

[90] V. M. Eyuboğlu and D. Forney, Jr., "Trellis precoding: combining coding, precoding, and shaping for intersymbol interference channels," *IEEE Trans. Inform. Theory*, Vol. 38, pp. 301–314, March 1992.

[91] V. M. Eyuboğlu and S. U. Qureshi, "Reduced-state sequence estimation with set partitioning and decision feedback," *IEEE Trans. Commun.*, Vol. 36, pp. 13–20, January 1988.

[92] G. Falciasecca, M. Frullone, G. Riva, M. Sentinelli, and A. M. Serra, "Investigation on a dynamic channel allocation for high capacity mobile radio systems," in *IEEE Vehicular Technology Conference*, Philadelphia, PA, pp. 176–181, 1988.

[93] D. D. Falconer and J. F. R. Magee, "Adaptive channel memory truncation for maximum likelihood sequence estimation," *Bell System Tech. J.*, Vol. 52, pp. 1541–1562, November 1973.

[94] D. D. Falconer and L. Ljung, "Application of fast Kalman estimation to adaptive equalization," *IEEE Trans. Commun.*, Vol. 26, pp. 1439–1446, October 1978.

[95] R. M. Fano, "A heuristic discussion of probabilistic decoding," *IEEE Trans. Inform. Theory*, Vol. 9, pp. 64–74, April 1963.

[96] K. Feher, *Advanced Digital Communications*. Englewood Cliffs, NJ: Prentice-Hall, 1987.

[97] W. Feller, *An Introduction of Probability Theory and Its Applications, Vol. I.* New York, NY: Wiley, 1968.

[98] L. F. Fenton, "The sum of log-normal probability distributions in scatter transmission systems," *IRE Trans. Commun.*, Vol. 8, pp. 57–67, March 1960.

[99] B. L. Floch, R. Halbert-Lassalle, and D. Castelain, "Digital sound broadcasting to mobile receivers," *IEEE Trans. Consumer Elect.*, Vol. 35, pp. 493–503, August 1989.

[100] G. D. Forney, Jr., "Maximum likelihood sequence estimation of digital sequence in the presence of intersymbol interference," *IEEE Trans. Inform. Theory*, Vol. 18, pp. 363–378, May 1972.

[101] G. D. Forney, Jr., "Coset codes – part I: introduction to geometrical classification," *IEEE Trans. Inform. Theory*, Vol. 34, pp. 1123–1151, September 1988.

[102] G. J. Foschini, "A reduced-state variant of maximum-likelihood sequence detection attaining optimum performance for high signal-to-noise ratio performance," *IEEE Trans. Inform. Theory*, Vol. 24, pp. 605–609, September 1977.

[103] E. A. Frech and C. L. Mesquida, "Cellular models and hand-off criteria," in *IEEE Veh. Technol. Conf.*, San Francisco, CA, pp. 128–135, May 1989.

[104] R. C. French, "Error rate predictions and measurements in the mobile radio data channel," *IEEE Trans. Veh. Technol.*, Vol. 27, pp. 214–220, August 1978.

[105] R. C. French, "The effect of fading and shadowing on channel reuse in mobile radio," *IEEE Trans. Veh. Technol.*, Vol. 28, pp. 171–181, August 1979.

[106] B. Friedlander, "Lattice filters for adaptive processing," *Proc. IEEE*, Vol. 70, pp. 829–867, August 1982.

[107] H. Furukawa and Y. Akaiwa, "A microcell overlaid with umbrella cell system," in *IEEE Veh. Technol. Conf.*, pp. 1455–1459, 1994.

[108] H. Furukawa and A. Yoshihiko, "Self-organized reuse partitioning, a dynamic channel assignment mehtod in cellular systems," in *IEEE Veh. Technol. Conf.*, Secaucus, NJ, pp. 524–527, May 1993.

[109] P. Gaasvik, M. Cornefjord, and V. Svenson, "Different methods of giving priority to handoff traffic in a mobile telephone system with directed retry," in *IEEE Veh. Technol. Conf.*, Saint Louis, MO, pp. 549–553, May 1991.

[110] W. F. Gabriel, "Adaptive processing array systems," *Proc. IEEE*, Vol. 80, pp. 152–162, January 1992.

[111] F. M. Gardner, ed., *Phaselock Techniques, second ed.* New York, NY: Wiley, 1979.

[112] G. J. Garrison, "A power spectral density analysis for digital FM," *IEEE Trans. Commun.*, Vol. 23, pp. 1228–1243, November 1975.

[113] E. A. Geraniotis and R. Mani, "Throughput analysis of a random access tree protocol for direct-sequence spread-spectrum packet radio networks," in *IEEE Military Commun. Conf.*, Washington, D. C., pp. 23.7.1–23.7.6, October 1987.

[114] E. A. Geraniotis, "Direct-sequence wpread-spectrum multiple-access communications over nonselective and frequency-selective Rician fading channels," *IEEE Trans. Commun.*, Vol. 34, pp. 756–764, August 1986.

[115] A. Gersho, "Adaptive equalization of highly dispersive channels," *Bell System Tech. J.*, Vol. 48, pp. 55–70, January 1969.

[116] A. Gersho and T. L. Lim, "Adaptive cancellation of intersymbol interference for data transmission," *Bell System Tech. J.*, Vol. 60, pp. 1997–2021, November 1981.

[117] K. S. Gilhousen, I. M. Jacobs, R. Padovani, A. J. Viterbi, L. A. Weaver, Jr., and C. W. III, "On the capacity of a cellular CDMA system," *IEEE Trans. Veh. Technol.*, Vol. 40, pp. 303–312, May 1991.

[118] K. S. Gilhousen, I. M. Jacobs, R. Padovani, L. A. Weaver, Jr., and C. W. III, "On the capacity of a cellular CDMA system," *IEEE Trans. Veh. Technol.*, Vol. 40, pp. 303–312, May 1991.

[119] A. A. Giordano and F. M. Hsu, eds., *Least Square Estimation with Applications to Digital Signal Processing.* New York, NY: Wiley, 1985.

[120] R. D. Gitlin, J. E. Mazo, and M. G. Taylor, "On the design of gradient algorithms for digitally implemented adaptive filters," *IEEE Trans. Circuit Theory*, Vol. 20, pp. 125–136, March 1973.

[121] R. D. Gitlin and S. B. Weinstein, "Fractionally-spaced equalization: an improved digital transversal equalizer," *Bell System Tech. J.*, Vol. 60, pp. 275–296, February 1981.

[122] R. Gold, "Optimum binary sequences for spread-spectrum multiplexing," *IEEE Trans. Inform. Theory*, Vol. 13, pp. 619–621, October 1967.

[123] A. J. Goldsmith, L. J. Greenstein, and G. J. Foschini, "Error statistics of real time power measurements in cellular channels with multipath and shadowing," in *IEEE Veh. Technol. Conf.*, Secaucus, NJ, pp. 108–110, May 1993.

[124] A. J. Goldsmith, L. J. Greenstein, and G. J. Foschini, "Error statistics of real-time power measurements in cellular channels with multipath and shadowing," *IEEE Trans. Veh. Technol.*, Vol. 43, pp. 439–446, August 1994.

[125] A. J. Goldsmith and L. J. Greenstein, "A Measurement-based model for predicting coverage areas of urban microcells," *IEEE J. Selec. Areas Commun.*, Vol. 11, pp. 1013–1023, September 1993.

[126] D. J. Goodman, S. A. Grandhi, and R. Vijayan, "Distributed dynamic channel assignment schemes," in *IEEE Veh. Technol. Conf.*, Secaucus, NJ, pp. 532–535, May 1993.

[127] D. N. Gordard, "Channel equalization using a Kalman filter for fast data transmission," *IBM Journal Research and Development*, Vol. 18, pp. 267–273, May 1974.

[128] I. Gradshteyn and I. Ryzhik, *Tables of Integrals, Series, and Products.* San Diego, CA: Academic Press, 1980.

[129] E. Green, "Path loss and signal variability analysis for microcells," in *5th Int. Conf. on Mobile Radio and Personal Commun.*, Coventry, UK, pp. 38–42, December 1989.

[130] O. Grimlund and B. Gudmundson, "Handoff strategies in microcellular systems," in *IEEE Veh. Technol. Conf.*, Saint Louis, MO, pp. 505–510, May 1991.

[131] J. L. Grubb, "The traveller's dream come true," *IEEE Commun. Mag.*, Vol. 29, pp. 48–51, November 1991.

[132] M. Gudmundson, "Analysis of handover algorithms," in *IEEE Veh. Technol. Conf.*, Saint Louis, MO, pp. 537–541, May 1991.

[133] M. Gudmundson, "Analysis of handover algorithms in cellular radio systems," Report No. TRITA-TTT-9107, Royal Institute of Technology, Stockholm, Sweden, April 1991.

[134] M. Gudmundson, "Correlation model for shadow fading in mobile radio systems," *Electronics Letters*, Vol. 27, pp. 2145–2146, November 1991.

[135] S. W. Halpern, "Reuse Partitioning in Cellular Systems," in *IEEE Veh. Technol. Conf.*, Toronto, Ontario, Canada, pp. 322–327, 1983.

[136] K. Hamied, *Advanced Radio Link Design and Radio Receiver Design for Mobile Communications*. Ph. D. thesis, Georgia Institute of Technology, Atlanta, GA, 1994.

[137] K. Hamied and G. L. Stüber, "A fractionally-spaced MLSE Receiver," in *IEEE Int. Conf. on Commun.*, Seattle , WA, pp. 7–11, 1995.

[138] K. Hanabe, V. Tetsuro, and T. Otsu, "Distributed adaptive channel allocation scheme with variable C/I threshold in cellular systems," in *IEEE Veh. Technol. Conf.*, Secaucus, NJ, pp. 164–167, May 1993.

[139] P. Harley, "Short distance attenuation measurements at 900 MHz and 1.8 GHz using low antenna heights for microcells," *IEEE J. Selec. Areas Commun.*, Vol. 7, pp. 5–11, January 1989.

[140] M. Hata and T. Nagatsu, "Mobile location using signal strength measurements in cellular systems," *IEEE Trans. Veh. Technol.*, Vol. 29, pp. 245–251, 1980.

[141] S. Haykin, ed., *Adaptive Filter Theory*. Englewood Cliff, NJ: Prentice-Hall, 1986.

[142] M-J. Ho and G. L. Stüber, "Co-channel interference of microcellular systems on shadowed Nakagami fading channels," in *IEEE Veh. Technol. Conf.*, Secaucus, NJ, pp. 568–571, May 1993.

[143] M-J. Ho and G. L. Stüber, "Capacity and Power Control for CDMA Microcells," *ACM/Baltzer Journal on Wireless Networks*, Vol. ??, p. ??, Oct 1995.

[144] P. Ho and D. Fung, "Error performance of interleaved trellis-coded PSK modulation in correlated Rayleigh fading channels," *IEEE Trans. Commun.*, Vol. 40, pp. 1800–1809, December 1992.

[145] P. Hoeher, "TCM on frequency-selective fading channels: a comparison of soft-output probabilistic equalizers," in *IEEE Global Commun. Conf.*, San Diego, CA, pp. 367–382, December 1990.

[146] J. M. Holtzman, "A simple, accurate method to calculate spread-spectrum multiple-access error probabilities," *IEEE Trans. Commun.*, Vol. 40, pp. 461–464, March 1992.

[147] J. Holtzman, "Adaptive measurement intervals for handoffs," in *IEEE Int. Conf. on Commun.*, Chicago, IL, pp. 1032–1036, June 1992.

[148] D. Hong and S. S. Rappaport, "Traffic model and performance analysis for cellular mobile radio telephone systems with prioritized and nonprioritized handoff procedures," *IEEE Trans. Veh. Technol.*, Vol. 35, pp. 77–92, August 1986.

[149] F. M. Hsu, A. A. Giordano, H. dePedro, and J. G. Proakis, "Adaptive equalization techniques for high speed transmission on fading dispersive HF channels," in *Nat. Telecommun. Conf.*, Houston, TX, pp. 58.1.1–58.1.7, November 1980.

[150] L-R. Hu and S. S. Rappaport, "Personal communication systems using multiple hierarchical cellular overlays," *IEEE J. Selec. Areas Commun.*, Vol. 13, pp. 406 – 415, February 1995.

[151] C-L. I, L. J. Greenstein, and R. Gitlin, "A microcell/macrocell cellular architecture for low- and high-mobility wireless users," *IEEE J. Selec. Areas Commun.*, Vol. 11, pp. 885 – 891, August 1993.

[152] K. Imamura and A. Murase, "Mobile communication control using multi-transmitter simul/sequential casting (MSSC)," in *IEEE Veh. Technol. Conf.*, Dallas, TX, pp. 334–341, May 1986.

[153] F. de Jager and C. B. Dekker, "Tamed Frequency modulation, a novel method to achieve spectrum economy in digital transmission," *IEEE Trans. Commun.*, Vol. 26, pp. 534–542, May 1978.

[154] W. C. Jakes, ed., *Microwave Mobile Communication.* New York, NY: IEEE Press, 1993.

[155] H. Jiang and S. S. Rappaport, "CBWL: A new channel assignment and sharing method for cellular communication systems," *IEEE Trans. Veh. Technol.*, Vol. 43, pp. 313–322, May 1994.

[156] M. Kaji and A. Akeyama, "UHF-band propagation characteristics for land mobile radio," in *Int. Symp. Ant. and Prop.*, Univ. of British Columbia, Canada, pp. 835–838, June 1985.

[157] T. Kanai, "Autonomous reuse partitioning in cellular systems," in *IEEE Veh. Technol. Conf.*, Denver, CO, pp. 782–785, May 1992.

[158] T. Kasami, "Weight distribution of Bose-Chaudhuri-Hocquenghem codes," in *Combinatorial Mathematics and its Applications*, University of North Carolina Press, Chapel Hill, NC, pp. 335–357, 1967.

[159] T. Kasami, S. Lin, and W. Peterson, "Some results on cyclic codes which are invariant under the affine group and their applications," *Information and Control*, Vol. 11, pp. 475–496, November-December 1968.

[160] E. Katz and G. Stüber, "Sequential sequence estimation for trellis-coded modulation on multipath fading ISI channels," *IEEE Trans. Commun.*, pp. 2882–2885, December 1995.

[161] M. Kavehrad and G. E. Bodeep, "Design and experimental results for a direct-sequence spread-spectrum radio using differential phase-shift keying modulation for indoor, wireless communications," *IEEE J. Selec. Areas Commun.*, Vol. 5, pp. 815–823, June 1987.

[162] M. Kavehrad and B. Ramamurthi, "Direct-sequence spread-spectrum with DPSK modulation and diversity for indoor wireless communications," *IEEE Trans. Commun.*, Vol. 35, pp. 224–236, February 1987.

[163] F. P. Kelly, *Reversibility and Stochastic Networks*. New York, NY: Wiley, 1979.

[164] F. P. Kelly, "Blocked probabilities in large circuit-switched networks," *Advances in Applied Probability*, Vol. 18, pp. 473–505, April-June 1986.

[165] J. Kennedy and M. C. Sullivan, "'Direction finding and smart antennas using software radio architecture," *IEEE Commun. Mag.*, Vol. 33, pp. 62–68, May 1995.

[166] S. Kozono, "Co-channel interference measurement method for mobile communication," *IEEE Trans. Veh. Technol.*, Vol. 36, pp. 7–13, January 1987.

[167] E. Kudoh and T. Matsumoto, "Effects of power control error on the system user capacity of DS/CDMA cellular mobile radios," *IEICE Transactions*, Vol. E75-B, pp. 524–529, June 1992.

[168] S. S. Kuek and W. C. Wong, "Ordered dynamic channel assignment scheme with reassignment in highway microcells," *IEEE Trans. Commun.*, Vol. 41, pp. 271–276, August 1992.

[169] G. Labedz, K. Felix, V. Lev, and D. Schaeffer, "Handover control issues in very high capacity cellular systems using small cells," in *Int. Conf. on Digital Land Mobile Radio Commun.*, Univ. of Warwick, Coventry, UK, 1987.

[170] W. H. Lam and R. Steele, "Spread-spectrum communications using diversity in an urban mobile radio environment," in *IEE Colloquium on Methods of Combating Multipath Effect in Wide Band Digital Cellular Mobile Systems*, London, England, pp. 6/1–6/11, October 1987.

[171] W. H. Lam and R. Steele, "Performance of direct-sequence spread-spectrum multiple-access systems in mobile radio," in *IEE Proc.*, pp. 1–14, February 1991.

[172] H. J. Larson and B. Schubert, *Probabilistic Models in Engineering Sciences, Vol. I.* New York, NY: Wiley, 1979.

[173] B. Larsson, B. Gudmundson, and K. Raith, "Receiver performance for the North American digital cellular system," in *IEEE Veh. Technol. Conf.*, Saint Louis, MO, pp. 1–6, May 1991.

[174] W. C. Y. Lee, *Mobile Communications Engineering.* New York, NY: McGraw Hill, 1982.

[175] W. C. Y. Lee, "Estimate of local average power of a mobile radio signal," *IEEE Trans. Veh. Technol.*, Vol. 34, pp. 22–27, February 1985.

[176] W. C. Y. Lee, *Mobile Communications Design Fundamentals.* Indianapolis, IN: Sams, 1986.

[177] W. C. Y. Lee, "Smaller cells for greater performance," *IEEE Communications Magazine*, pp. 19–30, November 1991.

[178] W. U. Lee and F. S. Hill, "A maximum-likelihood sequence estimator with decision-feedback equalization," *IEEE Trans. Commun.*, Vol. 25, pp. 971–979, June 1977.

[179] W. C. Y. Lee, "New cellular schemes for spectral efficiency," *IEEE Trans. Veh. Technol.*, Vol. 36, pp. 183–192, November 1987.

[180] W. C. Y. Lee, *Mobile Cellular Telecommunications Systems.* New York, NY: McGraw-Hill, 1989.

[181] W. C. Y. Lee, "Overview of cellular CDMA," *IEEE Trans. Veh. Technol.*, Vol. 40, pp. 291–302, May 1991.

[182] W. C. Y. Lee and Y. S. Yeh, "On the estimation of the seond-order statistics of log-normal fading in mobile radio environment," *IEEE Trans. Commun.*, Vol. 22, pp. 809–873, June 1974.

[183] J. S. Lehnert and M. B. Pursley, "Multipath diversity reception of spread-spectrum multiple-access communications," in *Proc. Conf. Inform. Sci. Syst.*, Johns Hopkins Univ., Baltimore, MD, March 1983.

[184] J. S. Lehnert and M. B. Pursley, "Error probability for binary direct-sequence spread-spectrum communications with random signature sequences," *IEEE Trans. Commun.*, Vol. 35, pp. 87–98, January 1987.

[185] J. S. Lehnert and M. B. Pursley, "Multipath diversity reception of spread-spectrum multiple-access communications," *IEEE Trans. Commun.*, Vol. 35, pp. 1189–1198, November 1987.

[186] A. Leon-Garcia, *Probability and Random Processes for Electrical Engineering.* Reading, MA: Addison-Wesley, 1989.

[187] J. Lin, F. Ling, , and J. Proakis, "Fading channel tracking properties of several adaptive algorithms for the North American digital cellular system," in *IEEE Veh. Technol. Conf.*, pp. 273–276, 1993.

[188] J. Lin, F. Ling, and J. Proakis, "Joint data and channel estimation for TDMA mobile channels," in *IEEE Int. Symp. on Pers., Indoor and Mobile Radio Commun.*, Boston, MA, pp. 235–239, 1992.

[189] S. Lin and D. J. Costello, Jr., *Error Control Coding: Fundamentals and Applications.* Englewood Cliffs, NJ: Prentice-Hall, 1983.

[190] F. Ling and J. G. Proakis, "A generalized multichannel least-squares lattice algorithm based on sequential processing stages," *IEEE Trans. Acoutics, Speech and Signal Proc.*, Vol. 32, pp. 381–389, April 1984.

[191] F. Ling and J. G. Proakis, "Adaptive lattice decision-feedback equalizers – their performance and application to time-variant multipath channels," *IEEE Trans. Commun.*, Vol. 33, pp. 348–356, April 1985.

[192] J-P. M. Linnartz, "Exact analysis of the outage probability in multiple-user mobile radio," *IEEE J. Selec. Areas Commun.*, Vol. 10, pp. 20–23, January 1992.

[193] Y. Liu, I. Oka, and E. Biglieri, "Error probability for digital transmission over nonlinear channels with applications to TCM," *IEEE Trans. Inform. Theory*, Vol. 36, pp. 1101–1110, September 1990.

[194] E. M. Long, *Decision-aided sequential sequence estimation for intersymbol interference channels.* Ph. D. thesis, Georgia Institute of Technology, 1989.

[195] E. M. Long and A. M. Bush, "Decision-aided sequential sequence estimation for intersymbol interference channels," in *IEEE Int. Conf. on Commun.*, Boston, MA, pp. 26.1.1–26.1.5, June 1989.

[196] F. Lotse and A. Wejke, "Propagation measurements for microcells in central Stockholm," in *IEEE Veh. Technol. Conf.*, Orlando, FL, pp. 539–541, May 1990.

[197] R. W. Lucky, "Automatic equalization for digital communication," *Bell System Tech. J.*, Vol. 44, pp. 547–588, April 1965.

[198] R. W. Lucky, "Techniques for adaptive equalization of digital communication systems," *Bell System Tech. J.*, Vol. 45, pp. 255–286, February 1966.

[199] R. Lucky, J. Salz, and E. Weldon, *Principles of Data Communication.* New York, NY: McGraw Hill, 1968.

[200] V. H. MacDonald, "The cellular concept," *Bell System Tech. J.*, Vol. 58, pp. 15–49, January 1979.

[201] F. R. Magee and J. G. Proakis, "Adaptive maximum-likelihood sequence estimation for digital signaling in the presence of intersymbol interference," *IEEE Trans. Inform. Theory*, Vol. 19, pp. 120–124, January 1973.

[202] A. Maloberti, "Radio transmission interface of the digital Paneuropean mobile system," in *IEEE Veh. Technol. Conf.*, San Francisco, CA, pp. 712–717, May 1989.

[203] M. Marsan and G. Hess, "Shadow variability in an urban land mobile radio environment," *Electronics Letters*, Vol. 26, pp. 646–648, May 1990.

[204] M. Marsan and G. C. Hess, "Cochannel isolation characteristics in an urban land mobile environment at 900 MHz," in *IEEE Veh. Technol. Conf.*, Saint Louis, MO, pp. 600–605, 1991.

[205] S. J. Mason, "Feedback theory: further properties of signal flow graphs," *IRE*, Vol. 44, pp. 920–926, July 1956.

[206] J. L. Massey, "Coding and modulation in digital communications," in *Int. Zurich Sem. Digital Communications*, Zurich, Switzerland, pp. E2(1)–E2(4), March 1974.

[207] J. E. Mazo, "On the independence theory of equalizer convergence," *Bell System Tech. J.*, Vol. 58, pp. 963–993, May 1979.

[208] J. E. Mazo, "Analysis of decision directed equalizer convergence," *Bell System Tech. J.*, Vol. 59, pp. 1857–1876, December 1980.

[209] W. R. Mende, "Evaluation of a proposed handover algorithm for the GSM cellular system," in *IEEE Veh. Technol. Conf.*, Orlando, FL, pp. 264–269, May 1990.

[210] L. B. Milstein, D. L. Schilling, R. L. Pickholtz, V. Erceg, M. Kullback, E. G. Kanterakis, D. Fishman, W. H. Biederman, and D. C. Salerno, "On the feasibility of a CDMA overlay for personal communications networks," *IEEE J. Selec. Areas Commun.*, Vol. 10, pp. 655–668, May 1992.

[211] S. Mockford and A. M. D. Turkmani, "Penetration loss into buildings at 900 MHz," in *IEE Colloquium on Propagation Factors and Interference Modeling for Mobile Radio Systems*, London, UK, pp. 1/1–1/4, November 1988.

[212] S. Mockford, A. M. D. Turkmani, and J. D. Parsons, "Local mean signal variability in rural areas at 900 MHz," in *IEEE Veh. Technol. Conf.*, Orlando, FL, pp. 610–615, May 1990.

[213] P. E. Mogensen, P. Eggers, C. Jensen, and J. B. Andersen, "Urban area radio propagation measurements at 955 and 1845 MHz for small and micro cells," in *IEEE Global Commun. Conf.*, Phoenix, AZ, pp. 1297–1302, December 1991.

[214] P. E. Mogensen and S. Petersen, "Antenna configuration measurements for direct microcells," in *IEEE Int. Symp. on Pers., Indoor and Mobile Radio Commun.*, Den Hague, The Netherlands, pp. 1075–1080, September 1994.

[215] M. A. Mokhtar and S. C. Gupta, "Capacity for cellular CDMA PCS's in Nakagami fading log-normal shadowing channels," in *IEEE Conf. Universal Personal Commun.*, pp. 190–194, 1992.

[216] P. Monsen, "Feedback equalization for fading dispersive channels," *IEEE Trans. Inform. Theory*, Vol. 17, pp. 56–64, January 1971.

[217] P. Monsen, "Theoretical and measured performance of DFE modem on a fading multipath channel," *IEEE Trans. Commun.*, Vol. 25, pp. 1144–1153, October 1977.

[218] P. Monsen, "Theoretical and measured performance of DFE modem on a fading multipath channel," *IEEE Trans. Commun.*, Vol. 32, pp. 5–12, January 1984.

[219] M. Morf, A. Vieira, and D. T. Lee, "Ladder forms for identification and speech processing," in *IEEE Conf. Dec. and Contr.*, New Orleans, LA, pp. 1074–1078, December 1977.

[220] R. K. Morrow, Jr. and J. S. Lehnert, "Bit-to-bit error dependence in slotted DS/CDMA packet systems with random signature sequences," *IEEE Trans. Commun.*, Vol. 37, pp. 1052–1061, October 1989.

[221] R. Muammar, "Co-channel interference in microcellular mobile radio system," in *IEEE Veh. Technol. Conf.*, Saint Louis, MO, pp. 198–203, May 1991.

[222] R. Muammar and S. C. Gupta, "Cochannel interference in high-capacity mobile radio systems," *IEEE Trans. Commun.*, Vol. 30, pp. 1973–1978, August 1982.

[223] M. S. Mueller, "Least-squares algorithms for adaptive equalizers," *Bell System Tech. J.*, Vol. 60, pp. 1905–1925, October 1981.

[224] M. S. Mueller and J. Salz, "A unified theory of data-aided equalization," *Bell System Tech. J.*, Vol. 60, pp. 2023–2038, November 1981.

[225] A. Murase, I. C. Symington, and E. Green, "Handover criterion for macro and microcellular systems," in *IEEE Veh. Technol. Conf.*, Saint Louis, MO, pp. 524–530, May 1991.

[226] K. Murota and K. Hirade, "GMSK modulation for digital mobile radio telephony," *IEEE Trans. Commun.*, Vol. 29, pp. 1044–1050, July 1981.

[227] Y. Nagata, "Analysis for spectrum efficiency in single cell trunked and cellular mobile radio," *IEEE Trans. Veh. Technol.*, Vol. 35, pp. 100–113, August 1987.

[228] A. F. Naguib, A. Paulraj, and T. Kailath, "Capacity improvement with base station antenna arrays in cellular CDMA," *IEEE Trans. Veh. Technol.*, Vol. 43, pp. 691–698, August 1994.

[229] M. Nakagami, "The m distribution; a general formula of intensity distribution of rapid fading," *Statistical Methods in Radio Wave Propagation*, W.G. Hoffman, ed., pp. 3–36, 1960.

[230] S. Nanda and D. J. Goodman, "Dynamic Resource Acquisition: distributed carrier allocation for TDMA cellular systems," *Third Generation Wireless Information Networks*, pp. 99–124, 1992.

[231] S. Nanda, "Teletraffic models for urban and suburban microcells: cell sizes and handoff rates," *IEEE Trans. Veh. Technol.*, Vol. 42, pp. 673–682, November 1993.

[232] R. W. Nettleton and G. R. Schloemer, "A high capacity assignment method for cellular mobile telephone systems," in *IEEE Veh. Technol. Conf.*, San Francisco, CA, pp. 359–367, May 1989.

[233] P. Newson and M. R. Heath, "The capacity of a spread spectrum CDMA system for cellular mobile radio with consideration of system imperfections," *IEEE J. Selec. Areas Commun.*, Vol. 12, pp. 673–683, 1994.

[234] H. Nyquist, "Certain topics in telegraph transmission theory," *Trans. American Inst. of Elect. Eng.*, Vol. 47, pp. 617–644, March 1928.

[235] H. Ochsner, "Direct-sequence spread-spectrum receiver for communication on frequency-selective fading channels," *IEEE J. Selec. Areas Commun.*, Vol. 5, pp. 188–193, February 1987.

[236] K. Okada, "A dynamic channel assignment strategy using information of speed and moving direction in microcellular systems," in *Int. Symp. Circuits and Systems*, Chicago, IL, pp. 2212–2215, May 1993.

[237] K. Okada and F. Kubota, "On dynamic channel assignment in cellular mobile radio systems," in *Int. Symp. Circuits and Systems*, Singapore, pp. 938–941, June 1991.

[238] K. Okada and F. Kubota, "A proposal of a dynamic channel assignment strategy with information of moving direction in microcellular systems," *Trans. IEICE*, Vol. E75-A, pp. 1667–1673, December 1992.

[239] K. Okanoue, A. Ushirokawa, H. Tomita, and Y. Furuya, "New MLSE receiver free from sample timing and input level controls," in *IEEE Veh. Technol. Conf.*, Secaucus, NJ, pp. 408–411, June 1993.

[240] Y. Okumura, E. Ohmuri, T. Kawano, and K. Fukuda, "Field strength and its variability in VHF and UHF land mobile radio service," *Rev. of the ECL*, Vol. 16, pp. 825–873, 1968.

[241] H. Panzer and R. Beck, "Adaptive resource allocation in metropolitan area cellular mobile radio systems," in *IEEE Veh. Technol. Conf.*, Orlando, FL, pp. 638–645, May 1990.

[242] Papoulis, *Probability, Random Variables, and Stochastic Processes*. New York, NY: McGraw-Hill, 1984.

[243] J. D. Parsons, *The Mobile Radio Propagation Channel.* New York, NY: Wiley, 1992.

[244] J. D. Parsons and J. G. Gardiner, *Mobile Communication Systems.* New York, NY: Halsted Press, 1989.

[245] S. Pasupathy, "Nyquist's third criterion," *Proc. IEEE*, Vol. 62, pp. 860–861, June 1974.

[246] J. M. Perl, A. Shpigel, and A. Reichman, "Adaptive receiver for digital communication over HF channels," *IEEE J. Selec. Areas Commun.*, Vol. 5, pp. 304–308, February 1987.

[247] R. L. Pickholtz, D. L. Schilling, and L. B. Milstein, "Theory of spread-spectrum communications – a tutorial," *IEEE Trans. Commun.*, Vol. 30, pp. 855 – 884, May 1982.

[248] A. R. Potter, "Implementation of PCNs using DCS1800," *IEEE Commun. Mag.*, Vol. 30, pp. 32–36, December 1992.

[249] R. Prasad and J. C. Arnbak, "Comments on "analysis for spectrum efficiency in single cell trunked and cellular mobile radio"," *IEEE Trans. Veh. Technol.*, Vol. 37, pp. 220–222, November 1988.

[250] R. Prasad and A. Kegel, "Effects of Rician faded and log-normal shadowed signals on spectrum efficiency in microcellular radio," *IEEE Trans. Veh. Technol.*, Vol. 42, pp. 274–281, August 1993.

[251] R. Prasad, A. Kegel, and J. C. Arnbak, "Analysis of system performance of high-capacity mobile radio," in *IEEE Veh. Technol. Conf.*, San Francisco, CA, pp. 306–309, May 1989.

[252] R. Prasad and A. Kegel, "Improved assessment of interference limits in cellular radio performance," *IEEE Trans. Veh. Technol.*, Vol. 40, pp. 412–419, May 1991.

[253] R. Prasad and A. Kegel, "Improved assessment of interference limits in cellular radio performance," *IEEE Trans. Veh. Technol.*, Vol. 40, pp. 412–419, May 1991.

[254] R. Prasad and A. Kegel, "Spectrum efficiency of microcellular systems," *Electronic Letters*, Vol. 27, pp. 423–425, February 1991.

[255] R. Price and P. E. Green, "A communication technique for multipath channels," *Proc. IEEE*, Vol. 46, pp. 555–570, March 1958.

[256] J. G. Proakis, *Digital Communications, 3rd ed.* New York, NY: McGraw-Hill, 1995.

[257] J. G. Proakis and J. Miller, "An adaptive receiver for digital signaling through channels with intersymbol interference," *IEEE Trans. Inform. Theory*, Vol. 15, pp. 484–497, July 1969.

[258] M. B. Pursley, F. D. Garber, and J. S. Lehnert, "Analysis of generalized quadriphase spread-spectrum communications," in *IEEE Int. Conf. on Commun.*, Seattle, WA, pp. 15.3.1–15.3.6, June 1980.

[259] S. U. Qureshi, "Adaptive equalization," *Proc. IEEE*, Vol. 73, pp. 1349–1387, September 1985.

[260] S. U. Qureshi and E. E. Newhall, "Analysis of maximum likelihood sequence estimation performance for quadrature amplitude modulation," *IEEE Trans. Inform. Theory*, Vol. 19, pp. 448–457, July 1973.

[261] R. Raheli, A. Polydoros, and C.-K. Tzou, "The principle of per-survivor processing: a general approach to approximate and adaptive MLSE," in *IEEE Global Commun. Conf.*, pp. 33.3.1–33.3.6, 1991.

[262] K. Raith and J. Uddenfelt, "Capacity of digital cellular TDMA systems," *IEEE Trans. Veh. Technol.*, Vol. 40, pp. 323–332, May 1991.

[263] B. Ramamurthi and M. Kavehrad, "Direct-sequence spread-spectrum with DPSK modulation and diversity for indoor wireless communications," in *IEEE Int. Conf. on Commun.*, Toronto, Ont., Canada, June 1986.

[264] T. Rappaport and L. Milstein, "Effects of path loss and fringe user distribution on CDMA cellular frequency reuse efficiency," in *IEEE Global Commun. Conf.*, San Diego, CA, pp. 500–506, December 1990.

[265] P. Raymond, "Performance analysis of cellular networks," *IEEE Trans. Commun.*, Vol. 39, pp. 1787–1793, December 1991.

[266] Research & Development Center for Radio Communications (RCR), "Digital cellular telecommunication systems," April 1991. RCR STD-27.

[267] D. O. Reudink, D. Jones, S. Meredith, and M. Reudink, "Narrow beam switched antenna experiment," First Workshop on Smart Antennas in Wireless Mobile Communications, 1994.

[268] S. O. Rice, "Noise in FM receivers," *Symposium Proceedings of Time Series Analysis*, pp. 395–422, 1963. M. Rosenblatt, ed.

[269] S. Rice, "Statistical properties of a sine wave plus noise," *Bell System Tech. J.*, Vol. 27, pp. 109–157, January 1948.

[270] M. Rouanne and D. Costello, "An algorithm for computing the distance spectrum of trellis codes," *IEEE J. Selec. Areas Commun.*, Vol. 7, pp. 929–940, August 1989.

[271] A. Rustako, N. Amitay, G. Owens, and R. Roman, "Radio propagation at microwave frequencies for line-of-sight microcellular mobile and personal communications," *IEEE Trans. Veh. Technol.*, Vol. 40, pp. 203–210, February 1991.

[272] J. Salz, "Optimum mean-square decision-feedback equalization," *Bell System Tech. J.*, Vol. 52, pp. 1341–1373, October 1973.

[273] A. Sampath and J. Holtzman, "Estimation of maximum Doppler frequency for hanodff decisions," in *IEEE Veh. Technol. Conf.*, Secaucus, NJ, pp. 859–862, May 1993.

[274] C. Sandeep and S. C. Gupta, "Performance of an adaptive multipath diversity receiver in a frequency selective Rayleigh fading channel," in *IEEE Veh. Technol. Conf.*, Philadelphia, PA, pp. 351–357, June 1988.

[275] E. H. Satorius and S. T. Alexander, "Channel equalization using adaptive lattice algorithms," *IEEE Trans. Commun.*, Vol. 27, pp. 899–905, June 1979.

[276] E. H. Satorius and J. D. Pack, "Application of least squares lattice algorithms to adaptive equalization," *IEEE Trans. Commun.*, Vol. 29, pp. 136–142, February 1981.

[277] D. Schleher, "Generalized Gram-Charlier series with application to the sum of lognormal variates," *IEEE Trans. Inform. Theory*, Vol. 23, pp. 275–280, March 1977.

[278] S. Schwartz and Y. S. Yeh, "On the distribution function and moments of power sums with log-normal components," *Bell System Tech. J.*, Vol. 61, pp. 1441–1462, September 1982.

[279] M. Serizawa and J. Murakami, "Phase tracking Viterbi demodulator," *Electronics Letters*, Vol. 40, pp. 792–794, 1989.

[280] N. Seshadri, "Joint data and channel estimation using fast blind trellis search techniques," in *IEEE Global Commun. Conf.*, San Diego, CA, pp. 1659–1663, 1990.

[281] W-H. Sheen and G. L. Stüber, "MLSE equalization and decoding for multipath-fading channels," *IEEE Trans. Commun.*, Vol. 39, pp. 1455–1464, October 1991.

[282] W-H. Sheen and G. L. Stüber, "Error probability for reduced-state sequence estimation," *IEEE J. Selec. Areas Commun.*, Vol. 10, pp. 571–578, April 1992.

[283] W-H. Sheen and G. L. Stüber, "Error probability of reduced-state sequence estimation for trellis-coded modulation on intersymbol interference channels," *IEEE Trans. Commun.*, Vol. 41, pp. 1265–1269, September 1993.

[284] W-H. Sheen and G. L. Stüber, "Error probability for maximum likelihood sequence estimation of trellis-coded modulation on ISI channels," *IEEE Trans. Commun.*, Vol. 42, pp. 1427–1430, April 1994.

[285] H. Shiino, N. Yamaguchi, , and Y. Shoji, "Performance of an adaptive maximum-likelihood receiver for fast fading multipath channel," in *IEEE Veh. Technol. Conf.*, Denver, CO, pp. 380–383, May 1992.

[286] M. K. Simon, J. K. Omura, R. A. Scholtz, and B. K. Levitt, *Spread Spectrum Communications.* Rockville, MD: Computer Science Press, 1985.

[287] N. R. Sollenberger, "Architecture and implementation of an efficient and robust TDMA frame structure for digital portable communications," *IEEE Trans. Veh. Technol.*, Vol. 40, pp. 250–260, January 1991.

[288] D. Stamatelos and A. Ephremides, "Multiple access capability of indoor wireless networks using spatial diversity," in *IEEE Int. Symp. on Pers., Indoor and Mobile Radio Commun.*, Den Hague, The Netherlands, pp. 1271–1275, September 1994.

[289] Stanford University, "First Workshop on Smart Antennas in Wireless Mobile Communications," 1994.

[290] Standford University, "Second Workshop on Smart Antennas in Wireless Mobile Communications," 1995.

[291] R. Steele, J. Whitehead, and W. C. Wong, "System aspects of cellular radio," *IEEE Commun. Mag.*, Vol. 33, pp. 80 – 86, January 1995.

[292] G. Stüber, "Soft decision direct-sequence DPSK receivers," *IEEE Trans. Veh. Technol.*, Vol. 37, pp. 151–157, August 1988.

[293] G. L. Stüber and C. Kchao, *Spread Spectrum Cellular Radio*. Georgia Institute of Technology, OCA Project No. E21-662 for Bell South Enterprises, December, 1990.

[294] H. Susuki, "A statistical model for urban radio propagation," *IEEE Trans. Commun.*, Vol. 25, pp. 673–680, July 1977.

[295] S. C. Swales, M. A. Beach, and D. J. Edwards, "Multi-beam adaptive base station antennas for cellular land mobile radio systems," in *IEEE Veh. Technol. Conf.*, San Francisco, CA, pp. 341–348, 1989.

[296] S. C. Swales, M. A. Beach, D. J. Edwards, and J. P. McGeehan, "The performance enhancement of multibeam adaptive base station antennas for cellular land mobile radio systems," *IEEE Trans. Veh. Technol.*, Vol. 39, pp. 56–67, February 1990.

[297] E. W. Swokowski, *Calculus with Analytical Geometry*. New York, NY: Prindle, Weber, and Schmidt, 1979.

[298] J. Tajima and K. Imamura, "A strategy for flexible channel assignment in mobile communication systems," *IEEE Trans. Veh. Technol.*, Vol. 37, pp. 92–103, May 1988.

[299] K. Takeo, M. Nishino, Y. Ameazwa, and S. Sato, "Adaptive traffic control scheme for non-uniform traffic distribution in microcellular mobile communication system," in *IEEE Veh. Technol. Conf.*, Orlando, FL, pp. 527–531, May 1990.

[300] S. Tekinay and B. Jabbari, "Handover and channel assignment in mobile cellular networks," *IEEE Commun. Mag.*, Vol. 29, pp. 42–46, 1991.

[301] H. Thaper, "Real-time application of trellis coding to high-speed voiceband data transmission," *IEEE J. Selec. Areas Commun.*, Vol. 2, pp. 648–658, September 1984.

[302] R. J. Tront, J. K. Cavers, and M. R. Ito, "Performance of Kalman decision-feedback equalization in HF radio modems," in *IEEE Int. Conf. on Commun.*, Toronto, Canada, pp. 1617–1621, June 1986.

[303] G. V. Tsoulos, M. A. Beach, and S. C. Swales, "Adaptive antennas for third generation DS-CDMA cellular systems," in *IEEE Veh. Technol. Conf.*, Chicago, IL, pp. 45–49, 1995.

[304] G. L. Turin, "Introduction to spread-spectrum antimultipath techniques and their application to urban digital radio," *Proc. IEEE*, Vol. 68, pp. 328–353, March 1980.

[305] G. L. Turin, "The effects of multipath and fading on the performance of direct-sequence spread-spectrum CDMA systems," *IEEE J. Selec. Areas Commun.*, Vol. 2, pp. 597–603, July 1984.

[306] A. M. D. Turkmani, "Probability of error for M-branch selection diversity," *IEE Proc. I.*, Vol. 139, pp. 71–78, February 1992.

[307] A. M. D. Turkmani, J. D. Parsons, F. Ju, and D. G. Lewis, "Microcellular radio measurements at 900,1500, and 1800 MHz," in *5th Int. Conf. on Mobile Radio and Personal Communications*, Coventry, UK, pp. 65–68, December 1989.

[308] W. H. Tuttlebee, "Cordless personal communications," *IEEE Commun. Mag.*, Vol. 30, pp. 42–53, December 1992.

[309] G. Ungerboeck, "Theory on the speed of convergence in adaptive equalizers for digital communication," *IBM Journal Research and Development*, Vol. 16, pp. 546–555, November 1972.

[310] G. Ungerboeck, "Adaptive maximum likelihood receiver for carrier-modulated data transmission systems," *IEEE Trans. Commun.*, Vol. 22, pp. 624–636, May 1974.

[311] G. Ungerboeck, "Fractional tap-spacing equalizer and consequences for clock recovery in data modems," *IEEE Trans. Commun.*, Vol. 24, pp. 856–864, August 1976.

[312] G. Ungerboeck, "Channel coding with multilevel phase signals," *IEEE Trans. Inform. Theory*, Vol. 28, pp. 55–67, January 1982.

[313] G. Ungerboeck, "Trellis coded modulation with redundant signal sets – part I: introduction," *IEEE Commun. Mag.*, Vol. 25, pp. 5 – 11, February 1987.

[314] G. Ungerboeck, "Trellis coded modulation with redundant signal sets – part II: state of the art," *IEEE Commun. Mag.*, Vol. 25, pp. 12 – 21, February 1987.

[315] G. M. Vachula and J. F. S. Hill, "On optimal detection of band-limited PAM signals with excess bandwidth," *IEEE Trans. Commun.*, Vol. 29, pp. 886–890, June 1981.

[316] F. L. Vermuelen and M. E. Hellman, "Reduced-state Viterbi decoding for channels with intersymbol interference," in *IEEE Int. Conf. on Commun.*, Minneapolis, MN, pp. 37B.1–37B.4, June 1974.

[317] R. Vijayan and J. Holtzman, "Sensitivity of handoff algorithms to variations in the propagation environment," in *IEEE Conf. Universal Personal Commun.*, Ottawa, Canada, pp. 158–162, October 1993.

[318] R. Vijayan and J. M. Holtzman, "A model for analyzing handoff algorithms," *IEEE Trans. Veh. Technol.*, Vol. 42, pp. 351–356, August 1993.

[319] R. Vijayan and J. Holtzman, "Analysis of handoff algorithms using nonstationary signal strength measurements," in *IEEE Global Commun. Conf.*, Orlando, FL, pp. 1405–1409, December 1992.

[320] A. J. Viterbi, *Principles of Coherent Communications.* New York, NY: Wiley, 1965.

[321] A. J. Viterbi, "Error bounds for convolutional codes and an asymptotically optimum decoding algorithm," *IEEE Trans. Inform. Theory*, Vol. 13, pp. 260–269, April 1967.

[322] A. J. Viterbi, "Convolutional codes and their performance in communication systems," *IEEE Trans. Commun.*, Vol. 19, pp. 751–772, October 1971.

[323] A. J. Viterbi, A. M. Viterbi, K. Gilhousen, and E. Zehavi, "Soft Handoff Extends CDMA Cell Coverage and Increases Reverse Channel Capacity," *IEEE J. Selec. Areas Commun.*, Vol. 12, pp. 1281–1288, October 1994.

[324] A. J. Viterbi, *CDMA Prinicples of Spread Spectrum Communication.* Reading, MA: Addison-Wesley, 1995.

[325] J.-F. Wagen, "Signal strength measurements at 881 MHz for urban microcells in dowtown Tampa," in *IEEE Global Commun. Conf.*, Phoenix, AZ, pp. 1313–1317, December 1991.

[326] E. H. Walker, "Penetration of radio signals into buildings in cellular radio environments," *Bell System Tech. J.*, Vol. 62, September 1983.

[327] L.-C. Wang, G. L. Stüber, and C.-T. Lea, "Architecture design, frequency planning, and performance analysis for a microcell/macrocell overlaying system," Submitted to IEEE Trans. on Vehicular Tech.

[328] L. F. Wei, "Trellis-coded modulation with multidimensional constellations," *IEEE Trans. Inform. Theory*, Vol. 33, pp. 483–501, July 1987.

[329] L. F. Wei, "Coded M-DPSK with built-in time diversity for fading channels," *IEEE Trans. Inform. Theory*, Vol. 39, pp. 1820–1839, November 1993.

[330] K. Wesolowski, "On the performance and convergence of the adaptive canceler of intersymbol interference in data transmission," *IEEE Trans. Commun.*, Vol. 33, pp. 425–432, May 1985.

[331] K. Wesolowski, "An efficient DFE & ML suboptimum receiver for data transmission over dispersive channels using two-dimensional signal constellations," *IEEE Trans. Commun.*, Vol. 35, pp. 336–339, March 1987.

[332] K. A. West and G. L. Stüber, "An aggressive dynamic channel assignment strategy for a microcellular environment," *IEEE Trans. Veh. Technol.*, Vol. 43, pp. 1027–1038, November 1994.

[333] B. Widrow, J. M. McCool, M. G. Larimore, and J. C. R. Johnson, "Adaptive switching circuits," in *IRE Wescon Conv.*, Los Angeles, CA, pp. 96–104, August 1960.

[334] B. Widrow, J. M. McCool, M. G. Larimore, and C. R. J. Jr., "Stationary and nonstationary learning characteristics of the LMS adaptive filter," in *Proc. IEEE*, pp. 1151–1162, August 1976.

[335] C. Wijffels, H. Miser, and R. Prasad, "A micro-cellular CDMA system over slow and fast Ricean fading channels with forward error correction," *IEEE Trans. Veh. Technol.*, Vol. 42, pp. 570–580, November 1993.

[336] A. Williamson, B. Egan, and J. Chester, "Mobile radio propagation in Auckland at 851 MHz," *Electronic Letters*, Vol. 20, pp. 517–518, June 1984.

[337] L. Wong and P. McLane, "Performance of trellis codes for a class of equalized ISI channels," *IEEE Trans. Commun.*, Vol. 36, pp. 1330–1336, December 1988.

[338] A. Wonjar, "Unknown bounds on performance in Nakagami channels," *IEEE Trans. Commun.*, Vol. 34, pp. 22–24, January 1986.

[339] J. M. Wozencraft and I. M. Jacobs, *Principles of Communication Engineering*. Prospect Heights, IL: Waveland Press, 1990.

[340] H. Xia, H. Bertoni, L. Maciel, and A. Landsay-Stewart, "Radio propagation measurements and modeling for line-of-sight microcellular systems," in *IEEE Veh. Technol. Conf.*, Denver, CO, pp. 349–354, May 1992.

[341] H. Xiang, "Binary code-division multiple-access systems operating in multipath fading, noisy channels," *IEEE Trans. Commun.*, Vol. 33, pp. 775–784, August 1985.

[342] F. Xiong, A. Zerik, and E. Shwedyk, "Sequential sequence estimation for channels with intersymbol interference of finite or infinite length," *IEEE Trans. Commun.*, Vol. 36, pp. 795–803, June 1990.

[343] H. Xue, R. Davies, M. Beach, and J. McGeehan, "'Linearity considerations in adaptive antenna array applications," in *IEEE Int. Symp. on Pers., Indoor and Mobile Radio Commun.*, Toronto, Canada, pp. 682–686, September 1995.

[344] Y. Yao and A. Sheikh, "Outage probability analysis for microcell mobile radio systems with cochannel interferers in Rician/Rayleigh fading environment," *Electronics Letters*, Vol. 26, pp. 864–866, June 1990.

[345] Y. Yao and U. Sheikh, "Investigation into cochannel interference in microcellular mobile radio systems," *IEEE Trans. Veh. Technol.*, Vol. 41, pp. 114–123, May 1992.

[346] Y. Yeh and S. C. Schwartz, "Outage probability in mobile telephony due to multiple log-normal interferers," *IEEE Trans. Commun.*, Vol. 32, pp. 380–388, April 1984.

[347] K. L. Yeung and T-S. P. Yum, "Compact pattern based channel assignment for cellular mobile systems," *IEEE Trans. Veh. Technol.*, Vol. 43, pp. 892–896, November 1994.

[348] S. Yoshida, A. Hirai, G. L. Tan, H. Zhou, and T. Takeuchi, "In-Service monitoring of multipath delay-spread and C/I for QPSK signal," in *IEEE Veh. Technol. Conf.*, Denver, CO, pp. 592–595, May 1992.

[349] T-S. P. Yum and W-S. Wong, "Hot-spot traffic relief in cellular systems," *IEEE J. Selec. Areas Commun.*, Vol. 11, pp. 934 – 940, August 1993.

[350] L. A. Zadeh, "Frequency analysis of variable networks," *Institute Radio Engineers*, Vol. 38, pp. 291–299, 1950.

[351] M. Zhang and T. Yum, "Comparisons of channel-assignment strategies in cellular mobile telephone systems," *IEEE Trans. Veh. Technol.*, Vol. 38, pp. 211–215, November 1989.

[352] N. Zhang and J. Holtzman, "Analysis of handoff algorithms using both absolute and relative measurements," in *IEEE Veh. Technol. Conf.*, Stockholm, Sweden, pp. 82–86, June 1994.

[353] N. Zhang and J. Holtzman, "Analysis of a CDMA Soft Handoff Algorithm," in *IEEE Int. Symp. on Pers., Indoor and Mobile Radio Commun.*, Toronto, Canada, pp. 819–823, September 1995.

[354] R. Ziemer and R. Peterson, *Digital Communications and Spread Spectrum Systems*. New York, NY: MacMillan, 1985.

[355] R. Ziemer and R. Peterson, *Introduction to Digital Communications*. New York, NY: MacMillan, 1992.

[356] M. Zorzi, "Simplified forward-link power control law in cellular CDMA," *IEEE Trans. Veh. Technol.*, Vol. 43, pp. 1088–1093, November 1994.

INDEX